ADVANCES IN EXPERIMENTAL MEDICINE AND BIOLOGY

Recent Volumes in this Series

Volume 460
MELATONIN AFTER FOUR DECADES: An Assessment of Its Potential
Edited by James Olcese

Volume 461
CYTOKINES, STRESS, AND DEPRESSION
Edited by Robert Dantzer, Emmanuelle E. Wollman, and Raz Yirmiya

Volume 462
ADVANCES IN BLADDER RESEARCH
Edited by Laurence S. Baskin and Simon W. Hayward

Volume 463
ENZYMOLOGY AND MOLECULAR BIOLOGY OF CARBONYL METABOLISM 7
Edited by Henry Weiner, Edmund Maser, David W. Crabb, and Ronald Lindahl

Volume 464
CHEMICALS VIA HIGHER PLANT BIOENGINEERING
Edited by Fereidoon Shahidi, Paul Kolodziejczyk, John R. Whitaker, Agustin Lopez Munguia, and Glenn Fuller

Volume 465
CANCER GENE THERAPY: Past Achievements and Future Challenges
Edited by Nagy A. Habib

Volume 466
CURRENT VIEWS OF FATTY ACID OXIDATION AND KETOGENESIS: From Organelles to Point Mutations
Edited by Patti A. Quant and Simon Eaton

Volume 467
TRYPTOPHAN, SEROTONIN, AND MELATONIN: Basic Aspects and Applications
Edited by Gerald Huether, Walter Kochen, Thomas J. Simat, and Hans Steinhart

Volume 468
THE FUNCTIONAL ROLES OF GLIAL CELLS IN HEALTH AND DISEASE: Dialogue between Glia and Neurons
Edited by Rebecca Matsas and Marco Tsacopoulos

Volume 469
EICOSANOIDS AND OTHER BIOACTIVE LIPIDS IN CANCER, INFLAMMATION, AND RADIATION INJURY, 4
Edited by Kenneth V. Honn, Lawrence J. Marnett, and Santosh Nigam

---

A Continuation Order Plan is available for this series. A continuation order will bring delivery of each new volume immediately upon publication. Volumes are billed only upon actual shipment. For further information please contact the publisher.

MELATONIN AFTER FOUR DECADES

An Assessment of Its Potential

Edited by

James Olcese

Institute for Hormone and Fertility Research
University of Hamburg
Hamburg, Germany

KLUWER ACADEMIC / PLENUM PUBLISHERS
New York, Boston, Dordrecht, London, Moscow

Library of Congress Cataloging-in-Publication Data

Melatonin after four decades : an assessment of its potential /
edited by James Olcese.
p. cm. -- (Advances in experimental medicine and biology ; v.
460)
"Proceedings of the Hanseatic Endocrine Conference on Melatonin
after Four Decades: an assessment of its potential, held August
27-30, 1998, in Hamburg, Germany"--T.p. verso
Includes bibliographical references and index.
ISBN 0-306-46134-X
1. Melatonin--Physiological effect Congresses. 2. Melatonin-
-Therapeutic use Congresses. I. Olcese, James. II. Hanseatic
Endocrine Conference on Melatonin after Four Decades (1998 :
Hamburg, Germany) III. Series.
[DNLM: 1. Melatonin--physiology Congresses. 2. Melatonin-
-therapeutic use Congresses. W1 AD559 v.460 1999 / WK 350 M5165
1999]
QP572.M44M435 1999
573.4--dc21
DNLM/DLC
for Library of Congress 99-31757
CIP

Proceedings of the Hanseatic Endocrine Conference on Melatonin after Four Decades: An Assessment of Its Potential, held August 27–30, 1998, in Hamburg, Germany

ISSN 0065-2598

ISBN 0-306-46134-X

233 Spring Street, New York, N.Y. 10013

10 9 8 7 6 5 4 3 2 1

A C.I.P. record for this book is available from the Library of Congress

Printed in the United States of America

PREFACE

The field of melatonin research, like many of its participants, has reached a respectable "middle-age" after many growth phases, identity crises and maturational experiences. First identified by Aaron Lerner and his colleagues at Yale, melatonin (or N-acetyl-5-methoxytryptamine as the chemists would prefer) was originally of some interest because of its exquisitely potent effects on melanophore pigment dispersion. Subsequently, its biochemistry was elucidated and within the first decade the unexpected and signature feature of its synthesis was discovered, viz its daily rhythmicity. Earlier clinical literature in which pubertal disorders were often associated with tumors of the pineal gland led researchers at about this time to the develop animal models for understanding melatonin's involvement in reproductive processes. Despite occasional frustration and frequent controversy, what eventually emerged in the second decade was the recognition that melatonin does indeed participate in regulating reproductive activities in many, but apparently not all species, and that this function is critically time-dependent. Put another way, melatonin's actions are often dissimilar when young versus old experimental animals are tested, but even more importantly, melatonin is only physiologically active when present in the organism's bloodstream (from where it diffuses quickly throughout the body) *at the appropriate time of day*. It didn't take long for this remarkable insight to permeate the field of chronobiology, out of which surprising new attributes for melatonin were found in the third decade. Even in the clinical field, melatonin was rapidly "transformed" into something more than a mere hormone. It thus became a key biological parameter for assessing circadian phase as well as a novel agent for modifying circadian clock function. Whereas for years the "secret" among well-travelled researchers was that melatonin could effectively mitigate the symptoms of jet lag, melatonin after its fourth decade is now being investigated in terms of innumerable expressions of circadian rhythmicity from sleep-wake cycles and body temperature rhythms to cerebral perfusion and cell proliferation.

A truly major breakthrough for the field of melatonin research was the cloning of the first melatonin receptor by Steve Reppert's laboratory some 5 years ago. Within only a few years many other species of melatonin receptor have been cloned, their tissue distribution mapped and signalling mechanisms identified. In this regard, the need for a logically consistent manner for assigning new receptors to the melatonin receptor subfamily led the IUPHAR earlier this year to recommend a new receptor nomenclature, which we have adopted in this book. It would be my personal plea to all current and future melatonin researchers that this classification be used, if for no other reason than to avoid the confusion similar to that which we all face with the variety of abbreviations currently used for melatonin (e.g., aMT, ML, MLT, MEL)!!

The conference from which this book derives its inspiration was organized as a tribute to the person of Aaron Lerner. His unique contribution four decades ago to the field of basic and applied biology, i.e., the discovery of melatonin, is finally beginning to receive the intellectual scrutiny that it deserves, as could be seen in the excellent lectures and posters presented at this conference. In evolving scientifically from a hormone with actions on the skin to a chronobiotic and, possibly oncostatic molecule, melatonin research has engaged many physiologists, biochemists, pharmacologists, psychiatrists and endocrinologists. The majority of these still active individuals were in attendance at our melatonin conference in August and have benefited the present book with their diverse and informative chapters. While this current "assessment of its potential" is necessarily broad and multi-disciplinary, one can anticipate an even greater interface with other disciplines in the future as the full spectrum of melatonin's biological actions become apparent.

I would be negligent not to take this opportunity to thank a number of individuals who made the Hanseatic Endocrine Conference 1998 and ultimately this book possible. For their untiring dedication to the many important organizational details I wish to thank A. Schade, P. Behring, P. Stegemann, Prof. R. Ivell, Dr. A. Mukhopadhyay, Prof. H. Schulte and Prof. F. Leidenberger, as well as the many members of the IHF whose time and support were tremendously helpful. Financial support from the Deutsche Forschungsgemeinschaft, GEFE e.V., Deutsche Gesellschaft für Endokrinologie, IHF, Lundbeck A/S, Bristol Myers-Squibb, the SmithKline Beecham Foundation, Schering AG, Becton-Dickinson, Boehringer Mannheim, Bühlmann Laboratories AG, Genecraft, Sigma-Aldrich, Dianova GmbH, IBL GmbH, Stockgrand Ltd and Servier is also gratefully acknowledged.

James Olcese

CONTENTS

1

MELATONIN—WITHOUT THE HYPE

Aaron B. Lerner

Department of Dermatology
Yale University School of Medicine
New Haven, Connecticut

The melatonin project was started at Yale by Dr. Yoshiyata Takahashi and me, and completed nearly 4 years later by Dr. James Case and me. In our investigations on normal and malignant pigment cells we were studying the action of humoral and neural factors on the dispersion and aggregation of melanosomes in frog melanocytes. My research was on pigmentation at basic and clinical levels. We had just completed the isolation of a melanocyte-stimulating hormone (MSH)—a dispersing factor—from the pituitary gland. We were intrigued by a paper by McCord and Allen in 1917 in which they claimed that extracts from bovine pineal glands lightened—that is, had an aggregating action on—the color of tadpoles. We had already studied the aggregating action of adrenaline, noradrenaline and acetylcholine. Was the agent in the pineal gland of these neural chemicals? We decided to find out. No one seemed to have been interested in the report by McCord and Allen. We had no competition.

In our isolation work the first step was to lyophilize fresh frozen pineal glands from cows. We could then remove debris, connective tissue, bone fragments, etc. so that we could work with clean glands. In Figure 1 a four-year-old boy has no difficulty holding a bag with a few thousand lyophilized pineal glands. In his other hand he is holding one gland. It looks like a single piece of puffed wheat or puffed rice. By the time the project was completed Armour had sent us more than 250,000 frozen glands and we were still complaining that we needed more.

Immediately after we established the structure of melatonin we realized that this relatively small moleculae has two unusual features viz., it has one N-acetyl group and one methoxy group. Melatonin was the first methoxyindole found in mammalian tissue. It was serotonin blocked at both ends. Biologic amines such as adrenaline, noradrenaline, serotonin and histamine, all antihistamines and many tranquilizers require a nitrogen atom carrying a positive charge to be active. When acetylated their charge and biologic activity are lost. But for pigmentation in frogs removing the charge on the nitrogen greatly enhances the potency. Melatonin is 100,000 times more potent than

Melatonin after Four Decades, edited by James Olcese.
Kluwer Academic / Plenum Publishers, New York, 2000.

Figure 1.

noradrenaline, serotonin or acetylcholine in lightening frog skin. The most potent agent to darken frog skin, α-MSH, also has an N-acetyl group.

The methoxy group was of special interest. It had been assumed that indoles with methoxy groups did not exist in mammals. Hydroxyl groups on aromatic rings have a negative charge but O-methylation removes the charge. Here again removing the charge of a group results in a more potent molecule. For potency the methoxy group on the ring is more important than the acetyl group on the nitrogen. Replacing the acetyl group with a hydrogen atom so that the molecule has an amino group instead of an N-acetyl one decreases the potency of the molecule on frog pigment cells by 5 fold. However, if one puts an hydroxy group on the ring instead of a methoxy one, activity is decreased by 20,000 fold.

If one were trying to isolate melatonin today instead of 44 years ago, the bioassay and chemical fractionations would most likely be similar to those we used. One always wonders about how many other biologically potent molecules exist, but are unknown because they are present in such small quantities and no sensitive bioassay is available to detect them.

Until 4 years ago everything went well. The research carried out by many of the participants of this meeting opened up new fields. They showed that melatonin is an important hormone especially in regard to its role related to the biologic actions of light in daily circadian and seasonal rhythms in human beings and in animals. There have been several international meetings. However in 1994 things changed. The United States Congress passed the Dietary Supplement Health and Education Act, which took herbs, vitamins, minerals, amino acids or their products out of control of the Food and Drug Administration (FDA). Also excluded from FDA regulation are compounds that are normally found in foods. Because tiny amounts of melatonin are present in foods such as bananas and rice, melatonin could be sold over the counter as a dietary supplement. In no time hype came into this field. Articles on melatonin for the lay public appeared everywhere. There were two major reports on melatonin in Newsweek magazine, including one in 1995 when it made the cover. Numerous books were published. In one year sales may have been as high as 200 million dollars. At least 10 million people have taken melatonin. Here was a hormone from a mysterious gland that could be purchased over the counter. But for what purpose? Without seeing a physician and without solid data, people could try taking melatonin for everything—to live longer, for arthritis, for immune disorders, for better sexual function, for neurologic disorders, etc. In spite of all the waste of money and of false hopes has anything good come from all this hype? Well, yes, melatonin is now inexpensive. It is relatively non-toxic. As known before this high promotion, melatonin is useful as a mild sedative and when taken properly to overcome jet-lag. For most of the other claims melatonin does not help. For a few, the jury is still out. But the mystery of the pineal gland is gone. Melatonin did help to start the modern studies of daily and seasonal biologic rhythms.

When I write about the history of melatonin I always want to say something about our research group. In the 4 years that it took to complete the melatonin project a total of only 7 people were involved, each for a period of approximately 2 years, except for the dishwasher (Kate Lehmann) and myself who were there for the entire time. Other than the dishwasher we had no technical help. By no help I mean zero. Two co-workers were postdoctoral fellows (Takahashi and Mori), one a resident (Case), one a medical student (Barkas) and one a part-time physician (Wright). Simply being a mentor for these wonderful and dedicated people was a reward in itself. Most were highly successful in academic careers. The late Dr. Yoshiyata Takahashi became Chairman of the Department of Internal Medicine at Gifu University as well as being the leading hepatologist in Japan. Dr. Wataru Mori rose from a fellow to being the Chairman of the Department of Pathology at the University of Tokyo. Later he became Dean of the Medical School and after that President of the University of Tokyo. He was only the second physician to become president of that University. He was the first person from Japan to be elected to the Institute of Medicine, which is part of the National Academy of Sciences. The late Dr. James Case was the first head of the Section of Dermatology at the University of Washington. Dr. Jack Barkas is at present Chairman of the Department of Psychiatry at Cornell. All got there with melatonin.

2

SEROTONIN *N*-ACETYLTRANSFERASE

A Personal Historical Perspective

D. C. Klein

Section on Neuroendocrinology
Laboratory of Developmental Neurobiology
National Institute of Child Health and Development
National Institutes of Health
Bethesda, Maryland

This chapter is written as part of a 40 year celebration of the discovery of melatonin and it seems most appropriate to this invited author to contribute something of a personal historical nature. Accordingly, the subject matter of the chapter is my relationship with serotonin *N*-acetyltransferase and the people who participated in the major advances in understanding this protein. Readers wanting a detailed up-to-date 1998 accounting of what's known about the enzyme are directed to other publications (1–11); earlier reviews will also be of interest (12–16).

1. THE PINEAL GLAND: A MELATONIN FACTORY ON A ~ 24-HOUR SCHEDULE

At the time I started working in this field around 1968, Aaron Lerner's discoveries of melatonin and the high concentrations of melatonin in the pineal gland were well known (17,18), as was the high amount of serotonin in the pineal gland (19). Julie Axelrod and his coworkers were generally interested in biogenic amine metabolizing enzymes, so it was not long before Lerner's findings led to their demonstration that the pineal gland had the enzymatic machinery required to convert serotonin to N-acetylserotonin and N-acetylserotonin to melatonin (20,21). This established the pineal gland as a melatonin factory.

The concept that there are 24-hour rhythmic changes in pineal indoles came from studies by Quay (22,23). He had developed fluorescent techniques to measure pineal melatonin and reported that there was a large day/night rhythm in serotonin and

Melatonin after Four Decades, edited by James Olcese.
Kluwer Academic / Plenum Publishers, New York, 2000.

melatonin. The large circadian changes in these compounds were intriguing and immediately raised questions in the minds of curious investigators about how changes in the concentrations of biogenic amines in neural tissues are regulated.

Interest in the regulation of melatonin levels first focused on the last enzyme in melatonin synthesis, hydroxyindole-O-methyltransferase (HIOMT). Axelrod and his coworkers characterized the protein and found that high levels were found only in the pineal gland, based on studies in mammals (12,24–27). In contrast, serotonin acetylation was thought to be mediated by a widely distributed enzyme, and therefore was not viewed as being a likely site of regulation (12). This acetyltransferase, arylamine *N*-acetyltransferase (E.C. 2.3.1.5), played an essential role in the discovery of acetyl CoA (28), and was well known at the time. Faced with the alternatives of either HIOMT or arylalkylamine N-acetyltransferase as the key regulator of melatonin production, Axelrod's team focused on the former—acetylation was ignored.

This hunch looked pretty good at first, because a positive relationship seemed to exist between environmental lighting, HIOMT and melatonin: HIOMT activity enzyme decreased following long-term exposure to light and decreased following long periods in continual lighting (12,25–27). In addition, melatonin was known to be high in the dark of night and low during the day, which matched up with the report that there was a large daily rhythm in HIOMT activity. Based on this, it seemed reasonable that HIOMT regulated the rhythm in melatonin production.

Over the years, the finding that HIOMT activity is elevated in the rat pineal gland following long term exposure to darkness and lower following light has been confirmed (29). However, the claim regarding HIOMT as a regulator of the melatonin rhythm didn't fare so well.

2. THE ROLE OF N-ACETYLTRANSFERASE IN REGULATING RHYTHMS IN PINEAL INDOLE METABOLISM

My entre into the pineal field came after hearing Russell Reiter give a seminar entitled "The Truth About the Pineal Gland", which I attended while both of us were at the University of Rochester. I came there in 1967 from Rice University after receiving my Ph.D. on the recently discovered thyroid hormone calcitonin. My goal in Rochester was to use *in vitro* techniques to study the hormonal control of bone metabolism as a fellow with Larry Raisz. After about a year of bone work, I realized that if I was going to make a contribution to science and establish a productive independent research program, the chances were slim that this could be done in the bone field. This area was filled with many well established, large and intensely competitive groups. I started to look around for new fields to conquer. I liked endocrinology and a look at my endocrinology text gave me my some indication of where I should go: the pineal gland or thymus. The reason? Not too much seemed to have been known about these glands since their chapters were the shortest in the book. Reiter's talk then helped turn me towards the pineal gland.

I learned more about melatonin primarily from articles by Axelrod's group, especially those that were coauthored by Dick Wurtman (12,25–27). These convinced me that there were lots of interesting open questions in the area. With Raisz's generous support and encouragement, I started to study melatonin.

The first step was to establish a pineal organ culture system, which proved to be easy—I simply replaced the bones with pineal glands. This worked and the glands seemed quite happy. They made melatonin from radiolabeled precursors and responded to norepinephrine (13,30), as had been reported by Axelrod and Wurtman, working with Harvey Schein (31). The melatonin response was robust, about 10-fold. My initial goal was to demonstrate that this was accompanied by an increase in HIOMT activity—which would confirm the then popular HIOMT hypothesis. However, norepinephrine treatment didn't increase HIOMT activity (13). Similarly, I discovered that dibutyryl cyclic AMP elevated melatonin production without increasing HIOMT activity (32), which provided more reason to question the role of HIOMT in regulating melatonin production.

The frustration with the *in vitro* results with HIOMT led me to reexamine the *in vivo* HIOMT rhythm. Again I was frustrated because I was unable to confirm the large nocturnal increase in activity that had been reported (21,25–27). I optimized the assay, following the directions of Angelo Notides, an excellent biochemist in the department, and also consulted with Axelrod and Wurtman. I tried everything. Still no luck—I was only able to find very small day/night difference at some ages (33), but was never able to find the large day/night rhythm that was in the literature.

I was faced with a puzzle. If HIOMT did not regulate the large change in melatonin production, what did? To get a better picture of what was taking place in the synthesis of melatonin I turned to Notides again for help and we developed a thin layer chromatography (TLC) system which separated serotonin oxidation and N-acetylation products (34). This was used to study effects of compounds which elevate melatonin production. The list now included the monoamine oxidase inhibitor harmine, in addition to norepinephrine and dibutyryl cyclic AMP (13,30,35). Glands were incubated with radiolabeled tryptophan or serotonin and a sample of gland homogenate or media extract was mixed with authentic standards and the mixture was resolved by TLC. In all cases the results were unequivocal—whenever there was an increase in the accumulation of radiolabeled melatonin production, the accumulation of radiolabeled N-acetylserotonin increased. I also found it was possible to increase melatonin production simply by elevating *N*-acetylserotonin. Based on these findings I suspected that the rate of melatonin production could increase simply by increasing the amount of *N*-acetylserotonin; and, that compounds which elevate melatonin production may do this by elevating the *N*-acetylation of serotonin.

In July of 1969 I moved to the National Institute of Child Health and Human Development (NICHD) of the National Institutes of Health (NIH). I had learned about the opportunity from Axelrod and was thrilled with the opportunity to work independently at the NIH.

I was given a tiny laboratory located in the basement of the living quarters of the women's branch of the U.S. Navy—the WAVES' Barracks. This was located in the National Naval Medical Center, which is separated from the NIH by a main thoroughfare. The labs were windowless (good for pineal work) and often flooded with heavy rains (bad for feet). These minor problems were easily ignored because waiting for me when I took the new job was a very nice surprise, an unusually capable and dedicated person, who shared my enthusiasm for pineal work, penchant for hard work and concern for detail—Joan Weller.

The pineal organ culture system was quickly reestablished and we started to work with it; we also tried to set up an assay to measure pineal serotonin N-acetyltransferase activity. This proved to be very difficult because we used daytime glands, which,—as we now know—have nearly undetectable levels of serotonin *N*-acetyltransferase activity. Hence, we were forced by circumstance to invent a very sensitive and specific assay so that we could selectively and reliably detect very low levels of serotonin *N*-acetyltransferase activity. The key element was to use TLC (34) to isolate products.

The first breakthrough with this assay came from the pineal organ culture system: treatment with norepinephrine or dibutyryl cyclic AMP treatment increased serotonin *N*-acetyltransferase activity (30,36). We also found that this was blocked by cycloheximide. I recall that the first results came on Thanksgiving Day.

The effects of norepinephrine or dibutyryl cyclic AMP on *N*-acetyltransferase activity were small. In 20–20 hindsight it is clear that our treatment period was not optimal—it was too long, allowing stimulated values to fall towards control levels. However, the response was sufficient to establish the fundamental principle that these compounds could elevate serotonin N-acetyltransferase activity and that this explained how they elevated melatonin production (30). Although these *in vitro* studies were important, they didn't provide enough evidence to prove what was happening *in vivo*. It was certainly possible that these *in vitro* results were not physiologically relevant.

I turned to *in vivo* studies and first tried to elevate serotonin *N*-acetyltransferase activity by keeping animals in the dark. This reasoning was based in part on the HIOMT studies. I was unfamiliar with the concepts of an endogenous clock and circadian rhythms, and only obtained tissue during normal working hours. I didn't know then that clock control of the enzyme would prevent an increase during the day. Soon, however, the concept of an endogenous clock would dawn on me.

The next big breakthrough came when I started to obtain tissue at night. There wasn't a proper animal room in the basement of the WAVES' Barracks, so I had to fix up a small closet with shelves for cages. I light-proofed the door and used a timer to control lighting. To obtain tissue, animals were taken down the hall to the laboratory and pineal glands were removed and placed in individual micro tubes—all this under room light.

When the glands were assayed, I was surprised and elated to see numbers I had never dreamed of seeing! Whereas I had become accustom to scintillation counter readouts of 100 or 200 DPM, I was now seeing some counts in the tens of thousands!!! I was still troubled, however, because there was enormous variation—some tubes had counts in the hundreds, some in the thousands and some in the tens of thousands.

Why the large variation? Two minor details were overlooked—the tubes were not numbered and I didn't record when animals were removed from the holding space and when the glands were put on dry ice. As a result, I had no way of determining if there was a correlation between enzyme activity and time in light. When I repeated the experiments I did a better job of record keeping and it became apparent that the highest DPMs came from animals that were in light for the least amount of time.

When I put this data together with daytime data, it was clear that I had discovered that *N*-acetyltransferase activity exhibited a very large day/night rhythm and that light exposure at night causes enzyme activity to drop very rapidly. So, in one single experiment I had the principles of two Science papers!! One reporting the day/night rhythm in serotonin N-acetyltransferase (37) and the second reporting light-induced rapid turn-off (38).

I was overwhelmed with excitement about the discoveries. I started to obtain night tissue without exposing animals to room light at all, by using dim red light. Soon I was able to demonstrate that the rhythm persisted in constant darkness, indicating it was driven by an endogenous clock (37).

The first publication of the rhythm received a lot of attention, because it redirected thinking in the field by establishing that serotonin *N*-acetyltransferase was the key regulator of the circadian rhythm in pineal melatonin and that the reciprocal changes in serotonin and melatonin were due to changes in the activity of serotonin N-acetyltransferase. It was proposed that serotonin N-acetylation limited the amount of melatonin made during the day; and, that at night melatonin production was increased by the increase in *N*-acetylserotonin which acted *via* mass action. The report also indicated that at night melatonin may be limited by HIOMT activity. This encompassing concept is valid for all vertebrates and has become a cornerstone of pineal biochemistry. For this reason the paper was identified as a Citation Classic and is one of the highest cited papers in the pineal and circadian literature (39). It is of incidental interest that another 1970 paper of mine, one on bone metabolism with Raisz, also became a Citation Classic (40). *What a great year!!!*

Although I never worked with Axelrod on the pineal gland, I consulted with him regularly about my work when I was at the NIH. At the time, no one in his laboratory was working on the pineal gland. Axelrod reviewed my manuscripts and showed me how to state things simply in easily understood language. The dramatic results he was reading about in my papers made him itchy to get back into the pineal field and as a result he eventually was part of highly productive work done with a series of coworkers, including Takeo Deguchi and Marty Zatz.

Although I quickly published the work on the acetyltransferase rhythm, the data on the rapid effect of light on serotonin N-acetyltransferase activity sat around for a while. I didn't realize how important it was until I looked at it side-by-side with a report from Helena Illnerová indicating that light exposure at night causes a very rapid increase in serotonin (41), which is otherwise very low at night. The rapid effect inhibitory effect of light on acetylation seemed to explain this. Photic suppression of N-acetyltransferase activity was of broad interest because of the rapidity of the effect, which suggested that remarkable regulatory mechanisms exist to trigger this. I wrote to Illnerová about my idea that the rapid light-induced increase in serotonin was due to the decrease in serotonin acetylation. She was in full agreement with my thinking and subsequently published extensively using serotonin N-acetyltransferase as an output marker for the suprachiasmatic nucleus (SCN) (42). I didn't meet Illnerová for several years. However, we finally met at a meeting on stress in Czechoslovakia in 1975. This meeting firmed up our scientific relationship and started a long friendship. It also was an important step towards establishing mechanisms through which Illnerová and one of her students, Jiri Vanecek, could come to the U.S. to work with me.

From the work done very early in the 70's it was clear that regulation of serotonin N-acetyltransferase was complex and important; and, that it begged for a molecular investigation. Although insufficient material was available for this type of work, it was possible to use the rhythm in serotonin N-acetyltransferase for other purposes. It played an important role as an output marker to describe the neural circuit which regulates the pineal gland (43). Of special note is the body of work which identified the SCN as the source of signals which control circadian rhythms in the pineal gland (44,45). This was accomplished in a collaborative study involving Bob Moore. This advance proved to be important on a broader scale because it helped establish the SCN as the "The Mind's Clock" (46).

Other important developments reflected some interesting pharmacological and biochemical observations. One came from Andy Parfitt, who found that depolarizing agents block adrenergic-cyclic AMP induction of serotonin N-acetyltransferase activity (47). David Sugden and I figured out that distinct β- and α-adrenergic components regulate the enzyme (48). "Namboo" M.A.A. Namboodiri established that serotonin

N-acetyltransferase was inactivated by protein thiol-disulfide exchange (49–51). And, the late Michel Buda found that a drop in cyclic AMP was the signal for the rapid decrease in enzyme activity (52).

Important advances were also made on the characterization of the enzyme. This started with communications with Wendell Weber, an expert on liver arylamine N-acetyltransferase. Both of us suspected the pineal enzyme with the rhythm in activity may not be arylamine N-acetyltransferase. We exchanged tissues and discovered that pineal homogenates which exhibited large differences in serotonin acetylation did not exhibit a day/night difference in arylamine N-acetyltransferase activity.

This convinced me that a detailed biochemical analysis of amine acetylation by the pineal gland might result in the characterization of a novel enzyme that was dedicated to melatonin synthesis. The effort was undertaken by Pierre Voisin and Namboodiri using both rat and sheep pineal glands (53). The work clearly established that the enzyme which exhibited the large day/night differences in serotonin acetylation was not arylamine N-acetyltransferase (EC 2.3.1.5), and that an arylalkylamine *N*-acetyltransferase (AANAT) could be chromatographically resolved from arylamine N-acetyltransferase in pineal homogenates. The enzymes also exhibited distinctly different biochemical characteristics. AANAT was given the designation EC 2.3. 1.87.

3. CLONING OF AANAT

Sometime after the identification of AANAT as a novel enzyme, it became apparent that the developing cloning technology would eventually allow us to obtain large amounts of AANAT and the tools necessary to study how it is regulated. At the start of this period, techniques were primitive and inefficient; certainly, molecular biology kits, reagents and commercial services were a thing of dreams. However, I was confident that sooner or later everything would fall into place and techniques required to clone AANAT would be available. I started to position ourselves accordingly by trying to do as much pineal-related molecular biology as possible. Joan Weller and I started with efforts aimed at purifying intact mRNA and making pineal cDNA libraries.

I also tried to identify relevant goals that would help us learn how to use molecular techniques. One of our first efforts was aimed at cloning an interesting protein found in both the pineal and retina—which was termed S-antigen at the time and now is called arrestin. I suspected that this could be cloned using an expression strategy in which protein was detected using antisera. At the time I was using Igal Gery's anti-S-antigen sera for other purposes and thought it might be used to clone S-antigen. I turned to Toshi Shinohara for help, because he had made and was using a bovine retinal cDNA library. We discussed cloning S-antigen and I brought him one of my postdoctoral fellows who wanted to learn molecular biology—Cheryl Craft. Bovine S-antigen was cloned shortly thereafter (54).

After getting our feet wet with the S-antigen project, we turned our attention to cloning other proteins of interest, including human HIOMT. This was done using a homology approach with probes based on the bovine sequence published by Takeo Deguchi's group (55). The first steps towards cloning human HIOMT were made by Joan Weller and Helena Illnerova; the work was completed by Susan Donohue who received help from Pat Roseboom (56). This was done entirely within our group and represented an important training experience.

Still looming over our program was the goal of cloning serotonin N-acetyltransferase. Several approaches had been tried over the years, all without success. Things started to happen in 1994. At that time the molecular biology experience and expertise in the laboratory was high. The program included Ruben Baler, who brought a sharp and nimble mind and excellent technical knowledge. Baler had just finished a challenging project on the identification and regulation of pineal Fos related protein-2 and was in full stride (57). Pat Roseboom had gained significant hands-on experience by cloning two interesting proteins, 14-3-3 ε and ζ (58), which in fact represented the unintended outcome of a failed attempt to clone AANAT.

Other members of the team were Steve Coon, who joined the program because of a common interest in adrenergic receptors and his desire to learn molecular biology; and, Marianne Bernard, who was strongly focused on the molecular regulation of HIOMT. The group was formidable in both expertise and intellectually combativeness. I convinced nearly everyone that it would be to each person's benefit if AANAT were to be cloned and it was agreed that other projects would be given lower priority, so that all possible approaches to cloning AANAT could be tried or retried.

The method that worked was one in which we had the least amount of confidence—an expression cloning strategy. We were encouraged by the success of Steve Reppert's program with this method, which was used to clone the first melatonin receptor (59). Immediately after hearing about this in February of 1996 at a Gordon Conference we set up the method. We thought that the enzyme was too unstable and the pineal gland was too specialized for this to occur in nonpineal cells. However, we were wrong.

The expression cloning strategy was spearheaded by Coon, who generated an ovine cDNA library and used a very sensitive AANAT activity assay that Baler and I had developed to detect expression. The method resolved products by TLC and visualized them with the PhosphoImager. Everything worked quite well and the start-to-finish time for this effort was remarkably short. Images verifying that AANAT had been cloned were faxed to me in May when I was in Italy attending a meeting. Immediately after cloning the ovine enzyme, the rat enzyme was cloned by Pat Roseboom. In July the paper describing the cloning of ovine serotonin N-acetyltransferase was submitted1. It included the findings that large rhythm in enzyme activity was accompanied by a large AANAT mRNA rhythm in the rat, but nearly no rhythm in the sheep. Analysis of the sequences indicated that AANAT and arylamine *N*-acetyltransferase were in different superfamilies—explaining why all previous efforts based on homology screening never worked.

This advance played an important role in the cloning of AANAT from other vertebrates and in the molecular analysis of the enzyme (8). Some of the most important advances made in my laboratory involved analysis of the AANAT promoter by Ruben Baler (6), which has led to rat transgenesis studies with David Carter. These have identified a sequence in the AANAT gene which directs tissue and time-of-day specific expression in transgenic rats. It might allow us to construct lines of rats that express genetic information of our choice selectively in the pineal gland.

Work by Pat Roseboom described regulation of AANAT mRNA in the rat (5) and work done by Marianne Bernard demonstrated that the expression of the AANAT gene in the chicken is regulated in a circadian manner and is driven by a pineal clock (7,56). This reflected collaborative projects done with Mike Iuvone and Marty Zatz. Work on the chicken pineal and retina is being extended by Nelson Chong and Iuvone.

The cloning of AANAT has led to a productive collaboration on the fish pineal gland with Válerie Bégay, Jack Falcón, Greg Cahill and Steve Coon (9). This established that the pineal clock drives a rhythm in AANAT mRNA in the pike and zebrafish, but that AANAT mRNA in the trout is expressed in a tonic manner. This fish effort also had the unexpected and surprising result of the identification of a second AANAT gene in pike. This may lead to identification of such a second gene in other species. Related to these studies on fish AANAT are efforts by Yoav Gothilf in my group to use AANAT as a marker gene to use to study pineal development in the zebrafish. He's detected AANAT mRNA in the embryonic pineal gland 22 hours postfertilization!

Use of the ovine AANAT sequence made it possible for Steve Coon and Ignacio Rodriguez to clone the human gene and to determine the chromosomal localization (3). This advance may provide a key step towards the identification of a molecular basis of low melatonin production in humans, as has been done by Pat Roseboom in the mouse. Roseboom found that low levels of AANAT activity occur in the C57/Bl mouse because of a mutation that causes missplicing and results in formation of a severely truncated protein (61).

One of the minor surprises which came out of the cloning of AANAT was that the gene is expressed in places and patterns that were not expected (8). Although it was expected that the retina would express the gene, Pat Roseboom discovered that rat retinal AANAT mRNA exhibits a day/night rhythm (5). And, Vince Cassone made the surprising finding that chick retina AANAT mRNA occurs in non-photoreceptor cells (7). Several efforts by Coon with others have resulted in the identification of AANAT transcripts outside the pineal gland and retina, including the pituitary gland, certain brain regions and the ovary (1–3,7,62). Others have discovered AANAT in the testis (63).

Soon after AANAT was cloned, we started to make antisera. Expressed protein and peptides were used as antigens. The first efforts to make anti-ovine AANAT by Joan Weller provided evidence that there was a close association between AANAT activity and protein. Jonathan Gastel extended this into studies in the rat, and made the very important discovery that the abundance of AANAT protein is regulated by proteasomal proteolysis, which explains how pineal AANAT activity rapidly disappears following exposure to light, at night (11). It is suspected that proteolysis involves a highly conserved lysine in the N-terminal region of the protein. It will be of interest to determine if this mechanism is conserved and explains photic suppression of AANAT activity in all vertebrates.

Kinetic analysis of AANAT was made possible by the preparation of a GST-expression construct by Jonathan Gastel. Using this construct, Phil Cole has been able to study the mechanism of acetylation and to predict that acetyl transfer involves the formation of an intermediate quarternary tryptamine-AcCoA complex (10). This is being supported by the results of structural analysis of AANAT by Alison Hickman and Fred Dyda (64). The structural work has used a highly soluble form of ovine AANAT to obtain crystals for X-ray crystallography. This reflects an important collaboration with Servier and their intention to develop AANAT-targeted drugs.

A striking outcome of the molecular analysis of AANAT regulation is the realization that there are distinct species-to-species differences in the mechanisms through which AANAT is regulated in vertebrates. However, it is clear that these represent "variations on a theme" and that in nearly all cases these mechanisms serve to insure that melatonin synthesis is only elevated at night in the dark. The apparent differences seem to serve to fine-tune the system so that it functions to optimize survival and behavior as required for each species.

4. THE FUTURE OF AANAT

In what directions will AANAT research go? It is reasonable to predict that continued investigations of AANAT will lead to a better understanding of clock control of gene expression in vertebrates; this should come form studies on the chicken and pike. Studies on rat AANAT promoters will shed light on the molecular basis of control of tissue specific expression of genes in the pineal gland and retina. Analysis of AANAT degradation should lead to a detailed understanding of this process in the pineal gland and will also reveal how proteolysis is regulated in the nervous system. Pharmacology studies are expected to lead to the identification of agents that modify AANAT activity in the pineal gland and elsewhere, with the possible evolution of such agents into AANAT-directed drugs that influence sleep and behavior. I expect that studies on the structure of AANAT will enhance our understanding of fundamental elements of enzyme action and will also provide a description of three dimensional surfaces that will be important in designing drugs and in understanding protein-protein interactions. There is good reason to suspect that the use of AANAT as a marker will enhance our understanding of embryological differentiation. I'm confident that these predictions will be realized. I am equally confident that there are lots of eager, imaginative and creative young scientists unknown to me who will make spectacular and surprising advances with AANAT in areas I haven't touched. Seeing how this develops will be interesting.

REFERENCES

1. Coon, S.L., Roseboom, P.H., Baler, R., Weller, J.L., Namboodiri, M.A.A., Koonin, E.V., and Klein, D.C. Pineal Serotonin N-acetyltransferase (EC 2.3.1.87): Expression cloning and molecular analysis. Science 270:1681–1683, 1995.
2. Roseboom, P.H. and Klein, D.C. Norepinephrine stimulation of pineal cyclic AMP response element-binding protein phosphorylation: Primary role of a β-adrenergic receptor/cyclic AMP mechanism. Mol Pharmacol 47:439–449, 1995.
3. Coon, S.L., Mazuruk, K., Bernard, B., Roseboom, P.H., Klein, D.C., and Rodriguez, I.R. The human serotonin N-acetyltransferase (EC 2.3.1.87) Gene: Structure, chromosomal localization and tissue expression. Genomics 34:76–84, 1996.
4. Klein, D.C., Roseboom, P.H., and Coon, S.L. New light is shining on the melatonin rhythm enzyme: The first post cloning view. Trends in Endocrinology and Metabolism 7:106–112, 1996.
5. Roseboom, P.H., Coon, S.L., Baler, R., McCune, S.K., Weller, J.L., and Klein, D.C. Melatonin synthesis: Regulation of the more than 150-fold nocturnal increase in serotonin N-acetyltransferase messenger RNA in the rat pineal gland. Endocrinol 137:3033–3044, 1996.
6. Baler, R., Covington, S., and Klein, D.C. The rat arylalkylamine *N*-acetyltransferase gene promoter: cAMP activation via a CRE/CCAAT complex. J Biol Chem 272:6979–6985, 1997.
7. Bernard, M., Iuvone, P.M., Cassone, V.M., Roseboom, P.H., Coon, S. L., and Klein, D.C. Melatonin synthesis: Photic and circadian regulation of serotonin *N*-acetyltransferase mRNA in the chicken pineal gland and retina. J Neurochem 68:213–222, 1997.
8. Klein, D.C., Coon, S.L., Roseboom, P.H., Weller, J.L., Bernard, M., Gastel, J.A., Zatz, M., Iubone, M., Rodriquez, I.R., Begay, V., Falcon, J., Cahill, G., Cassone, V.M., and Baler, R. The melatonin rhythm generating enzyme: Molecular regulation of serotonin *N*-acetyltransferase in the pineal gland. Rec Prog Hormone Res 52:307–358, 1997.

9. Bégay, V., Falcón, J., Cahill, G., Klein, D.C., and Coon, S. Transcripts encoding two melatonin synthesis enzymes in the teleost pineal organ: Circadian regulation in pike and zebrafish, but not in trout. Endocrinol 139:905–912, 1998.
10. DeAngelis, J., Gastel, J.A., Klein, D.C., and Cole, P.A. Kinetic mechanism of serotonin *N*-acetyltransferase (EC 2.3.1.87). J Biol Chem 273:3045–3050, 1998.
11. Gastel, J.A., Roseboom, P.H., Rinaldi, P.A., Weller, J.L., and Klein, D.C. Melatonin production: Proteasomal proteolysis in serotonin *N*-acetyltransferase regulation. Science 279:1358–1360, 1998.
12. Wurtman, R.J. and Axelrod, J. The formation metabolism, and physiological effects of melatonin. Adv Pharmacol 6A:141–149, 1968.
13. Klein, D.C. and Berg, G.R. Pineal gland: Stimulation of melatonin production by norepinephrine involves cyclic AMP mediated stimulation of N-acetyltransferase. Adv Biochem Psychopharmacol 5:241–263, 1970.
14. Klein, D.C. The pineal gland: A model of neuroendocrine regulation. In: The Hypothalamus. Reichlin, S., Baldessarini, R.J., and Martin, B. (Eds.) Raven Press, New York, pp. 303–327, 1978.
15. Klein, D.C. Photoneural regulation of the mammalian pineal gland. (ed.) D. Evered and S. Clark. In: Photoperiodism, Melatonin and the Pineal. Pitman, London (Ciba Foundation Symposium 117) pp. 38–56, 1985.
16. Klein, D.C., Schaad, N., Namboodiri, M.A.A., Yu, L., and Weller, J.L. Regulation of pineal serotonin N-acetyltransferase activity. Biochem Soc Trans 20:299–304, 1992.
17. Lerner, A.B., Case, J.D., and Heinzelman, R.V. Structure of melatonin. J Am Chem Soc 81:6084–6085, 1959.
18. Lerner, A.B., Case, J.D., Takahashi, Y., Lee, T.H., and Mori, W. Isolation of melatonin, the pineal gland factor that lightens melanocytes. J Am Chem Soc 80:2587–2592, 1958.
19. Quay, W.B. and Halevy, A. Experimental modification of the rat pineal's content of serotonin and related indole amines. Physiol Zool 35:1–7, 1962.
20. Weissbach, H., Redfield, B.G., and Axelrod, J. Biosynthesis of melatonin; Enzymic conversion of serotonin to N-acetylserotonin. Biochim Biophys Acta 43:352–353, 1960.
21. Axelrod, J. and Weissbach, H. Enzymatic O-methylation of N-acetylserotonin to melatonin. Science 131:1312, 1960.
22. Quay, W.B. Circadian rhythm in rat pineal serotonin and its modification of estrous cycle and photoperiod. Gen Compl Endocrinol 3:1473–1479, 1963.
23. Quay, W.B. Circadian and estrous rhythms in pineal melatonin and 5-hydroxyindole-3-indole acetic acid. Proc Soc Exp Biol Med 115:710–714, 1964.
24. Axelrod, J., MacLean, R.D., Albers, R.W., and Weissbach, H. Regional distribution of methyl transferase enzymes in the nervous system and glandular tissues. Regional Neurochemistry, S.S. Kety and J. Elkes, eds., Pergamon Press, Oxford, pp. 307–311, 1961.
25. Wurtman, R.J., Axelrod, J., and Fischer, J.E. Melatonin synthesis in the pineal gland: Effect of light mediated by the sympathetic nervous system. Science 143:1328–1329, 1964.
26. Axelrod, J., Wurtman, R.J., and Snyder, S.H. Control of hydroxyindole-O-methyltransferase activity in the rat pineal gland by environmental lighting. J Biol Chem 240:949–953, 1965.
27. Wurtman, R.J., Axelrod, J., and Philips, L.W. Melatonin synthesis in the pineal gland: Control by light Science 142:1071–1072, 1963.
28. Buda, M. and Klein, D.C. N-acetylation of biogenic amines. In: Biochemistry and function of amine enzymes. Usdin, E., Weiner, N., and Youdim, M. (eds.) M. Dekker, New York, pp. 527–544, 1978.
29. Sugden, D. and Klein, D.C. Regulation of rat pineal hydroxyindole-O-methyltransferase in neonatal and adult rats. J Neurochem 40:1647–1653, 1983.
30. Klein, D.C., Berg, G.R., and Weller, J.L. Melatonin synthesis: Adenosine 3′,5′-monophosphate and norepinephrine stimulate N-acetyltransferase. Science 168:979–980, 1970.
31. Axelrod, J., Shein, H.M., and Wurtman, R.J. Stimulation of C^{14}-tryptophan by norepinephrine in rat pineal in organ culture. Proc Natl Acad Sci USA 62:544–549, 1969.
32. Klein, D.C., Berg, G.R., Weller, J.L., and Glinsmann, W. Pineal gland: Dibutyryl cyclic adenosine monophosphate stimulation of labeled melatonin production. Science 167:1738–1740, 1970.
33. Klein, D.C. and Lines, S.V. Pineal hydroxyindole-O-methyltransferase activity in the growing rat. Endocrinol 84:1523–1525, 1969.
34. Klein, D.C. and Notides, A. Thin-layer chromatographic separation of pineal gland derivatives of serotonin-^{14}C. Anal Biochem 31:480–483, 1969.
35. Klein, D.C. and Rowe, J. Pineal gland in organ culture. I. Inhibition by harmine of serotonin-^{14}C oxidation, accompanied by stimulation of melatonin-^{14}C production. Mol Pharmacol 6:164–171, 1970.

36. Berg, G.R. and Klein, D.C. Pineal gland in organ culture. II. Role of adenosine 3′,5′-monophosphate in the regulation of radiolabeled melatonin production. Endocrinol 89:453–464, 1971.
37. Klein, D.C. and Weller, J.L. Indole metabolism in the pineal gland: A circadian rhythm in N-acetyltransferase. Science 169:1093–1095, 1970.
38. Klein, D.C. and Weller, J.L. Rapid light-induced decrease in pineal serotonin N-acetyltransferase activity. Science 177:532–533, 1972.
39. Klein, D.C. Citation Classic: Commentary on Klein, D.C. and Weller J.L. Indole metabolism in the pineal gland: A circadian rhythm in N-acetyltransferase. Science 169:1093–1095 (1970) Current Contents 31:16, 1988.
40. Klein, D.C. Citation Classic: Commentary on Klein, D.C. and Raisz L.G. Prostaglandins: Stimulation of bone resorption in tissue culture. Endocrinology 86:1436–1440 (1970) Current Contents 30:17, 1988.
41. Illnerová, H. Effect of light on the serotonin content of the pineal gland. Life Sci 10:955–961, 1971.
42. Illnerová, H. The suprachiasmatic nucleus and rhythmic pineal melatonin production In: *Suprachiasmatic Nucleus* (ed. Klein D.C., Moore R.Y., and Reppert S.M.) Oxford Press, New York, pp. 197–216, 1991.
43. Klein, D.C., Weller, J.L., and Moore, R.Y. Melatonin metabolism: Neural regulation of pineal serotonin: acetyl coenzyme A N-acetyltransferase activity. Proc Natl Acad Sci USA 68:3107–3110, 1971.
44. Moore, R.Y. and Klein, D.C. Visual pathways and the central neural control of a circadian rhythm in pineal serotonin N-acetyltransferase activity. Brain Res 71:17–33, 1974.
45. Klein, D.C. and Moore, R.Y. Pineal N-acetyltransferase and hydroxyindole-O-methyltransferase: control by the retinohypothalamic tract and the suprachiasmatic nucleus. Brain Res 174:245–262, 1979.
46. Klein, D.C., Moore, R.Y., and Reppert, S.M. (eds.) Suprachiasmatic Nucleus: The Mind's Clock. Oxford University Press, NY, 1991.
47. Parfitt, A., Weller, J.L., Klein, D.C., Sakai, K.K., and Marks, B.H. Blockade by ouabain or elevated potassium ion concentration of the adrenergic and adenosine cyclic 3′,5′-monophosphate-induced stimulation of pineal serotonin N-acetyltransferase activity. Mol Pharmacol 11:241–255, 1975.
48. Klein, D.C., Sugden, D. and Weller, J.L. Postsynaptic α-adrenergic receptors potentiate the β-adrenergic stimulation of pineal serotonin N-acetyltransferase. Proc Natl Acad Sci USA 80:599–603, 1983.
49. Namboodiri, M.A.A., Weller, J.L., and Klein, D.C. Evidence of inactivation of pineal indoleamine N-acetyltransferase by protein thiol-disulfide exchange. J Biol Chem 255:6032–6035, 1980.
50. Namboodiri, M.A.A., Favilla, J.T., and Klein, D.C. Pineal N-acetyltransferase is inactivated by disulfide-containing peptides: Insulin is the most potent. Science 213:571–573, 1981.
51. Klein, D.C. and Namboodiri, M.A.A. Control of the circadian rhythm in pineal serotonin N-acetyltransferase activity: possible role of protein thiol: disulfide exchange. Trends in Biochemical Science 7:98–102, 1982.
52. Klein, D.C., Buda, M., Kapoor, C., and Krishna, G. Pineal serotonin N-acetyltransferase activity: Abrupt decrease in adenosine 3′,5′-monophosphate may be signal for "turnoff". Science 199:309–311, 1978.
53. Voisin, P., Namboodiri, M.A.A., and Klein, D.C. Arylamine N-acetyltransferase and arylalkylamine N-acetyltransferase in the mammalian pineal gland. J Biol Chem 259:10913–10918, 1984.
54. Shinohara, T., Craft, C.M., Stein, P., Ziegler Jr., J.S., Wistow, G., Katial, A., Gery, I., and Klein, D.C. Isolation of cDNAs for bovine S-antigen. In: Pineal and Retinal Relationships. P.J. O'Brien and D.C. Klein (eds.) Academic Press, Orlando pp. 331–343, 1986.
55. Ishida, I., Obinata, M., and Deguchi, T. Molecular cloning and nucleotide sequence of cDNA encoding hydroxyindole-O-methyltransferase of bovine pineal glands. J Biol Chem 262:2895–2899, 1987.
56. Donohue, S.J., Roseboom, P.H., Illnerová, H., Weller, J.L., and Klein, D.C. Human hydroxyindole-O-methyltransferase: Presence of LINE-1 fragment in a cDNA clone and pineal mRNA. DNA and Cell Biol 12:715–727, 1993.
57. Baler, R. and Klein, D.C. Circadian expression of transcription factor Fra 2 in the rat pineal gland. J Biol Chem 270:27319–27325, 1995.
58. Roseboom, P.H., Weller, J.L., Babila, T., Aiken, A., Sellers, L.A., Moffett, J.R., Namboodiri, M.A.A., and Klein, D.C. Cloning and characterization of the ε and ζ isoforms of the rat 14-3-3 proteins. DNA and Cell Biol 13:629–640, 1994.
59. Ebisawa, T., Karne, S., Learner, M.R., and Reppert, S.M. Expression cloning of a high-affinity melatonin receptor from *Xenopus* dermal melanophores. Proc Natl Acad Sci USA 91:6133–6137, 1994.
60. Bernard, M., Klein, D.C., and Zatz, M. Chick pineal clock regulates serotonin *N*-acetyltransferase mRNA rhythm in culture. Proc Natl Acad Sci USA 94:304–309, 1997.

61. Roseboom, P.H., Namboodiri, M.A.A., Zimonjic, D.B., Popescu, N.C., Rodriguez, I.R., Gastel, J.A., and Klein, D.C. Natural melatonin knockdown in C57BL/6J mice: Rare mechanism truncates serotonin *N*-acetyltransferase. Mol Brain Res 63:189–197, 1998.
62. Fleming, J.V., Barrett, P., Coon, S.L., Klein, D.C., and Morgan, P.J. Ovine arylalkylamine N-acetyltransferase in the pineal and pituitary glands: differences in function and regulation. Endocrinology 140:972–978, 1999.
63. Borjigen, J., Wang, M.M., and Snyder, S.H. Diurnal variation in mRNA encoding serotonin N-acetyltransferase in pineal gland. Nature 378:783–785, 1995.
64. Hickman, A.B., Klein, D.C., and Dyda, F. Melatonin biosynthesis: The structure of serotonin N-acetyltransferase at 2.5 Å resolution suggests a catalytic mechanism. Mol Cell 3:23–32, 1999.

3

EVOLUTION OF MELATONIN-PRODUCING PINEALOCYTES

Horst-W. Korf

Dr. Senckenbergische Anatomie
Anatomisches Institut II
Universitätsklinikum der Johann Wolfgang Goethe-Universität
Theodor-Stern-Kai 7, 60590 Frankfurt/Main

1. INTRODUCTION

In all vertebrate species, melatonin is produced during darkness and its formation is inhibited by light. Melatonin can thus be considered as the hormone of darkness. Because of its lipophilic nature, melatonin reaches all parts of an organism and plays an important role in several physiological processes throughout all vertebrate classes from lamprey to man (1). In view of melatonin's widespread effects, the organs and cells producing melatonin and their fate during phylogeny are of considerable neurobiological interest. As far as vertebrates are concerned it is generally accepted that melatonin is made in two organs, the retina and the pineal. Obviously melatonin made in the retina serves as an intrinsic signaling substance and does not enter the general circulation under normal conditions. By contrast melatonin made in the pineal serves as a neurohormone which is released into the pineal capillaries and accounts for the melatonin levels in the bloodstream. The retina and pineal organ share several common features. 1) They are derivatives of the diencephalon and as such parts of the central nervous system. 2) They are the two major photoreceptor organs of vertebrates. 3) Depending on the species, both organs are capable of generating circadian (endogenous) rhythms, which persist in the absence of environmental cues (so called zeitgebers). All these aspects underline the firm evolutionary and functional link between melatonin production, generation of circadian rhythms and photoreception/phototransduction.

Tel.: +69 6301 6040; Fax: +69 6301 6017; e-mail: korf@em.uni-frankfurt.de

Melatonin after Four Decades, edited by James Olcese.
Kluwer Academic / Plenum Publishers, New York, 2000.

The phylogenetic development of the two melatonin-producing organs is strikingly different: whereas the overall organization of the retina is well conserved and very similar in all vertebrate classes, the pineal organ has undergone a complex transformation in the course of evolution which affects the structural and functional features of the principal cell type of the pineal parenchyma, the pinealocytes, as well as the neuronal connections. In poikilothermic vertebrates, the pineal organ is neuronally connected to the brain via prominent pinealofugal pathways conveying synaptic signals to a variety of brain areas. These connections are gradually reduced during phylogeny. Conversely, the sympathetic innervation of the pineal complex originating from the superior cervical ganglion develops progressively in the course of evolution and is most prominent in mammals, but absent in lampreys and teleosts (34,43). This contribution will touch upon some evolutionary aspects of the melatonin-producing cells in the pineal organ, the pinealocytes.

2. TYPES OF PINEALOCYTES

According to structural and ultrastructural criteria vertebrate pinealocytes have been subdivided into three major categories: true pineal photoreceptors, modified pineal photoreceptors and pinealocytes sensu stricto (11,59; cf. 43). True pineal photoreceptors are restricted to anamniotes, modified pineal photoreceptors are predominantly found in lacertilian and avian species and pinealocytes sensu stricto form the parenchyma of the mammalian and also the ophidian pineal gland.

2.1. True Pineal Photoreceptors

True pineal photoreceptors bear an outer segment that protrudes into the pineal lumen and consists of numerous disks produced by successive basoapical invaginations of the plasma membrane (Figure 1). Depending on the species, the number of outer segment disks varies between 10 and 300. The outer segment is connected to the inner segment via a cilium of the $9 \times 2 + 0$ type. Opposite to the outer segment, the pineal photoreceptor gives rise to a basal process originating from the perikaryon and contributing to prominent intrapineal neuropil formations. Its enlarged terminals contain numerous electron lucent synaptic vesicles intermingled with synaptic ribbons and scattered dense core vesicles (Fig. 1). Via these terminals, the true pineal photoreceptors establish synapses with intrapineal second-order neurons. Adjacent photoreceptor cells are connected via gap junctions, suggesting that they are electrically coupled (cf. 19).

By means of immunocytochemical and biochemical investigations it was shown that pineal photoreceptors contain molecules of the phototransduction cascade which are very closely related to or even identical with those expressed by retinal photoreceptors. Thus, immunoreaction for rod-opsin, the proteinous component of the rod visual pigment rhodopsin has been found in the outer segments of many pineal photoreceptors in lamprey (77), teleosts (85), frogs (84) and certain reptiles (81). In the pineal of the clawed toad and the agamid lizard (*Uromastix hardwicki*) very few or no rod-opsin immunoreactive pineal outer segments were found (26,32). These species, however, have pineal photoreceptors that react with antibodies raised against chicken cone-opsin. Cone-opsin immunoreactive outer segments are also found in ranid frogs, but they are less frequent than the rod-opsin immunoreactive outer segments. To date it has not yet been investigated whether a subpopulation of true pineal photoreceptors

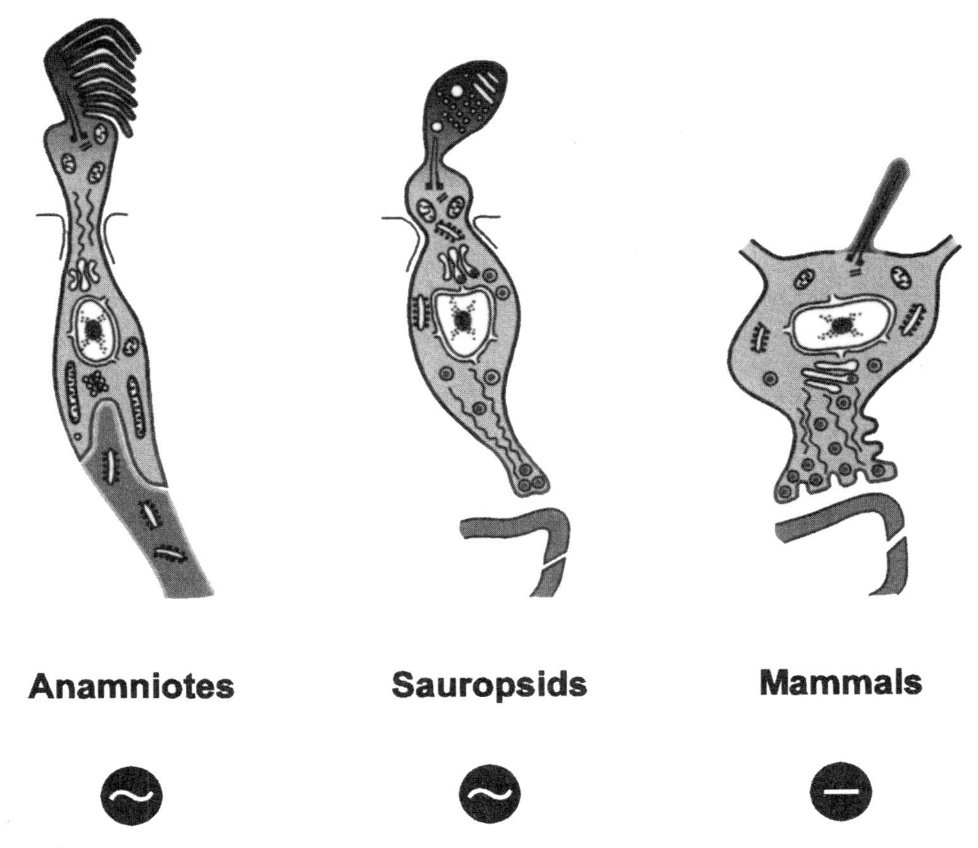

Figure 1. The three principal types of pinealocytes. Classical pineal photoreceptors of anamniotes, modified photoreceptors of sauropsids and pinealocytes sensu stricto of mammals. Classical and modified photoreceptors are directly light-sensitive and may harbor an endogenous oscillator (~). Pinealocytes of adult mammals do not contain an oscillator (–) and have lost the direct light sensitivity.

contains the pineal-specific photopigment pinopsin, which has been cloned from the chicken pineal organ. This novel photopigment is different from all known retinal photopigments, but has 45% homology with the opsin of blue cones (47,58). Recently, Blackshaw and Snyder (6) cloned an opsin which is specifically expressed in the pineal complex (pineal and parapineal organs) of the channel catfish. The parapinopsin amino acid sequence is 40–45% identical to chicken pinopsin, but it does not appear as an orthologue of chicken pinopsin. A high level of parapinopsin expression was found in the majority of parapineal cells, but it occurs only in a limited number of pineal photoreceptors which appear concentrated in the proximal portion of the pineal where they make up 8% of the cells. Parapinopsin was not found in retinal photoreceptors and probes for catfish rhodopsin and red cone pigment only labeled retinal, but no pineal photoreceptors. Taken together, all findings indicate that multiple types of true pineal photoreceptors exist; some have pineal-specific photopigments, others are closely related to retinal rods or cones.

Microspectrographic and electrophysiological experiments have also revealed the presence of multiple photopigments (19,51,71). For further and more comprehensive discussion on pineal photopigments and neurophysiological properties of true pineal photoreceptors, see Ekström and Meissl (19). Many true pineal photoreceptors share additional molecular features with their retinal counterparts; they express immunoreactive alpha-transducin (80), S-antigen (40) and recoverin (42).

The concept that true pineal photoreceptors belong to the neuronal cell lineage is supported by the demonstration of neuronal markers in these cells (neuron-specific enolase: 60; neurofilament 200 kDa: 8). Interestingly, also retinal photoreceptors were shown to contain neurofilament immunoreaction (8). In both retinal and pineal photoreceptors, the immunoreactive neurofilament is confined to the axoneme which connects the outer with the inner segment. The observations suggest that neurofilaments form a part of the photoreceptor cytoskeleton.

The neurotransmitter employed by pineal photoreceptors has not yet been precisely identified. Electrophysiological studies in frogs have suggested that the pineal photoreceptors—like their retinal counterparts—utilize excitatory amino acids, i.e., glutamate or aspartate as transmitter (52,53). In frogs and reptiles, glutamate and aspartate have been shown in pineal photoreceptors of the pineal organ and parietal eye, respectively, by means of immunocytochemical techniques (82,83). These findings conform to earlier biochemical results showing that glutamate and aspartate are present in the trout (54) and goldfish pineal (48).

There are several lines of evidence that true photoreceptor cells are the source of melatonin produced in the anamniote vertebrate pineal organ. Thus, hydroxyindole-*O*-methyltransferase, the ultimate enzyme of the melatonin biosynthesis, has been immunocytochemically identified in virtually all pineal pineal photoreceptors of teleosts (21). Also, serotonin, the precursor of melatonin has been localized to true pineal photoreceptors by means of immunocytochemistry, but, depending on the species, there are conspicous differences from cell to cell (cf. 18,77). In the lamprey, *Lampetra japonica*, the pineal photoreceptors which bear long rod-opsin immunoreactive outer segments were shown to lack serotonin immunoreactivity that could, however, be demonstrated in two other types of pinealocytes (77). One of these had a short rod-opsin immunoreactive outer segment and displays features of the modified photoreceptor cell; the other cell type was located in the proximal portion of the pineal, the atrium. It lacked an outer segment and contained only a weak S-antigen immunoreaction, thus resembling the pinealocyte sensu stricto found in the mammalian pineal organ. These results speak in favor of cell-to-cell differences in the indoleamine metabolism. They fully conform to earlier electron microscopic and fluorometric results (50) and support the notion that multiple types of pinealocytes are present at an early evolutionary state. Also in the pineal organ of the pike, true pineal photoreceptors were found to coexist with modified photoreceptor cells (20). These findings are relevant for considerations on the cytoevolution of melatonin-producing pinealocytes. One idea is that mammalian pinealocytes represent the final stage of a gradual transformation of true pineal photoreceptors characteristic of anamniotes. The occurrence of all three types of pinealocytes (i.e., true and modified photoreceptors as well as pinealocytes of the mammalian type) in the pineal organ of the lamprey (a most basic vertebrate) suggests that they may have evolved in parallel.

2.2. Modified Pineal Photoreceptors

Modified pineal photoreceptors are endowed with a rudimentary outer segment which is less regular than that of true pineal photoreceptors. Some modified photoreceptors have only a bulbous cilium lacking membrane disks (Fig. 1). The basal process of the cells contains synaptic ribbons intermingled with clear vesicles and dense-core granules. The number of the latter varies considerably on a species-to-species basis but, in general, these granules are much more numerous in modified than in true pineal photoreceptors. The dense-core granules are taken as an indication of a high secretory activity of the modified photoreceptor cells, although the content of these granules has not been clarified. According to current concepts, melatonin is not stored within the pinealocyte, but immediately released after its formation. Thus, it may be speculated that these granules contain a neuroactive substance that is different from melatonin. Since they closely resemble the granules of peptidergic neurons, one may assume that they harbor a peptide that may be co-released with melatonin. In this context, it is interesting that modified photoreceptors, but also true pineal photoreceptors and pinealocytes sensu stricto react with antibodies directed against secretory proteins of the subcommissural organ (66). The basal processes of modified photoreceptors terminate adjacent to the basal lamina or are apposed to basal processes of other modified photoreceptors. Obviously the basal processes of modified photoreceptors do not form synaptic contacts with intrapineal neurons.

As shown by immunocytochemistry, modified photoreceptors of the avian pineal organ contain photoreceptor-specific proteins, but again conspicuous species differences are evident. Several rod-opsin immunoreactive outer segment remnants or whole cells were found in the pineal of the pigeon (81) and Pekin duck (35). In contrast, only very few rod-opsin positive elements were found in the Japanese quail (22) and chicken (33). Most probably, the latter finding can be explained by the fact that in the chicken pineal organ, pinopsin is the predominating photopigment (28,47,58) and that the rod-opsin antibody used did not cross-react with pinopsin that has 45% homology with the opsin of blue cones in the retina. Interestingly, also the S-antigen immunoreaction, which is the marker most widely distributed in all types of vertebrate pinealocytes, is absent from the chicken pineal organ, although the S-antigen immunoreaction is expressed by photoreceptors in the chicken retina. These results suggest that the arrestin (S-antigen) molecule of the pinopsin phototransduction cascade differs from the arrestin of the rhodopsin phototransduction cascade. Many strongly S-antigen immunoreactive modified pineal photoreceptors were found in the quail, duck and pigeon. The latter species were also shown to contain alpha-transducin-, recoverin-and interstitial retinol binding protein (IRBP) immunoreactive cells (22,42,80).

The immunocytochemical results conform to studies showing the direct light sensitivity of the chicken pineal organ (13,76,91) and the presence of 11-cis retinal, the prosthetic group of any vertebrate photopigment, and its light-dependent stereoisomerization in the quail pineal organ (23). As shown by in vitro experiments, light stimuli elicit an acute inhibition of the melatonin biosynthesis in cultured chicken pineal organs. Thus, the directly light-sensitive cells in the chicken pineal organ appear to translate the photic stimulus into a neuroendocrine response and can be classified as a photoneuroendocrine cell. The capacity of the modified photoreceptor cells to produce melatonin has been corroborated by immunocytochemical demonstration of serotonin and HIOMT in these cells (5,86). Moreover, in the chicken pineal some multipolar cells (interfollicular cells) displayed immunoreactive HIOMT.

2.3. Pinealocytes Sensu Stricto

Pinealocytes sensu stricto (Fig. 1) form the main cellular component of the mammalian and also of the ophidian pineal organ. In some species, e.g., the hamster and the opossum, certain pinealocytes are directly exposed to the cerebrospinal fluid and can be classified as CSF-contacting pinealocytes (39,40). Mammalian pinealocytes lack outer and inner segments, but contain cilia of the 9 + 0 type presumably representing remnants of the receptor pole. Although one type of pinealocyte has been shown to possess long axonlike processes leaving the pineal organ and penetrating into the brain (40,41), most pinealocytes bear processes terminating within the pineal parenchyma at the basal lamina of the perivascular space. These processes contain synaptic ribbons and a varying number of clear and dense core vesicles (87).

There is general agreement that mammalian pinealocytes produce melatonin. However, it is not clear whether melatonin biosynthesis occurs in all or only in certain cells. Immunocytochemical demonstration of melatonin has revealed conflicting results and is considered unreliable by most immunocytochemists. Immunocytochemical demonstration of serotonin, the precursor of melatonin, has shown an even distribution in the mouse pineal organ, but in the gerbil pineal the serotonin immunoreactivity displays a conspicuous cell-to-cell variation. In the bovine, not all but only certain pinealocytes are labeled with an antibody against hydroxyindole-*O*-methyltransferase (cf. 43). After the successful cloning of the NAT, some novel antibodies became available. Preliminary findings obtained with these new tools suggest that—in the sheep—the majority of pinealocytes express NAT; the intensity of the immunoreaction, however, varies considerably on a cell-to-cell basis.

The close phylogenetic relationship between mammalian pinealocytes and pineal photoreceptors is illustrated by the fact that mammalian pinealocytes display immunoreactions for photoreceptor-specific proteins, such as rod-opsin (29,37), S-antigen (38,39,40,41) and recoverin (42,69). Biochemical investigations indicate the presence of rhodopsin kinase (72) and a cone cyclic GMP phosphodiesterase in rat pinealocytes (10). The bovine pineal was shown to express both rod-type and cone-type cyclic nucleotide-gated channels by means of RT-PCR (14). The number of cells immunoreactive for photoreceptor-specific proteins varies with the species and the antibodies applied. In the rodent and cat pineal organ, the majority of pinealocytes are S-antigen immunoreactive, whereas in the human pineal organ, the S-antigen immunoreactive cells make up only 5–10% of the total population. The number of rod-opsin immunoreactive pinealocytes is constantly smaller than that of the S-antigen immunoreactive cells. No rod-opsin immunoreactive pinealocyte is found in the pineal organ of adult albino rodents. In man, a small subpopulation of pinealocytes (3–5%) bind the rod-opsin antibody. Approximately 25–30% of the pinealocytes are rod-opsin immunoreactive in adult pigmented mice (wild type and C57BL). The most interesting pattern of immunoreactions for "photoreceptor-specific" proteins has been observed in the "blind" mole rat, *Cryptomys damarensis*, which is endowed with a characteristic patch of white hairs in the parietal region of the skull. In this species, approximately 50% of pinealocytes are rod-opsin immunoreactive. A considerable number of cells also displays immunoreactive alpha-transducin, which has never been shown with certainty in the pineal organ of adult individuals of any other mammlian species (36,69). All these results imply the existence of functionally different subtypes of pinealocytes some of which may share molecular features with retinal rods or cones whereas others are unrelated to retinal photoreceptors and may belong to the family of pinopsin- or parapinopsin containing cells.

Immunochemical and in situ hybridization histochemical investigations have provided evidence that the rod-opsin-, S-antigen and recoverin immunoreactions are indeed elicited by the authentic proteins of the phototransduction cascade. Three bands of approximately 40, 75 and 110 kDa were found to bind the rod-opsin antibody in immunoblots of the pineal organ of the pigmented mouse (44). Corresponding bands were found in the mouse retina. Moreover, the S-antigen and recoverin antibodies labeled bands of identical molecular mass in the retina and the pineal organ of various mammalian species (cf. 33, for review). Also, the in situ hybridization revealed S-antigen expression in the retina and pineal organ (36).

Ontogenetic studies with in situ hybridization have shown that the majority of neonatal rat pinealocytes expresses principal components needed to reconstitute a functional phototransduction pathway (7). Rhodopsin expression was found in virtually all pinealocytes with highest levels at postnatal day 2 and a rapid decline at postnatal day 12. Blue cone pigment expression was considerably higher than rhodopsin expression and did not decline. Accordingly, high expression of cone transducin and cone phosphodiesterase was observed. Rhodopsin kinase expression was found in pinealocytes at all ontogenetic stages, whereas recoverin levels were low throughout development. Expression of S-antigen (rod arrestin) was found to be fairly robust. The results suggest that rat pinealocytes may use a mixture of rod- and cone-specific phototransduction elements. The results conform to earlier observations indicating that the neonatal rat pineal organ is directly photosensitive (92). The functional significance of photoreceptor-specific proteins persisting in the pinealocytes of adult mammals remains to be established. It is generally accepted that these cells have lost the direct light sensitivity. This concept is confirmed by studies showing that the mammalian pineal organ does not contain 11-cis retinal, the prosthetic group essential for a functioning photopigment (24,44).

Mammalian pinealocytes display an enriched glutamate immunoreactivity (49,63). Moreover, microvesicles isolated from the bovine pineal gland have been shown to accumulate L-glutamate against a concentration gradient by means of a vesicular L-glutamate transporter (56,57). These results suggest that microvesicles of mammalian pinealocytes accumulate glutamate and that glutamate is secreted from pinealocytes upon exocytosis of microvesicles. Thus, similarities between mammalian pinealocytes and true pineal photoreceptor cells appear to exist also in regard to the transmitter content.

The close relationship between mammalian pinealocytes and neurons is stressed by the fact that the vast majority of mammalian pinealocytes is immunoreactive to synaptophysin, neurofilaments (29,60,64), synaptotagmin I, synaptobrevin II, syntaxin I (62), and rab 3 (61). Interestingly, mammalian pinealocytes lack synapsin as do retinal photoreceptors.

3. FUNCTIONAL ASPECTS OF THE PHYLOGENETIC TRANSFORMATION OF PINEALOCYTES

As can be expected from the structural transformation of pinealocytes in the course of evolution, also the receptor mechanims and signal transduction cascades regulating melatonin biosynthesis show striking interspecific variation. In the pineal organ of poikilothermic vertebrates and birds, melatonin biosynthesis is regulated through light stimuli perceived by true or modified pineal photoreceptors. In most of these nonmammalian species (e.g., lamprey, zebra fish, pike, house sparrow) pineal melatonin biosynthesis is also under control of an endogenous oscillator which resides in the pineal itself and is capable of generating endogenous (circadian) rhythms in the absence of any environmental cue. The pineal organ of mammals contains—at least in adult individuals—neither functional photoreceptors nor endogenous rhythm generators. Here, the melatonin biosynthesis is regulated through signals from the retina as photoreceptor organ and the suprachiasmatic nucleus as endogenous oscillator. These signals are transmitted to the mammalian pineal organ via a complex neuronal chain whose final pathway is the postganglionic sympathetic innervation of the pineal organ. Accordingly, the generation and regulation of the rhythm in pineal melatonin biosynthesis of mammals depends on the reception and transduction of noradrenergic signals via alpha- and beta-adrenergic receptors.

In all vertebrate species investigated thus far, cyclic AMP and calcium ions have been identified as the second messengers that stimulate NAT and melatonin biosynthesis, but as can be expected from the evolutionary transformations, the regulation of these second messengers and their importance for melatonin biosynthesis differs from species to species (trout: 3, 45, 55, 79; chicken: 15, 16, 17, 27, 65, 88, 89, 90, 91; rat: 30, 70, 46; Maronde et al., this volume). After the successful cloning of a cDNA for the serotonin-N-acetyltransferase (9,12,31) in several vertebrate species it became readily evident that also the mechanisms regulating this key enzyme of the melatonin biosynthesis vary on a species to species basis. In the sheep the level of NAT mRNA varies between day and night by a factor of 1.5, but the enzyme activity is increased 7 fold in the night (12). In the rainbow trout (whose pineal is directly photosensitive, but does not contain an endogenous oscillator) the NAT mRNA levels are virtually kept constant during day and night, whereas the NAT activity is rapidly regulated by light and darkness (4). These findings indicate the predominance of posttranscriptional control mechanisms in sheep and trout. By contrast, the regulation of NAT in rat comprises an important transcriptional component: the 100-fold nocturnal increase of NAT activity in the rat pineal organ is accompanied by a more than 150-fold increase in NAT mRNA levels at night time (9,68). Interestingly, posttranscriptional processes are also involved in determining the activity of NAT in rat pineal: as shown by Gastel et al. (25) the acute inhibition of NAT activity upon exposure of rats to light stimuli in the middle of the night does not involve transcriptional and translational control mechanisms, but is mediated by rapid and reversible control of selective proteasomal proteolysis.

In the rat, the transcriptional and translational upregulation, maintenance and downregulation of NAT is essentially coupled to the cyclic AMP pathway, whose activation may be potentiated by the concomitant elevation of the intracellular calcium concentration. Norepinephrine-induced activation of the cyclic AMP pathway causes phosphorylation of the activating transcription factor CREB (67,78) and stimulates expression of the inhibitory transcription factor ICER (73,74,75) both acting upon genes containing a CRE in their promoters. In the rat, these two transcription factor appear as a crucial link transforming activation of the second messenger systems to activation and inhibition of the transcription of the NAT gene (see Maronde et al., this volume), whose promoter was shown to contain a CRE (2). Interestingly, the levels of p-CREB did not show day/night differences in the trout pineal organ (S. Kroeber, E. Maronde and H.-W. Korf, unpublished results). These results are in line with the rather constant levels of NATmRNA in the trout pineal organ.

4. CONCLUSION

In this contribution some evolutionary and regulatory aspects of melatonin-producing pinealocytes are reviewed. The results suggest that the final neuroendocrine pathway of the pinealocytes, i.e., the nocturnal production and secretion of melatonin is well conserved in the course of evolution. In contrast, the receptor mechanisms and signal transduction cascades regulating production and secretion of the neurohormone display conspicuous species-to-species differences. This scenario may also hold true for other neurosecretory cell types.

ACKNOWLEDGMENT

Our experimental studies have been supported by the Deutsche Forschungsgemeinschaft. I am grateful to Dr. Erik Maronde for his help with the preparation of this manuscript.

REFERENCES

1. Arendt, J. Melatonin and the mammalian pineal gland. London, Chapman and Hall, pp. 1–331, 1995.
2. Baler, R., Covington, S., and Klein, D.C. The rat arylalkylamine *N*-acetyltransferase gene promoter. J. Biol. Chem. 272:6979–6985, 1997.
3. Bégay, V., Bois, P., Collin, J.P., Lenfant, J., and Falcon, J. Calcium and melatonin production in dissociated trout pineal photoreceptor cells in culture. Cell Calcium 16:37–46, 1994.
4. Bégay, V., Falcón, J., Cahill, G.M., Klein, D.C., and Coon, S. Transcripts encoding two melatonin synthesis enzymes in the teleost pineal organ: circadian regulation in pike and zebrafish, but not in trout. Endocrinology 139:905–912, 1998.
5. Bernard, M., Voisin, P., Guerlotte, J., and Collin, J.P. Molecular and cellular aspects of hydoxyindole-*O*-methyltransferase expression in the developing chick pineal gland. Dev. Brain Res. 59:75–81, 1991.
6. Blackshaw, S. and Snyder, S.H. Parapinopsin, a novel catfish opsin localized to the parapineal organ, defines a new gene family. J. Neurosci. 17:8083–8092, 1997.
7. Blackshaw, S. and Snyder, S.H. Development expression pattern of phototransduction components in mammalian pineal implies a light-sensing function. J. Neurosci. 17:8074–8082, 1997.
8. Blank, H.M., Müller, B., and Korf, H.W. Comparative investigations of the neuronal apparatus in the pineal organ and retina of the rainbow trout: immunocytochemical demonstration of neurofilament 200-kDa and neuropeptide Y, and tracing with DiI. Cell Tissue Res. 288:417–425, 1997.
9. Borjigin, J., Wang, M.M., and Snyder, S.H. Diurnal variation in mRNA encoding serotonin-N-acetyltransferase in pineal gland. Nature 378:783–785, 1995.
10. Caracamo, B., Hurwitz, M.Y., Craft, C.M., and Hurwitz, R.L. The mammalian pineal expresses the cone but not the rod cyclic GMP phosphodiesterase. J. Neurochem. 65:1085–1092, 1995.
11. Collin, J.P. Differentiation and regression of the cells of the sensory line in the epiphysis cerebri. In: Wolstenholme, G.E.W. and Knight, J. (eds.) The pineal gland. Churchill-Livingstone, Edinburgh, pp. 79–125, 1971.
12. Coon, S.L., Roseboom, P.H., Baler, R., Weller, J.L., Namboodiri, M.A.A., Koonin, E.V., and Klein, D.C. Pineal serotonin-N-acetyltransferase: expression cloning and molecular analysis. Science 270:1681–1683, 1995.
13. Deguchi, T. Rhodopsin-like photosensitivity of isolated chicken pineal gland. Nature 290:706–707, 1981.
14. Distler, M., Biel, M., Flockerzi, V., and Hofmann, F. Expression of cyclic nucleotide-gated cation channels in non-sensory tissues and cells. Neuropharmacology 33:1275–1282, 1994.
15. Dryer, S.E. and Henderson, D. A cyclic GMP-activated channel in dissociated cells of the chick pineal gland. Nature 353:756–758, 1991.
16. D'Souza, T. and Dryer, S.E. Intracellular free Ca^{2+} in dissociated cells of the chick pineal gland: regulation by membrane depolarization, second messengers and neuromodulators, and evidence for release of intracellular Ca^{2+} stores. Brain Res. 656:85–94, 1994.
17. D'Souza, T. and Dryer, S.E. A cationic channel regulated by a vertebrate intrinsic circadian oscillator. Nature 382:165–167, 1996.

18. Ekström, P. and Meissl, H. Electron microscopic analysis of S-antigen- and serotonin-immunoreactive neural and sensory elements in the photosensory pineal organ of the salmon. J. Comp. Neurol. 292:73–82, 1990.
19. Ekström, P. and Meissl, H. The pineal organ of teleost fishes. Rev. Fish Biol. Fisheries 7:199–284, 1997.
20. Falcon, J. Identification et propriétés des cellules photoneuronedocrines de l'organe pinéal. Thesis, University of Poitiers, 1984.
21. Falcon, J., Begay, V., Goujon, J.M., Voisin, P., Guerlotte, J., and Collin, J.P. Immunocytochemical localization of hydroxyindole-O-methyltransferase in pineal photoreceptor cells of several fish species. J. Comp. Neurol. 341:559–566, 1994.
22. Foster, R.G., Korf, H.W., and Schalken, J.J. Immunocytochemical markers revealing retinal and pineal but not hypothalamic photoreceptor systems in the Japanese quail. Cell Tissue Res. 248:161–167, 1987.
23. Foster, R.G., Schalken, J.J., Timmers, A.M., and De Grip, W.J. A comparison of some photoreceptor characteristics in the pineal and retina. I. The Japanese quail (*Coturnix coturnix*). J. Comp. Physiol. [A] 165:553–563, 1989a.
24. Foster, R.G., Timmers, A.M., Schalken, J.J., and de Grip, W.J. A comparison of some photoreceptor characteristics in the pineal and retina. II. The Djungarian hamster (*Phodopus sungorus*). J. Comp. Physiol. [A] 165:565–572, 1989b.
25. Gastel, J.A., Roseboom, P., Rinaldi, P.A., Weller, J.L., and Klein, D.C. Melatonin production: proteasomal proteolysis in serotonin N-acetyltransferase regulation. Science 279:1358–1360, 1998.
26. Hafeez, M.A., Korf, H.W., and Oksche, A. Immunocytochemical and electron microscopic investigations of the pineal organ in adult agamid lizards, *Uromastix hardwicki*. Cell Tissue Res. 250:571–578, 1987.
27. Harrison, N. and Zatz, M. Voltage-dependent calcium channels regulate melatonin output from cultured chick pineal cells. J. Neurosci. 9:2462–2467, 1989.
28. Hirunagi, K., Ebihara, S., Okano, T., Takanaka, Y., and Fukuda, Y. Immunoelectron-microscopic investigation of the subcellular localization of pinopsin in the pineal organ of chicken. Cell Tissue Res. 289:235–241, 1997.
29. Huang, S.K., Klein, D.C., and Korf, H.W. Immunocytochemical demonstration of rod-opsin, S-antigen, and neuron-specific proteins in the human pineal gland. Cell Tissue Res. 267:493–498, 1992.
30. Klein, D.C. Photoneural regulation of the mammalian pineal gland. In: Evered, D. and Clark, S. (eds.) Photoperiodism, melatonin and the pineal gland. London: Pitman, pp. 38–56, 1985.
31. Klein, D.C., Coon, S.L., Roseboom, P.H., Weller, J.L., Bernard, M., Gastel, J.A., Zatz, M., Iuvone, M., Rodriguez, I.R., Begay, V., Falcon, J., Cahill, G.M., Cassone, V.M., and Baler, R. The melatonin rhythm generating enzyme: molecular regulation of serotonin N-acetyltransferase in the pineal gland. Rec. Progr. Horm. Res. 52:307–358, 1997.
32. Korf, B., Rollag, M.D., and Korf, H.W. Ontogenetic development of S-antigen- and rod-opsin immunoreactions in retinal and pineal photoreceptors of *Xenopus laevis* in relation to the onset of melatonin-dependent color-change mechanisms. Cell Tissue Res. 258:319–329, 1989.
33. Korf, H.W. The pineal organ as a component of the biological clock. Ann. N.Y. Acad. Sci. 719:13–42, 1994.
34. Korf, H.W. Innervation of the pineal gland. In: Burnstock, G. (ed.) Series on the autonomic nervous system, vol 10. Autonomic-endocrine interactions (Unsicker, K., ed.). Harwood, Amsterdam, pp. 129–180, 1996.
35. Korf, H.W. and Vigh-Teichmann, I. Sensory and central nervous elements in the avian pineal organ. Ophthalmic Res. 16:96–101, 1984.
36. Korf, H.W. and Wicht, H. Receptor and effector mechanisms in the pineal organ. Prog. Brain Res. 91:285–297, 1992.
37. Korf, H.W., Foster, R.G., Ekström, P., and Schalken, J.J. Opsin-like immunoreaction in the retinae and pineal organs of four mammalian species. Cell Tissue Res. 242:645–648, 1985a.
38. Korf, H.W., Møller, M., Gery, I., Ziegler, J.S., and Klein, D.C. Immunocytochemical demonstration of retinal S-antigen in the pineal organ of four mammalian species. Cell Tissue Res. 239:81–85, 1985b.
39. Korf, H.W., Oksche, A., Ekström, P., Veen, T., van Zigler, J.S., Gery, I., Stein, P., and Klein, D.C. S-antigen immunocytochemistry. In: O'Brien, P. and Klein, D.C. (eds.) Pineal and retinal relationships. Academic Press, New York, pp. 343–355, 1986a.
40. Korf, H.W., Oksche, A., Ekström, P., Zigler, J.S., Gery, I., and Klein, D.C. Pinealocyte projections into the mammalian brain revealed with S-antigen antiserum. Science 231:735–737, 1986b.

41. Korf, H.W., Sato, T., and Oksche, A. Complex relationships between the pineal organ and the medial habenular nucleus-pretectal region of the mouse as revealed by S-antigen immunocytochemistry. Cell Tissue Res. 261:493–500, 1990.
42. Korf, H.W., White, B.H., Schaad, D.C., and Klein, D.C. Recoverin in pineal organs and retinae of various vertebrate species including man. Brain Res. 595:57–66, 1992.
43. Korf, H.W., Schomerus, C., and Stehle, J. The pineal organ, its hormone melatonin and the photoneuroendocrine system. Adv. Anat. Embryol. Cell Biol. 146:1–100, 1998.
44. Kramm, C.M., De Grip, W.J., and Korf, H.W. Rod-opsin immunoreaction in the pineal organ of the pigmented mouse does not indicate the presence of a functional photopigment. Cell Tissue Res. 274:71–78, 1993.
45. Kroeber, S., Schomerus, C., and Korf, H.W. Calcium oscillations in a subpopulation of S-antigen-immunoreactive pinealocytes of the rainbow trout (*Oncorhynchus mykiss*). Brain Res. 744:68–76, 1997.
46. Maronde, E., Schomerus, C., Stehle, J., and Korf, H.W. Control of CREB phosphorylation and its role for induction of melatonin synthesis in rat. Biol. Cell 89:505–511, 1997.
47. Max, M., McKinnon, P.J., Seidenman, K.J., Barrett, R.K., Applebury, M.L., Takahashi, J.S., and Margolskee, R.F. Pineal opsin: A nonvisual opsin expressed in chick pineal. Science 267:1502–1506, 1995.
48. McNulty, J., Rathbun, W.E., and Druse, M.J. Ultrastructural and biochemical responses of photoreceptor pinealocytes to light and dark in vivo and in vitro. Life Sci. 43:845–850, 1988.
49. McNulty, J.A., Kus, L., and Ottersen, O.P. Immunocytochemical and circadian biochemical analysis of neuroactive amino acids in the pineal gland of the rat: effect of superior cervical ganglionectomy. Cell Tissue Res. 269:515–523, 1992.
50. Meiniel, A. New aspects of the phylogenetic evolution of sensory cell lines in the vertebrate pineal complex. In: Oksche, A. and Pévet, P. (eds.) The pineal organ: photobiology-biochronometry-endocrinology. Elsevier, Amsterdam, pp. 27–48, 1981.
51. Meissl, H. and Ekström, P. Extraretinal photoreception by pineal systems: a tool for photoperiodic time measurement? Trends Comp. Biochem. Physiol. 1:1223–1240, 1993.
52. Meissl, H. and George, S.R. Electrophysiological studies on neuronal transmission in the frog's photosensory pineal organ. The effect of amino acids and biogenic amines. Vision Res. 24:1727–1734, 1984a.
53. Meissl, H. and George, S.R. Photosensory properties of the pineal organ. Microiontophoretic application of excitatory amino acids onto pineal neurons. Ophthalmic Res. 16:114–118, 1984b.
54. Meissl, H., Donley, C.S., and Wissler, J.H. Free amino acids and amines in the pineal organ of the rainbow trout (*Salmo gairdneri*): influence of light and dark. Comp. Biochem. Physiol. 61C:401–405, 1978.
55. Meissl, H., Kroeber, S., Yáñez, J., and Korf, H.W. Regulation of melatonin production and intracellular calcium concentrations in the trout pineal organ. Cell Tissue Res. 286:315–323, 1996.
56. Moriyama, Y. and Yamamoto, A. Microvesicles isolated from bovine pineal gland specifically accumulate L-glutamate. FEBS Lett. 367:233–236, 1995a.
57. Moriyama, Y. and Yamamoto, A. Vesicular L-glutamate transporter in microvesicles from bovine pineal glands. J. Biol. Chem. 270:22314–22320, 1995b.
58. Okano, T., Yoshizawa, T., and Fukada, Y. Pinopsin is a chicken pineal photoreceptive molecule. Nature 372:94–96, 1994.
59. Oksche, A. Sensory and glandular elements of the pineal organ. In: Wolstenholme, G.E.W. and Knight, J. (eds.) The pineal gland. Churchill-Livingstone, Edinburgh, pp. 127–146, 1971.
60. Oksche, A., Korf, H.W., and Rodríguez, E.M. Pinealocytes as photoneuroendocrine units of neuronal origin: Concepts and evidence. In: Fraschini, F. and Reiter, R.J. (eds.) Advances in pineal research. Vol. 2. London: John Libbey, pp. 1–18, 1987.
61. Redecker, P. The ras-like rab3A protein is present in pinealocytes of the gerbil pineal gland. Neurosci. Lett. 184:117–120, 1995.
62. Redecker, P. Synaptotagmin I, synaptobrevin II, and syntaxin I are coexpressed in rat and gerbil pinealocytes. Cell Tissue Res. 283:443–454, 1996.
63. Redecker, P. and Veh, R.W. Glutamate immunoreactivity is enriched over pinealocytes of the gerbil pineal gland. Cell Tissue Res. 278:579–588, 1994.
64. Redecker, P., Grube, D., and Jahn, R. Immunohistochemical localization of synaptophysin (p38) in the pineal gland of the Mongolian gerbil (*Meriones unguiculatus*). Anat. Embryol. 181:433–440, 1990.
65. Robertson, L.M. and Takahashi, J.S. Circadian clock in cell culture: I. Oscillation of melatonin release from dissociated chick pineal cells in flow-through microcarrier culture. J. Neurosci. 8:12–21, 1988.

66. Rodríguez, E.M., Korf, H.W., Oksche, A., Yulis, C.R., and Hein, S. Pinealocytes immunoreactive with antisera against secretory glycoproteins of the subcommissural organ: a comparative study. Cell Tissue Res. 254:469–480, 1988.
67. Roseboom, P.H. and Klein, D.C. Norepinephrine stimulation of pineal cyclic AMP response element-binding protein phosphorylation: involvement of a β-adrenergic/cyclic AMP mechanism. Mol. Pharmacol. 47:439–449, 1995.
68. Roseboom, P.H., Coon, S.L., Baler, R., McCune, S.K., Weller, J.L., and Klein, D.C. Melatonin synthesis: analysis of the more than 150-fold nocturnal increase in serotonin-N-acetyltransferase messenger ribonucleotide acid in the rat pineal gland. Endocrinology 137:3033–3044, 1996.
69. Schomerus, C., Ruth, P., and Korf, H.W. Photoreceptor-specific proteins in the mammalian pineal organ: immunocytochemical data and functional considerations. Acta Neurobiol. Exp. 54 (Suppl.):9–17, 1994.
70. Schomerus, C., Laedtke, E., and Korf, H.W. Calcium responses of isolated, immunocytochemically identified rat pinealocytes to noradrenergic, cholinergic and vasopressinergic stimulations. Neurochem. Int. 27:163–175, 1995.
71. Solessio, E. and Engbretson, G.A. Antagonistic chromatic mechanisms in photoreceptors of the parietal eye of lizards. Nature 364:442–445, 1993.
72. Somers, R.L. and Klein, D.C. Rhodopsin kinase activity in the mammalian pineal gland and other tissues. Science 226:182–184, 1984.
73. Stehle, J.H. Pineal gene expression: dawn in a dark matter. J. Pineal Res. 18:179–190, 1995.
74. Stehle, J.H., Foulkes, N.S., Molina, C.A., Simonneaux, V., Pévet, P., and Sassone-Corsi, P. Adrenergic signals direct rhythmic expression of transcriptional repressor CREM in the pineal gland. Nature 356:314–320, 1993.
75. Stehle, J.H., Foulkes, N.S., Pévet, P., and Sassone-Corsi, P. Developmental maturation of pineal gland function: Synchronized CREM inducibility and adrenergic stimulation. Mol. Endocrinol. 9:706–716, 1995.
76. Takahashi, J.S., Murakami, N., Nikaido, S., Pratt, B., and Robertson, L.M. The avian pineal—a vertebrate model system of the circadian oscillator: cellular regulation of circadian rhythms by light, secondary messengers, and macromolecular synthesis. Rec. Prog. Horm. Res. 45:279–352, 1989.
77. Tamotsu, S., Korf, H.W., Morita, Y., and Oksche, A. Immunocytochemical localization of serotonin and photoreceptor-specific proteins (rod-opsin, S-antigen) in the pineal complex of the river lamprey, *Lampetra japonica*, with special reference to photoneuroendocrine cells. Cell Tissue Res. 262:205–216, 1990.
78. Tamotsu, S., Schomerus, C., Stehle, J.H., Roseboom, P.H., and Korf, H.W. Norepinephrine-induced phosphorylation of the transcription factor CREB in isolated rat pinealocytes: an immunocytochemical study. Cell Tissue Res. 282:219–226, 1995.
79. Thibault, C., Falcón, J., Greenhouse, S.S., Lowery, A., Gern, W.A., and Collin, J.P. Regulation of melatonin production by pineal photoreceptor cells: role of cyclic nucleotides in the trout (*Oncorhynchus mykiss*). J. Neurochem. 61:332–339, 1993.
80. Van Veen, T., Östholm, T., Gierschik, P., Spiegel, A., Somers, R., Korf, H.W., and Klein, D.C. Alpha-transducin immunoreactivity in retinae and sensory pineal organs of adult vertebrates. Proc. Natl. Acad. Sci. USA 83:912–916, 1986.
81. Vigh, B., Vigh-Teichmann, I., Röhlich, P., and Aros, B. Immunoreactive opsin in the pineal organ of reptiles and birds. Z. Mikrosk. Anat. Forsch. 96:113–129, 1982.
82. Vigh, B., Vigh-Teichmann, I., Debreceni, K., and Takacs, J. Similar fine-structural localization of immunoreactive glutamate in the pineal complex and retina of frogs. Arch. Histol. Cytol. 58:37–44, 1995.
83. Vigh, B., Debreceni, K., Fejer, Z., and Vigh-Teichmann, I. Immunoreactive excitatory amino acids in the parietal eye of lizards, a comparison with the pineal organ and retina. Cell Tissue Res. 287:275–283, 1997.
84. Vigh-Teichmann, I. and Vigh, B. Opsin: immunocytochemical characterization of different types of photoreceptors in the frog pineal organ. J. Pineal Res. 8:323–333, 1990.
85. Vigh-Teichmann, I., Korf, H.W., Oksche, A., and Vigh, B. Opsin-immunoreactive outer segments and acetylcholinesterase-positive neurons in the pineal complex of *Phoxinus phoxinus* (Teleostei, Cyrinidae). Cell Tissue Res. 262:205–216, 1982.
86. Voisin, P., Guerlotte, J., and Collin, J.P. An antiserum against chicken hydroxyindole-O-methyltransferase reacts with the enzyme from pineal gland and retina and labels pineal modified photoreceptors. Mol. Brain Res. 4:53–61, 1988.

87. Vollrath L. Synaptic ribbons of a mammalian pineal gland. Circadian changes. Z. Zellforsch. 145:171–183, 1973.
88. Zatz, M. Agents that affect calcium influx can change cyclic nucleotide levels in cultured chick pineal cells. Brain Res. 583:304–307, 1992.
89. Zatz, M. and Mullen, D.A. Norepinephrine, acting via adenylate cyclase, inhibits melatonin output but does not phase-shift the pacemaker in cultured chick pineal cell. Brain Res. 450:137–143, 1988a.
90. Zatz, M. and Mullen, D. Photoendocrine transduction in cultured chick pineal cells II. Effects of forskolin, 8-bromocyclic AMP, and 8-bromocyclic GMP on the melatonin rhythm. Brain Res. 453:51–62, 1988b.
91. Zatz, M., Mullen, D.A., and Moskal, J.R. Photoendocrine transduction in cultured chick pineal cells: effects of light, dark, and potassium on the melatonin rhythm. Brain Res. 438:199–215, 1988.
92. Zweig, M., Snyder, S.H., and Axelrod, J. Evidence for a nonretinal pathway of light to pineal gland in newborn rats. Proc. Natl. Acad. Sci. USA 56:515–520, 1966.

4

MELATONIN BIOSYNTHESIS IN CHICKEN RETINA

Regulation of Tryptophan Hydroxylase and Arylalkylamine *N*-Acetyltransferase

P. M. Iuvone,[1] N. W. Chong,[2] M. Bernard,[2] A. D. Brown,[1] K. B. Thomas,[1] and D. C. Klein[2]

[1] Department of Pharmacology
Emory University School of Medicine
Atlanta, Georgia
[2] Section on Neuroendocrinology
Laboratory of Developmental Neurobiology
National Institute of Child Health and Development
NIH, Bethesda, Maryland

1. SUMMARY

Melatonin is synthesized from the amino acid tryptophan (tryptophan → 5-hydroxytryptophan → serotonin → N-acetylserotonin → melatonin). In the chick retina, mRNA levels and activities of two enzymes in this pathway, tryptophan hydroxylase (TPH) and arylalkylamine *N*-acetyltranferase (AA-NAT), display circadian rhythms. The dramatic nocturnal increase in melatonin production in this tissue in part reflects circadian clock-driven increases in mRNA encoding both enzymes. This appears to be translated into increases in both TPH and AA-NAT protein. In the case of AA-NAT, however, this translation is strongly dependent upon environmental lighting. Light acts through post-transcriptional mechanisms to regulate AA-NAT activity; a hypothetical mechanism is proteasomal proteolysis that is otherwise inhibited in the dark by second messengers. Accordingly, melatonin production in the retina, as in the chicken pineal gland, is thought to be regulated by two mechanisms. One is clock-driven changes in TPH and AA-NAT mRNAs, which in turn drive changes in the synthesis of the corresponding encoded proteins. The second is light-induced post-transcriptional degradation of AA-NAT. These mechanisms insure that retinal

Melatonin after Four Decades, edited by James Olcese.
Kluwer Academic / Plenum Publishers, New York, 2000.

melatonin production follows a precise schedule that reflects daily changes in the environmental lighting.

2. MELATONIN BIOSYNTHESIS IN CHICKEN RETINA

The retinae of many vertebrates have the capacity to synthesize melatonin (44). However, this source of melatonin does not appear to contribute substantially to circulating melatonin, which is pineal derived. Retinal melatonin appears to act locally as a paracrine neuromodulator and regulator of rhythmic retinal physiology (21,23). Retinal melatonin is synthesized primarily in photoreceptor cells (2,10,17,18,42) and its production is regulated by a local, retinal circadian clock (4,10,41).

Melatonin modulates the release of several neurotranmitters in the retina. It inhibits the release of dopamine (6,13) and acetylcholine (29) from amacrine cells. Melatonin stimulates the release of glutamate (15), the photoreceptor neurotransmitter. The release of dopamine and acetylcholine are stimulated by light (e.g., 5,27), while glutamate is released from photoreceptors in darkness (e.g., 34). Thus, melatonin mimics, and may partiially mediate, the effects of darkness on neurotranmitter release. The effect of melatonin on dopamine release appears particularly important, as dopamine is a mediator of circadian changes in visual sensitivity (8,26).

Melatonin promotes dark-adaptive photomechanical movements in photoreceptors and retinal pigment epithelial (RPE) cells (11,30–32,35). Melatonin activates rod photoreceptor disk shedding and phagocytosis by RPE cells (3). Disk shedding and phagocytosis occur in a circadian fashion in most vertebrates, and melatonin may be a neurohumoral link between the retinal circadian clock and the rhythmic turnover of photosensitive outer segment membranes.

Melatonin has effects on photoreceptor survival. Its administration prior to exposure to high intensity light enhances photoreceptor degeneration in albino rats (7,25,43). In contrast, intraocular injection of the melatonin antagonist, luzindole, promotes photoreceptor survival and maintenance of function in light-damaged retinas (36). The mechanisms responsible for these effects are unknown, but may reflect the inhibitory effect of melatonin on dopamine release and loss of a neuroprotective action of dopamine (22). Alternatively, it may reflect direct effects of melatonin on photoreceptor-RPE interactions or a disruption of retinal circadian physiology.

The chicken has become an important experimental model for studying retinal melatonin biosynthesis and actions, because its retina has a relatively high density of melatonin receptors (14,33) and robustly synthesizes melatonin under the influence of a circadian clock (19). Chicken retina contains all of the enzymes of the melatonin biosynthetic pathway and readily converts [^{14}C]trytophan to [^{14}C]melatonin (39). Melatonin levels in chick retina are low during the day, but show a dramatic nocturnal increase in chicks housed under a light-dark cycle or in constant darkness (19). Acute light exposure at night rapidly reduces retinal melatonin levels. This review will focus on the regulation of two enzymes of the melatonin biosynthetic pathway, TPH and AA-NAT, and on their roles in regulating circadian melatonin biosynthesis and the acute inhibitory effect of light in the chick retina.

3. REGULATION OF TPH AND AA-NAT EXPRESSION AND ACTIVITY

3.1. Localization of TPH and AA-NAT mRNAs in Retina

AA-NAT mRNA is localized primarily to the photoreceptor layer of the chick retina, with a lower level of expression in the ganglion cell layer (2); expression was not detected in the inner nuclear layer. TPH mRNA expression is also strong in the photoreceptor layer (12). In addition, TPH mRNA is found in the ganglion cell and inner nuclear layers. The localization of TPH mRNA alone (without AA-NAT mRNA) in the inner nuclear layer corresponds to the previously described serotonin-immunoreactive amacrine cells (28,39). Expression of both mRNAs in photoreceptors is consistent with this cell type as the primary source of retinal melatonin.

Localization of TPH and AA-NAT mRNAs in the ganglion cell layer is unexpected, as serotonin immunoreactivity has not been observed in this cell layer of chick retina. This observation suggests that some ganglion cells may synthesize melatonin or *N*-acetylserotonin. However, the available evidence suggests that melatonin synthesis in ganglion cells is not significant relative to that formed in photoreceptor cells. This is because destruction of most ganglion cells by kainic acid does not decrease melatonin production nor damage photoreceptor cells (39).

3.2. Circadian Rhythms of TPH and AA-NAT mRNAs and Activities

The daily rhythms of TPH and AA-NAT activities are very similar in retinas of chicks exposed to a 12h light—12h dark cycle (LD) (Figure 1). Activities of both enzymes are low during the day and show robust increases early in the dark phase of the light-dark cycle. These rhythms of activity appear to be generated, at least in part, by rhythmic expression of TPH and AA-NAT mRNAs (2,12). The daily rhythms of mRNA level and activity are driven by a circadian oscillator and persist in constant light (LL) and constant darkness (DD). The rhythms of mRNA level under constant conditions are very similar for both enzymes, showing peaks in the middle of the subjective dark phase. Thus, expression of the TPH and AA-NAT genes may be regulated by the same clock driven mechanism.

In DD, the rhythms of TPH activity and NAT activity are similar to one another and appear to reflect the expression patterns of their respective mRNAs (2,12,38). In LL, high amplitude TPH activity rhythms occur (Figure 2). Nocturnal activity is only slightly reduced on the first day of LL, but gradually declines over successive days. LL has similar effects on the pattern of TPH mRNA expression (12). In contrast, AA-NAT activity is greatly suppressed (~85%) at night in LL, resulting in a markedly reduced amplitude of the daily rhythm (Figs. 2, 3). The amplitude of the AA-NAT mRNA rhythm is also reduced, but to a much smaller extent (~50%) (Figure 3). This observation indicates that light has a suppressive effect on AA-NAT activity, but not on TPH activity, that may reflect differences in post-transcriptional regulation.

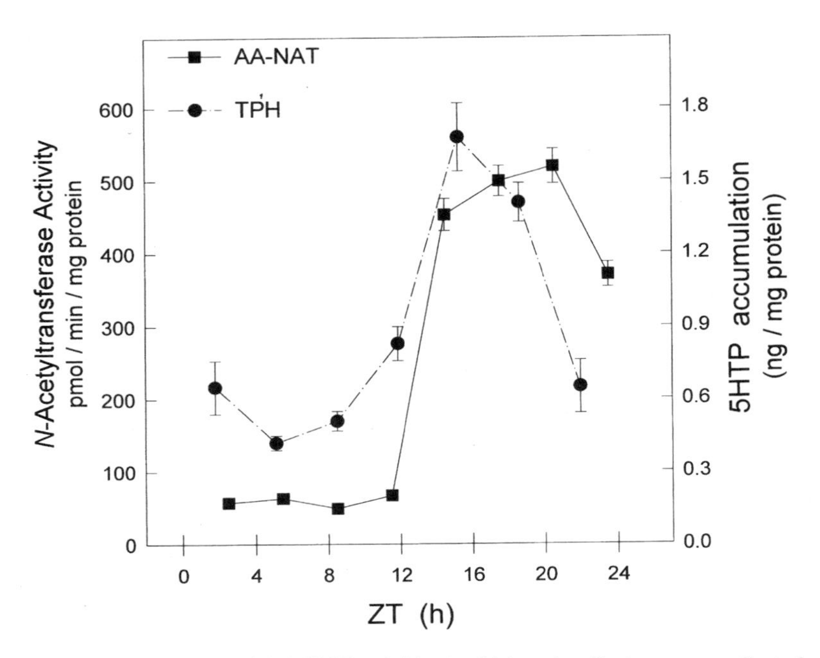

Figure 1. Daily rhythms of TPH and AA-NAT activities in chick retina. Retinas were collected at the zeitgeber times (ZT) indicated during the 12 h light—12 h dark cycle (LD, lights on at ZT 0). TPH activity was estimated from the *in situ* accumulation of 5-hydroxytryptophan (5HTP) 30 min following inhibition of aromatic L-amino acid decarboxylase activity with *m*-hydroxybenzylhydrazine (*m*HBH)(38). AA-NAT activity was measured in retinal homogenates using tryptamine and acetyl coenzyme A as substrates (40). Adapted from Thomas and Iuvone (38) and Iuvone and Alonso-Gomez (23).

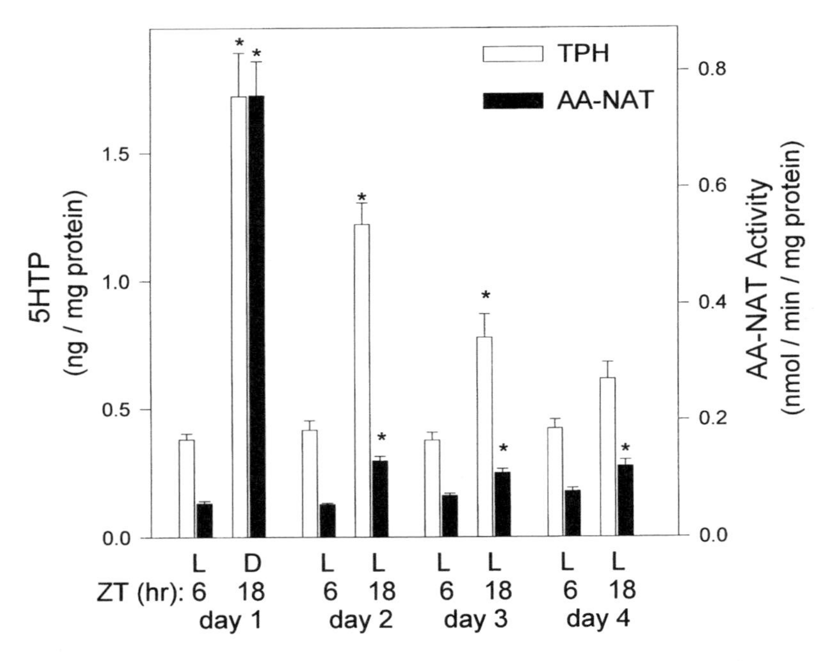

Figure 2. Effect of constant light on the rhythms of TPH and AA-NAT activities. Retinas were dissected 30 min following injection of *m*HBH at midday (ZT 6) and midnight (ZT 18). On day 1, animals were exposed to the entrained LD cycle. Constant light (LL) began on the second day. *$p < 0.01$ *vs* ZT6. Adapted from Thomas and Iuvone (38).

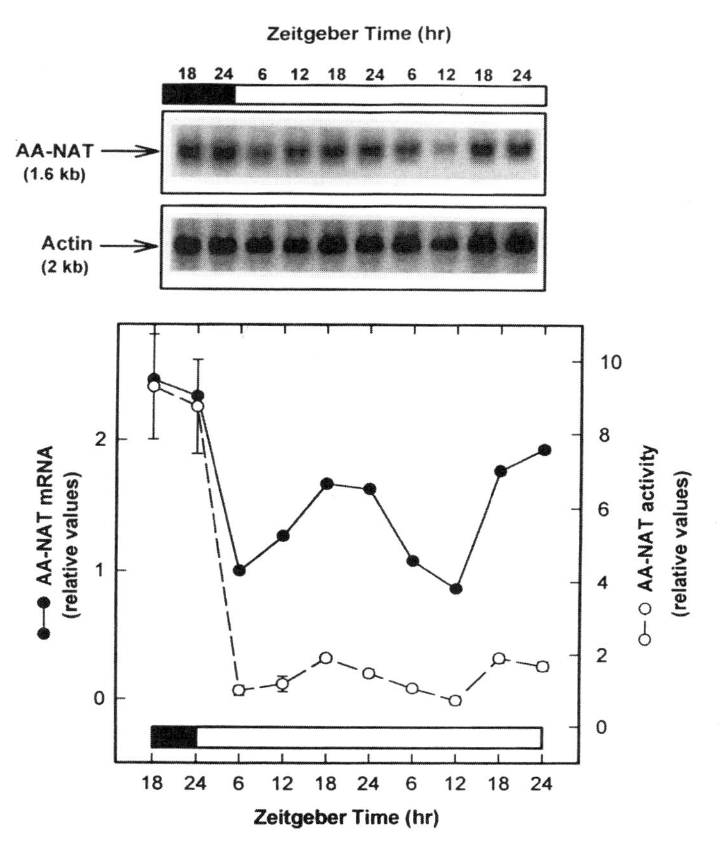

Figure 3. Rhythms of AA-NAT mRNA and activity in LL. Chickens were housed in LD and were then transferred to LL. The open horizontal bar indicates when lights were on and the vertical lines indicate the subjective day/night transitions. Retinas were taken at the indicated times and processed for RNA analysis. The top panels show representative Northern blot analyses of AA-NAT and actin mRNAs. Each lane contains 20 µg of total RNA. The bottom panels represent the quantitative analysis of the Northern blots (•–•) and the levels of AA-NAT activity (○–○) in each experimental group. The abundance of the AA-NAT transcript has been normalized to actin mRNA, to correct for variations in loading. All values are expressed relative to the first ZT 6 time point values. AA-NAT activity at ZT 6 was 26 ± 3 pmoles/min/mg protein. Adapted from Bernard et al. (2).

3.3. Effects of Acute Light Exposure on TPH and AA-NAT mRNAs and Enzyme Activities

Light is known to suppress melatonin production by suppressing AANAT activity in both the pineal gland and retina (20,24). In the chicken, it has been found that light exposure during the middle of the dark phase of the LD cycle reduces retinal melatonin levels to daytime values within 1 hour (19). This treatment elicits only small decreases (~30%) of TPH and AA-NAT mRNAs (2,12). Acute light exposure also has little inhibitory effect (~30%) on TPH activity (Figure 4). In contrast, AA-NAT activity is dramatically reduced (~80%) by acute light exposure (Figures 4 and 5). Thus, the effects of acute light exposure on TPH activity appear to reflect mainly changes of mRNA level, while that on AA-NAT activity is elicited by

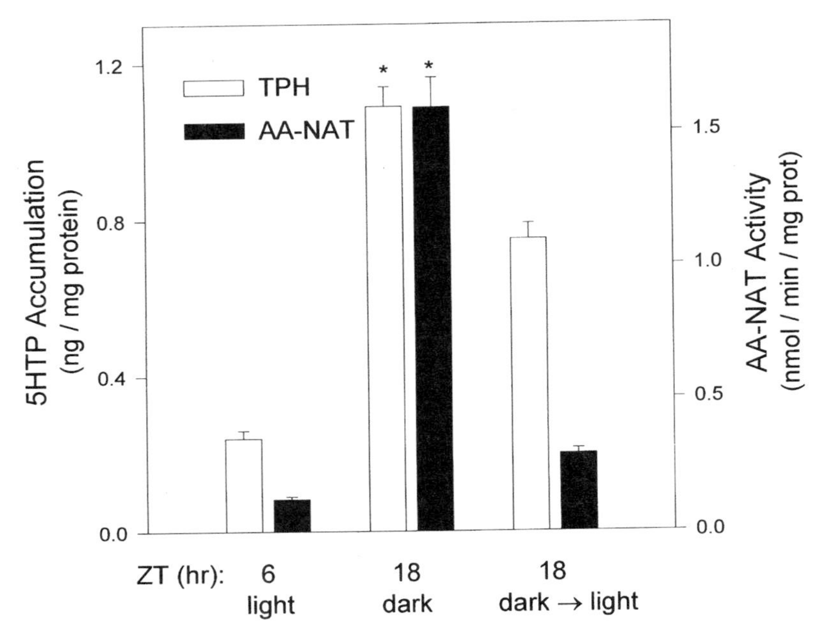

Figure 4. Differential effects of acute light exposure on nocturnal TPH and AA-NAT activities. Animals were exposed to fluorescent room light ($2 \times 10^{-4}\,W/cm^2$) during middark phase (ZT 18) for 60 min (Dark → Light). After the initial 30 min of light exposure, they were injected with *m*HBH. Retinas were dissected in light 30 min after injection. Control animals (dark) were treated identically except that they were not exposed to light. Retinas were also obtained from *m*HBH-treated animals during midlight phase (ZT 6). Adapted from Thomas and Iuvone (38).

both decreased mRNA levels and post-transcriptional regulation of enzyme activity. The post-transcriptional regulatory mechanism may involve decreased cyclic AMP levels in the photoreceptor cell resulting in a destabilization of AA-NAT (1). By analogy to the rat pineal gland (16), the destabilization of AA-NAT may result in its proteolysis.

4. ROLES OF TPH AND AA-NAT IN REGULATING MELATONIN BIOSYNTHESIS

The dramatic nocturnal increase of melatonin biosynthesis involves the induction of both TPH and AA-NAT. Similar to the situation in *Xenopus* retina (9), administration of the serotonin precursor, 5-hydroxytryptophan (5HTP), at night greatly increases chick retinal melatonin levels (Figure 6). Administration of tryptophan has no effect. This observation indicates that AA-NAT is not saturated with substrate *in situ*, and that the induction of TPH at night contributes to high rates of melatonin biosynthesis by increasing the supply of serotonin to AA-NAT. In contrast, adminstration of 5HTP during the daytime, when AA-NAT activity is low, elicits a relatively small stimulation of melatonin level (37). Similarly, little increase of melatonin level is observed following 5HTP administration after acute light exposure at night, which suppresses AA-NAT activity (Fig. 6). Accordingly, it is clear that AA-NAT plays a primary role in regulating melatonin production, and that large changes in serotonin synthesis do not appear sufficient to produce large changes in melatonin production.

The acute inhibitory effect of light exposure on AA-NAT activity appears to result in an accumulation of serotonin sufficient to result in an increase in its oxidation by monoamine oxidase to 5-hydroxyindoleacetic acid (38) (Figure 7). Serotonin continues to be synthesized during light exposure due to high TPH activity, but little of it is *N*-acetylated. The rapid decline in melatonin levels are primarily due to the decrease in AANAT activity, not to a decrease in serotonin, as indicated by the increase in 5-HT and oxidation products that typically results from light exposure.

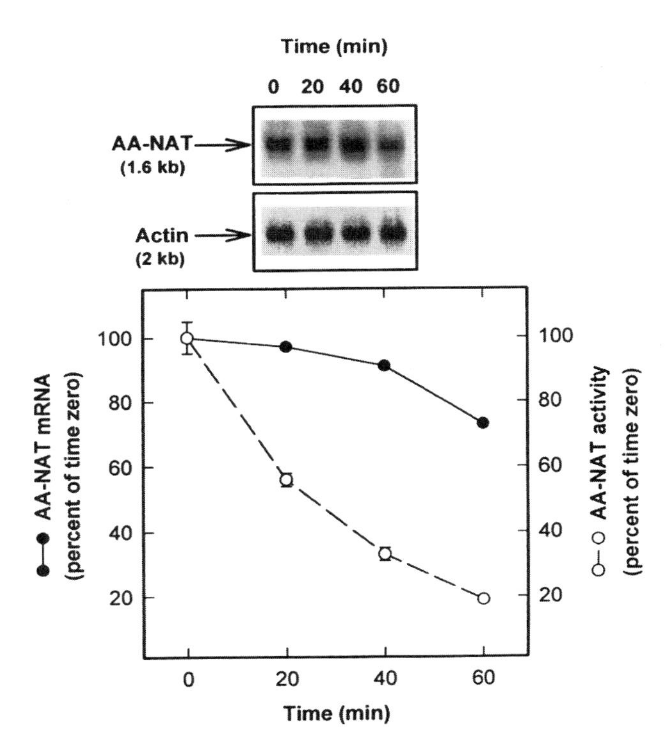

Figure 5. Acute effect of light at night on levels of AA-NAT mRNA and activity. Chickens were housed in LD. Starting at ZT 18 (midnight), the animals were exposed to fluorescent white light for the indicated times. The top panels show representative Northern blot analyses of AA-NAT and actin mRNAs. Each lane contains 20 μg of total RNA. The bottom panels represent the quantitative analysis of the Northern blots (•–•) and the levels of AA-NAT activity (○–○) in each experimental group. The abundance of the AA-NAT transcript has been normalized to actin mRNA. All values represent the mean of duplicate determinations (RNA) or the mean ± s.e.m. (activity; n = 8 retinas per group) and are expressed as percent of the t = 0 time point (ZT 18). Individual mRNA values were within 18% of the mean. AA-NAT activity at ZT 18 was 494 ± 67 pmoles/min/mg protein. Adapted from Bernard et al. (2).

5. CONCLUSION

These studies are consistent with the hypothesis that induction of both TPH and AA-NAT is essential for the large nocturnal increase of melatonin biosynthesis. The induction of the two enzymes appear to be coordinately regulated by the circadian clock. The activity of TPH is controlled primarily by changes in TPH mRNA level, whereas that of AA-NAT reflects a combination of transcriptional and post-transcriptional regulation. The post-transcriptional regulatory mechanisms appear to be largely responsible for the acute inhibitory effects of light on AA-NAT activity and melatonin levels in the retina.

ACKNOWLEDGMENTS

Research in the laboratory of PMI was supported in part by NIH grant RO1-EY04864.

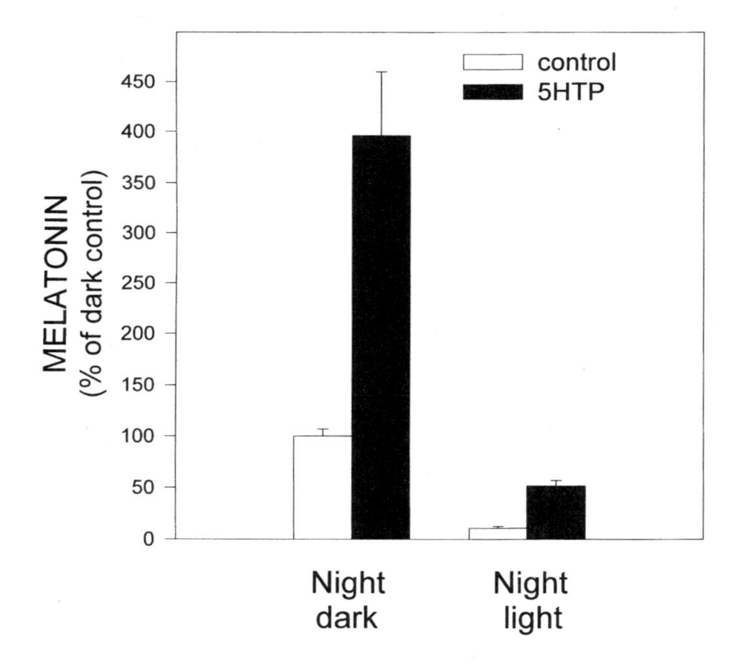

Figure 6. Effect of 5HTP administration on nocturnal melatonin levels in darkness and during acute light exposure. Animals were injected with 5HTP (100 mg/kg ip) during dark phase (ZT15). Retinas were dissected 90 minutes later in darkness (night/dark) or following 90 min exposure to light (night/light). Melatonin content of the retinas was determined by radioimmunoassay. N = 6–8/group. Light exposure significantly reduced nocturnal melatonin level ($p < 0.05$). 5HTP administration significantly increased melatonin levels relative to vehicle controls in both darkness and during acute light exposure, but the increase was much smaller in light-treated retinas. Tryptophan administration (500 mg/kg ip) had no effect on nocturnal melatonin levels. From unpublished data of A.D. Brown, K.B. Thomas, and P.M. Iuvone.

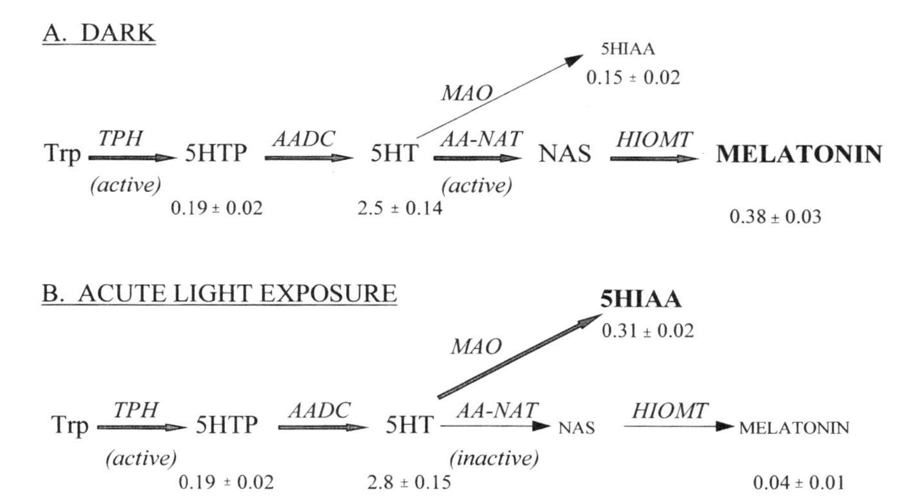

Figure 7. Nocturnal tryptophan metabolism in darkness and following light exposure. A. At night, in darkness, TPH and AA-NAT activities are high, resulting in significant metabolism of trptophan (Trp) to melatonin. B. Following acute light exposure, TPH activity remains high but AA-NAT is inactivated. As a consequence, metabolism in this pathway is shunted from production of melatonin to 5-hydroxyindolacetic acid (5HIAA), a product of serotonin (5HT) oxidation by monoamine oxidase (MAO).Other abbreviations: AADC = aromatic L-amino acid decarboxylase; HIOMT = hydroxyindole *O*-methyltransferase; NAS = *N*-acetylserotonin. Numerical values below 5HTP, 5HT, 5HIAA, and melatonin are steady-state levels expressed in ng/mg protein; values for 5HTP, 5HIAA, and 5HT are from Thomas and Iuvone (38); melatonin values are unpublished data of A.D. Brown and P.M. Iuvone.

REFERENCES

1. Alonso-Gómez, A.L. and Iuvone, P.M. Melatonin biosynthesis in cultured chick retinal photoreceptor cells: calcium and cyclic AMP protect serotonin N-acetyltransferase from inactivation in cycloheximide-treated cells. J Neurochem 65:1054–1060, 1995.
2. Bernard, M., Iuvone, P.M., Cassone, V.M., Roseboom, P.H., Coon, S.L., and Klein, D.C. Avian melatonin synthesis: photic and circadian regulation of serotonin *N*-acetyltransferase mRNA in the chicken pineal gland and retina. J Neurochem 68:213–224, 1997.
3. Besharse, J.C. and Dunis, D.A. Methoxyindoles and photoreceptor metabolism: activation of rod shedding. Science 219:1341–1343, 1983.
4. Besharse, J.C. and Iuvone, P.M. Circadian clock in Xenopus eye controlling retinal serotonin N-acetyltransferase. Nature 305:133–135, 1983.
5. Boatright, J.H., Hoel, M.J., and Iuvone, P.M. Stimulation of endogenous dopamine release and metabolism in amphibian retina by light and K+-evoked depolarization. Brain Res 482:164–168, 1989.
6. Boatright, J.H., Rubim, N.M., and Iuvone, P.M. Regulation of endogenous dopamine release in amphibian retina by melatonin: the role of GABA. Vis Neurosci 11:1013–1018, 1994.
7. Bubenik, G.A. and Purtill, R.A. The role of melatonin and dopamine in retinal physiology. Can J Physiol Pharmacol 58:1457–1462, 1980.
8. Buelow, N., Kelly, M.E., Iuvone, P.M., Underwood, H., and Barlow, R.B., Jr. Circadian modulation of retinal function in the Japanese quail: role of dopamine. Soc Neurosci Abs 18:768, 1992.
9. Cahill, G.M. and Besharse, J.C. Circadian regulation of melatonin in the retina of *Xenopus laevis*: Limitation by serotonin availability. J Neurochem 54:716–719, 1990.
10. Cahill, G.M. and Besharse, J.C. Circadian clock functions localized in Xenopus retinal photoreceptors. Neuron 10:573–577, 1993.

11. Cheze, G. and Ali, M.A. Role de l'epiphyse dans la migration du pigment epithelial retinien chez quelques Teleosteens. Can J Zool 54:475–481, 1976.
12. Chong, N.W., Cassone, V.M., Bernard, M., Klein, D.C., and Iuvone, P.M. Circadian expression of tryptophan hydroxylase mRNA in chicken retina. Mol Brain Res 61:243–250, 1998.
13. Dubocovich, M.L. Melatonin is a potent modulator of dopamine release in the retina. Nature 306:782–784, 1983.
14. Dubocovich, M.L. and Takahashi, J.S. Use of 2-[^{125}I]-iodomelatonin to characterize melatonin binding sites in chicken retina. Proc Natl Acad Sci, USA 84:3916–3920, 1987.
15. Faillace, M.P., Sarmiento, M.I.K., and Rosenstein, R.E. Melatonin effect on [^{3}H]glutamate uptake and release in the golden hamster retina. J Neurochem 67:623–628, 1996.
16. Gastel, J.A., Roseboom, P.H., Rinaldi, P.A., Weller, J.L., and Klein, D.C. Melatonin production: Proteasomal proteolysis in serotonin N-acetyltransferase regulation. Science 279:1358–1360, 1998.
17. Green, C.B., Cahill, G.M., and Besharse, J.C. Tryptophan Hydroxylase in *Xenopus laevis* retina: Localization to photoreceptors and circadian rhythm *in vitro*. Invest Ophthalmol 35:1701, 1994.
18. Guerlotté, J., Grève, P., Bernard, M., Grechez-Cassiau, A., Morin, F., Collin, J.P., and Voisin, P. Hydroxyindole-*O*-methyltransferase in the chicken retina: Immunocytochemical localization and daily rhythm of mRNA. Eur J Neurosci 8:710–715, 1996.
19. Hamm, H.E. and Menaker, M. Retinal rhythms in chicks—circadian variation in melatonin and serotonin N-acetyltransferase. Proc Natl Acad Sci, USA 77:4998–5002, 1980.
20. Hamm, H.E., Takahashi, J.S., and Menaker, M. Light-induced decrease of serotonin N-acetyltransferase activity and melatonin in the chicken pineal gland and retina. Brain Res 266:287–293, 1983.
21. Iuvone, P.M. Cell biology and metabolic activity of photoreceptor cells: light-evoked and circadian regulation, In *Neurobiology and Clinical Aspects of the Outer Retina*. Djamgoz, M.B.A., Archer, S., and Vallerga, S. (eds.), pp. 25–55, Chapman & Hall, London, 1995.
22. Iuvone, P.M. Circadian rhythms of melatonin biosynthesis in retinal photoreceptor cells: Signal transduction, interactions with dopamine, and speculations on a role in cell survival, In *Retinal Degeneration and Regeneration*. Kato, S., Osborne, N.N., and Tamai, M. (eds.), pp. 3–13, Kugler Publications, Amsterdam/New York, 1996.
23. Iuvone, P.M. and Alonso-Gómez, A.L. Melatonin in the vertebrate retina, In *Retine, Luminiere, et Radiations*. Christen, Y., Doly, M., and Droy-Lefaix, M.-T. (eds.), pp. 49–62, Irvinn, Paris, 1998.
24. Klein, D.C. and Weller, J.L. Rapid light-induced decrease in pineal serotonin N-acetyltransferase activity. Science 177:532–533, 1972.
25. Leino, M., Aho, I.-M., Kari, E., Gynther, J., and Markkanen, S. Effects of melatonin and 6-methoxytetrahydro-β-carboline in light induced retinal damage: a computerized morphometric method. Life Sci 35:1997–2001, 1984.
26. Mangel, S.C. Dopamine mediates circadian rhythmicity and light adaptation in the retina. Exp Eye Res 67 Suppl. 1:S190, 1998.
27. Massey, S.C. and Neal, M.J. The light evoked release of acetylcholinefrom the rabbit retina *in vivo* and its inhibition by gamma-aminobutyric acid. J Neurochem 32:1327–1329, 1979.
28. Millar, T.J., Winder, C., Ishimoto, I., and Morgan, I.G. Putative serotonergic bipolar and amacrine cells in the chicken retina. Brain Res 439:77–87, 1988.
29. Mitchell, C.K. and Redburn, D.A. Melatonin inhibits ACh release from rabbit retina. Visual Neurosci 7:479–486, 1991.
30. Pang, S.F. and Yew, D.T. Pigment aggregation by melatonin in the retinal pigment epithelium and choroid of guinea pigs, *Cavia pocellus*. Experientia 35:231–233, 1979.
31. Pierce, M.E. and Besharse, J.C. Circadian regulation of retinomotor movements. I. Interaction of melatonin and dopamine in the control of cone length. J Gen Physiol 86:671–689, 1985.
32. Pierce, M.E. and Besharse, J.C. Melatonin and rhythmic photoreceptor metabolism: melatonin-induced cone elongation is blocked by high light intensity. Brain Res 405:400–404, 1987.
33. Reppert, S.M., Weaver, D.R., Cassone, V.M., Godson, C., and Kolakowski, L.F., Jr. Melatonin receptors are for the birds: Molecular analysis of two receptor subtypes differentially expressed in chick brain. Neuron 15:1003–1015, 1995.
34. Schmitz, Y. and Witkovsky, P. Glutamate release by the intact light-responsive photoreceptor layer of the *Xenopus* retina. J Neurosci Methods 68:55–60, 1996.
35. Stenkamp, D.L., Iuvone, P.M., and Adler, R. Photomechanical movements of cultured embryonic photoreceptors: Regulation by exogenous neuromodulators and by a regulable source of endogenous dopamine. J Neurosci 14:3083–3096, 1994.
36. Sugawara, T., Sieving, P.A., Iuvone, P.M., and Bush, R.A. The melatonin antagonist luzindole protects retinal photoreceptors from light damage in the rat. Invest Ophthalmol Vis Sci 39:2458–2465, 1998.

37. Thomas, K.B. A circadian rhythm of tryptophan hydroxylase activity in the retina and pineal gland. Doctoral Dissertation, Emory University, 1991.
38. Thomas, K.B. and Iuvone, P.M. Circadian rhythm of tryptophan hydroxylase activity in chicken retina. Cell Mol Neurobiol 11:511–527, 1991.
39. Thomas, K.B., Tigges, M., and Iuvone, P.M. Melatonin synthesis and circadian tryptophan hydroxylase activity in chicken retina following destruction of serotonin immunoreactive amacrine and bipolar cells by kainic acid. Brain Res 601:303–307, 1993.
40. Thomas, K.B., Zawilska, J., and Iuvone, P.M. Arylalkylamine (serotonin) N-acetyltransferase assay using high performance liquid chromatography with fluorescence or electrochemical detection of N-acetyltryptamine. Anal Biochem 184:228–234, 1990.
41. Tosini, G. and Menaker, M. Circadian rhythms in cultured mammalian retina. Science 272:419–421, 1996.
42. Wiechmann, A.F. and Craft, C.M. Localization of mRNA encoding the indolamine synthesizing enzyme, hydroxyindole-*O*-methyltransferase, in chicken pineal gland and retina by in situ hybridization. Neurosci Lett 150:207–211, 1993.
43. Wiechmann, A.F. and O'Steen, W.K. Melatonin increases photoreceptor susceptibility to light-induced damage. Invest Ophthalmol Vis Sci 33:1894–1902, 1992.
44. Zawilska, J.B. and Nowak, J.Z. Regulatory mechanisms in melatonin biosynthesis in retina. Neurochem Int 20:23–36, 1992.

ULTRAVIOLET LIGHT SUPPRESSES MELATONIN BIOSYNTHESIS IN CHICK PINEAL GLAND

Jolanta B. Zawilska, Jolanta Rosiak, and Jerzy Z. Nowak

Department of Biogenic Amines
Polish Academy of Sciences
POB-225, Łódź-1, 90-950 Poland

1. INTRODUCTION

The avian pineal gland synthesizes melatonin (MLT) in a circadian rhythm generated by an endogenous pacemaker. The rhythmic production of MLT appears to be regulated primarily by distinct changes in the activity of serotonin N-acetyltransferase (NAT). Visible light is a predominant environmental factor controlling MLT biosynthesis and as such it has two distinct effects on the hormone production. Thus, exposure to light at night results in acute decrease in NAT activity, MLT content and release. Light also resets the phase of the free-running circadian oscillator generating the rhythm of MLT production (1). It has recently been demonstrated that, like visible light, ultraviolet radiation (with wavelengths in the range of 320–400 nm; UV-A) is an inhibitory factor controlling MLT biosynthesis in pineal gland of rodents (2,3). The aim of this study was to analyze effects of UV-A light on the nocturnal NAT activity in chick pineal gland.

2. MATERIALS AND METHODS

Experiments were performed on two and seven weeks old chicks maintained under a 12 h light: 12 h dark cycle (LD; lights on at 22.00) for a minimum of 8 days prior to the study. During the fourth or the ninth hour of the dark phase of the LD cycle the animals were exposed to UV-A light (325–390 nm total bandwidth, λ_{max} = 365 nm, an irradiance of 10 μW/cm^2 at the heads' level). Ultraviolet light was produced by 16 W VL-204 BLB lamp (Vilber Lourmat, Marne la Vallee, France) and passed through a

Melatonin after Four Decades, edited by James Olcese.
Kluwer Academic / Plenum Publishers, New York, 2000.

UFS-6 filter. NAT activity was measured in supernatants of pineal homogenates by the radioisotopic method (4).

3. RESULTS AND DISCUSSION

Acute exposure of dark-adapted chicks to UV-A light produced a marked suppression of nighttime NAT activity of the pineal gland. The magnitude of the observed decline in the enzyme activity was dependent on the duration of the light pulse and age of chicks, with two weeks old birds being more sensitive than seven weeks old animals (Figure 1).

In another set of experiments, after a 5 min pulse of UV-A light chicks were transferred to darkness and a time course of NAT reactivation was determined for the pineal gland. When the exposure took place at the beginning of the fourth hour of the dark phase, i.e., when levels of both NAT-mRNA and NAT activity are increasing (5), the small UV-A light-evoked decline in the enzyme activity gradually deepened during the first 40 min treatments of the birds with darkness, from 9% (at 0 min) to 43% (at 40 min). After 40 min of darkness, NAT activity of the chick pineal gland began to rise, reaching complete restoration within 2 hours (Figure 2A). When the UV-A pulse of the same duration was applied in the middle of the ninth hour of the dark phase, i.e., during a declining portion of circadian oscillation in NAT-mRNA and NAT activity of the chick pineal gland (5), it produced a significant (by 23%) but short-lasting (up to 15 min) decline in the pineal NAT activity (Figure 2B).

Earlier studies have demonstrated that noradrenaline acting on postsynaptic α_2-adrenergic receptors plays an important role in the regulation of MLT biosynthesis in the chick pineal gland (6). Stimulation of these receptors have been shown to mediate, at least partially, the suppressive action of white light on pineal NAT activity and MLT level. Therefore, in the present studies we examined a possible involvement of an α_2-noradrenergic signal in the action of UV-A light on NAT activity of the chick pineal

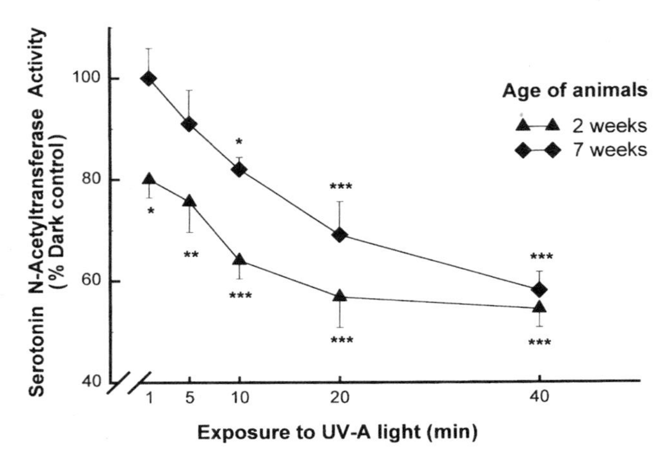

Figure 1. Effects of near-ultraviolet (UV-A) light on the nocturnal serotonin N-acetyltransferase activity of the chick pineal gland. Data shown are means ± SEM. N = 5–10/group. The enzyme activity in the dark control group (in nmol/hr/pineal gland): two weeks old chicks, 17.52 ± 0.93; seven weeks old chicks, 33.21 ± 2.83. *$P < 0.05$ vs dark control, **$P < 0.01$ vs dark control. ***$P < 0.001$ vs dark control.

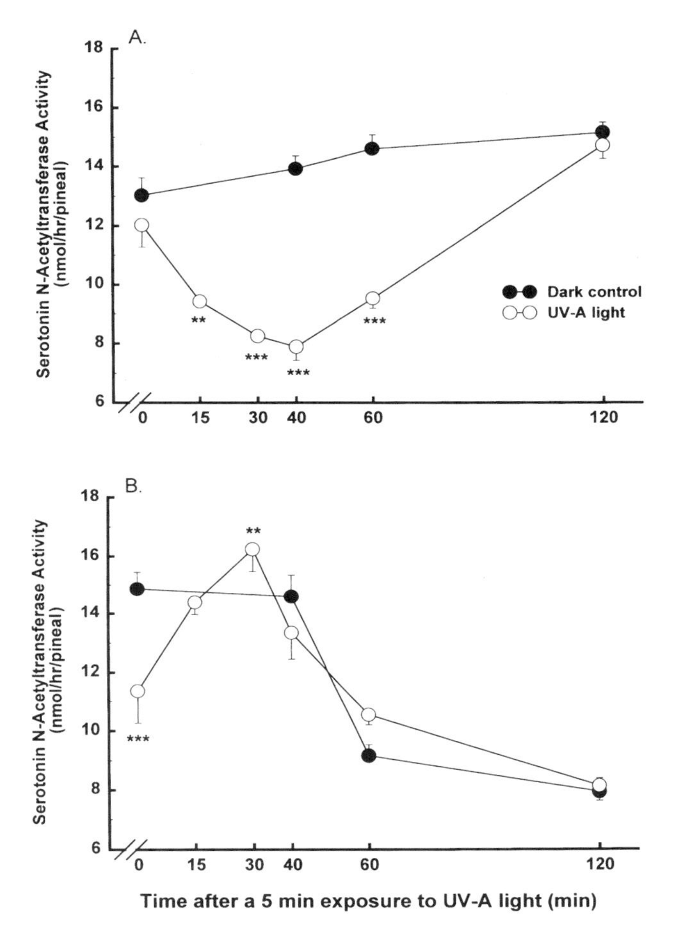

Figure 2. Time course of reactivation of serotonin N-acetyltransferase activity of the chick pineal gland after a short exposure to UV-A light. At the beginning of the fourth hour (A) or in the middle of the ninth hour (B) of the dark phase of the LD cycle the animals were exposed to a 5 min pulse of UV-A light, and then returned to darkness. Dark control refers to the enzyme activity in pineal glands isolated from chicks kept under dark conditions throughout the experiment. Data shown are means ± SEM. N = 5–15/group. $^{**}P < 0.01$ vs dark control, $^{***}P < 0.001$ vs dark control.

gland. Pretreatment of chicks with α-methyl-p-tyrosine, an effective in vivo inhibitor of tyrosine hydroxylase, significantly reduced the UV-A light-evoked decline in the chick pineal NAT activity (7). Furthermore, yohimbine (2 mg/kg, ip), a blocker of α_2-adrenergic receptors, antagonized the suppressive action of UV-A radiation on the enzyme activity [NAT activity (in nmol/hr/pineal): dark, 16.15 ± 1.13; dark + yohimbine, 18.22 ± 1.71; dark + 20 min UV-A light pulse, 10.11 ± 0.95[a]; dark + yohimbine + UV-A light, 17.28 ± 1.23[b]. N = 6–10/group. [a]$P < 0.001$ vs dark, [b]$P < 0.001$ vs dark + UV-A light].

In summary, results of our study indicate that UV-A radiation, similar to visible light, potently suppresses the nocturnal NAT activity of the chick pineal gland with the α_2-noradrenergic signal playing a role of an intermediate in this action.

ACKNOWLEDGMENT

Supported by the KBN grant No.4 P05A 074 12.

REFERENCES

1. Takahashi J.S., Murakami N., Nikaido S.S., Pratt B.L., and Robertson L.M. Rec. Prog. Horm. Res. 1989, 5:279–352.
2. Brainard G.C., Barker F.M., Hoffman R.J., Stetson M.H., Hanifin J.P., Podolin P.L., and Rollag M.D. Vision Res. 1994, 34:1521–1533.
3. Zawilska J.B., Rosiak J., and Nowak J.Z. Neurosci. Lett. 1998, 243:49–52.
4. Nowak J.Z., Żurawska E., and Zawilska J. Neurochem. Int. 1989, 14:397–406.
5. Bernard M., Iuvone P.M., Cassone V.M., Roseboom P.H., Coon S.L., and Klein D.C. J. Neurochem. 1997, 68:213–224.
6. Zawilska J. and Iuvone P.M. J. Pharmacol. Exp. Ther. 1989, 250:86–92.
7. Zawilska J.B., Rosiak J., Trzepizur K., and Nowak J.Z. The effects of near-ultraviolet light on serotonin N-acetyltransferase activity in the chick pineal gland. J. Pineal Res. 1999, 26:122–127.

6

EFFECTS OF VASOACTIVE INTESTINAL PEPTIDE AND HISTAMINE ON MELATONIN AND cAMP PRODUCTION IN CHICK EMBRYO PINEAL CELLS

Martina Macková and Dalma Lamošová

Institute of Animal Biochemistry and Genetics
SASci, Ivanka pri Dunaji, Slovakia

1. INTRODUCTION

The avian pineal gland, which seems to be a major source of circulating melatonin, contains a circadian oscillator that controls the rhythm of melatonin synthesis and release (1). Melatonin production is rhythmic in chick pineal cells. The key regulator of melatonin synthesis is cAMP. Agents which increase cAMP levels, such as vasoactive intestinal peptide (VIP), increase melatonin production (2). One of the widely reported actions of histamine (HA) is stimulation of cAMP production. The effect of HA on cAMP synthesis in the chick pineal gland was studied under in vivo and in vitro conditions and the gland was found to be highly sensitive to this amine (3,4).

A rhythmic profile of melatonin production was detected already during embryonic life, firstly in 16-day-old chick embryos (5). The important role of cAMP in melatonin synthesis in the chick embryo we investigated by application of agents which increase and/or decrease the level of cAMP (6).

Therefore, the aim of the present work was to study the effects of VIP and HA treatment on melatonin and cAMP production in pineal cells isolated from post-hatched chicks and chick embryos.

2. MATERIALS AND METHODS

Eggs of broiler hens were incubated in a forced draught incubator with temperature maintained at 37.5 ± 0.3°C under a light: dark cycle of 12:12 (light from 06.00

Melatonin after Four Decades, edited by James Olcese.
Kluwer Academic / Plenum Publishers, New York, 2000.

hour). Light intensity was at a range of 200 to 400 lux measured at the egg surface. Embryos at days 16 and 19 of development and post-hatched chicks at day 5 were decapitated during the light phase and pineal glands were processed as described previosly (5,6).

Pineal cell cultures from embryos of both ages and post-hatched chicks were used in investigating effects of VIP and HA alone and in combination. VIP and HA (10^{-7} mol l^{-1} and 10^{-4} mol l^{-1} resp., Sigma) were applicated on day 3 of pineal cell culture during the daytime, for 6 hours after the onset of light (for melatonin production) and for 1 hour in the middle of the light phase (for cAMP efflux and accumulation). Melatonin and cAMP production were measured by direct RIAs. Effects of drug administrations were evaluated by *t*-test.

3. RESULTS

VIP and HA alone and in combination evoked an increase of cAMP and melatonin levels in the embryonic period (Table 1). Results showed that stimulatory effect of VIP was stronger on melatonin production while HA influenced mainly efflux and accumulation of cAMP. However, the effect of HA alone didn't increase significantly melatonin production in the postembryonic period.

4. DISSCUSION

Our experimental results showed the stimulatory effect of VIP on melatonin and cAMP production in chick embryo pineal cells and confirmed published data from the postnatal period (2). Melatonin production and cAMP levels in embryonic pineal cells were increased after treatment with HA in agreement with experimental data obtained in post-hatched chicks (4). However, similar to the study of Nowak et al. (4), the stimulatory effect on melatonin production was only a marginal one. Melatonin and cAMP production after treatment of pineal cells with both agents in combination were

Table 1. Effects of VIP and HA on melatonin and cAMP production in pineal cells isolated from chick embryos and post-hatched chicks*

		16-day-old embryos	*P*	19-day-old embryos	*P*	5-day-old chicks	*P*
	control	2.74 ± 0.81		5.0 ± 1.11		8.1 ± 0.33	
Efflux cAMP	VIP	105.4 ± 6.20	***	68.12 ± 6.5	***	728.1 ± 111	***
(nmol/ml)	HA	127.3 ± 28.9	**	130.96 ± 9.8	***	1168.6 ± 65	***
	VIP + HA	354.4 ± 77.1	**	248.6 ± 12.4	***	1886.5 ± 87	***
	control	55.0 ± 6.5		54.4 ± 2.3		181.9 ± 4.0	
Accumulation	VIP	124.1 ± 20.2	*	76.9 ± 6.9	*	344.1 ± 17.0	***
cAMP	HA	129.3 ± 13.6	**	104.6 ± 7.6	**	435.9 ± 32.2	***
(nmol/ml)	VIP + HA	220.7 ± 22.4	***	105.7 ± 4.8	***	656.9 ± 50.3	***
	control	0.58 ± 0.16		9.94 ± 0.42		43.1 ± 2.6	
Melatonin	VIP	1.35 ± 0.19	*	20.3 ± 2.64	*	70.9 ± 6.4	**
production	HA	1.10 ± 0.12	*	11.8 ± 0.64	*	47.0 ± 4.6	
(ng/ml)	VIP + HA	1.50 ± 0.04	**	18.6 ± 1.00	***	68.1 ± 6.5	**

*Data are expressed as mean ± SEM. * $P < 0.05$, ** $P < 0.01$, *** $P < 0.001$ vs. control.

increased in both the embryonic and postembryonic period. There was an additional effect of HA and VIP on cAMP levels (efflux and accumulation), while the stimulatory effect of combined action of both agents on melatonin production is equal to action of VIP alone.

Results presented in this paper suggest that the biochemical pathways involving cAMP and VIP, which control melatonin production during the postnatal period, are developed before hatching and VIP stimulated melatonin production already in embryonic stages.

ACKNOWLEDGMENT

This work was supported by a grant from the Scientific Agency of Slovak Republic (2/5044/98).

REFERENCES

1. Takahashi J.S., Murakami N., Nikaido S.S., Pratt B.L., and Robertson L.M. (1989) The avian pineal, a vertebrate model system of the circadian oscillator: cellular regulation of circadian rhythms by light, second messengers and macromolecular synthesis. Recent Prog. Horm. Res. 45:279–352.
2. Pratt B.L. and Takahashi J.S. (1989) Vasoactive intestinal polypeptide and α_2-adrenoreceptor agonists regulate adenosine 3′,5′-monophosphate accumulation and melatonin release in chick pineal cell cultures. Endocrinol. 125:2375–2384.
3. Nowak J.Z., Sek B., D'Souza T., and Dryer S. (1995) Does histamine stimulate cyclic AMP formation in the avian pineal gland via a novel (non-H1, non-H2, non-H3) histamine receptor subtype. Neurochem. Int. 27:519–526.
4. Nowak J.Z., Zawilska J.B., Woldan-Tambor A., Sek B., Voisin P., Lintunen M., and Panula P. (1997) Histamine in the chick pineal gland: Origin, metabolism, and effects on the pineal function. J. Pineal Res. 22:26–32.
5. Lamošová D., Zeman M., Macková M., and Gwinner E. (1995) Development of rhythmic melatonin synthesis in cultured pineal glands and pineal cells isolated from chick embryo. Experientia 51:970–975.
6. Macková M., Lamošová D., and Zeman M. (1998) Regulation of rhythmic melatonin production in pineal cells of chick embryo by cyclic AMP. Cell. Mol. Life Sci. 54:461–466.

7

CERAMIDE INHIBITS L-TYPE CALCIUM CHANNEL CURRENTS IN RAT PINEALOCYTES

C. L. Chik,* B. Li, T. Negishi, E. Karpinski, and A. K. Ho

Department of Physiology and Department of Medicine*
Faculty of Medicine
University of Alberta
Edmonton, Alberta, Canada T6G 2H7

1. INTRODUCTION

We have recently identified the Ca^{2+} channels expressed in the rat pineal gland (1,2). Electrophysiological studies have demonstrated that rat pinealocytes express the L-type but not the T- or N-type Ca^{2+} channel. RT-PCR analysis of the Ca^{2+} channel subunit mRNAs revealed that rat pinealocytes express the α_{1D}, α_{2b}, β_2 and β_4 subunits (2). These channels can be inhibited by activation of the adrenergic mechanism through the involvement of cAMP- and cGMP-dependent protein kinases (1,2). Activation of insulin growth factor-1 receptors also has an inhibitory effect on these channels through a tyrosine kinase-dependent mechanism (3). Furthermore, these channels are under tonic phosphorylation control through phosphoprotein phosphatase 1 (4).

Another signalling pathway which has received much attention recently is the sphingomyelin cycle (5,6) (Figure 1). Ceramide, which is produced following sphingomyelin hydrolysis, has been shown to regulate intracellular enzymes such as protein kinase C (7), tyrosine kinases (8), diacylglycerol kinase (9), phosphatases (10) and phospholipases (11). Since both kinases and phosphatases can modulate the cyclic nucleotide responses (12,13) as well as the L-type Ca^{2+} channel in the rat pinealocyte, we investigated whether the sphingomyelin cycle could modulate pineal function through its effect on cyclic nucleotide accumulation and/or the L-type Ca^{2+} channel.

Melatonin after Four Decades, edited by James Olcese.
Kluwer Academic / Plenum Publishers, New York, 2000.

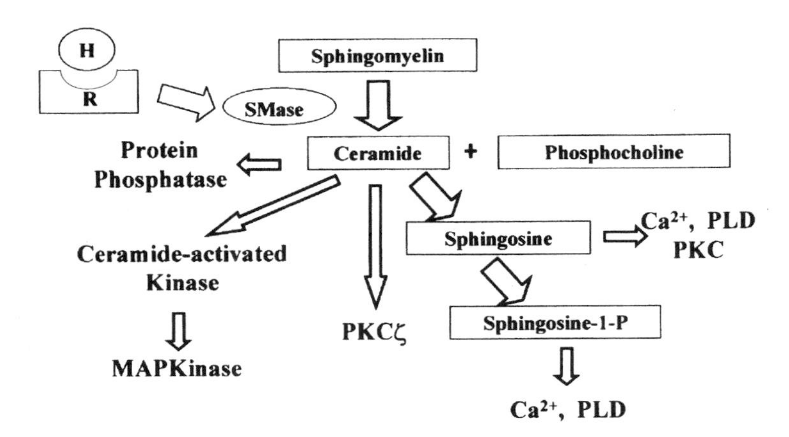

Figure 1. A schematic representation of the sphinogomyelin cycle. H, hormone or other ligand; MAPKinase, mitogen-activated protein kinase; PKC, protein kinase C; PKCζ, the ζ isozyme of PKC; PLD, phospholipase D; R, receptor; SMase, sphingomyelinase.

2. EFFECT OF CERAMIDE ON PINEAL CYCLIC NUCLEOTIDE RESPONSES

Pinealocytes were prepared from male Sprague-Dawley rats (150 gm, University of Alberta Animal unit) by trypsinization as previously described (14,15). The cells were suspended in Dulbecco's modified Eagle's medium containing 10% fetal calf serum and maintained at 37°C for 24 hr in a gas mixture of 95% air and 5% CO_2 before experimental treatment. Aliquots of cells (1.5×10^4 cells/0.4 ml) were treated with drugs which had been prepared in concentrated solutions in water or dimethylsulfoxide. The duration of the drug treatment period was 15 min for cAMP and cGMP accumulation. Cellular cAMP and cGMP contents were determined using a radioimmunoassay procedure in which samples were acetylated prior to analysis (16).

2.1. Effect of Ceramide on Adrenergic-Stimulated cAMP and cGMP Responses (17)

Treatment with isoproterenol (ISO, 1 μM) alone caused a significant increase in cAMP and cGMP accumulation; the addition of a depolarizing concentration of K^+ (30 mM KCl) potentiated these responses by 8–10 fold (Figure 2). Pretreatment with C2-ceramide for 5 min had no effect on basal or ISO-stimulated cAMP or cGMP accumulation. However, C2-ceramide inhibited the ISO + KCl-stimulated responses in a dose-dependent manner. Significant inhibition was observed at 10 μM of C2-ceramide and at 100 μM the reduction was more than 70% for both cAMP and cGMP responses. C6- and C8-ceramide were also effective in reducing the ISO + KCl-stimulated cAMP and cGMP responses (Figure 3). In contrast, C2-dihydroceramide, an inactive analogue, had no effect.

2.2. Effects of Ceramide on Cyclic Nucleotide Accumulation Stimulated by ISO in the Presence of Other Potentiating Agents

The inhibitory effect of C2-ceramide was dependent on the treatments used to potentiate the β-adrenergic-stimulated cAMP and cGMP accumulation (17). Pre-

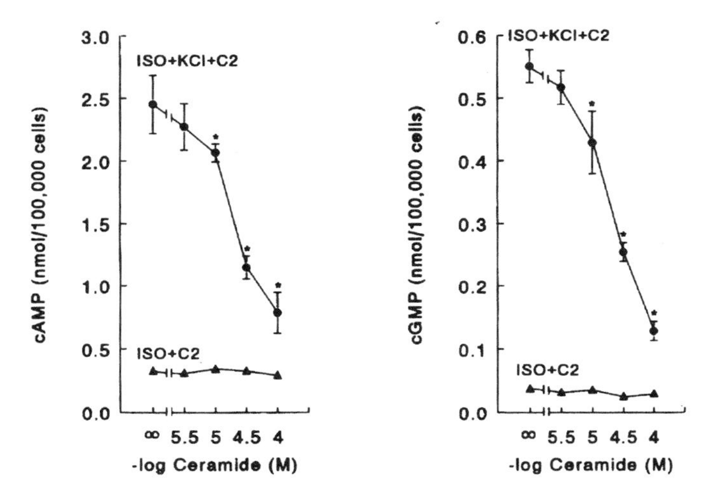

Figure 2. Effect of C2-ceramide on ISO- and ISO + KCl-stimulated cAMP and cGMP accumulation in rat pinealocytes. Pinealocytes (1.5×10^4 cells/ 400 µl) were incubated in DMEM with 10% fetal bovine serum and pre-treated with or without C2-ceramide (C2, 3–100 µM) for 5 min. The cells were then stimulated with ISO (1 µM) in the presence or absence of KCl (30 mM) for an additional 15 min. Each value represents the mean ± SEM of determinations done in duplicate on three samples of cells. *Significantly different from treatment with ISO + KCl.

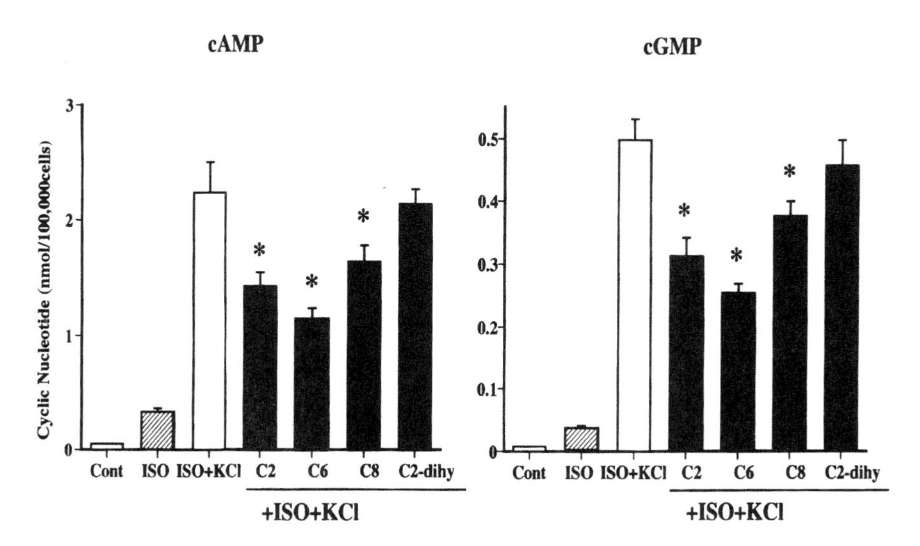

Figure 3. Effect of ceramides on the potentiating effect of K^+ on ISO-stimulated cAMP and cGMP responses. Pinealocytes (1.5×10^4 cells/ 400 µl) were incubated in DMEM with 10% fetal bovine serum and pre-treated with or without C2-, C2-dihydro(C2-dihy)-, C6- and C8-ceramide (30 µM) for 5 min. The cells were then stimulated with ISO (1 µM) and KCl (30 mM) for an additional 15 min. Each value represents the mean ± SEM of determinations done in duplicate on three samples of cells. *Significantly different from treatment with ISO + KCl.

treatment with C2-ceramide for 15 min had no effect on the potentiation by phenylephrine [an α-adrenergic agonist] (18), 4β-phorbol 12-myristate 13-acetate [a protein kinase C activator] (18–20), genistein [a tyrosine kinase inhibitor] (21,22), calyculin A [a serine/threonine phosphatase inhibitor] (13) or isobutylmethylxanthine [a phosphodiesterase inhibitor]. These results suggest that the inhibitory effect of C2-ceramide may be specific to intracellular Ca^{2+} elevating agents.

Additional studies showed that the inhibitory effect of C2-ceramide was dependent on the mechanism by which intracellular Ca^{2+} was elevated. Pre-treatment with C2-ceramide for 5 min inhibited only the BayK 8644 (an L-type Ca^{2+} channel agonist)-mediated potentiation of the ISO-stimulated cAMP and cGMP responses (17). In contrast, C2-ceramide had no effect on the potentiating effects of ionomycin [a Ca^{2+} ionophore] (23) or thapsigargin [an intracellular Ca^{2+}-ATPase inhibitor] (24).

Treatment with 1-phenyl-2-hexadecanoylamino-3-morpholino-1-propanol (PPMP), a glucosylceramide synthase inhibitor, has been shown to increase cellular ceramide levels (25,26). Pre-treatment with PPMP for 15 min had an effect similar to that of C2-ceramide: PPMP had no effect on basal or ISO-stimulated cAMP and cGMP accumulation, but selectively inhibited the ISO + KCl- or ISO + BayK 8644-stimulated cAMP and cGMP accumulation (Figure 4).

Taken together, these results indicate that ceramide selectively inhibits the potentiation mediated by elevation of intracellular Ca^{2+}. Ceramide has no effect on the potentiation mediated by an activator of protein kinase C or inhibitors of tyrosine kinase and phosphatases. Furthermore, this inhibitory effect of ceramide is highly specific and depends on the mechanism through which intracellular Ca^{2+} is elevated. For example,

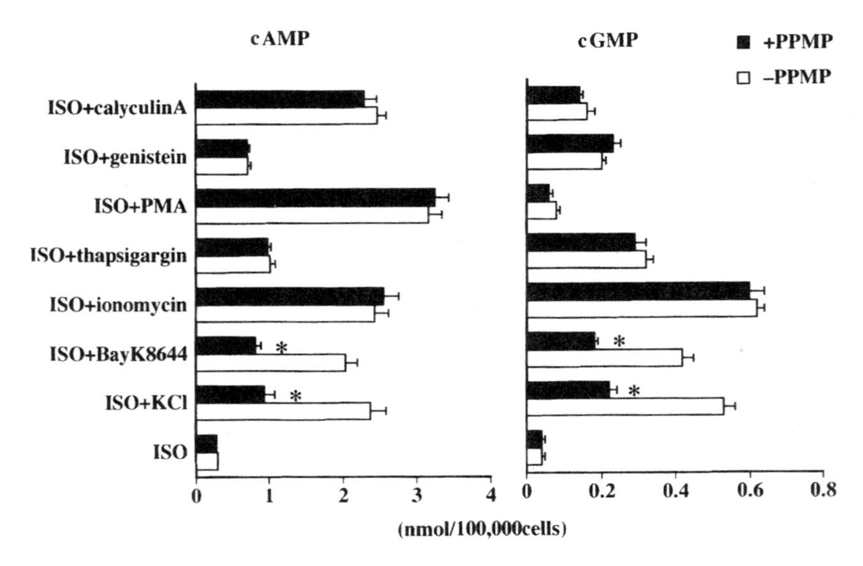

Figure 4. Effect of PPMP on the cAMP and cGMP responses stimulated by ISO in the absence or presence of potentiating agents. Pinealocytes (1.5×10^4 cells/ 400 μl) were incubated in DMEM with 10% fetal bovine serum and pre-treated with or without PPMP (10 μM) for 15 min. The cells were then stimulated with ISO (1 μM) in the presence or absence of potentiating agents as indicated, for an additional 15 min. Each value represents the mean ± SEM of determinations done in duplicate on three samples of cells. *Significantly different from corresponding treatment without PPMP.

ceramide inhibits only the potentiation caused by agents such as depolarizing concentrations of K^+ or BayK 8644 which elevate intracellular Ca^{2+} through activation of the L-type Ca^{2+} channel. Ceramide, however, has no effect on the potentiation mediated by ionomycin or thapsigargin, two agents which elevate intracellular Ca^{2+} in rat pinealocytes (24). Our results therefore suggest that an increase in ceramide levels significantly inhibits L-type Ca^{2+} channels in rat pinealocytes.

3. EFFECT OF CERAMIDE ON THE L-TYPE CA^{2+} CHANNELS

3.1. Patch Clamp Analysis

Ca^{2+} channel current recordings were obtained using the whole cell version of the patch clamp technique (1–4,27). Patch electrodes were filled with a solution containing (in mM): 70 Cs_2-aspartate, 20 HEPES, 11 EGTA, 1 $CaCl_2$, 5 $MgCl_2$. $6H_2O$, 5 glucose, 5 ATP-Na_2 and 5 K-succinate. Creatine phosphokinase (50 U/ml) and phosphocreatine-Na_2 (20 mM) were added to the pipette solution to reduce current run down. The bath solution contained (in mM): 105 Tris Cl, 0.8 $MgCl_2$. $6H_2O$, 5.4 KCl, 20 $BaCl_2$, 0.02 tetrodotoxin and 10 HEPES. 20 mM Ba^{2+} was used as the charge carrier. The membrane currents were measured using an Axopatch 1B whole cell patch clamp amplifier (Axon Instruments, Foster City, CA). The data were sampled using pClamp software (pClamp 5.7) and a Digidata 1200 analogue-to-digital interface (Axon Instruments). Analysis was performed using the pClamp software. At a holding potential of −50 mV and with Cs^+ in the intracellular solution, hyperpolarizing pulses did not activate any currents. Data are presented as the mean ± SEM percentages of control values. The pretreatment I–V relationship was plotted and used as a control.

3.2. Effect of Ceramide on the L-Type Ca^{2+} Channel Current in Rat Pinealocytes

As reported previously, the only voltage-dependent Ca^{2+} channel current found in dissociated pinealocytes is the dihydropyridine-sensitive L-type Ca^{2+} channel current (1–4). The effect of C2-ceramide on the L-type Ca^{2+} channel current is shown in Figure 5A, 5B. C2-ceramide (30 μM) decreased the peak amplitude of the L-type Ca^{2+} channel current in a pinealocyte by about 25% (Figure 5A). It did not shift the peak inward current along its voltage axis, as shown in the I–V relationship (Figure 5A). The onset of inhibition caused by C2-ceramide occurred within 8 to 10 min and the maximal inhibition was observed between 15 and 20 min. The inhibition by C2-ceramide persisted in the presence of BayK 8644 (Figure 5B). Similar to C2-ceramide, C6-ceramide also inhibited basal and BayK 8644-stimulated the L-type Ca^{2+} channel current (data not shown). In contrast, C2-dihydroceramide, an inactive analogue, had no effect on the basal and BayK 8644-stimulated L-type Ca^{2+} channel current.

A similar inhibitory effect on the L-type Ca^{2+} channel current was observed when endogenous ceramide level was increased by PPMP, a glucosylceramide synthase inhibitor (Figure 5B). The onset of inhibition caused by PPMP occurred within 8 to 10 min and the maximal inhibition was observed between 15 and 20 min. PPMP also inhibited the BayK 8644-mediated increase in the L-type Ca^{2+} channel current (Figure 5B).

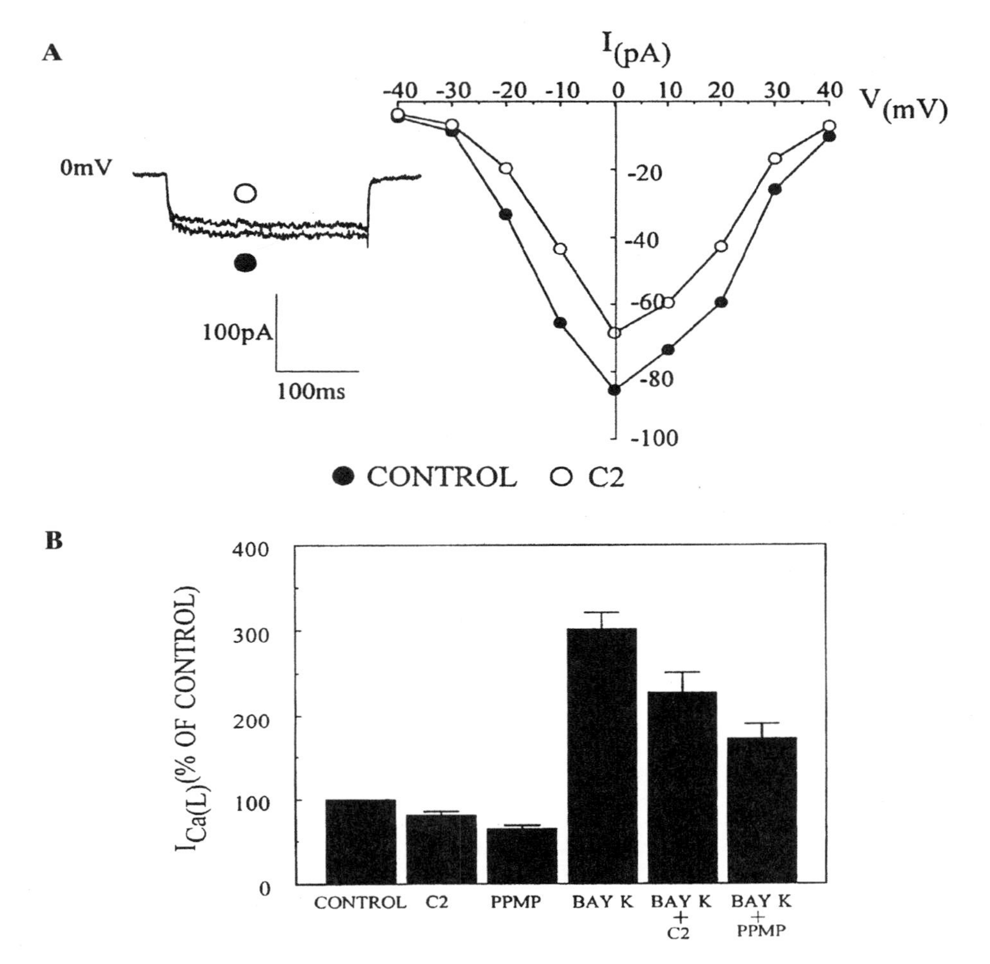

Figure 5. Effect of C2-ceramide on the L-type Ca^{2+} channel current in rat pinealocytes. (A) The L-type Ca^{2+} channel current is activated by depolarizing a pinealocyte from –40 mV to 10 mV in the absence (•) or presence (○) of C2-ceramide (C2, 30 μM). The I–V relationships obtained from the same cell are shown on the right. (B) The combined data showing the effect of C2-ceramide (30 μM) and PPMP (10 μM) in the absence or presence of BayK 8644 (Bay K, 1 μM), n = 5 for each treatment group.

3.3. Interleukin 1β Inhibits the L-Type Ca^{2+} Channel Current

Several cytokines, including interleukin 1β, tumor necrosis factor-α, and interferon-γ are known to induce sphingomyelin hydrolysis to ceramide (6). Since abundant expression of interleukin 1β and its specific receptor has been demonstrated in the rat pineal gland (28), the effect of interleukin 1β on the L-type Ca^{2+} channel current was examined. Similar to ceramides, interleukin 1β (10 ng/ml) inhibited the amplitude of the L-type Ca^{2+} channel current by about 23% (data not shown). The onset of inhibition caused by interleukin 1β occurred within 8 to 10 min and the maximal inhibition was observed between 15 and 20 min. Interleukin 1β also inhibited the BayK 8644-mediated increase in the L-type Ca^{2+} channel current.

4. DISCUSSION

Studies of ceramide on cyclic nucleotides indicate that it selectively inhibits the potentiation of cyclic nucleotide accumulation in the rat pinealocyte by BayK 8644 or a depolarizing concentration of K^+, suggesting that ceramide may inhibit the L-type Ca^{2+} channel current. Using the whole cell version of the patch clamp technique, we confirm that ceramide has an inhibitory effect on the L-type Ca^{2+} channel current. This represents a first report of an inhibitory effect of ceramide on the α_{1D} subtype of Ca^{2+} channel current which is commonly found in neuroendocrine cells. This observation suggests that the sphingomyelin cycle may play an important role in hormone secretion.

Both the inhibitory effects of ceramide on cyclic nucleotides and the L-type Ca^{2+} channel current appear specific: because 1) C2- and C6-ceramide produce similar effects on the nucleotides and the current; 2) C2-dihydroceramide, an inactive analogue of ceramide, is without effect; and 3) PPMP, an inhibitor of ceramide metabolism which elevates ceramide levels in other cell types, produces the same effect as exogenous ceramide. Indeed, our data on PPMP indicate that there may be a high rate of turnover of sphingomyelin in the rat pineal gland since 15 min of treatment with PPMP is sufficient to generate enough ceramide endogenously to produce a similar inhibitory effect to that observed after exogenous ceramide.

Although we have previously shown that the L-type Ca^{2+} channel current in rat pinealocytes is regulated by cAMP- and cGMP-dependent protein kinases, the observed effect of ceramide is independent of elevation of cyclic nucleotides. Neither ceramides nor PPMP has an effect on basal cAMP and cGMP accumulation. The effect of ceramide also appears to be selective for the voltage-gated Ca^{2+} channels since it has no effect on the norepinephrine-mediated increase in intracellular Ca^{2+} (our unpublished observation). Whether the effect of ceramide represents a direct effect on the channel or mediated through an intermediate second messenger remains unknown. Possible second messengers include mitogen-activated protein kinase and protein kinase C, a known modulator of these channels in rat pinealocytes (3). However, the protein kinase C isozyme(s) activated by ceramide appear different from those that are involved in potentiating the pineal cyclic nucleotide responses as ceramide had no effect on the β-adrenergic-stimulated cyclic nucleotide accumulation.

Our results, together with the known involvement of ceramide in the inhibitory effect of cytokines on the L-type Ca^{2+} channel current in rat ventricular myocytes (29), suggest that the sphingomyelin cycle likely plays an important role in the regulation of L-type Ca^{2+} channels in general. One probable endogenous activator of the sphingomyelin cycle in the rat pineal gland is interleukin 1β. This is based on our observation that interleukin 1β has an inhibitory effect on the L-type Ca^{2+} channel current and there is high expression of this cytokine in the rat pineal gland (28). Our results also add to the growing evidence that the L-type Ca^{2+} channel plays an important role in the control of pineal function. Recently, these channels have been shown to be involved in the acetylcholine-mediated L-glutamate release, hence indirectly regulates norepinephrine-dependent melatonin synthesis (30,31). At present, the role of the sphingomyelin cycle in the control of pineal N-acetyltransferase and melatonin synthesis remains to be determined.

ACKNOWLEDGMENTS

A. K. Ho is a scholar of the Alberta Heritage Foundation for Medical Research (AHFMR). This work was supported by grants from the Medical Research Council of Canada and the University of Alberta Hospital Foundation. The authors would like to thank Dr. Albert Baukal (NICHD, NIH) for the supply of antisera for the RIA.

REFERENCES

1. Chik, C.L., Liu, Q.-Y., Li, B., Karpinski, E., and Ho, A.K. Cyclic GMP inhibits L-type Ca^{2+} channel currents through protein phosphorylation in rat pinealocytes. J. Neurosci. 15:3104–3109, 1995.
2. Chik, C.L., Liu, Q.-Y., Li, B., Klein, D.C., Zylka, M., Chin, H., Kim, D.S., Karpinski, E., and Ho, A.K. α_{1D} L-type Ca^{2+} channel currents: inhibition by a β-adrenergic agonist and PACAP in rat pinealocytes. J. Neurochem. 68:1078–1087, 1997.
3. Chik, C.L., Li, B., Karpinski, E., and Ho, A.K. Insulin and insulin-like growth factor-1 inhibit the L-type Ca^{2+} channel current in rat pinealocytes. Endocrinology 138:2033–2042, 1997.
4. Chik, C.L., Li, B., Karpinski, E., and Ho, A.K. Regulation of the L-type Ca^{2+} channel current in rat pinealocytes: role of basal phosphorylation. J. Neurochem., 72:73–80, 1999.
5. Hannun, Y.A. and Bell, R.M. Functions of phospholipids and sphingolipid breakdown products in cellular regulation. Science 243:500–507, 1989.
6. Hannun, Y.A. The sphingomyelin cycle and the second messenger function of ceramide. J. Biol. Chem. 267:3125–3128, 1994.
7. Lee, J.Y., Hannun, Y.A., and Obeid, L.M. Ceramide inactivates cellular protein kinase Cα. J. Biol. Chem. 271:13169–13174, 1996.
8. Raines, M.A., Kolesnick, R.N., and Golde, D.W. Sphingomyelinase and ceramide activate mitogen-activated protein kinase in myeloid HL-60 cells. J. Biol. Chem. 268:14572–14575, 1993.
9. Younes, A., Kahn, D.W., Besterman, J.M., Bittman, R., Byun, H.-S., and Kolesnick, R.N. Ceramide is a competitive inhibitor of diacylglycerol kinase *in vitro* and in intact human leukemia (HL-60) cells. J. Biol. Chem. 267:842–847, 1992.
10. Dobrowsky, R.T. and Hannun, Y.A. Ceramide stimulates a cytosolic protein phosphatase. J. Biol. Chem. 267:5048–5051, 1992.
11. Abousalham, A., Liossis, C., O'Brien, L., and Brindley, D.N. Cell-permeable ceramides prevent the activation of phospholipase D by ADP-ribosylation factor and RhoA. J. Biol. Chem. 272:1069–1075, 1997.
12. Chik, C.L. and Ho, A.K. Multiple receptor regulation of cyclic nucleotides in rat pinealocytes. Prog. Biophys. Molec. Biol. 53:197–203, 1989.
13. Ho, A.K. and Chik, C.L. Phosphoprotein phosphatase inhibitors potentiate adrenergic-stimulated cAMP and cGMP production in rat pinealocytes. Am. J. Physiol. 268:E458–E466, 1995.
14. Buda, M. and Klein, D.C. A suspension culture of pinealocytes: regulation of N-acetyltransferase activity. Endocrinology 103:1483–1493, 1978.
15. Chik, C.L., Ho, A.K., and Klein, D.C. α_1-Adrenergic potentiation of vasoactive intestinal peptide stimulation of rat pinealocyte adenosine 3′,5′-monophosphate and guanosine 3′,5′-monophosphate: evidence for a role of calcium and protein kinase C. Endocrinology 122:702–708, 1988.
16. Harper, J.F. and Brooker, G. Femtomole sensitive radioimmunoassay for cyclic AMP and cyclic GMP after 2′0 acetylation by acetic anhydride in aqueous solution. J. Cyclic Nucleotide Res. 1:207–218, 1975.
17. Negishi, T., Chik, C.L., and Ho, A.K. Ceramide selectively inhibits calcium-mediated potentiation of β-adrenergic-stimulated cyclic nucleotide accumulation in rat pinealocytes. Biochem. Biophys. Res. Commun. 244:57–61, 1998.
18. Sugden, D., Vanecek, J., Klein, D.C., Thomas, T.P., and Anderson, W.B. Activation of protein kinase C potentiates isoprenaline-induced cyclic AMP accumulation in rat pinealocytes. Nature 314:359–361, 1985.
19. Ho, A.K., Chik, C.L., and Klein, D.C. Protein kinase C is involved in the adrenergic stimulation of pineal cGMP accumulation. J. Biol. Chem. 262:10059–10064, 1987.
20. Ho, A.K., Thomas, T.P., Chik, C.L., Anderson, W.B., and Klein, D.C. Protein kinase C: subcellular redistribution by increased Ca^{2+} influx. J. Biol. Chem. 263:9292–9297, 1988.

21. Ho, A.K., Wiest, R., Ogiwara, T., Murdoch, G., and Chik, C.L. Potentiation of adrenergic-stimulated cyclic AMP accumulation by tyrosine kinase inhibitors in rat pinealocytes. J. Neurochem. 65:1597–1603, 1995.
22. Ogiwara, T., Murdoch, G., Chik, C.L., and Ho, A.K. Tyrosine kinase inhibitors enhance cGMP production in rat pinealocytes. Biochem. Biophys. Res. Commun. 207:994–1002, 1995.
23. Sugden, A.L., Sugden, D., and Klein, D.C. Essential role of calcium influx in the adrenergic regulation of cAMP and cGMP in rat pinealocytes. J. Biol. Chem. 261:11608–11612, 1986.
24. Ho, A.K., Ogiwara, T., and Chik, C.L. Thapsigargin modulates agonist-stimulated cAMP responses through calcium-dependent and -independent mechanisms in rat pinealocytes. Mol. Pharmacol. 49:1104–1112, 1996.
25. Inokuchi, J. and Radin, N.S. Preparation of the active isomer of 1-phenyl 2-decanoylamino-3-morpholino-1-propanol, inhibitor of murine glucocerebroside synthetase. J. Lipid Res. 28:565–571, 1987.
26. Posse de Chaves, E., Bussiere, M., Vance, D.E., Campenot, R.B., and Vance, J.E. Elevation of ceramide within distal neurites inhibits neurite growth in cultured rat sympathetic neurons. J. Biol. Chem. 272:3028–3035, 1997.
27. Hamill, O.P., Marty, A., Neher, E., Sakmann, B., and Sigworth, F.J. Improved patch-clamp techniques for high-resolution current recording from cells and cell-free membrane patches. Pflugers Arch. 391:85–100, 1981.
28. Wong, M.-L., Bongiorna, P.B., Rettori, V., McCann, S.M., and Licino, J. Interleukin (IL) 1β, IL-1 receptor antagonist, IL-10, and IL-13 gene expression in the central nervous system and anterior pituitary during systemic inflammation: pathophysiological implications. Proc. Natl. Acad. Sci. USA 94:227–232, 1997.
29. Schreur, K.D. and Liu, S. Involvement of ceramide in inhibitory effect of IL-1β on L-type Ca^{2+} current in adult rat ventricular myocytes. Am. J. Physiol. 272:H2592–2598, 1997.
30. Letz, B., Schomerus, C., Maronde, E., Korf, H.W., and Horbmacher, C. Stimulation of a nicotinic ACh receptor causes depolarization and activation of L-type Ca^{2+} channels in rat pinealocytes. J. Physiol. 499:329–340, 1997.
31. Yamada, H., Ogura, A., Koizumi, S., Yamaguchi, A., and Moriyama, Y. Acetylcholine triggers L-glutamate exocytosis via nicotinic receptors and inhibits melatonin synthesis in rat pinealocytes. J. Neurosci. 18:4946–4952, 1998.

8

EXPRESSION OF MELATONIN RECEPTORS AND 2-[^{125}I]IODOMELATONIN BINDING SITES IN THE PITUITARY OF A TELEOST FISH

Pascaline Gaildrat and Jack Falcón

Département des Neurosciences
Laboratoire de Neurobiologie Cellulaire
CNRS UMR 6558
Université de Poitiers
40 avenue du Recteur Pineau
86022 Poitiers Cedex France

ABSTRACT

The mechanisms of the photoperiodic control of fish physiology (growth, reproduction) and behavior (locomotor activity) are far from being understood. We show here that 2-[^{125}I]iodomelatonin binds specifically to membrane preparations from pike (*Esox lucius*, L.) pituitaries (K_D: 556 pM; Bmax: 2.8 fmol/mg proteins). Radioautography indicated that the binding was restricted to a part of the pituitary only. Using polymerase chain reaction from pike genomic DNA we were able to isolate, subclone and sequence two fragments. The so-called P4 and P8 fragments displayed homology with, respectively, the Mel_{1a} and Mel_{1b} receptor subtypes. The P4 and P8, probes allowed detection of mRNAs corresponding to these receptors in different areas of the brain, including the pituitary. This is the first evidence that melatonin receptors and binding sites are expressed in the pituitary of a non-mammalian species. We suggest that in fish the melatonin-mediated photoperiodic control of neuroendocrine functions might involve a direct effect on the pituitary.

Tel: 33 (0)549.45.39.77; Fax: 33 (0)549.45.40.51; e-mail: Jack.Falcon@campus.univ-poitiers.fr

Melatonin after Four Decades, edited by James Olcese.
Kluwer Academic / Plenum Publishers, New York, 2000.

1. INTRODUCTION

Light is a major environmental factor regulating the daily and seasonal rhythms of melatonin production by the pineal organ (9,26). In all vertebrate species examined thus far, melatonin production is high during night-time and low during day-time. The daily pattern changes on an annual basis as a consequence of the seasonal variations in day length. Melatonin is involved in the control of numerous daily and seasonal rhythms and may thus be regarded as an internal "Zeitgeber". Its effects are mediated through specific receptors. Melatonin binding sites were first characterized and mapped using the melatonin agonist 2-[^{125}I]iodomelatonin (^{125}I-melatonin); high (pM range) and low (nM range) affinity binding sites were evidenced (5,25). Three melatonin receptor subtypes—the Mel_{1a}, Mel_{1b} and Mel_{1c}—have been cloned in the zebrafish, *Xenopus* and chicken, whereas only the first two are found in mammals (7,18,27,28). The binding and cloning studies indicated that the receptors belong to the 7 transmembrane family coupled to guanine nucleotide (G) binding proteins (25,27). They are usually coupled to inhibition of the adenylyl cyclase, but other transduction pathways are being investigated.

Part of the circadian effects of melatonin are mediated through its action on the hypothalamic circadian clocks (the suprachiasmatic nuclei [SCN] in mammals, and the visual (v)SCN in birds) (2,13,14,19,24,29,32). The reproductive effects of melatonin have been clearly demonstrated in mammals. For example, in seasonally breeding species, short photoperiod or long photoperiod plus melatonin treatment may induce either gonadal regression, or reactivation of the reproductive cycle, in long day and short day breeders, respectively; conversely increasing day length will induce opposite effects (20,26). The targets mediating the reproductive effects of melatonin include the premammillary and mediobasal areas of the hypothalamus (to modulate LH secretion) and the pars tuberalis of the hypophysis (involved in the seasonal control of prolactin secretion) (13,17,20,23,29). The mechanisms by which melatonin generates its reproductive effects are only partly known, and the data concern mainly mammals.

Many studies have investigated the part of the pineal and its hormone melatonin in the regulation of reproduction in ectotherms and birds (22,34). To date, the available data differ with gender, photoperiod and temperature administration, and reproductive stage, so that no clear-cut picture arises. This is further complicated by the observation that the ^{125}I-melatonin binding sites display a widespread distribution in nonmammals (2,6,8,21,30,33). One study reports that intra-ventricular melatonin administration increased plasma gonadotropin hormone (GtH II) in the Atlantic croaker, and that physiological concentrations of melatonin stimulated GtH II release from cultured pituitary fragments (15). These results contrast with those reporting no binding of melatonin in the pituitary of the fish and birds investigated thus far (2,11). The present study was designed to determine whether the pituitary of a teleost (pike) expresses melatonin receptors. For this purpose, we looked at the expression of melatonin receptor transcripts, using the polymerase chain reaction (PCR), and at the binding of ^{125}I-melatonin on crude membrane preparations and frozen tissue sections.

2. MATERIALS AND METHODS

2.1. Animals and Tissue Specimens

Wild pikes (*Esox lucius* L., teleost) of both sexes (500 to 2000 g) were obtained during the winter from ponds at Poitou-Charentes (France). In the laboratory, fish were maintained for 24 h in oxygenated and filtered tap-water under conditions resembling their natural habitat with respect to temperature and illumination (1000 lux intensity at the water surface). They were killed by decapitation. Nervous tissues, including different brain areas, and the retina, the pituitary as well as peripheral organs, including liver, gonads and intestine, were removed, frozen in dry ice and stored at –78°C. Pituitaries were also frozen at –50°C in isopentane for *in situ* binding studies.

2.2. Isolation of Two Putative cDNA Melatonin Receptor Fragments

Polymerase chain reaction (PCR) amplification was accomplished using degenerate primers based on the peptide sequences INRYCYIC and MAYFNSC, which are located near the 3rd and in the 7th transmembrane domains, respectively, and are conserved among the *Xenopus* and the mammalian (Mel_{1a} and Mel_{1b}) melatonin receptors (27–29). Genomic DNA was subjected to 30 cycles of PCR amplification (94°C, 45 sec; 45°C, 2 min; 72°C, 2 min) using 200 nM of each degenerate primer, and 1.25 U of TaqDNA polymerase (Appligene). PCR products were run on 1.5% agarose gel. The amplified DNA was purified and cloned into pGEM-T vector (Promega). The nucleotide sequences of recombinant clones were determined using "ABIPRISM™ Dye Terminator Cycle Sequencing Ready Reaction Kit" (Perkin Elmer) with AmpliTaq DNA Polymerase, FS. Primers were synthetic oligonucleotides that were either vector specific or derived from sequence information.

After sequence determination, specific primers (Figure 1) were designed to amplify a fragment to be used as a probe in northern and Southern blot experiments. For this purpose, the plasmid DNA, containing the PCR-generated insert of a pike putative melatonin receptor (see above), was subjected to 35 cycles of PCR amplification (94°C, 45 sec; 60°C, 45 sec; 72°C, 30 sec).

2.3. Expression of Two Putative Melatonin Receptors in the Pike

2.3.1. Northern Blot Analysis. Total RNA was isolated from different tissues using the TRIzol method (TRIzol™, GibcoBRL). Poly(A)$^+$ RNA was then purified using the Oligotex mRNA kit (QIAGEN), subjected to electrophoresis trough a 1% agarose-formaldehyde gel, and finally blotted overnight in 10X SSC (1.5 M NaCl/0.15 M sodium citrate, adjusted to pH 7 with HCl) on nitrocellulose sheets. The blot membrane was dried at 80°C and non-specific sites were blocked for 2 hr at 42°C with hybridization buffer containing 40% formamide, 10% dextran sulfate, 4X SSC, 1X Denhardt's solution (100X Denhardt's: 2% w/v polyvinylpyrrolidone, 2% w/v bovine serum albumin, 2% w/v Ficoll 400), 20 mM Tris-HCl (pH 7.4) and 0.3 mg/ml denatured calf thymus DNA. The membrane was hybridized with the probes labeled by random priming with [α-^{32}P]dCTP (3000Ci/mmol; Amersham) (10). Hybridization was performed overnight at 42°C with 120 ng of labeled probe in 5 ml of hybridization buffer. The blot membrane was then washed (3 times 15 min) in 2X SSC/0.1% SDS (sodium dodecyl sulfate) at room temperature, followed by incubation (2 times 30 min) in 0.1X SSC/0.1% SDS at 52°C. The membrane was then exposed to Hyperfilm (Kodak) at –80°C using intensifying screens. After deshybridization, the blot was hybridize with a human β actin cDNA probe obtained by PCR amplification of lymphocyte cDNA, covering a 207 bases sequence of exons 4 and 5 (4).

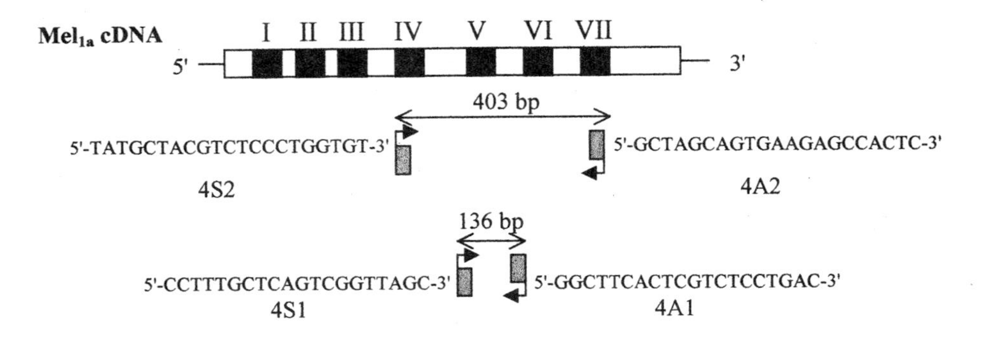

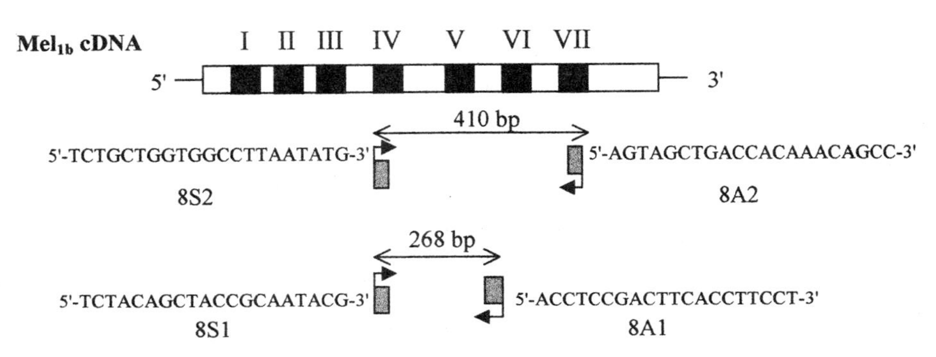

Figure 1. Oligonucleotide sequences of the specific primers used in the polymerase chain reactions (PCR). As explained in the text, PCR amplification of genomic DNA was accomplished using degenerate primers based on the peptide sequences INRYCYIC and MAYFNSC, located in conserved regions near the third (black boxe III) and in the seventh (black boxe VII) transmembrane domains of the receptor. After cloning and sequence determination, this allowed to obtain the P4 and P8 fragments which displayed homologies, respectively, with the Mel_{1a} and Mel_{1b} receptor subtypes. For each fragment, two specific primer pairs were designed. The box-arrows indicate the schematic positions of the PCR primers and their extension by the Taq polymerase. The letters S and A mean sense and anti-sense. Total RNAs from the tissues under investigation were reverse transcribed and the cDNA product were submitted to PCR amplification using the 4S2-4A2 or 8S2-8A2 primers pairs (see Fig. 4). The 4S1-4A1 or 8S1-8A1 primer pairs were used to generate specific probes from the cloned P4 and P8 fragments.

2.3.2. Reverse Transcription (RT) and Polymerase Chain Reaction (PCR). The first-strand cDNA was synthesized on Dynabeads oligo$(dT)_{25}$ (Dynal). The RNA (3 µg) from each tissue was reverse transcribed with 5×10^6 beads. Each 20 µl cDNA synthesis reaction contained reverse transcription buffer (50 mM Tris-HCl [pH 8.3], 75 mM KCl, 3 mM $MgCl_2$), 10 mM DTT (dithioteitrol), 0.5 mM of each dNTP and 200 U of M-MLV reverse transcriptase (GibcoBRL). A control group in which reverse transcriptase was omitted was processed in parallel. The RT reaction was carried out for 2.5 hr at 42°C, and single-strand cDNA was then purified. An aliquot of the reaction mixture (1.5 µl) was subjected to 35 cycles of PCR amplification (94°C, 45 sec; 60°C, 45 sec; 72°C, 30 sec) using specific oligonucleotide primers (Figure 1). The RT-PCR fragments were also sequenced. Another aliquot was subjected to 30 cycles of PCR amplification (94°C, 45 sec; 54°C, 45 sec; 72°C, 1 min) using the actin primers (5′-TGAAGCCCAGAGCAAAAGAG-3′ and 5′-TCTCCTTGATGTCACGCACA-3′) and the product was run on 1.5% agarose gel.

2.3.3. Southern Blotting. One percent of the RT-PCR amplified products (see above) was electrophoresed in a 1.5% agarose gel in 1X TAE buffer (40 mM Tris, 20 mM sodium acetate, 1 mM EDTA), transferred to a nitrocellulose membrane. Hybridization was performed as described above. Blots were then exposed to Fuji RX X ray films at room temperature.

2.4. Binding Assays on Tissue Sections

Pituitary serial coronal sections (20 μm thick) were obtained on a cryostat at –20°C; they were thaw-mounted onto gelatin-coated slides, and kept at –20°C. Autoradiographic binding procedure was performed as previously described (12). Sections were pre-incubated at 4°C for 15 min in Tris buffer (100 mM, pH 7.4) containing 4 mM $CaCl_2$, and then incubated in the same buffer containing 100 pM ^{125}I-melatonin (Dupont NEN, France) with or without an excess of cold melatonin (1 μM) for 60 min at room temperature. Sections were washed twice for 30 sec in assay buffer, followed by a 30 sec wash in ice cold distilled water. Sections were air-dried and exposed to Biomax MR film (Kodak) for 10 days. Six pike pituitaries were used for this experimental series, which were processed in 3 independent experiments.

2.5. Binding Assays on Pituitary Membranes

Membranes were prepared, and binding assays were performed as previously described (11). In brief, 60 pooled pituitaries were suspended in 10 volumes (v/v) of ice-cold $CaCl_2$/Tris buffer (50 mM), and homogenized by means of an Ultraturax. The homogenate was centrifuged at 800 g (for 10 min at +4°C) and the pellet was discarded. The supernatant was then centrifuged at 80,000 g (for 20 min at +4°C). The resulting pellet was re-suspended by sonication, washed in 10 ml of Tris buffer, and re-centrifuged as above indicated. The final pellet was re-suspended in Tris buffer, at a final concentration of 2–5 mg proteins/ml. Membranes were stored at –78°C. Proteins were determined using the method of Bradford (1) using bovine serum albumin for the preparation of standards.

Aliquots containing 50 to 100 μg of membrane proteins were incubated in a final volume of 60 μl of Tris–HCl buffer in the presence of increasing concentrations of ^{125}I-melatonin. Nonspecific binding was determined by the addition of an excess of unlabelled melatonin (50 μM in the final reactive volume). After one hour incubation at +21°C, membranes were collected by rapid vacuum filtration through Whatman GF/C glass fiber filters. The filters were washed 3 times with 3 ml of ice-cold incubation buffer and radioactivity was measured (Wallak γ counter). Each curve corresponds to data obtained from sixty pooled pituitaries; each plot corresponds to triplicate determinations; all experiments were duplicated. Data were fitted to the equation of a rectangular hyperbola (16).

3. RESULTS

3.1. Isolation of Two Putative Melatonin Receptor Fragments

The PCR amplification of pike genomic DNA using degenerate primers allowed to obtain fragments of the appropriate length (504 bp), which were cloned and sequenced. Of 20 clones analyzed, 2 putative melatonin receptor fragments—P4 and P8—were identified which are 70% identical (Figure 2). Among all the cloned melatonin receptors, P4 is most similar to the zebrafish Mel_{1a} melatonin receptor (Z1.4), with which it shares 84% amino acid identity. P8 displays similarities with the zebrafish Mel_{1b} melatonin receptor (Z2.6), with which it shares 79% amino acid identity.

3.2. Expression of P4 and P8 mRNA

P4 and P8 mRNA distribution was examined by northern blot analysis in several brain areas, including the pituitary as well as in the liver and retina. Hybridization with the P8 probe revealed the presence of one transcript (approximately 5.4 kb) highly expressed in the optic tectum (Figure 3). No hybridization signal was apparent in other brain areas, pituitary or liver, either with the P8 or with the P4 probe (not shown).

The expression of the P4 and P8 genes was also examined using a RT-PCR assay (Figure 4). No amplification was detected in the controls, thereby indicating that genomic DNA was not amplified. The P4 labeled mRNA was expressed in several brain areas and the retina. Actin served as a control to verify the amount of template for each sample. Although no precise quantification could be done, our results suggest high levels of expression in the optic tectum and diencephalon. Noticeable expression was also found in the olfactory bulbs, telencephalon and retina whereas threshold levels were detected in the pituitary and the cerebellum. The P8 transcript was detected in the same tissues but the relative pattern of expression was apparently different. A high expression was found in the optic tectum and telencephalon, and a much lower expression was apparent in the pituitary, olfactory bulbs, diencephalon and cerebellum. A very low hybridization signal was obtained in the retina. No expression was detected in the ovaries, liver and intestine with either probe.

3.3. Binding of ^{125}I-Melatonin in Pituitary Sections and Membranes

In vitro radioautography on pituitary sections indicated that the labeling did not cover the whole surface of the gland (Figure 5). Rather, it appeared as a moon quarter in the antero-ventral part. Alternative sections incubated in the presence of an excess (1 μM) of melatonin displayed absolutely no binding.

To study the binding of ^{125}I-melatonin on pituitary membranes we used the parameters set up for the binding on membrane preparations from the brain (11). Under these conditions, ^{125}I-melatonin bound in a saturable fashion to pituitary membranes (Figure 6). Scatchard re-plot of the data indicated that Bmax was 2.8 fmol/mg prot. and K_D was 556 pM. Nonspecific binding increased linearly with increasing ^{125}I-melatonin concentrations.

4. DISCUSSION

This study reports the cloning of two receptor fragments (P4 and P8) from the pike that contain structural motifs typical of the G protein-coupled melatonin receptor family. The comparison of their deduced amino-acid sequences with the sequences from the zebrafish melatonin receptors (29) supports the view that the P4 and P8 fragments correspond, respectively, to the Mel_{1a} and Mel_{1b} receptor subtypes. Northern blot analysis indicated that expression of the Mel_{1b} was high in the optic tectum. No

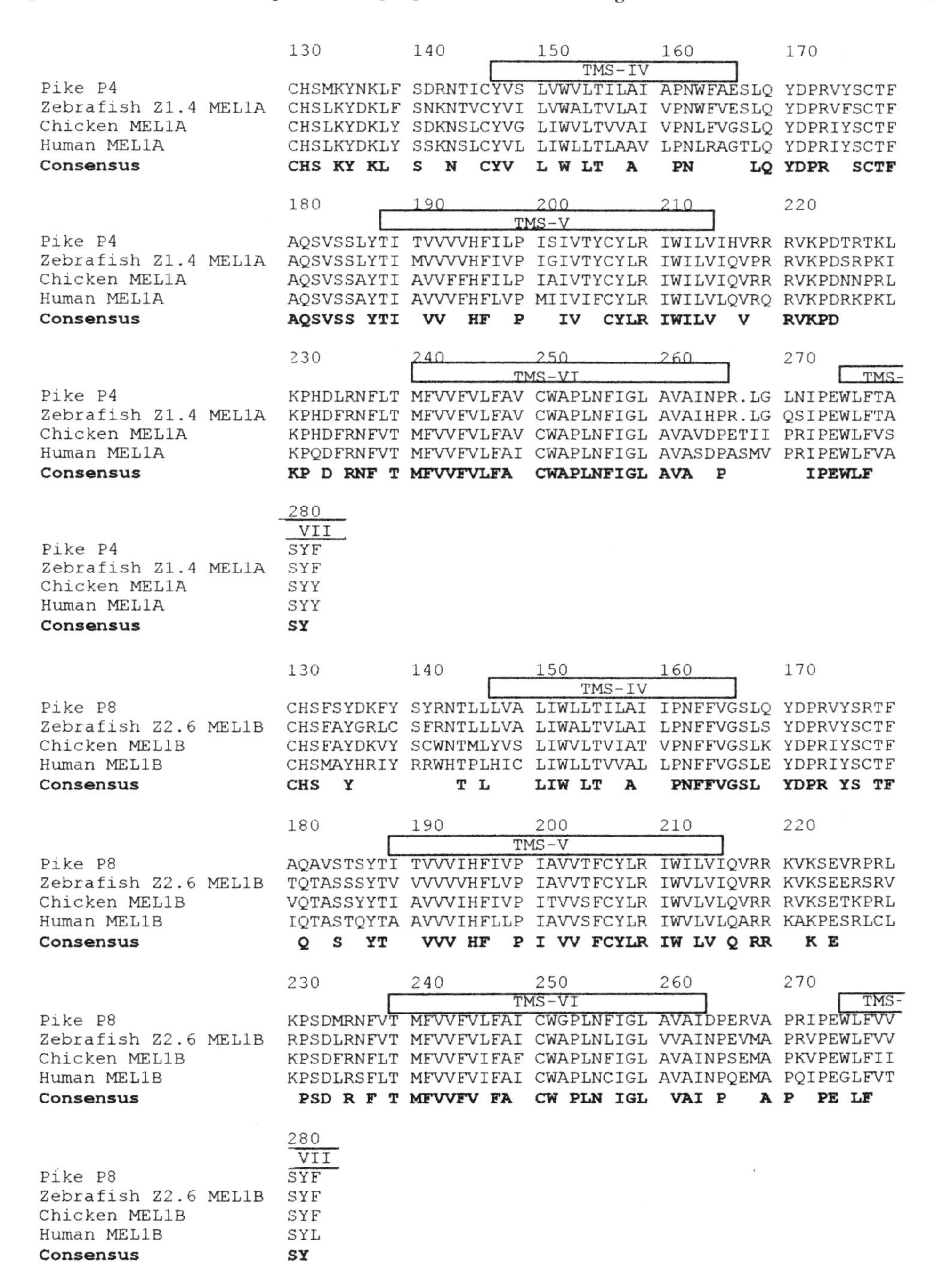

Figure 2. Deduced amino acid sequences of pike P4 and P8 fragments and their comparison with the Zebrafish, Chicken and Human melatonin receptors. Genbank accession numbers are U31823 (Zebrafish Z1.4), U31824 (Zebrafish Z2.6), U31820 (Chicken Mel_{1a}), U30609 (Chicken Mel_{1b}), U14108 (Human Mel_{1a}) and U25341 (Human Mel_{1b}). TMS: transmembrane domain.

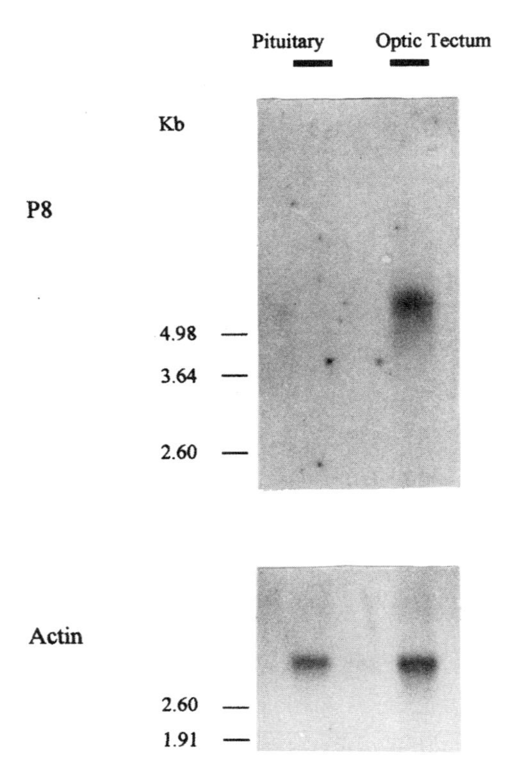

Figure 3. Expression of P8 putative (Mel_{1b}) melatonin receptor mRNA assessed by northern blot analysis. Each lane contained 2.5 μg of poly(A)$^+$RNA. The lower portion of the blot depicts the hybridization pattern obtained with an actin probe to verify equal loading of lanes. The blot was exposed to X-ray film for one week at –80°C.

hybridization signal was detected in other brain areas with either probe, suggesting that the levels of expression were low. However, RT-PCR analysis, using specific primers, indicated that all the brain areas expressed the Mel_{1a} and Mel_{1b} subtypes. Radioautographic studies also report a widespread distribution of the ^{125}I-melatonin binding sites in the brain of the goldfish (21), the trout (3), the salmon (8) and the pike (not shown). Our data agree with previous investigations which showed that ^{125}I-melatonin could bind to membrane preparations from pike brain (11). Binding was saturable reversible and sensitive to GTP, suggesting the presence of G-protein coupled melatonin receptors. In addition, displacement experiments suggested the presence of two components with affinities in the femtomolar and nanomolar range of concentrations respectively (11).

The interesting finding of this study was that the P8 probe and, to a lesser extent, the P4 probe allowed to evidence some expression in the pituitary. To determine whether this expression was correlated with the presence of melatonin binding sites, we investigated the binding of ^{125}I-melatonin to sections and membranes. Indeed, ^{125}I-melatonin bound in a saturable manner to pike pituitary membranes. The dissociation constant was within the range of that found with the brain membranes (11). In contrast, the number of binding sites was 10- to 15-fold lower in the pituitary than in the brain. This resulted in a low signal-to-noise ratio in the pituitary. Tissue sections from the pituitary exhibited a specific binding, but localized only in the antero-ventral part of the gland. Our results contrast with previous investigations indicating that no melatonin binding sites are present in the pituitary of nonmammalian vertebrates

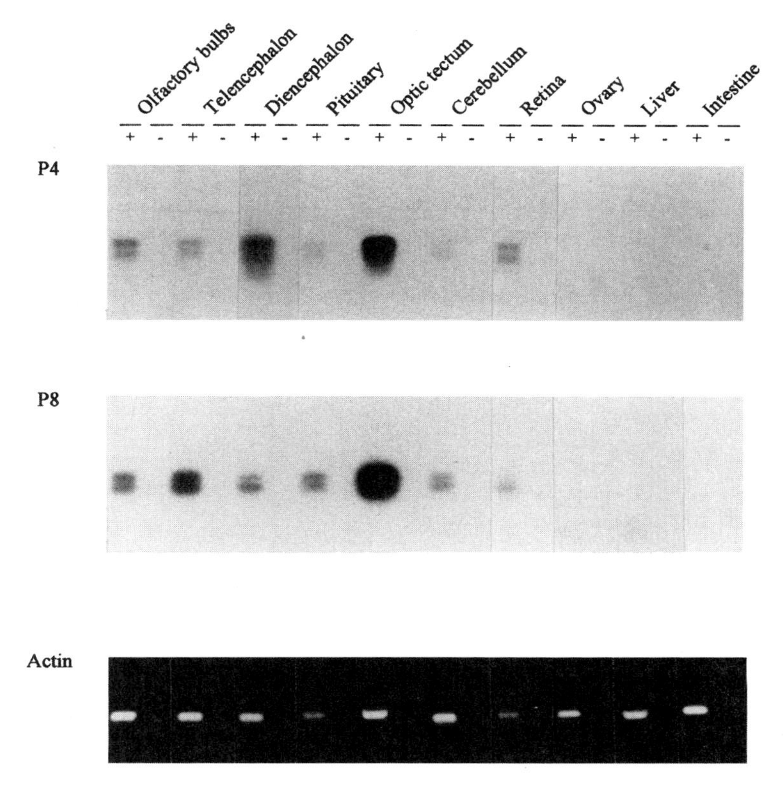

Figure 4. Tissue-specific distribution of P4 and P8 melatonin receptor mRNA assessed by RT-PCR. Sample (3 µg) of each tissue total RNA was subjected to reverse transcriptase using Dynabeads oligo$(dT)_{25}$ followed by PCR using two additional oligonucleotides (4S2-4A2 or 8S2-8A2) as described in Materials and Methods and in Fig. 1. One percent of the amplified cDNA products were Southern-blotted and probed with a ^{32}P-labeled specific melatonin receptor subtype fragment internal to the PCR fragment (+). For each tissue, a control group, in which reverse transcriptase was omitted, was processed in parallel (–). Actin served as a control to verify the amount of template for each sample.

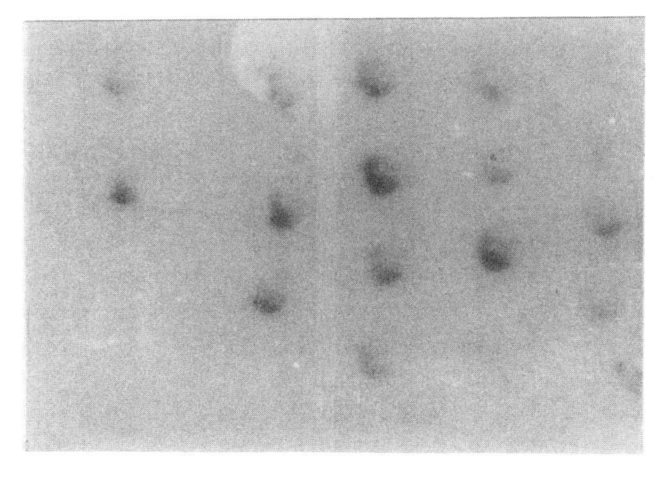

Figure 5. 2-[^{125}I]iodomelatonin binding sites in the pituitaries revealed by *in vitro* radioautography. Images were produced on Xray films by 20 µm coronal tissue sections through pituitaries incubated with 100 pM of 2-[^{125}I]iodomelatonin. Sections incubated in the presence of an excess (1 µM) of melatonin were devoid of any labeling (not shown).

including fish (refs in 2,11). The difficulty to evidence melatonin binding sites in the pituitary might result from a number of factors: species investigated, low levels of expression, age of the animals, possible nycthemeral and/or circannual rhythms of expression.

Altogether, our data speak in favor of the presence of melatonin receptors in the pituitary of the pike. Preliminary investigations indicate that melatonin modulates cyclic AMP levels in this organ, suggesting that these are functional receptors. A comparison of the radioautography sections with histological sections from freshly fixed pike pituitaries suggests that the binding is associated to an area containing gonadotrophs as well as prolactin and growth hormone cells.

In conclusion, the present paper is the first to report the presence of melatonin receptors in the pituitary of a non-mammalian vertebrate. This opens new lines of

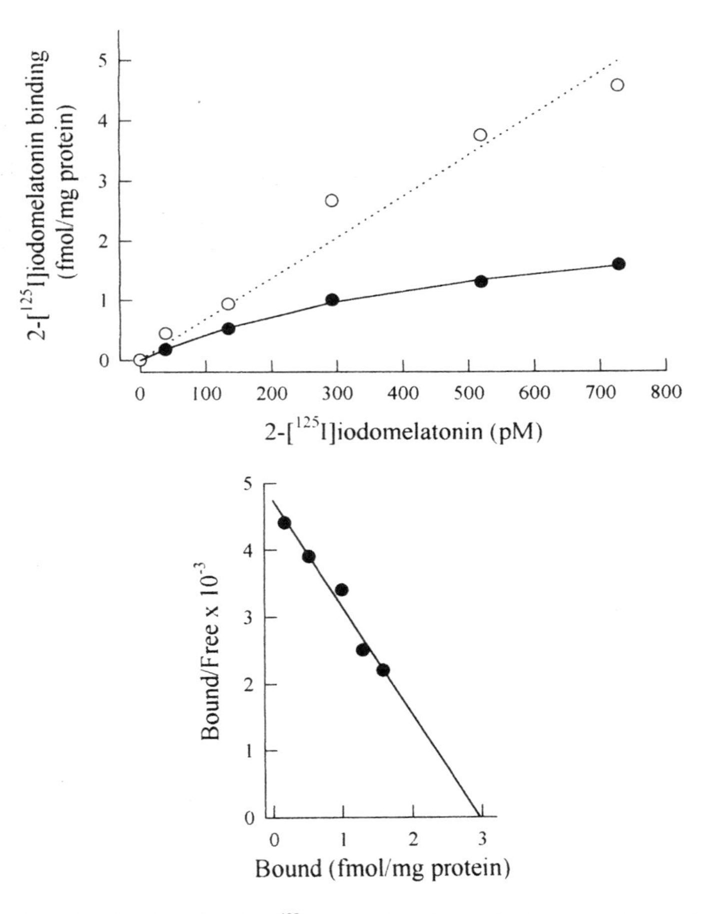

Figure 6. Equilibrium saturation binding of 2-[^{125}I]iodomelatonin to pike brain membranes. Membranes were prepared and binding performed as described in Materials and Methods, in the presence of increasing concentration of 2-[^{125}I]iodomelatonin. In the top graph, the specific binding (solid circles) is defined as total binding minus non-specific binding (open circles), determined in the presence of 50μM of melatonin. Scatchard plot of the experimental data with the best least-squares regression line is shown in the bottom graph. Each experiment used pooled homogenates from 60 pituitaries. Means ± SEM (n = 3). One representative experiment out of 2.

investigation related to the melatonin-mediated photoperiodic control of neuroendocrine functions in fish. Future investigations will aim to identify which cell type(s) and which hormonal output(s) is (are) modulated by melatonin in the pike pituitary. Interestingly enough, physiological concentrations of melatonin modulated 1) GtH II release from cultured pituitary fragments of the Atlantic croaker (15), and 2) gonadotropin release by neonatal rat gonadotrophs challenged with LH-RH (31). The fish pituitary offers interesting perspectives to study molecular and cellular events related to the transduction of the melatonin signal in ectotherms.

REFERENCES

1. Bradford, M.M. A rapid and sensitive method for the quantitation of microgram quantities of protein utilizing the principle of protein-dye binding. Anal. Biochem. 72:248–254, 1976.
2. Cassone, V.M., Brooks, D.S., and Kelm, T.A. Comparative distribution of 2[^{125}I]iodomelatonin binding in the brains of diurnal birds: outgroup analysis with turtles. Brain Behav. Evol. 45:241–256, 1995.
3. Davies, B., Hannah, L.T., Randall, C.F., Bromage, N., and Williams, L.M. Central melatonin binding sites in rainbow trout (*Onchorhynchus mykiss*). Gen. Comp. Endocrinol. 96:19–26, 1994.
4. Delfau, M.H., Kerckaert, J.P., d. Collyn, M., Fenaux, P., Lai, J.L., Jouet, J.P., and Grandchamp, B. Detection of minimal residual disease in chronic myeloid leukemia patients after bone marrow transplantation by polymerase chain reaction. Leukemia, 4:1–5, 1990.
5. Dubocovich, M.L. Melatonin receptors: are there multiple subtypes? Trends. Pharmacol. Sci. 16:50–56, 1995.
6. Dubocovich, M.L. and Takahashi, J.S. Use of 2-[^{125}I]iodomelatonin to characterize melatonin binding sites in chicken retina. Proc. Natl. Acad. Sci. USA 84:3916–3920, 1987.
7. Ebisawa, T., Karne, S., Lerner, M.R., and Reppert, S.M. Expression cloning of a high-affinity melatonin receptor from *Xenopus* dermal melanophores. Proc. Natl. Acad. Sci. USA 91:6133–6137, 1994.
8. Ekstrom, P. and Vanecek, J. Localization of 2-[^{125}I]iodomelatonin binding sites in the brain of the Atlantic salmon, *Salmo salar* L. Neuroendocrinology. 55:529–537, 1992.
9. Falcón, J. Cellular circadian clocks in the pineal. Prog. Neurobiol. 58:121–162, 1999.
10. Feinberg, A.P. and Vogelstein, B. A technique for radiolabeling DNA restriction endonuclease fragments to high specific activity. Anal. Biochem. 132:6–13, 1983.
11. Gaildrat, P., Ron, B., and Falcón, J. Daily and circadian variations in 2-[125I]-iodomelatonin binding sites in the pike brain (Esox lucius). J. Neuroendocrinol. 10:511–517, 1998.
12. Gauer, F., Masson-Pévet, M., and Pévet, P. Pinealectomy and constant illumination increase the density of melatonin binding sites in the pars tuberalis of rodents. Brain Res. 575:32–38, 1992.
13. Gauer, F., Masson-Pévet, M., Skene, D.J., Vivien-Roels, B., and Pévet, P. Daily rhythms of melatonin binding sites in the rat pars tuberalis and suprachiasmatic nuclei; evidence for a regulation of melatonin receptors by melatonin itself. Neuroendocrinology. 57:120–126, 1993.
14. Gwinner, E., Hau, M., and Heigl, S. Melatonin: generation and modulation of avian circadian rhythms. Brain Res. Bull. 44:439–444, 1997.
15. Khan, I.A. and Thomas, P. Melatonin influences gonadotropin II secretion in the Atlantic croaker (Micropogonias undulatus). Gen. Comp. Endocrinol. 104:231–242, 1996.
16. Limbird, L.E. *Cell surface receptors: a short course on theory and methods*. Martnus-Nijhoff, Boston, 1986.
17. Lincoln, G.A. and Clarke, I.J. Photoperiodically-induced cycles in the secretion of prolactin in hypothalamo-pituitary disconnected rams: evidence for translation of the melatonin signal in the pituitary gland. J. Neuroendocrinol. 6:251–260, 1994.
18. Liu, F., Yuan, H., Sugamori, K.S., Hamadanizadeh, A., Lee, F.J., Pang, S.F., Brown, G.M., Pristupa, Z.B., and Niznik, H.B. Molecular and functional characterization of a partial cDNA encoding a novel chicken brain melatonin receptor. FEBS Lett. 374:273–278, 1995.
19. Liu, C., Weaver, D.R., Jin, X., Shearman, L.P., Pieschl, R.L., Gribkoff, V.K., and Reppert, S.M. Molecular dissection of two distinct actions of melatonin on the suprachiasmatic circadian clock. Neuron, 19:91–102, 1997.
20. Malpaux, B., Daveau, A., Maurice-Mandon, F., Duarte, G., and Chemineau, P. Evidence that melatonin acts in the premammillary hypothalamic area to control reproduction in the ewe: presence of binding

sites and stimulation of luteinizing hormone secretion by in situ microimplant delivery. Endocrinology, 139:1508–1516, 1998.
21. Martinoli, M.G., Williams, L.M., Kah, O., Titchener, L.T., and Pelletier, G. Distribution of central melatonin binding sites in the goldfish (*Carassius auratus*). Mol. Cell. Neurosci. 2:78–85, 1991.
22. Mayer, I., Bornestaf, C., and Borg, B. Melatonin in non-mammalian vertebrates: Physiological role in reproduction? Comp. Biochem. Physiol. [A] 118:515–531, 1997.
23. Maywood, E.S., Bittman, E.L., and Hastings, M.H. Lesions of the melatonin- and androgen-responsive tissue of the dorsomedial nucleus of the hypothalamus block the gonadal response of male Syrian hamsters to programmed infusions of melatonin. Biol. Reprod. 54:470–477, 1996.
24. McArthur, A.J., Gillette, M.U., and Prosser, R.A. Melatonin directly resets the rat suprachiasmatic circadian clock in vitro. Brain Res. 565:158–161, 1991.
25. Morgan, P.J., Barrett, P., Howell, H.E., and Helliwell, R. Melatonin receptors: localization, molecular pharmacology and physiological significance. Neurochem. Int. 24:101–146, 1994.
26. Reiter, R.J. Melatonin: the chemical expression of darkness. Mol. Cell Endocrinol. 79:153–158, 1991.
27. Reppert, S.M., Weaver, D.R., and Ebisawa, T. Cloning and characterization of a mammalian melatonin receptor that mediates reproductive and circadian responses. Neuron, 13:1177–1185, 1994.
28. Reppert, S.M., Godson, C., Mahle, C.D., Weaver, D.R., Slaugenhaupt, S.A., and Gusella, J.F. Molecular characterization of a second melatonin receptor expressed in human retina and brain: the Mel1b melatonin receptor. Proc. Natl. Acad. Sci. USA 92:8734–8738, 1995.
29. Reppert, S.M., Weaver, D.R., Cassone, V.M., Godson, C., and Kolakowski, L.F., Jr. Melatonin receptors are for the birds: molecular analysis of two receptor subtypes differentially expressed in chick brain. Neuron, 15:1003–1015, 1995.
30. Rivkees, S.A., Cassone, V.M., Weaver, D.R., and Reppert, S.M. Melatonin receptors in chick brain: characterization and localization. Endocrinology, 125:363–368, 1989.
31. Vanecek, J. and Klein, D.C. Melatonin inhibition of GnRH-induced LH release from neonatal rat gonadotroph: involvement of Ca^{2+} not cAMP. Am. J. Physiol. 269:85–90, 1995.
32. Vanecek, J., Pavlik, A., and Illnerova, H. Hypothalamic melatonin receptor sites revealed by autoradiography. Brain Res. 435:359–362, 1987.
33. Wiechmann, A.F. and Wirsig-Wiechmann, C.R. Localization and quantification of high-affinity melatonin binding sites in *Rana pipiens* retina. J. Pineal Res. 10:174–179, 1991.
34. Zachmann, A., Ali, M.A., and Falcón, J. Melatonin and its effects in fishes: an overview. In: *Rhythms in fishes*. M.A. Ali, ed. Plenum press, New York, pp. 149–165, 1992.

9

MELATONIN RELEASE FROM THE PINEALS OF TWO SPARIDS

Sparus aurata and *Acanthopagrus bifasciatus*

Benny Ron[1] and Darren K. Okimoto[2]

[1] National Center for Mariculture
Israel Oceanographic & Limnological Research Ltd.
North Shore, P.O. Box 1212, Eilat 88112
Israel
[2] Psychology Department
University of Delaware
Newark, Delaware, 19716

INTRODUCTION

We present here data on the photic regulation of the melatonin-generating system in two representatives of the family Sparidae, the gilthead seabream, *Sparus aurata*, and the double-bar seabream, *Acanthopagrus bifasciatus*. The commercially-important *S. aurata* is a native of the Mediterranean Sea and an excellent model species since the fish is hardy, breeds readily in captivity, and can be easily maintained in high densities in the laboratory. *A. bifasciatus*, which is native to the Red Sea and the Western Pacific Ocean, has been recently domesticated, spawned, and grown in our laboratory at the National Center for Mariculture in Eilat, Israel, as a potential new species for the aquaculture industry. The objectives of this study were to: a) investigate the properties of the circadian oscillator *in vitro* in *S. aurata* and *A. bifasciatus*; b) examine the phylogenetic consistency in either the presence or absence of a pineal clock in the *S. aurata* and *A. bifasciatus*; and c) characterize the development of the melatonin-generating system during ontogeny in the *S. aurata*.

Melatonin after Four Decades, edited by James Olcese.
Kluwer Academic / Plenum Publishers, New York, 2000.

METHODS

Utilizing a static culture methodology, we investigated the photic entrainment of pineal organs that were obtained from post larval (40 days posthatching) and juvenile *S. aurata* and juvenile *A. bifasciatus* exposed to 12 hr light: 12 hr dark for 3–4 days (lights on 08:30 hr) and constant temperature (24°C). To test for the presence of a pineal clock in these selected representatives of the Sparidae family, pineal glands from juvenile *S. aurata* and juvenile *A. bifasciatus* were placed in perifusion culture, and exposed to an LD 12–12 cycle for 1 day (lights on 08:30 hr) followed by dark:dark (DD) for 3 days and constant temperature (24°C). Melatonin concentrations in culture medium were measured by radioimmunoassay.

RESULTS

Rhythmic melatonin release under 12 hr light: 12 hr dark was detected in both species and in *S. aurata* at the post larval stage of development, with higher melatonin release occurring during the dark period (Figures 1, 2 and 3). Interestingly, we observed a species difference in melatonin release during the first LD cycle, suggesting the existence of interspecies variations in the melatonin-generating system in the Sparidae family (Figure 1 vs. Figure 2). An endogenous rhythm reflecting a free-running oscillator was seen under DD (Figures 4 and 5) supported the existence of a pineal clock in both *S. aurata* and *A. bifasciatus*. Periodogram analysis (Fourier transform, S-Plus) of the melatonin rhythm under DD has revealed that their pineal clock is circadian in nature (Tau of about 24 hr). More precisely, *S. aurata* pineal 1, 2 and 3, Tau values were 23.5, 24.6 and 23.1 correspondingly, while computation of Tau for *A. bifasciatus* pineal 1, 2 and 3, produced 25.0, 24.9 and 24.9 values correspondingly.

CONCLUSIONS

These data suggest that the pineals of *S. aurata* and *A. bifasciatus* are photosensitive (e.g., contain photoreceptors) and clocked and that photic entrainment is present

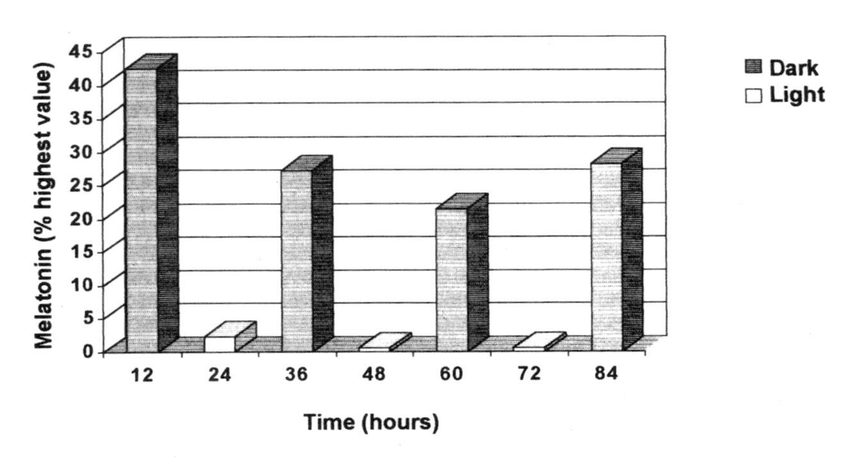

Figure 1. *Sparus aurata* (n = 6).

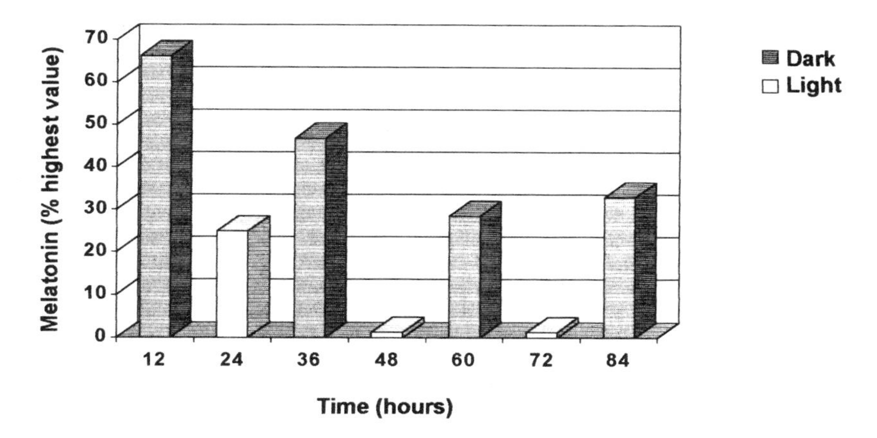

Figure 2. *Acanthopagrus bifasciatus* (n = 6).

at the post-larval stage of development in *S. aurata*. Specifically, the appearance of rhythmic melatonin release in response to an LD cycle indicates the development of photosensitivity, whereas rhythmicity under DD suggests the presence of a functional pineal clock. The presence of rhythmic melatonin release from pineals of different Sparidae species under LD and DD supports the notion that a pineal clock occurs at the family level of taxonomic organization. The significant difference of melatonin concentrations between the L and D periods, during the first LD cycle, suggests the existence of inter-species variations in the melatonin-generating system within the Sparidae family. The melatonin-generating system of the gilthead seabream, *Sparus aurata*, is fully functional at the post-larva stage. Further characterization of the melatonin-generating systems of these two sparids is in progress.

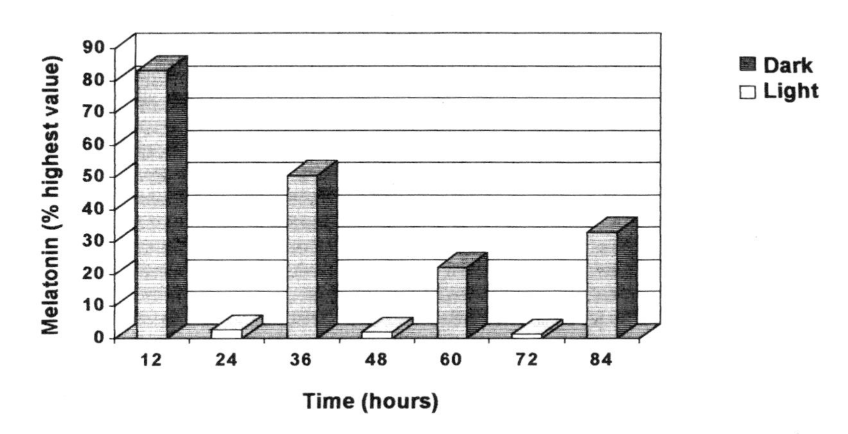

Figure 3. *Sparus aurata* larvae (n = 6).

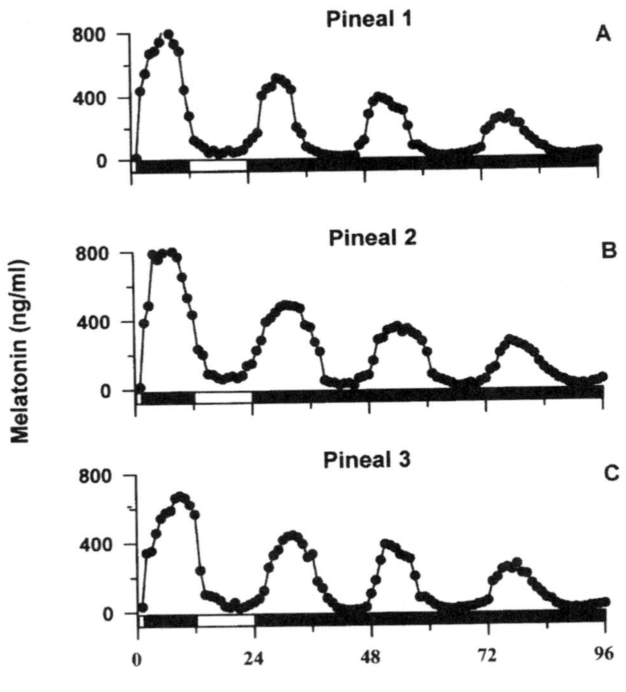

Figure 4. *Sparus aurata.*

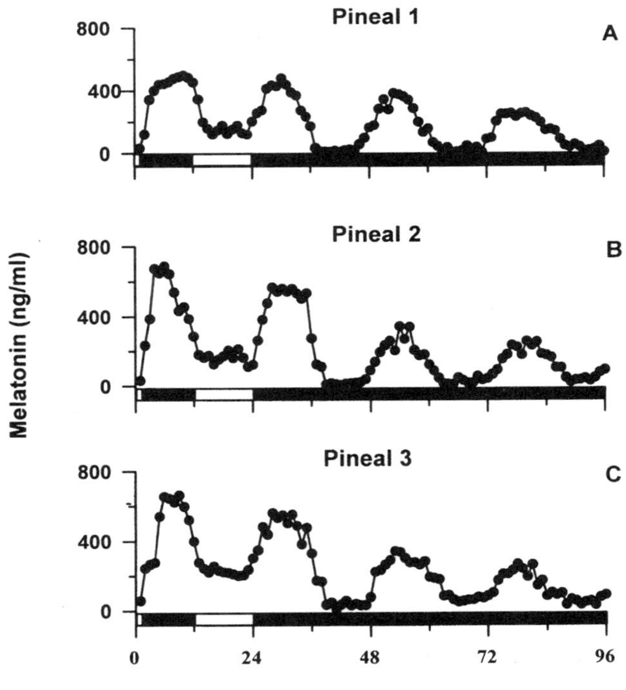

Figure 5. *Acanthopagrus bifasciatus.*

REFERENCES

Falcon, J., Thibault, C., Begay, V., Zachmann, A., and Collin, J-P. Regulation of rhythmic melatonin secretion by fish pineal photoreceptor cells. In "Rhythm in Fishes" (M.A. Ali, ed.), NATO ASI series. Series A, Life Sciences, Vol. 236, pp. 167–198. Plenum Press, New York, 1992.

Iigo, M., Kezuka, H., Aida, K., and Hanyu, I. Circadian rhythms of melatonin secretion from superfused goldfish (*Carassius auratus*) pineal glands *in vitro*. Gen. Comp. Endocrinol. 83, 152–158, 1991.

Zachmann, A., Ali, M.A., and Falcon, J. Melatonin and its effect in fishes: an overview. In "Rhythms in Fishes" (M.A. Ali, ed.), NATO ASI series. Series A, Life Sciences, Vol. 236, pp. 149–165. Plenum Press, New York, 1992.

10

PHOTOENDOCRINE SIGNAL TRANSDUCTION IN PINEAL PHOTORECEPTORS OF THE TROUT

Role of cGMP and Nitric Oxide

B. Zipfel, H. A. Schmid, and H. Meissl

Max-Planck-Institute for Physiological and Clinical Research
W.G. Kerckhoff-Institute, Parkstr. 1
D-61231 Bad Nauheim, Germany

1. SUMMARY

This study describes the presence and distribution of cGMP-immunoreactivity and of the nitric oxide (NO) synthesizing enzyme, NO synthase (NOS), as demonstrated by use of the NADPH-diaphorase technique in directly light sensitive pineal organ of the trout. Cyclic GMP immunohistochemistry revealed immunoreactivity in pineal photoreceptor cells that were identified by double-labeling with S-antigen, whereas NADPH-positive structures were located adjacent to these photoreceptor cells. Since NO is known to stimulate synthesis of cGMP, these results indicate a role for NO in pineal function, eg. in cGMP related events in the phototransduction process as well as in the light-dark control of melatonin synthesis.

2. INTRODUCTION

Photoreceptor cells of the pineal organ of fish and other lower vertebrates transduce photic information about the 24-h light-dark cycle into neuronal and endocrine messages. The pineal hormone melatonin is synthesized and released rhythmically with elevated levels at night and low levels at daytime and its synthesis may be regulated either by the ambient photoperiod or by an endogenous circadian pacemaker that is entrained by the photoperiod (2,3).

Melatonin after Four Decades, edited by James Olcese.
Kluwer Academic / Plenum Publishers, New York, 2000.

In contrast to the intensively investigated retinal phototransduction cascade, the current knowledge about this process in pineal organs is fragmentary and the coupling between the phototransduction cascade and melatonin synthesis is unresolved (7). Whereas cyclic guanosine monophosphate (cGMP) is suggested to be a second messenger in retinal and pineal phototransduction, the knowledge about its role in melatonin synthesis is unknown. It was shown that light induces a decrease in cGMP production in the fish pineal organ (3), but cGMP analogs did not influence serotonin-N-acetyltransferase (NAT) activity in the trout pineal (9). Pineal organs of the pike kept in constant darkness exhibit a circadian pattern of the cGMP content with a maximum in the beginning of the subjective night (4).

In this study we investigated the presence and regulation of cGMP and NO synthase in the photosensitive pineal organ of a fish, the rainbow trout *Oncorhynchus mykiss*. Special emphasis was laid on the possible role of nitric oxide, which is believed to be an intercellular/intracellular messenger in several areas of the central and peripheral nervous system and which plays a major role in the stimulation of cGMP synthesis (6).

3. MATERIAL AND METHODS

3.1. Immunocytochemistry and NADPH-Diaphorase Staining

For the immunohistological studies, pineal organs were incubated for 2 hours either in darkness or light in 0.5 ml of RPMI 1640 containing 1 mM 3-isobutyl-1-methylxantine (IBMX) and the nitric oxide (NO) donor S-nitroso-N-penicillamine (SNAP). After the incubation period the organs were immediately fixed for two hours by adding 1 ml of 4% paraformaldehyde and then processed for immunohistochemistry or NADPH-diaphorase staining. A semiquantitative evaluation of immunostained slices was conveyed from photographs using an image analysis system RAG 200 (Biocom, France).

NADPH-diaphorase staining was perfomed using a modified protocol of Östholm et al. (8). For a direct comparison with S-antigen and cGMP-immunolabeling, the same sections were also processed for NADPH histochemistry.

3.2. Perifusion Culture

Pineal organs were rapidly removed from decapitated animals and transferred to flow-through organ culture chambers. Each chamber was continuously perifused with modified RPMI 1640 medium at a flow rate of 0.5 ml/hr and a temperature of 17°C. The perifusion medium was collected at the outlet of the experimental chambers in time intervals of 1 h. This medium was used for the determination of hormone content. Melatonin was separated by high performance liquid chromatography as previously described (10). Drugs were added directly to the perifusion medium.

4. RESULTS

4.1. Immunocytochemistry and NADPH-Diaphorase

Immunolabeling of the pineal organ with a highly specific antibody against cGMP revealed predominantely stained photoreceptor-like cells. Double-labeling of the same sections with an antiserum against S-antigen, a specific marker for retinal and pineal

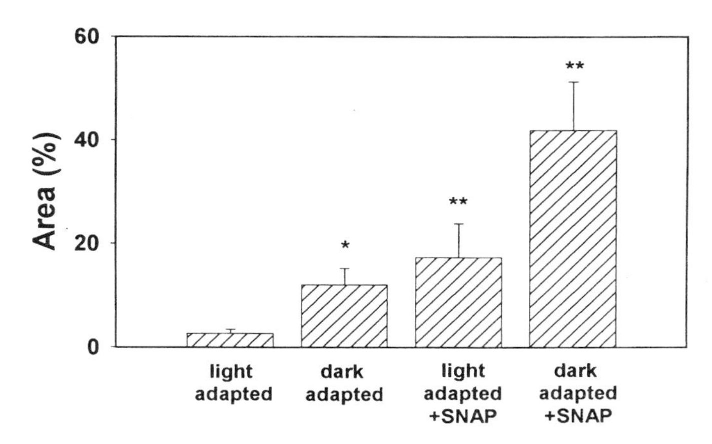

Figure 1. Semiquantitative measurement of the intensity of cGMP-immunofluorescence after treatment of the light- and dark-adapted pineal organs with 100μM S-nitroso-N-penicillamine (SNAP) for 5 minutes. The intensity of FITC-immunostained slices was determined from photographs using an image analysis system RAG 200 (Biocom/France) and expressed in % area of the entire pineal area with an optical density >100 (arbitrary units). **$P < 0.001$ SNAP treated pineal organs vs coresponding groups (light-or dark-adapted); *$P < 0.01$ dark-adapted vs. light-adapted group.

photoreceptors, showed that most cells that are immunopositive to cGMP represent true photoreceptor cells.

Dark-adapted pineal organs showed a stronger cGMP-immunoreaction compared to light-adapted organs. Preincubation of pineal organs with the nitric oxide (NO) donor SNAP resulted in a strong increase in cGMP-immunoreactivity. These results were confirmed by a semiquantitative evaluation of FITC-immunostained slices using an image analysis system RAG 200 (Biocom, France) (Figure 1).

Staining of the same section of the pineal with the NADPH-diaphorase technique demonstrated that NADPH-diaphorase-positive cells were located in close proximity to cGMP- and S-antigen-positive cells.

4.2. Melatonin Synthesis

Melatonin production of the explanted trout pineal organ maintained in perifusion culture is directly regulated by the ambient illumination with low levels in the photopic range and high levels in the scotopic range and in darkness. The effect of light is clearly reduced in the presence of IBMX, a non-specific inhibitor of phosophodiesterases (PDE), showing maximal melatonin production in dim light. The nitric oxide donor SNAP increases melatonin production in IBMX-treated pineal organs.

5. DISCUSSION

The present study, using a highly selective antibody against cGMP, demonstrates that pinealocytes contain cGMP-immunoreactivity with an intensity that was dependent on the natural stimulus, light or darkness, and on the presence of specific NO donors like S-nitroso-N-penicillamine, SNAP. Using a double-immunofluorescence method for the demonstration of cGMP and S-antigen in combination with NADPH diaphorase histochemistry, which is a reliable method for the detection of NO synthase activity (1), we could demonstrate that most pinealocytes which were immunolabeled for cGMP represent S-antigen-positive photoreceptors and that NADPH diaphorase positive cells or fibers are located adjacent to the immunolabeled cells. These data raise new questions about a possible influence of a NO synthase/NO system in the regulation of cGMP in pineal photoreceptors as it was recently suggested for retinal photoreceptors (5). Since the phosphodiesterase inhibitors IBMX and zaprinast affect melatonin synthesis, cGMP possesses at least an indirect influence on the melatonin generating system. However, it remains to be elucidated whether nitric oxide stimulated cGMP-synthesis is responsible for the control of the endocrine, neuronal or both pineal messages.

REFERENCES

1. Dawson, T.M., Bredt, D.S., Fotuhi, M., Hwang, P.M., and Snyder, S.H. Nitric oxide synthase and neuronal NADPH diaphorase are identical in brain and peripheral tissues. Proc. Natl. Acad. Sci. USA 88:7797–7801, 1991.
2. Ekström, P. and Meissl, H. The pineal organ of teleost fishes. Rev. Fish Biol. Fisheries 7:199–284, 1997.
3. Falcón, J., Thibault, C., Begay, V., Zachmann, A., and Collin, J.-P. Regulation of the rhythmic melatonin secretion by fish pineal photoreceptor cells. In: Ali, M.A. (ed.) Rhythms in Fishes, Plenum Press, New York, London, 1992.
4. Falcón, J. and Gaildrat, P. Variations in cyclic adenosine 3′,5′-monophosphate and cyclic guanosine 3′,5′-monophosphate content and efflux from the photosensitive pineal organ of the pike in culture. Pflügers Arch. Eur. J. Physiol. 433:336–342, 1997.
5. Goldstein, I.M., Ostwald, P., and Roth, S. Nitric oxide: a review of its role in retinal function and desease. Vision Res. 36:2979–2994, 1996.
6. Knowles, R.G., Palacios, M., Palmer, R.M.J., and Moncada, S. Formation of nitric oxide from L-arginine in the central nervous system: a transduction mechanism for the stimulation of soluble guanylate cyclase. Proc. Natl. Acad. Sci. USA 86:5159–5162, 1989.
7. Meissl, H. Photic regulation of pineal function—Analogies between retinal and pineal photoreception. Biol. Cell 89:549–554, 1997.
8. Östholm, T., Holmqvist, B.I., Alm, P., and Ekström, P. Nitric oxide synthase in the CNS of the atlantic salmon. Neurosci. Lett. 168:233–237, 1994.
9. Thibault, C., Falcón, J., Greenhouse, S.S., Lowery, C.A., Gern, W.A., and Collin, J.-P. Regulation of melatonin production by pineal photoreceptor cells: role for cyclic nucleotides in the trout (*Oncorhynchus mykiss*) J. Neurochem. 61:332–339, 1993.
10. Yáñez, J. and Meissl, H. Secretion of the methoxyindoles melatonin, 5-methoxytryptophol, 5-methoxyindoleacetic acid, and 5-methoxytryptamine from trout pineal organs in superfusion culture: effects of light intensity. Gen. Comp. Endocrinol. 101:165–172, 1996.

11

INTRINSIC GLUTAMINERGIC SYSTEM NEGATIVELY REGULATES MELATONIN SYNTHESIS IN MAMMALIAN PINEAL GLAND

Yoshinori Moriyama, Hiroshi Yamada, Mitsuko Hayashi, and Shouki Yatsushiro

Department of Cell Membrane Biology
Institute of Scientific and Industrial Research (ISIR)
Osaka University
Ibaraki Osaka 567-0047, Japan

The mammalian pineal gland is a photoneuroendocrine transducer that rhythmically synthesizes and secretes melatonin at night in response to photoperioic stimuli and signals from endogenous circadian oscillators (1–3). In the rat, this process is controlled by the suprachiasmatic nuclei (SCN) through sympathetic neurons projecting into the glands. These neurons secrete norepinephrine (NE) which binds to adrenergic receptors and activates adenylate cyclase through a heterotrimeric guanine nucleotide-binding protein (Gs). The resultant increase of cAMP causes transcriptional activation of the serotonin N-acetyltransferase (NAT), resulting in increased melatonin output (4). Thus, the role of sympathetic innervation as a positive regulatory mechanism for melatonin synthesis has been firmly established. However, the daily change of NAT mRNA and its activity are not completely correlated with the rate of decrease of NAT activity, suggesting the presence of regulatory mechanism(s) other than adrenergic one in pineal gland.

From the studies on the structure and function of synaptic-like microvesicles (or microvesicle; MVs) in bovine pineal gland (5–7), we have learned that pinealocytes are glutaminergic neuroendocrine cells, and use L-glutamate as a negative regulator for melatonin synthesis (Figure 1). In this article, we describe briefly the mechanism by which the glutaminergic system inhibits melatonin synthesis in pineal gland.

Melatonin after Four Decades, edited by James Olcese.
Kluwer Academic / Plenum Publishers, New York, 2000.

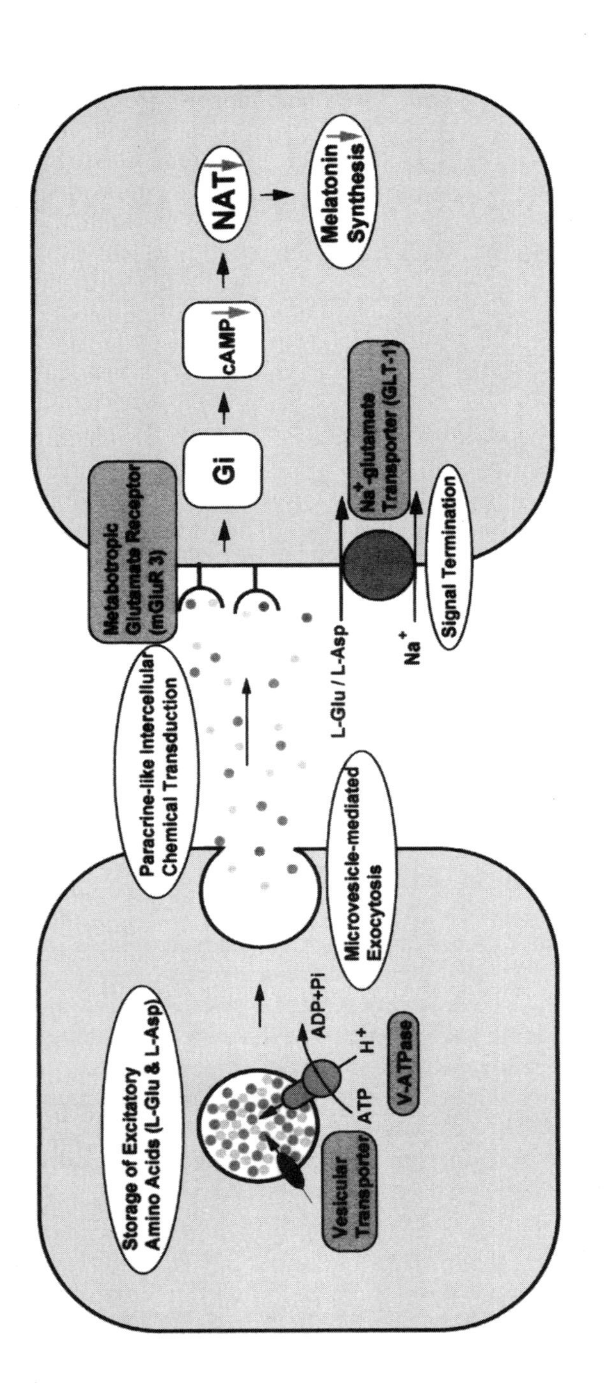

Figure 1. Glutaminergic systems in rat pineal gland.

1. MICROVESICLES-MEDIATED GLUTAMATE EXOCYTOSIS (GLUTAMATE SIGNAL OUTPUT)

Mammalian pinealocytes possess a large number of MVs, which are morphologically similar to synaptic vesicles (7,8). Some role in exocytic process has been postulated, since the MVs are especially rich in process terminal region, the site for speculative secretory events in pinealocytes (9,10). We have established a procedure for isolation of MVs from bovine pineal glands with almost 20-fold enrichment over homogenate as determined by immunoreactivity against anti-synaptophysin antibodies (5). Isolated MVs are small organelles (average diameter being 50–70 nm) with relatively electron transluent contents and contain several membrane proteins for exocytosis as well as accumulation of neurotransmitters. These include synaptotagmin (5), SV2B (11), *N*-ethylmaleimide-sensitive fusion protein (NSF) (12), and vacuolar H^+-ATPase (5) (also see in Hayashi et al., this series). Upon the addition of ATP, the isolated MVs form an electrochemical gradient of protons and take up L-glutamate against the concentration gradient, indicating the presence of active transport system of L-glutamate in MVs. The properties of the glutamate uptake are essentially the same as those of neuronal synaptic vesicles (5,6), suggesting that vesicular glutamate transporter functions in MVs (Figure 1). This is the first example for the presence of a vesicular glutamate transporter outside of neurons. In cultured pinealocytes, glutamate immunoreactivities are concentrated within MVs, indicating that the accumulation of glutamate in MVs occurs *in vivo*.

The glutamate in MVs can be secreted to the extracellular space through exocytosis from cultured pinealocytes (0.18 nmoles L-glutamate released per min $\times 10^6$ cells). The glutamate secretion is triggered upon treatment of KCl or A23187 in the presence of Ca^{2+}. The release of glutamate is dependent on extracellular Ca^{2+}, inhibited by either L-type Ca^{2+} channel blockers such as nifedipine and nitrendipine (13) or treatment of Bothlinum neurotoxin type E accompanied with specific cleavage of synaptic vesicle associated protein 25 (SNAP25) (14). These results indicated that entry of Ca^{2+} through L-type Ca^{2+} channel directly triggered the glutamate secretion. Regulated exocytosis, similar to various secretory granule or synaptic vesicle, may occur in pinealocytes. It is noteworthy that L-aspartate but not D-form, another excitatory amino acid, is accumulated in MVs and also secreted through the same mechanism (15). The mechanism for accumulation of L-aspartate in MVs is unknown at present. Participation of the vesicular glutamate transporter is not likely, since L-aspartate is not the substrate for the pineal vesicular glutamate transporter (5,6).

Taken together, we concluded that pinealocytes can fire glutamate signals with a similar mechanism as in synapses. The internal concentration of glutamate in MVs reaches around 50 mM, which might be high enough to inhibit melatonin synthesis (see below) even after suitable dilution through exocytosis.

2. GLUTAMATE SIGNAL INPUT

Exogenous L-glutamate reversibly inhibited both NE-stimulated melatonin synthesis and NAT activity with the concentration required for 50% inhibition being 85 μM (16). The inhibition was not observed with D-glutamate or its metabolites such as γ-aminobutyrate (GABA) (14), but observed with D, L-aspartate (17,18), supporting that receptor-mediated glutamate signaling is involved in the inhibition. To define the role of glutamate in pinealocytes, we investigated a signal transduction pathway by which glutamate inhibits melatonin synthesis in rat pinealocytes.

Pharmacological analyses with agonists for various glutamate receptors (GluRs) were performed to investigate participation of GluRs in this putative glutamate signaling in pinealocytes. It was found that that 1 mM 1(1S,3R)-aminocyclopentane-1,3-dicarboxylic acid (tACPD), (2S,1′S,2′S)-2-(carboxycyclopropyl)-glycine (L-CCG-I) and (2S,1′S,2′R,3′R)-2-(2′,3′)dicarboxycyclopropyl-glycine (DCGIV), agonists for class II metabotropic GluRs (mGluRs) (19–22), inhibited melatonin synthesis about 50% as was the case of L-glutamate. No other GluR agonists including N-methyl-D-aspartate (NMDA) and quisqualate were effective (16). These results strongly suggest that class II mGluRs are involved in glutamate signaling. Because this class of mGluRs is known to be negatively coupled to adenylate cyclase (19,23–25), we hypothesized that upon binding of glutamate, the receptor triggers an G_i cascade resulting in decreased cAMP concentration and subsequently, decreased NAT activity and melatonin synthesis.

To test this hypothesis, we investigated whether class II mGluRs are expressed in pineal glands. There are two known isoforms of class II mGluRs, mGluR2 and mGluR3 (19,23–25). RT-PCR analysis and Northern blotting of pineal RNA demonstrated the expression of mRNA for mGluR3. No expression of the mRNAs for mGluR3 was observed in other tissues tested. There was no detectable hybridization of the mGluR2 probe to pineal gland RNA, indicating that the level of mGluR2 transcript was below detection limit of our Northern analysis. Furthermore, immunohistochemistry with antibodies specific for class II mGluRs demonstrated that the antigen was expressed in synaptophysin-positive cells, an indicator of pinealocytes (10,5,14). No antigen was present in non-pinealocyte cells. $G_{i1\alpha}$, the major subunit of the G_i protein that is linked to mGluR3, was also detected immunologically in pinealocytes but not in other cells types (16). These results confirmed that both mGluR3 and G_i proteins are present in the same pinealocytes.

Both mGluR3 and adenylate cyclase seem to be functionally coupled in pinealocytes as following reasons: the effect of dibutylyl cAMP, a non-hydrolyzable cAMP analogue, and pertussis toxin (PTX), a specific uncoupler of adenylate cyclase and G_i, which were expected to eliminate agonist-evoked decrease of cAMP concentration and NAT activity, restored NAT activity and melatonin synthesis, both of which were inhibited by L-glutamate and mGluR3 agonists. PTX also blocked the glutamate-dependent decrease of cAMP. Furthermore, under these conditions, PTX was found to block ADP-ribosylation of $G_{i\alpha}$ (16). Thus, mGluR3 and adenylate cyclase are coupled through G_i in pinealocytes. These results indicate that the inhibitory cAMP cascade is involved in the glutamate-evoked inhibition of melatonin synthesis.

3. GLUTAMATE SIGNAL TERMINATION

It is important point out that presence of signal termination system(s) could be necessary for the functional operation of the glutamate-evoked signaling cascade shown above (Figure 1). In the synaptic cleft, released L-glutamate is rapidly sequestered by Na^+-dependent transporters (26–28). The reuptake system is responsible for termination of the glutamate signal in chemical transduction and a reduction of glutamate cytotoxicity to neuronal tissues (27,28). Therefore, we wondered whether a similar mechanism to that in nervous tissues is present in pineal glands.

We identified and characterized the plasma membrane-type glutamate transporter in rat pinealocytes. The radiolabeled glutamate uptake by the cultured pinealocytes was driven by extracellular Na^+, saturated with glutamate concetration, and significantly inhibited by L- or D-aspartate at 100 μM, β-threo-hydroxyaspartate at 50 μM, pyrrolidine dicarboxylate at 50 μM and L-cysteine sulfinate at 10 μM, inhibitors for plasma membrane glutamate transporter to 95, 77, 79, and 47%, at maximum, respectively. Consistently, clearance of extracellular glutamate measured by HPLC was dependent on Na^+, and inhibited by β-threo-hydroxyaspartate and L-cysteine sulfinate. Immunological studies with site-specific antibodies against the three isoforms of the Na^+-dependent glutamate transporter (GLT-1, GLAST and EAAC1) indicated the expression of only the GLT-1 type transporter in the pineal glands. Expression of the GLT-1 type transporter in pineal glands was further demonstrated by RT-PCR combined Northern blotting with specific DNA probes (29). Immunohistochemical analysis further indicated that the immunological counterpart(s) of the GLT-1 is localizaed in pinealocytes (29). These results demonstrated that the GLT-1 type Na^+-dependent transporter is expressed and functions as a reuptake system for glutamate in rat pinealocytes. Since GLT-1 is known to be localized to astrocytes (30–32), this is the first observation of the presence of GLT-1-like transporter in endocrine cells.

4. PROPOSED AUTONOMIC REGULATION OF MELATONIN SYNTHEIS BY GLUTAMINERGIC SYSTEMS

Based on these observations, we concluded that rat pinealocytes are equipped with machineries for input, output and termination of the glutamate signals (Fig. 1). Since the same pinealocytes express all these glutaminergic elements, the glutamate may be used as autocrine- or paracrine-like signals in the endocrine organ: upon stimulation, pinealocytes secrete L-glutamate so as to inhibit NE-stimulated melatonin synthesis. Excess amount of glutamate is rapidly taken up by the reuptake system(s). Thus, melatonin synthesis in pinealocytes is negatively regulated by the glutaminergic systems. The glutamate-evoked regulation of melatonin synthesis seems to be a novel type of hormonal regulation, because a classical neurotransmitter is involved in the hormonal synthesis. Now, we are trying to reveal the elements of the pineal glutaminergic systems at the molecular level. The glutamate-evoked inhibition of melatonin synthesis is observed even in the presence of NE, indicating that the glutaminergic signal dominates adrenergic control. It is likely that the glutaminergic system functions as an autonomic regulatory mechanism against neuronal control in the pineal gland (Figure 2).

Characterization of the glutaminergic systems shown here causes other important questions to be solved. One is the role of nicotinic acetylcholine receptor (nAchR) in pinealocytes. Recently, we found that acetylcholine triggers glutamate exocytosis via nAchR and inhibits NE-dependent melatonin synthesis through an inhibitory cAMP cascade (33) (Figure 1). Thus, *in vivo* a stimulant for glutamate exocytosis may be acetylcholine. Our results suggest the very fascinating hypothesis that parasympathetic neurons negatively control melatonin synthesis by way of endogenous glutaminergic systems in pineal glands (Figure 2). We are now performing a series of experiments to test the hypothesis.

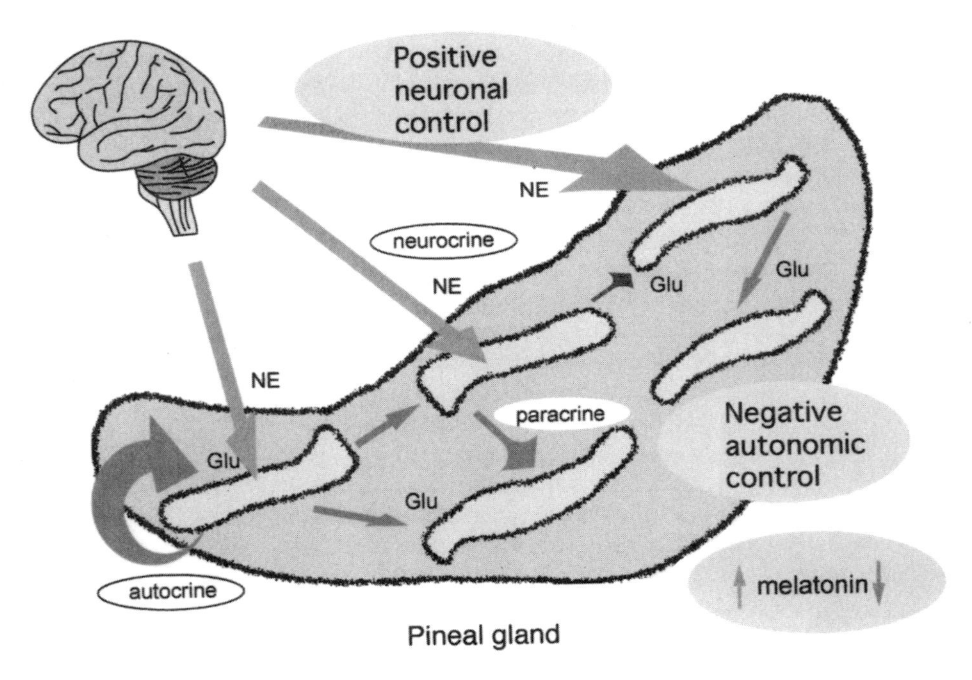

Figure 2. Autonomic regulation of endocrine function in pineal gland.

Another important question is the role of D-aspartate in pinealocytes. This amino acid was recently found to be present in pinealocytes at extraordinarily high concentration. We found that the D-aspartate is released from pinealocytes by non-vesicular mechanisms other than exocytosis, and it inhibits strongly the NE-stimulated melatonin synthesis (15,17,18). This is the first example for putative physiological functions of the amino acid that its mode of action has been revealed (17,18). The mechanism by which pinealocytes use D-aspartate as a signal transducing molecule may open a new field in pineal physiology.

ACKNOWLEDGMENT

This work was supported in part by grants from the Japanese Ministry of Education, Science and Culture, CREST, Japan Science and Technology Corpolation, the Terumo Science Foundation, the Novartis Pharma Foundation (Japan) for the Promotion of Science, the Takeda Science Foundation, and Salt Science Foundation. We also thank Prof. M. Futai, and Prof. A. Yamaguchi for their encouragement and supply during the works.

REFERENCES

1. Axelrod J. The pineal gland: a neurochemical transducer. Science 184:1341–1348 (1974).
2. Klein D.C. Photoneural regulation of the mammalian pineal gland, in Ciba Foundation Symposium

117: Photoperiodism, melatonin, and the pineal gland (Evered D. and Clark S., eds.): pp. 38–56 Pitman London (1985).

3. Reiter R.J. Pineal melatonin: cell biology of its synthesis and of its physiological interactions. Endocrine Rev. 12:151–180 (1991).
4. Foulkes N.S., Borjigin J., Snyder S.H., and Sassone-Corsi P. Rhythmic transcription: the molecular basis of circadian melatonin synthesis. Trends. Neurosci. 20:487–492 (1997).
5. Moriyama Y. and Yamamoto A. Microvesicles isolated from bovine pineal gland specifically accumulate L-glutamate. FEBS Lett. 367:233–236 (1995).
6. Moriyama Y. and Yamamoto A. Vesicular L-glutamate transporter in microvesicles from bovine pineal glands: driving force, mechanism of chloride anion-activation, and substrate specificity. J. Biol. Chem. 270:22314–22320 (1995).
7. Moriyama Y., Yamamoto A., Yamada H., Tashiro Y., and Futai M. Role of endocrine cell microvesicles in intercellular chemical transduction. Biol. Chem. 377:155–165 (1996).
8. Thomas-Reetz A. and De Camilli P. A role for synaptic vesicles in non-neuronal cells: clues from pancreatic β cells and from chromaffin cells. FASEB J. 8:209–216 (1994).
9. Reiter R.J. The mammalian pineal gland: structure and function. Am. J. Anat. 162:287–313 (1981).
10. Redecker P. and Bargsten G. Synaptophysin—a common constituent of presumptive secretory microvesicles in the mammalian pinealocyte: a study of rat and gerbil pineal glands. J. Neurosci. Res. 34:79–96 (1993).
11. Hayashi M., Yamamoto A., Yatsushiro S., Yamada H., Futai M., Yamaguchi A., and Moriyama Y. Synaptic protein SV2B, but not SV2A, is predominantly expressed and associated with microvesicles in rat pinealocytes. J. Neurochem. 71:356–365 (1998).
12. Moriyama Y., Yamamoto A., Tagaya M., Tashiro Y., and Michibata H. Localization of *N*-ethylmaleimide-sensitive fusion protein in pinealocytes. NeuroReport 6:1757–1760 (1995).
13. Yamada H., Yamamoto A., Takahashi M., Michibata H., Kumon H., and Moriyama Y. The L-type Ca^{2+} channel is involved in microvesicle-mediated glutamate exocytosis from rat pinealocytes. J. Pineal Res. 21:165–174 (1996).
14. Yamada H., Yamamoto A., Yodozawa S., Kozaki S., Takahashi M., Michibata H., Morita M., Furuichi T., Mikoshiba K., and Moriyama Y. Microvesicle-mediated exocytosis of glutamate is a novel paracrine-like signal transduction mechanism and inhibits melatonin secretion in rat pinealocytes. J. Pineal Res. 21:175–191 (1996).
15. Yatsushiro S., Yamada H., Kozaki S., Kumon H., Michibata H., Yamamoto A., and Moriyama Y. L-Aspartate but not the D form is secreted through microvesicle-mediated exocytosis and is sequestered through Na^+-dependent transporter in rat pinealocytes. J. Neurochem. 69:340–347 (1997).
16. Yamada H., Yastushiro S., Ishio S., Hayashi M., Nishi T., Yamamoto A., Futai M., Yamaguchi A., and Moriyama Y. Metabotropic glutamate receptors negatively regulate melatonin synthesis in rat pinealocytes. J. Neuroscience 18:2056–2062 (1998).
17. Yamada H., Yamaguchi A., and Moriyama Y. L-Aspartate-evoked inhibition of melatonin production in rat pineal glands. Neurosci. Lett. 228:103–108 (1997).
18. Ishio S., Yamada H., Hayashi M., Yatsushiro S., Noumi T., Yamaguchi A., and Moriyama Y. D-Aspartate modulates melatonin synthesis in rat pinealocytes. Neurosci. Lett. 249:143–146 (1998).
19. Tanabe Y., Masu M., Ishii T., Shigemoto R., and Nakanishi S. A family of metabotropic glutamate receptors. Neuron 8:169–179 (1992).
20. Hayashi Y., Tanabe Y., Aramori I., Masu M., Shimamoto K., Ohfune Y., and Nakanishi S. Agonist analysis of 2-(carboxycyclopropyl) glycine isomers for cloned metabotropic glutamate receptor subtypes expressed in Chinese hamster ovary cells. Br. J. Pharmacol. 107:539–543 (1992).
21. Hayashi Y., Momiyama A., Takahashi T., Ohishi H., Ogawa-Meguro R., Shigemoto R., Mizuno N., and Nakanishi S. Role of a metabotropic glutamate receptor in synaptic modulation in the accessory olfactory bulb. Nature 366:687–690 (1993).
22. Hayashi Y., Sekiyama N., Nakanishi S., Jane D.E., Sunter D.C., Birse E.F., Udvarhelyi P.M., and Watkins J.C. Analysis of agonist and antagonist activities of phenylglycine derivatives for different cloned metabotropic glutamate receptor subtypes. J. Neuroscience 14:3370–3377 (1994).
23. Tanabe Y., Nomura A., Masu M., Shigemoto R., Mizuno N., and Nakanishi S. Signal transduction, pharmacological properties, and expression patterns of two rat metabotropic glutamate receptors, mGluR3 and mGluR4. J. Neuroscience 13:1372–1378 (1993).
24. Nakanishi S. Molecular diversity of glutamate receptors and implications for brain function. Science 258:597–603 (1992).

25. Riedel G. Function of metabotropic glutamate receptors in learning and memory. Trends in Neuroscience 19:219–223 (1996).
26. Kanner B. and Schuldiner S. Mechanism of transport and storage of neurotransmitters. CRC Crit. Rev. Biochem. 22:1–38 (1987).
27. Kanner B.I. Glutamate transporters from brain. A novel neurotransmitter transporter family. FEBS Lett. 325:95–99 (1993).
28. Kanai Y., Smith C.P., and Hediger M.A. A new family of neurotransmitter transporters: the high affinity glutamate transporters. FASEB. J. 7:1450–1459 (1993).
29. Yamada H., Yatsushiro S., Yamamoto A., Nishi T., Futai M., Yamaguchi A., and Moriyama Y. Functional expression of GLT-1 type Na^+-dependent glutamate transporter in rat pinealocytes. J. Neurochem. 69:1491–1498 (1997).
30. Danbolt N.C., Storm-Mathisen J., and Kanner B.I. An [Na^+ + K^+] coupled L-glutamate transporter purified from brain is located in glial cell processes. Neuroscience 51:295–310 (1992).
31. Rothstein J.D., Martin L., Levey A.I., Dykes-Hoberg M., Jin L., Wu D., Nash N., and Kuncl R.W. Localization of neuronal and glial glutamate transporter. Neuron 13:713–725 (1994).
32. Lehre K.P., Levy L.M., Ottersen O.P., Storm-Mathisen J., and Danbolt N.C. Differential expression of two glial glutamate transporters in the rat brain: quantitative and immunocytochemical observations. J. Neuroscience 15:1835–1853 (1995).
33. Yamada H., Ogura A., Yamaguchi A., and Moriyama Y. Acetylcholine triggers L-glutamate exocytosis via nicotinic receptors and inhibits melatonin synthesis in rat pinealocytes. J. Neuroscience 18:4946–4952 (1998).

12

SYNAPTIC VESICLE PROTEIN SV2B, BUT NOT SV2A, IS PREDOMINANTLY EXPRESSED AND ASSOCIATED WITH MICROVESICLES IN RAT PINEALOCYTES

Mitsuko Hayashi,[1] Shouki Yatsushiro,[2] Hiroshi Yamada,[2] Akitsugu Yamamoto,[2] Masamitsu Futai,[1] Akihito Yamaguchi,[2] and Yoshinori Moriyama[1]

[1] Departments of Cell Membrane Biology and Molecular Cell Biology
Institute of Scientific and Industrial Research (ISIR)
Osaka University
CREST, Japan Science and Technology Corporation
Ibaraki, Osaka 567, Japan
[2] Department of Physiology
Kansai Medical University
Moriguchi, Osaka 570, Japan

We have shown that rat pinealocytes are glutaminergic endocrine cells, and secrete L-glutamate through exocytosis (reviewed in 1). The secreted L-glutamate binds to class II metabotropic glutamate receptor, and inhibits N-acetyltransferase activity through an inhibitory cAMP cascade (2). It is quite likely that pinealocytes use L-glutamate as a negative regulator for melatonin synthesis through paracrine- or autocrine-like intercellular signal transduction.

Microvesicles (MVs), endocrine counterparts of neuronal synaptic vesicles, are present in mammalian pinealocyte (3–5) and are responsible for L-glutamate exocytosis (6). Studies on the structure and function of the pineal MVs are important to understand the molecular events on the mechanism of the glutamate signal output. Our recent studies indicate that pineal MVs contain synaptophysin, synaptotagmin, VAMP2, and N-ethylmaleimide-sensitive fusion protein (NSF), but devoid of synapsins (Figure 1) (1). Furthermore, the MVs also possess vacuolar H^+-ATPase (V-ATPase) as a main constituent (approximately 10% of total protein), and vesicular glutamate transporter. The V-ATPase and vesicular glutamate transporter are energetically coupled: L-glutamate transport is driven by membrane potential (inside positive) established by

Melatonin after Four Decades, edited by James Olcese.
Kluwer Academic / Plenum Publishers, New York, 2000.

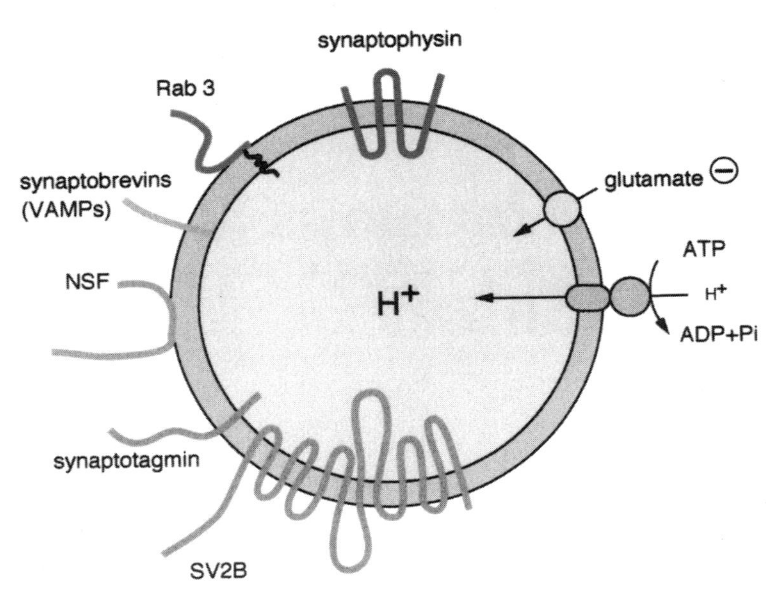

Figure 1. Membrane Proteins of MVs from Mammalian Pinealocytes.

V-ATPase (Figure 1) (3,4). L-Glutamate inside MVs is exocytosed Ca^{2+}-dependently and appears in extracellular space (6). The exocytosis seem to occur with similar mechanism to neuronal synaptic vesicles and endocrine secretory granules. However, the rate of MVs exocytosis is about 1/10 of known regulated exocytosis (6,7), and detailed studies on the mechanism of exocytosis will be important to reveal the mechanistic similarity and difference of the known exocytosis.

To characterize the molecular organization of microvesicles in more detail, we investigated in this study the expression and localization of synaptic vesicle protein 2 (SV2), in rat pinealocytes. SV2, originally identified in cholinergic vesicles from the electric fish, *Discopyge ommanta* (8), is a glycosylated synaptic vesicle membrane protein comprising 12 transmembrane regions (9–11). Two isoforms, abbreviated as SV2A and SV2B, have been identified in mammals: the former consists of 742 amino acids and the latter of 683 amino acids, with 65% sequence homology (12,13).

Reverse transcription polymerase chain reaction (RT-PCR) analysis indicated that transcripts specific for two isoforms, SV2A, a ubiquitous form present in neuronal and endocrine cells, and SV2B, a neuron-specific form, are amplified in pineal RNAs. Northern blotting with specific transcripts indicated that the mRNA for SV2B is predominantly expressed, whereas that for SV2A is below the detection limit. Site-specific antibodies against SV2B recognized a single 72 kDa polypeptide in the pineal membrane fraction, whereas anti-SV2A antibodies did not recognize any polypeptides. Immunohistochemical analysis of cultured cells indicated that SV2B is expressed in pinealocytes but not in other types of cells. SV2B is present in somata and is especially rich in processes, which are filled with microvesicles. SV2B is co-localized with synaptophysin and synaptotagmin, markers for microvesicles. Immunoelectron-microscopy indicated that SV2B is associated with microvesicles. These results indicated that SV2B, a neuronal specific form, but not SV2A, a ubiquitous form, is unexpectedly expressed in pinealocytes and associated with MVs. Pinealocytes are the first example of stable expression of SV2B in non-neuronal cells. Since SV2B is also expressed in cultured αTC6, clonal pancreatic α cells, SV2B is not a specific protein for neurons.

REFERENCES

1. Moriyama Y., Yamamoto A., Yamada H., Tashiro Y., and Futai M. (1996) Role of endocrine cell microvesicles in intercellular chemical transduction. Biol. Chem. **377**, 155–165.
2. Yamada H., Yatsushiro S., Ishio S., Hayashi M., Nishi T., Yamamoto A., Futai M., Yamaguchi A., and Moriyama Y. (1998) Metabotropic glutamate receptors negatively regulate melatonin synthesis in rat pinealocytes. J. Neurosci. **18**, 2056–2062.
3. Moriyama Y. and Yamamoto A. (1995a) Microvesicles isolated from bovine pineal gland specifically accumulate L-glutamate. FEBS Lett. **367**, 233–236.
4. Moriyama Y. and Yamamoto A. (1995b) Veciular L-glutamate transporter in microvesicles from bovine pineal glands: driving force, mechanism of chloride anion-activation, and substrate specificity. J. Biol. Chem. **270**, 2314–22320.
5. Redecker P. and Bargsten G. (1993) Synaptophysin—a common constituent of presumptive secretory microvesicles in the mammalian pinealocyte: a study of rat and gerbil pineal glands. J. Neurosci. Res. **34**, 79–96.
6. Yamada H., Yamamoto A., Yodozawa S., Kozaki S., Takahashi M., Michibata H., Morita M., Furuichi T., Mikoshiba K., and Moriyama Y. (1996a) Microvesicle mediated exocytosis of glutamate is a novel paracrine-like chemical transduction mechanism and inhibits melatonin secretion in rat pinealocytes. J. Pineal Res. **21**, 175–191.
7. Yamada H., Yamamoto A., Takahashi M., Michibata H., Kumon H., and Moriyama Y. (1996b) The L-type Ca^{2+} channel is involved in the microvesicle-mediated exocytosis of glutamate in rat pinealocytes. J. Pineal Res. **21**, 165–174.
8. Buckley K. and Kelly R.B. (1985) Identification of a transmembrane glycoprotein specific for secretory vesicles of neural and endocrine cells. J. Cell Biol. **100**, 1284–1294.
9. Bajjalieh S.M., Peterson K., Linial M., and Scheller R.H. (1992) SV2, a brain synaptic vesicle protein homologous to bacterial transporters. Science **257**, 1271–1273.
10. Feany M.B., Lee S., Edwards R.H., and Buckley K.M. (1992) The synaptic vesicle protein 2 is a novel type of transmembrane transporter. Cell **70**, 861–867.
11. Gringrich J.A., Andersen P.H., Tiberi M., Mestikawy S., Jorgensen P.N., Fremeau R.T. Jr., and Caron M.G. (1992) Identification, characterization, and molecular cloning of a novel transporter-like protein localized to the central nervous system. FEBS lett. **312**, 115–122.
12. Bajjalieh S.M., Peterson K., Linial M., and Scheller R.H. (1993) Brain contains two forms of synaptic vesicle protein 2. Proc. Natl. Acad. Sci. USA **90**, 2150–2154.
13. Bajjalieh S.M., Frantz G.D., Weimann J.M., McConnel S.K., and Scheller R.H. (1994) Differential expression of synaptic vesicle protein 2 (SV2) isoforms. J. Neurosci. **14**, 5223–5235.

13

NEUROPEPTIDE Y (NPY) AND NPY RECEPTORS IN THE RAT PINEAL GLAND

Jens D. Mikkelsen,[1] Frank Hauser,[1] and James Olcese[2]

[1] Neurobiology Department
H. Lundbeck A/S, Valby-Copenhagen
Denmark
[2] Institute for Hormone and Fertility Research
Hamburg, Germany

1. SUMMARY

NPY is considered to play an important role in pineal function, because it is co-stored with the dominant pineal transmitter noradrenaline. However, little evidence from the literature suggests that NPY alone is a strong regulator of melatonin synthesis or secretion and it is therefore more likely that NPY modulates noradrenergic neurotransmission in the rat pineal gland. The purpose of the present studies was to determine the nature and origin of NPYergic inputs to, and the type of specific NPY receptor subtypes in, the rat pineal gland. Gel filtration and immunocytochemistry using region-specific antisera revealed that all proNPY present in intrapineal nerve fibres is cleaved to amidated NPY and a C-terminal flanking peptide of NPY (CPON). The vast majority of NPY content in the pineal gland was found to be of sympathetic origin. Receptor autoradiography showed that only a few NPY specific binding sites were present in the superficial pineal gland. A reverse transcriptase polymerase chain reaction detected sequences of only NPY receptor subtype Y1 and not other NPY receptor subtypes in pineal extracts. These results together with the available literature imply that NPY under certain conditions is co-released with noradrenaline and exerts its actions either presynaptically or on the pinealocyte through a Y1 receptor. The available data indicate that NPY has no effect alone, but acts in concert with noradrenaline. A presynaptic action regulating noradrenaline neurotransmission is also possible. NPY has been reported only to act on melatonin secretion *in vitro*, and it remains to be established what function NPY plays in the pineal gland *in vivo*. This

Melatonin after Four Decades, edited by James Olcese.
Kluwer Academic / Plenum Publishers, New York, 2000.

paper discuss possible modulatory actions of NPY being a predominant sympathetic transmitter.

2. INTRODUCTION

The mammalian pineal gland is densely innervated by noradrenergic nerve fibres originating from the superior cervical ganglia (1) that plays an essential role in synthesis of melatonin (2). It is well established that NPY occurs together with noradrenaline in the postganglionic sympathetic fibres (3), and several studies have shown that the NPY-containing nerve fibres in the pineal gland, at least partly, originate from noradrenergic neurons in the superior cervical ganglia (4–6). NPY and noradrenaline are co-released from the nerve terminal (7,8) and it is therefore tempting to speculate how these transmitters interact postsynaptically on the pinealocyte.

Noradrenaline is the major neurotransmitter regulating melatonin synthesis and release and it is considered to be responsible for activation of cAMP-dependent signalling pathways and for the circadian variation in melatonin secretion and release. A possible role for NPY may be either to modulate the actions of noradrenaline in the early night and/or to regulate the decline in melatonin secretion in the late night. The purpose of this report is to cover the literature and extend with our own studies a clearer definition of the nature of NPY, the origin of NPYergic innervation, the NPY receptors, and the activity of NPY in the rat pineal gland.

As illustrated in Figure 1, the NPY precursor undergoes cleavage at a single dibasic site Lys38-Arg39 resulting in the formation of 1–39 amino acid NPY and a C-flanking peptide of NPY (CPON). The NPY fragment is further processed successively by carboxypeptidase-like and peptidylglycine alpha-amidating monooxygenase enzymes. The nature of NPY-immunoreactivity has not been examined in the pineal gland, thus gel filtration and site specific antisera were used to determine the processing of the NPY precursor.

Some intrapineal NPY fibres remain after superior cervical ganglionectomy (6,9). NPY is present in ocular nerves originating from the parasympathetic system (10) though it remains to be shown that the non-sympathetic nerves in the rat pineal gland also originate from a parasympathetic ganglion. Hence, while the NPYergic input in

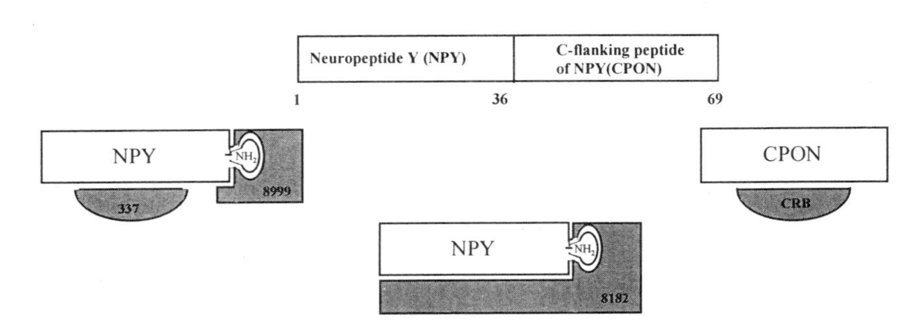

Figure 1. Schematic illustration of the used antisera and their specificity for distinct epitopes of the NPY precursor, NPY, or CPON.

the pineal is considered to be sympathetic, a central input may also play a role in some species. In order to understand what systems innervate the pineal gland, histological and radioimmunological analyses were carried out in normal and superior cervical ganglionectomised rats.

NPY exerts its actions through at least six different receptors subtypes, designated Y1–Y6 (11,12). These receptor subtypes are differently expressed in rat tissues. Y1, Y2 and Y5 are expressed in the brain, whereas Y4 has been found in some circumventricular organs (13–16). Y1 and Y2 receptors are activated by different NPY peptide analogues and considered to be post- and presynaptic, respectively (17). Y4 and Y5 receptors are postsynaptic, and shows affinity for pancreatic polypeptide. Binding of NPY has been shown on pinealocytes and the Y1 mRNA shown to be expressed in rat pineal tissue (18,19). Since these studies were reported, additional NPY specific receptors have been cloned and pharmacologically characterised (10). We have analysed the level of specific binding of NPY in the rat pineal gland and used reverse transcriptase polymerase chain reaction to determine which NPY receptors that are expressed in rat pineals.

NPY modulates the noradrenergic transmission in the rat pineal gland at both pre- and postsynaptic levels (18,19). Unequivocal results have been reported on the role of NPY alone (18–20), and little evidence for NPY activating transcriptional events is available. Given that NPY activates the pinealocyte through Y1 receptors, several possible signalling pathways may be activated that are important in pineal function, and in particular related to how NPY may interact with noradrenergic signalling.

3. MATERIALS AND METHODS

3.1. Extraction of Pineals, Gel Filtration, and Radioimmunoassays

For gel filtration, tissue extracts obtained from superficial pineal glands (pooled from 50 adult Wistar rats) were applied to Sephadex G-50 superfine columns (1.6×95 cm) eluted at 4°C, and aliquots subjected to radioimmunoassays as earlier described (21,22). As illustrated in Figure 1, four different antisera directed against different epitopes of ProNPY and its products were used. Antiserum code 337 crossreacts approximately 75% with proNPY, the antiserum code #8999 recognised NPY only in its amidated form, and antiserum #8182 recognised both (21–24). Finally, an antiserum against CPON obtained from Cambridge Research Biochemicals was used for immunohistochemistry.

In order to measure the relative content of NPY from sympathetic and non-sympathetic sources, superficial pineal glands were removed 4 hrs after light onset (12:12) from nine normal and twelve ganglionectomised rats and frozen immediately on dry ice. Individual glands were then homogenised and extracts assayed for NPY as earlier described (25).

3.2. Fixation and Immunohistochemistry

Naive, sham-operated, and superior cervical ganglionectomised male Wistar rats were fixed by perfusion in 4% paraformaldehyde in 0.1 M phosphate buffer (pH 7.4).

The brains and pineals were sectioned in a cryostat, thawed onto pre-gelatinised glass slides and incubated with the above mentioned antisera all of which were diluted 1:1000. The immunoreactions were performed using the biotin-avidin method and diamonobenzidine was used as chromagen (see 6,22).

For dual-immunocytochemistry sections were incubated with both anti-NPY (#8182) and a monoclonal mouse antiserum against tyrosine hydroxylase (Immunonuclear, Stillwater) diluted 1:4000. A dual fluorescence method using biotinylated rabbit-anti mouse (Jackson Immunores Labs, West Grove) and FITC-conjugated donkey anti-rabbit IgG (Nordic Immunology, Tilburg, The Netherlands) diluted 1:400 and 1:100, respectively was applied. Finally, the sections were incubated with streptavidin Texas Red-complexes diluted 1:100, coverslipped and examined in a Nikon Eclipse fluorescence microscope.

3.3. Receptor Autoradiography

Six frozen rat brains and pineals were sectioned in 20 μm thick serial coronal sections, thawed onto gelatin-coated slides, and kept at –80°C until the autoradiography was performed. The sections were washed in 25 mM HEPES (pH 7.4), 2.5 mM $CaCl_2$, 1 mM $MgCl_2$, 0.5 mg/ml bacitracin and 0.5 g/l BSA, and then incubated for 60 minutes at room temperature in the same buffer to which 0.1 nM [^{125}I][Tyr-36] monoiodo PYY (Amersham, Birkerød, Denmark) and 1 mM of phenylmethylsulfonyl flouride were added (13,26). One series of sections was incubated with radioactive ligand alone, another with the radioactive ligand and 1 μM NPY (non-specific). The slides were washed in the preincubation buffer for 4×10 minutes at room temperature, air-dried in a stream of cold air and exposed on a Kodak D19 film for 3 days before being developed in Kodak D19 developer.

3.4. Reverse Transcriptase-Polymerase Chain Reaction (RT-PCR)

A total number of 50 male Wistar rats (200 g) were sacrificed during the day. All superficial pineal glands, anterior pituitaries and four hippocampi were removed and frozen at –80°C. Total RNA was isolated from pooled pineals (55 mg wet weight), pituitaries (90 mg) and hippocampus (130 mg) using a RNAeasy total RNA isolation kit (Qiagen) according the instructions of the manufacturer, treated with 1 unit RNase-free Dnase I (Promega) for 30 minutes at 37°C and then repurified using the same RNA isolation kit in a final volume of 50 μl. RT-PCR was performed using RT-PCR beads (Pharmacia). First-strand cDNA was synthesised using poly(dT)$_{12-18}$ and 1 μl total RNA for 30 minutes at 42°C. Thereafter, primer pairs specific for the rat Y1 receptor (5′-TGCGGCGTTCAAGGACAAGTATG and 5′-GTGGCA GAGCAGGAACAGCAG ATT, resulting in a 352 bp product), rat Y2 receptor (5′-GGCCTGGCGTGGGTGTC and 5′-TGCCTTCGCTGATGGTAGTGGTC, resulting in a 284 bp product), rat Y4 receptor (5′-CCCTTCCTGGCCAATAGCATC and 5′-GCAGGGGCAGCCAGAG AAC, resulting in a 328 bp product) or rat Y5 receptor (5′-GGCAAACCATGGCTACTTCCTGAT and 5′-CTGGGGGTTTTTGC CTGGTTC, resulting in a 363 bp product) were added. Cycling parameters were 94°C for 1 minute, 58°C for 1 minute and 72°C for 1 minute for a total of 32 cycles. 1/5 of the PCR products were electrophoresed in a 1.5 % agarose gel and stained with EtBr. Experiments without reverse transcription were performed as control.

4. RESULTS

4.1. The NPY Precursor is Processed to Amidated NPY and CPON

As shown in Figure 2, gel filtration elution profiles of NPY immunoreactivity from the pineal gland as measured both by antiserum 337 and 8999 demonstrated a single peak with an apparent molecular size corresponding to that of synthetic porcine NPY. Filtration, collections and assays of pineal extracts resulted in similar elution positions of desamidoNPY- (identified with antiserum 337) and NPYamide-immunoreactivity (identified with antiserum 8999) indicating that the vast majority, if not all, of NPY-

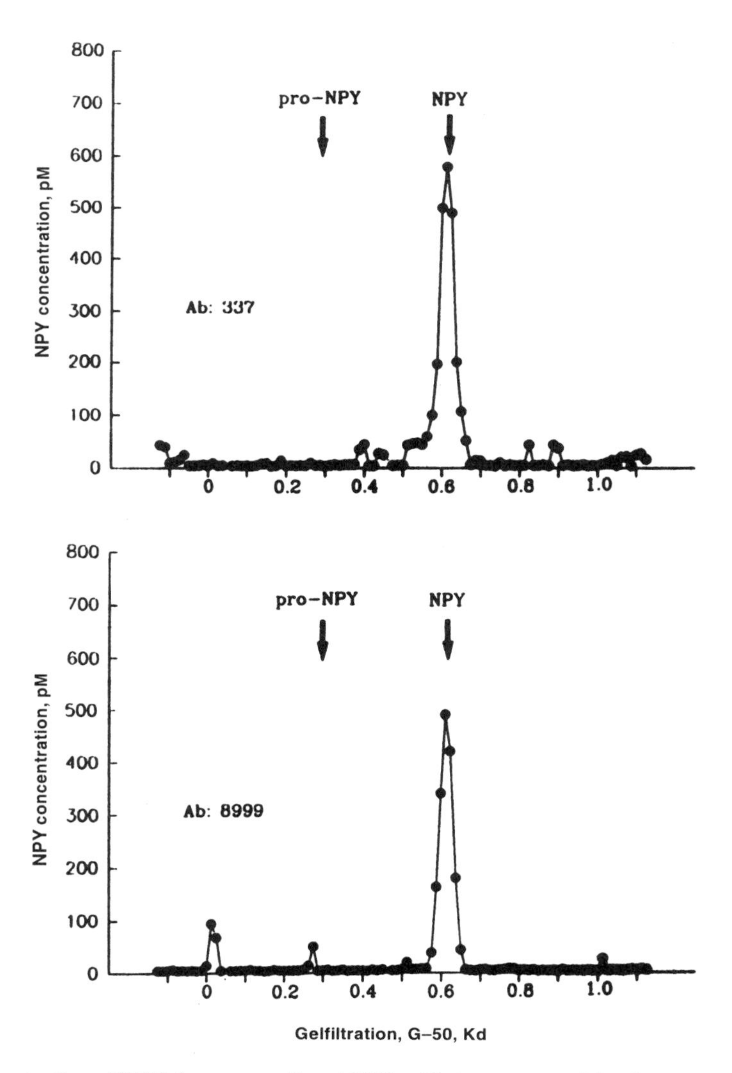

Figure 2. Concentration of NPY- (upper panel) and NPYamide-immunoreactivity (lower panel) in fractions extracted from rat superficial pineals using Sephadex G-50 gel filtration. The two profiles are comparable and show that NPY is present only in an amidated form.

immunoreactivity is amidated NPY. No molecular species corresponding to proNPY was identified. Furthermore, CPON is found in intrapineal nerve fibres (Figure 3), which suggests that proNPY is cleaved to the two peptide fragments, and the relative concentration of the two fragments is only dependent on their degradation rate and not synthesis or processing.

NPY/CPON-immunoreactive nerve fibres were found in the meningeal tissue, in fibres surrounding pial arteries, in the choroid plexus adjacent to the deep pineal gland, and in the connective tissue connected to the superficial pineal gland (Figures 3 and 4). Positive nerve fibres were found close to the intrapineal vessels and intermingled between groups of pinealocytes.

The rostral part of the pineal complex, comprising the body of the deep pineal gland together with the pinealocytes attached to the pineal stalk contained several NPY/CPON-immunoreactive fibres.

4.3. NPY in Ganglionectomised Rats

Radioimmunoassays revealed that the content of NPY was 1181.2 ± 83.2 fmol/gland in normal pineals, and in the pineal of ganglionectomised rats 27.4 ± 3.5 fmol/gland (Figure 5). This was in accordance with the results of Zhang et al. (1991) showing that the number of NPY-containing fibres in the superficial pineal was considerably lower in bilateral ganglionectomised animals than in normal animals.

In the deep pineal and the pineal stalk, the number of NPY-immunoreactive fibres was only moderately decreased after superior cervical ganglionectomy (6). These fibres are likely to originate from the brain. Dual-labelling showed that NPY in the deep pineal gland of ganglionectomised rats were not co-stored with tyrosine hydroxylase (Figure 6) indicating that these fibres originate from those central NPY cells that are not catecholaminergic.

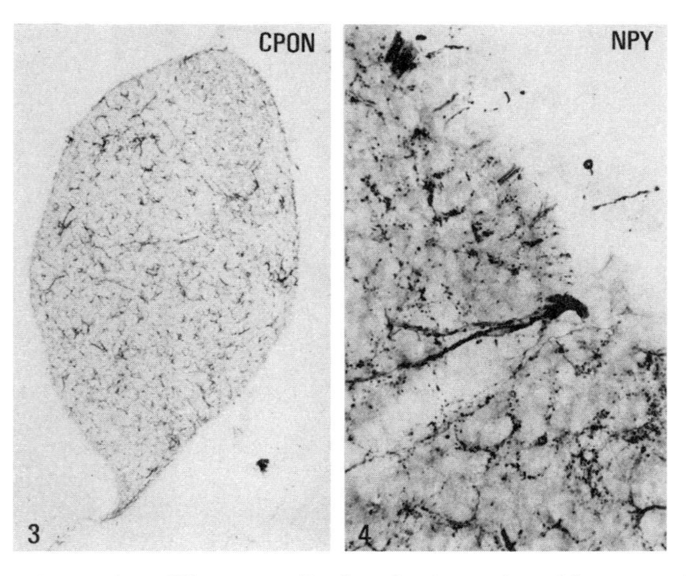

Figures 3–4. Photomicrographs at different magnifications showing sections of the rat superficial pineal gland immunostained for CPON (Figure 3) and NPY (Figure 4).

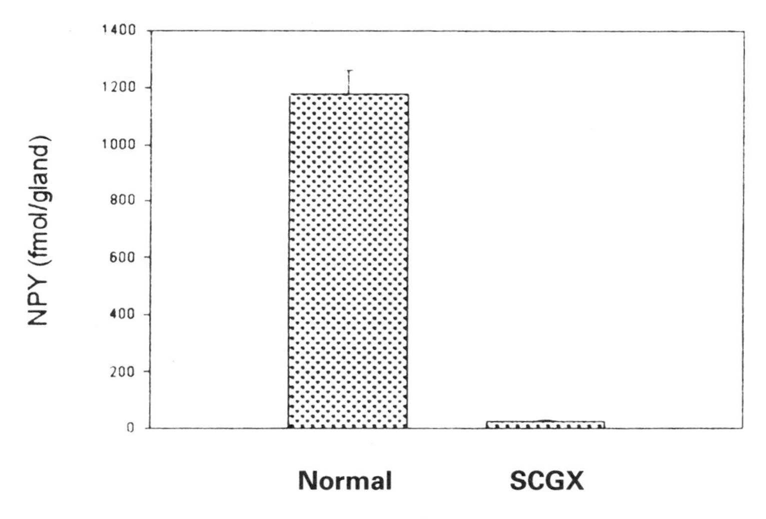

Figure 5. The content of NPY in extracts of the superficial pineal gland measure with radioimmunoassay. Effects of superior cervical ganglionectomy on content of NPY-immunoreactivity in the pineal gland.

4.4. NPY Receptors in the Pineal

Binding of sections with iodinated peptide YY (PYY) revealed the presence of a low concentration of binding sites in the rat superficial pineal gland (Figure 7). High levels of binding was observed in the hippocampus and the cerebral cortex.

RT-PCR analysis of pineal total RNA revealed expression of Y1 mRNA, but virtually no expression of Y2, Y4 or Y5 mRNA (Figure 8). In the anterior pituitary, expres-

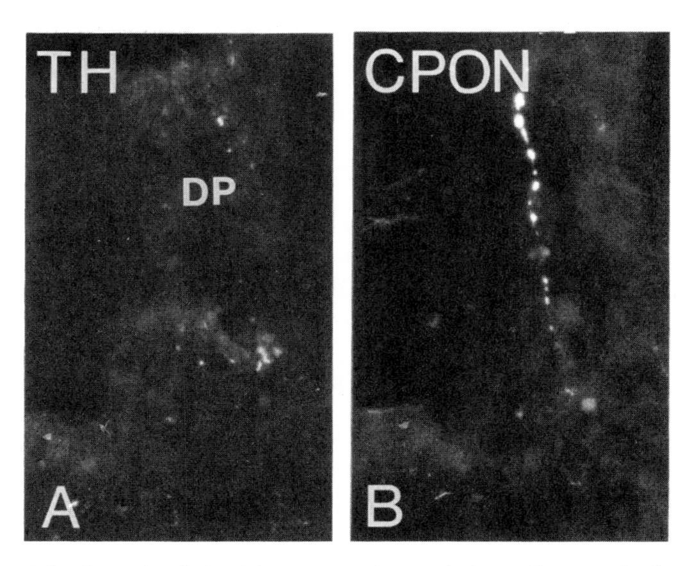

Figure 6. Section of the deep pineal gland from a superior cervical ganglionectomised rat immunoreacted for both tyrosine hydroxylase (TH, A) and CPON (B). A positive CPON-fibre not co-storing noradrenaline is present in the pinealopetal projection originating from the brain.

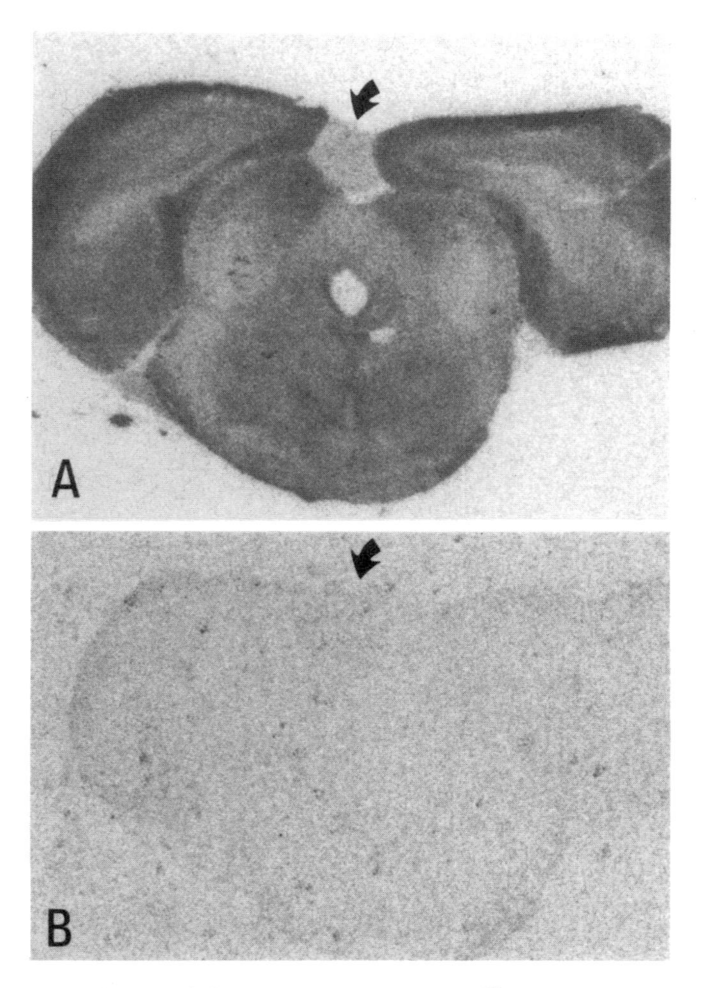

Figure 7. Receptor autoradiograms (A) showing the distribution of ^{125}I-PYY binding in the rat pineal gland (arrows). The content of binding in the pineal gland is relatively low compared to the brain. The lower figure (B) shows an adjacent section incubated with an excess of NPY.

sion of Y2 mRNA was dominant and in the hippocampus all cloned receptors apart from Y4 were expressed (Figure 8). This shows that the Y1 subtype is the only post-synaptic receptor amongst the NPY receptors subtypes known today that is expressed in the pineal gland.

5. DISCUSSION

5.1. The NPY Precursor is Cleaved and NPY is Amidated

The studies showed that the nerve fibres in the pineal gland contain only NPY in a final and processed form. Termination of processing has been observed in other tissues

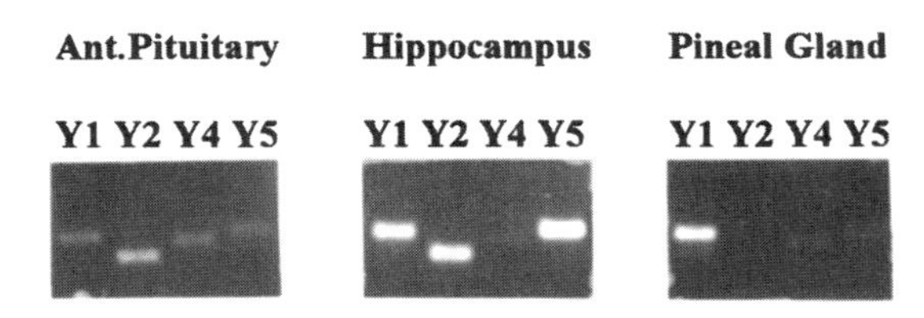

Figure 8. Expression of Y receptor mRNA in the pineal gland, the anterior pituitary and the hippocampus using a reverse transcriptase-polymerase chain reaction. The reactions were performed in the three tissues simultanously. It shows that Y1 is the dominant, if not the only postsynaptic receptor that is expressed in the rat pineal.

such as the suprachiasmatic nucleus (21,22). In contrast to other peptide precursors, ProNPY is fully processed to active molecules. Because CPON has not been associated with any function in the organs studied, we conclude that NPY in the nerve fibres are amidated NPY.

5.2. NPY Originates Predominantly from the Sympathetic Superior Cervical Ganglia

It is known that the superficial pineal gland of the rat is densely innervated by nerve fibres containing NPY (4,6). Further, it was revealed that the vast majority of the NPY-immunoreactivity (97%) originates from the superior cervical ganglia and a minority from elsewhere.

Notably, the NPYergic innervation of the pineal gland is found to be species dependent. High amounts of NPY nerve fibres have been observed in the rat and sheep (4,6,9), whereas little innervation is seen in the mink and primate (27,28). Further, while pineal NPYergic fibres originate mainly from a sympathetic source in rat and sheep, NPY originates to a large extent also from neurones in the brain in the mink and monkey.

The presence of NPY in the central input of the rat pineal is supported by the observation that most of the NPY/CPON-immunoreactive fibres remaining after ganglionectomy was seen in that part of the pineal complex that is known to receive a central innervation, namely the deep pineal gland. Also a parasympathetic innervation cannot be excluded (29–31).

5.3. NPY in the Deep Pineal and the Central Innervation

The origin of the NPY/CPON-immunoreactive neurons contained in the central pinealopetal projection have not been identified yet with certainty. It is shown that the majority of the central input contains NPY and not noradrenaline, which excludes catecholaminergic neurons in the brain stem costoring NPY (32). A NPYergic projection originating from the intergeniculate leaflet of the lateral geniculate nucleus to the suprachiasmatic nucleus has been demonstrated (33–36) and this projection has been shown to mediate phase advances to the pacemaker (37,38). Since the intergeniculate leaflet projects to the pineal gland (24), it is likely that this projection supplies the deep pineal with NPY/CPON-positive afferents, and thereby connects the pineal gland directly with the circadian timing system.

5.4. The Pinealocyte Expresses Only Y1 Receptors

The NPY receptor subtypes are differently expressed in the organism. In the pineal gland, only the Y1 receptor subtype was found to be expressed. This finding is in accordance with a previous study report in binding of NPY to a Y1 preferring binding site in a primary culture of pinealocytes (18). The level of total binding in the pineal is low compared to other tissues. Other known NPY receptors were not expressed in the rat pineal gland.

Noteworthy, postsynaptic Y4 and Y5 receptors were not found in the pineal. The Y4 receptor is expressed in the periphery and in some circumventricular organs considered to be a target for blood borne pancreatic polypeptide (39). The Y5 receptor is expressed in the hypothalamus and not in peripheral tissues (16) and is apparently not playing any role in pineal. However, presynaptic Y receptors expressed in the sympathetic cell body or in other neurons innervating the pineal are probably present (see below) and undetectable in the used mRNA assay.

5.5. NPY Neurotransmission in the Rat Pineal Complex: Functional Considerations

A major question related to these observations is how NPY via the postsynaptic receptor inhibits melatonin release. It has been demonstrated that NPY inhibits β-adrenergic, VIPergic and cholera toxin-induced cAMP accumulation in pineal cultures, through a pertussin toxin sensitive G_i protein (40). This raises interesting perspectives, because two transmitters in the same nerve terminal, or even in the same vesicle, activate and inhibit the pinealocyte. Whether impulse frequency affect the proportional release of NPY and noradrenaline from sympathetic endings in the pineal gland remains to be established.

The role of NPY in other organs has been attributed to its co-transmission with catecholamines, because they are costored in the same postganglionergic nerves (3). In peripheral sympathetic nerve endings, noradrenaline is stored in two different types of vesicles, the large and the small dense core vesicles (41). Single electrical stimulation of other sympathetic nerves, such as splenic and mesenteric nerves (at frequencies lower than 20 Hz), produced release of both noradrenaline and NPY from the nerve endings that increased with the frequency of stimulation, but the ratio of noradrenaline to NPY remained constant (7,8). It is not known under what conditions noradrenaline is released alone, and if similar mechanisms occur in pineal nerves. However, the co-localisation of NPY noradrenaline implies that NPY is released together with noradrenaline.

The sympathetic nervous system is rhythmic in many organs (42). Its role in pineal function is unclear—in particular the mechanisms involved in the inhibition of the melatonin secretion in the late dark-phase are poorly understood. It is speculated that altered impulse frequencies during the night could affect the release of noradrenaline and NPY in such a way that noradrenaline had a major effect in the beginning of the dark phase, and NPY plays a role later resulting in the decline of N-acetyl transferase activity and subsequent of melatonin secretion.

The pineal tissue shares the characteristics of a sympathetic input with many other peripheral organs. Thus, there may be a functional component of this input that relates to regulation of the vascular system, and another related to the neuroendocrine regulation. In general, NPY contracts isolated peripheral blood vessels, but the responses

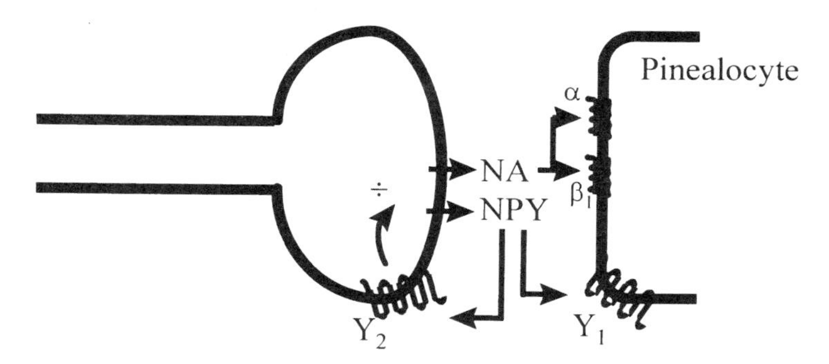

Figure 9. Schematic illustrations of possible effect of NPY and noradrenaline (NA) at the synaptic level. NPY acts either via a post-synaptic receptor in concert with NA to regulate melatonin synthesis and secretion or pre-synaptically to inhibit further release of NA to the extracellular space.

in terms of dose-response relationships are different from organ to organ (43). Using the radioactive microsphere technique for measurements of regional blood flow, the effects of NPY was among many organs studied also in the pineal gland of the rabbit (44). After 2 minutes of NPY, there was a marked blood flow reduction in the pineal gland (44).

In summary, NPY seems to have at least three effects on the sympathetic neuroeffector junction (Figure 9). There is a presynaptic effect (expressed as inhibition of norepinephrine release), and two postsynaptic effects; a direct response, and a potentiation/inhibition of the noradrenergic response (17,45,46). Apart from the direct effect, the same appears to be true for the pineal gland. However, it remains to be understood how this input can drive the circadian rhythm, which is unique for pineal physiology.

ACKNOWLEDGMENTS

Tine Ø. Gerlén, Pia M. Carstensen are thanked for technical assistance, and Tina Bott for secretarial assistance. This study was supported by the Danish Medical Research Council.

REFERENCES

1. Bowers C.W., Dahm L.M., and Zigmond R.E. The number and distribution of sympathetic neurons that innervate the rat pineal gland. Neuroscience 13:87–96, 1984.
2. Reiter R.J. Pineal melatonin: cell biology of its synthesis and of its physiological interactions. Endocr Rev 12:151–180, 1991.
3. Lundberg J.M., Terenius L., Hökfelt T., Martling C.R., Tatemoto K., Mutt V., Polak J., Bloom S., and Goldstein M. Neuropeptide Y (NPY)-like immunoreactivity in peripheral noradrenergic neurons and effects of NPY on sympathetic function. Acta Physiol Scand 116:477–480, 1982.
4. Schon F., Allen J.M., Yeats J.C., Allen Y.S., Ballesta J., Polak J.M., Kelly J.S., and Bloom S.R. Neuropeptide Y innervation of the rodent pineal gland and cerebral blood vessels. Neurosci Lett 57:65–71, 1985.
5. Reuss S. and Moore R.Y. Neuropeptide Y-containing neurons in the rat superior cervical ganglion: projections to the pineal gland. J Pineal Res 6:307–316, 1989.

6. Zhang E.T., Mikkelsen J.D., and Møller M. Tyrosine hydroxylase- and neuropeptide Y-immunoreactive nerve fibers in the pineal complex of untreated rats and rats following removal of the superior cervical ganglia. Cell Tissue Res 265:63–71, 1991.
7. De Potter W.P., Partoens P., Schoups A., Llona I., and Coen E.P. Noradrenergic neurons release both noradrenaline and neuropeptide Y from a single pool: the large dense cored vesicles. Synapse 25:44–55, 1997.
8. Donoso M.V., Brown N., Carrasco C., Cortes V., Fournier A., and Huidobro-Toro J.P. Stimulation of the sympathetic perimesenteric arterial nerves releases neuropeptide Y potentiating the vasomotor activity of noradrenaline: involvement of neuropeptide Y-Y1 receptors. J Neurochem 69:1048–1059, 1997.
9. Cozzi B., Mikkelsen J.D., Ravault J.P., and Møller M. Neuropeptide Y (NPY) and C-flanking peptide of NPY in the pineal gland of normal and ganglionectomized sheep. J Comp Neurol 316:238–250, 1992.
10. Jones M.A. and Marfurt C.F. Peptidergic innervation of the rat cornea. Exp Eye Res 66:421–435, 1998.
11. Blomqvist A.G. and Herzog H. Y-receptor subtypes—how many more? Trends Neurosci 20:294–298, 1997.
12. Herzog H., Hort Y.J., Ball H.J., Hayes G., Shine J., and Selbie L.A. Cloned human neuropeptide Y receptor couples to two different second messenger systems. Proc Natl Acad Sci USA 89:5794–5798, 1992.
13. Larsen P.J., Sheikh S.P., Jakobsen C.R., Schwartz T.W., and Mikkelsen J.D. Regional distribution of putative NPY Y1 receptors and neurons expression Y1 mRNA in forebrain areas of the rat central nervous system. Eur J Neurosci 5:1622–1637, 1993.
14. Larsen P.J. and Kristensen P. The neuropeptide Y (Y4) receptor is highly expressed in neurons of the rat dorsal vagal complex. Molecular Brain Research 48:1–6, 1997.
15. Gustafson E.L., Smith K.E., Durkin M.M., Walker M.W., Gerald C., Weinshank R., and Branchek T.A. Distribution of the neuropeptide Y Y2 receptor mRNA in rat central nervous system. Brain Res Mol Brain Res 46:223–235, 1997.
16. Dumont Y., Fournier A., and Quirion R. Expression and characterization of the neuropeptide Y Y5 receptor subtype in the rat brain. J Neurosci 18:5565–5574, 1998.
17. Wahlestedt C., Yanaihara N., and Håkanson R. Evidence for different pre- and post-junctional receptors for neuropeptide Y and related peptides. Regul Pept 13:307–318, 1986.
18. Olcese J. Neuropeptide Y: an endogenous inhibitor of norepinephrine-stimulated melatonin secretion in the rat pineal gland. J Neurochem 57:943–947, 1991.
19. Simonneaux V., Ouichou A., Craft C., and Pevet P. Presynaptic and postsynaptic effects of neuropeptide Y in the rat pineal gland. J Neurochem 62:2464–2471, 1994.
20. Vacas M.I., Sarmiento M.I., Pereyra E.N., Etchegoyen G.S., and Cardinali D.P. In vitro effect of neuropeptide Y on melatonin and norepinephrine release in rat pineal gland. Cell Mol Neurobiol 7:309–315, 1987.
21. O'Hare M.M. and Schwartz T.W. Expression and precursor processing of neuropeptide Y in human and murine neuroblastoma and pheochromocytoma cell lines. Cancer Res 49:7015–7019, 1989.
22. Mikkelsen J.D., Larsen P.J., O'Hare M.M., and Wiegand S.J. Gastrin releasing peptide in the rat suprachiasmatic nucleus: an immunohistochemical, chromatographic and radioimmunological study. Neuroscience 40:55–66, 1991.
23. Blinkenberg M., Kruse-Larsen C., and Mikkelsen J.D. An immunohistochemical localization of neuropeptide Y (NPY) in its amidated form in human frontal cortex. Peptides 11:129–137, 1990.
24. Mikkelsen J.D. and Møller M. A direct neural projection from the intergeniculate leaflet of the lateral geniculate nucleus to the deep pineal gland of the rat, demonstrated with Phaseolus vulgaris leucoagglutinin. Brain Res 520:342–346, 1990.
25. Jessop D.S., Larsen P.J., Mikkelsen J.D., Lightman S.L., and Chowdrey H.S. Effects of adrenalectomy and hypertonic saline on neuropeptide Y content in the posterior pituitary of the rat. Neuroendocrinology 57:416–421, 1993.
26. Greisen M.H., Sheikh S.P., Bolwig T.G., and Mikkelsen J.D. Reduction of neuropeptide Y binding sites in the rat hippocampus after electroconvulsive stimulations. Brain Res 776:105–110, 1997.
27. Larsen P.J., Enquist L.W., and Card J.P. Characterization of the multisynaptic neuronal control of the rat pineal gland using viral transneuronal tracing. Eur J Neurosci 10:128–145, 1998.
28. Mikkelsen J.D. and Mick G. Neuropeptide Y (NPY)-immunoreactive nerve fibers in the pineal gland of the macaque (Macaca fascicularis). J Neuroendocrinol 4:681–688, 1992.
29. Shiotani Y., Yamano M., Shiosaka S., Emson P.C., Hillyard C.J., Girgis S., and MacIntyre I. Distribution and origins of substance P (SP)-, calcitonin gene-related peptide (CGRP)-, vasoactive

intestinal polypeptide (VIP)-and neuropeptide Y (NPY)-containing nerve fibers in the pineal gland of gerbils. Neurosci Lett 70:187–192, 1986.
30. Landis S.C., Jackson P.C., Fredieu J.R., and Thibault J. Catecholaminergic properties of cholinergic neurons and synapses in adult rat ciliary ganglion. J Neurosci 7:3574–3587, 1987.
31. Gibbins I.L. and Morris J.L. Co-existence of immunoreactivity to neuropeptide Y and vasoactive intestinal peptide in non-noradrenergic axons innervating guinea pig cerebral arteries after sympathectomy. Brain Res 444:402–406, 1988.
32. Sawchenko P.E., Swanson L.W., Grzanna R., Howe P.R., Bloom S.R., and Polak J.M. Colocalization of neuropeptide Y immunoreactivity in brainstem catecholaminergic neurons that project to the paraventricular nucleus of the hypothalamus. J Comp Neurol 241:138–153, 1985.
33. Harrington M.E., Nance D.M., and Rusak B. Double-labeling of neuropeptide Y-immunoreactive neurons which project from the geniculate to the suprachiasmatic nuclei. Brain Res 410:275–282, 1987.
34. Harrington M.E., Nance D.M., and Rusak B. Neuropeptide Y immunoreactivity in the hamster geniculo-suprachiasmatic tract. Brain Res Bull 15:465–472, 1985.
35. Card J.P. and Moore R.Y. Organisation of lateral geniculate-hypothalamic connections in the rat. J Comp Neurol 284:135–147, 1989.
36. Mikkelsen J.D. Projections from the lateral geniculate nucleus to the hypothalamus of the Mongolian gerbil (Meriones unguiculatus): an anterograde and retrograde tracing study. J Comp Neurol 299:493–508, 1990.
37. Albers H.E. and Ferris C.F. Neuropeptide Y: role in light-dark cycle entrainment of hamster circadian rhythms. Neurosci Lett 50:163–168, 1984.
38. Rusak B., Meijer J.H., and Harrington M.E. Hamster circadian rhythms are phase-shifted by electrical stimulation of the geniculo-hypothalamic tract. Brain Res 493:283–291, 1989.
39. Whitcomb D.C., Taylor I.L., and Vigna S.R. Characterization of saturable binding sites for circulating pancreatic polypeptide in rat brain. Am J Physiol 259:G687–691, 1990.
40. Harada Y., Okubo M., Yaga K., Kaneko T., and Kaku K. Neuropeptide Y inhibits beta-adrenergic agonist- and vasoactive intestinal peptide-induced cyclic AMP accumulation in rat pinealocytes through pertussis toxin-sensitive G protein. J Neurochem 59:2178–2183, 1992.
41. Fried G., Terenius L., Hökfelt T., and Goldstein M. Evidence for differential localization of noradrenaline and neuropeptide Y in neuronal storage vesicles isolated from rat vas deferens. J Neurosci 5:450–458, 1985.
42. Malpas S.C. The rhythmicity of sympathetic nerve activity [In Process Citation]. Prog Neurobiol 56:65–96, 1998.
43. Edvinsson L., Ekblad E., Håkanson R., and Wahlestedt C. Neuropeptide Y potentiates the effect of various vasoconstrictor agents on rabbit blood vessels. Br J Pharmacol 83:519–525, 1984.
44. Nilsson S.F. Neuropeptide Y (NPY): a vasoconstrictor in the eye, brain and other tissues in the rabbit. Acta Physiol Scand 141:455–467, 1991.
45. Håkanson R., Wahlestedt C., Ekblad E., Edvinsson L., and Sundler F. Neuropeptide Y: coexistence with noradrenaline. Functional implications. Prog Brain Res 68:279–87:279–287, 1986.
46. Potter E.K., Mitchell L., McCloskey M.J., Tseng A., Goodman A.E., Shine J., and McCloskey D.I. Pre- and postjunctional actions of neuropeptide Y and related peptides. Regul Pept 25:167–177, 1989.

14

SIGNAL TRANSDUCTION IN THE RODENT PINEAL ORGAN

From the Membrane to the Nucleus

Erik Maronde,[1] Martina Pfeffer,[1] Charlotte von Gall,[1] Faramarz Dehghani,[1] Christof Schomerus,[1] Helmut Wicht,[1] Susanne Kroeber,[1] James Olcese,[2] Jörg H. Stehle,[1] and Horst-Werner Korf[1]

[1] Dr. Senckenbergische Anatomie
Anatomisches Institut II
Klinikum der Johann Wolfgang Goethe-Universität
Theodor-Stern-Kai-7, D-60590 Frankfurt
Germany
[2] Institut für Hormon- und Fortpflanzungsforschung
Universität Hamburg, D-22529 Hamburg
Germany

SUMMARY

The rodent pineal organ transduces a photoneural input into a hormonal output. This photoneuroendocrine transduction leads to highly elevated levels of the hormone melatonin at nightime which serves as a message for darkness. The melatonin rhythm depends on transcriptional, translational and posttranslational regulation of the arylalkylamine-*N*-acetyltransferase, the key enzyme of melatonin biosynthesis. These regulatory mechanisms are fundamentally linked to two second messenger systems, namely the cAMP- and the Ca^{2+}-signal transduction pathways. Our data gained by molecular biology, immunohistochemistry and single-cell imaging demonstrate a time- and substance-specific activation of these signaling pathways and provide a framework for the understanding of the complex signal transduction cascades in the rodent pineal gland which in concert not only regulate the basic profile but also finetune the circadian rhythm in melatonin synthesis.

Melatonin after Four Decades, edited by James Olcese.
Kluwer Academic / Plenum Publishers, New York, 2000.

1. INTRODUCTION

The mammalian pineal organ is a photoneuroendocrine transducer which rhythmically produces and secretes melatonin during night time in response to photoperiodic stimuli and signals from endogenous circadian oscillators (32,39). The rhythm in melatonin is essential for seasonal reproduction and maternal-fetal communication, influences activity and sleep, and modulates the function of the endogenous rhythm generator which plays an important role also in human physiology and pathology (sleep-wake cycle, shift-work, rapid travel across several time zones, seasonal affective disorders) (2).

Melatonin biosynthesis starts with the uptake of circulating tryptophan into pinealocytes and subsequent 5-hydroxylation by tryptophan hydroxylase. 5-Hydroxytryptophan is transformed into serotonin by the aromatic L-amino acid decarboxylase. The next step is the formation of *N*-acetylserotonin catalyzed by arylalkylamine-*N*-acetyltransferase (AANAT). Finally, *N*-acetylserotonin is *O*-methylated and converted into melatonin by means of the hydroxyindole-*O*-methyltransferase. According to current concepts, the lipophilic melatonin is not stored within the pinealocytes but is released into pineal capillaries immediately after its formation.

The large rhythmic changes in synthesis (and release) of melatonin are based upon an increase in AANAT activity during night time which is 150-fold in the rat pineal organ (54). This rate-limiting enzyme of the melatonin biosynthesis has recently been cloned from rat and sheep pineal libraries (12,18,54).

The regulation of AANAT expression and activity in the mammalian pineal organ depends on a master circadian oscillator located outside the pineal organ in the suprachiasmatic nucleus (SCN) of the hypothalamus. The SCN receives photoperiodic information from the retina via the retinohypothalamic tract which serves the synchronization of the endogenous rhythm with the daily light-dark changes. The SCN transmits its signals to the pineal organ via a multisynaptic neural pathway comprising the paraventricular nucleus of the hypothalamus, the intermediolateral column of the thoracic spinal cord and the superior cervical ganglion. The latter gives rise to the very prominent sympathetic innervation of the mammalian pineal organ (for review and references, see 34). The SCN not only drives the AANAT rhythm, but is also a major target for melatonin. Thus, the SCN and the pineal organ appear to be connected by a feedback loop.

Norepinephrine (NE), the primary neurotransmitter of the sympathetic fibers, is released from the intrapineal nerve terminals in large quantities at the onset of darkness (21) and acts upon α_1- and β-adrenergic receptors in the pinealocyte membrane (32). Activation of β-adrenergic receptors appears as the major event through which NE stimulates melatonin biosynthesis during darkness. The β-adrenergic effects of NE are mediated via the GTP-binding protein G_s and lead to an activation of adenylate cyclase, which causes a tenfold increase in cAMP. Stimulation of α_1-adrenergic receptors alone has no effect on cAMP accumulation, but causes a dramatic increase in the intracellular concentration of free calcium ions $[Ca^{2+}]_i$ and potentiates the β-adrenergic effects on cAMP levels: Simultaneous activation of α_1- and β-adrenoreceptors causes a 100-fold increase in cAMP. The cAMP pathway is crucially linked to the transcriptional and translational up-regulation and maintenance of NAT activity (35,54) and was suggested to be involved also in NAT down-regulation (63,24).

This paper will provide a review on NE-regulated signal transduction which finally results in up- and downregulation of gene expression in the rodent pineal organ. The model species were the Wistar rat and the melatonin proficient C3H mouse.

2. EXPERIMENTAL PROCEDURES

2.1. Animals and Preparation of Pineal Glands

All experiments were conducted in accordance with guidelines on the care of experimental animals as approved by the European Communities Council Directive (86/609/EEC). For cell and organ culture, pineal glands were removed from adult, male Wistar rats or C3H mice decapitated in the morning (between 08:00 and 12:00 h). For analysis of light/dark variations, pineal glands were directly removed from adult, male Wistar rats or C3H mice decapitated at different times of the day (LD 12:12) (*ex vivo*). Pineal glands taken during the night were dissected under dim red light illumination. For isolation of pinealocytes, the glands were immediately dissociated by papain digestion and repetitive pipetting as previously described (67). Briefly, the dispersed cells were resuspended in Dulbecco's Modified Eagle Medium (DMEM) supplemented with fetal calf serum (10%), HEPES (10 mM), ascorbic acid (100 μg/ml), penicillin (100 U/ml), streptomycin (100 μg/ml), glutamine (2 mM) plus glucose (7 mg/ml), immobilized on coated coverslips and incubated for up to 72 h. The viability of the cells was found to be approximately 95% as assessed by trypan blue exclusion.

2.2. Determination of $G_{s\alpha}$- and β_1-ADR-mRNA by *In Situ* Hybridization

Sense and antisense oligonucleotide probes complementary to amino acids 379 to 394 of rat $G_{s\alpha}$ were ^{35}S-labeled by the terminal transferase reaction. A partial rat β_1-ADR-cDNA was generated by PCR using primers annealing upstream of the third transmembrane domain and downstream of the fifth transmembrane domain spanning 506 bp of the coding region (40). PCR fragments of the expected size were electroeluted after size selection, subcloned into the vector pBS SK^- (Stratagene) and analyzed by terminal sequencing as described (62). ^{35}S-labeled β_1-ADR *sense* and *antisense* ribonucleotide probes were generated by *in vitro* transcription with T3 or T7 polymerase (Boehringer Mannheim) after plasmid linearization with *XhoI* or *EcoRI*, respectively. *In situ* hybridization of $G_{s\alpha}$ and β_1-ADR was performed according to published procedures (63,64,65,50,51).

2.3. Fura-2 Calcium Imaging of Rat and Mouse Pinealocytes

The intracellular concentration of free calcium ions ($[Ca^{2+}]_i$) of isolated rat and C3H pineal cells was analyzed with an Attofluor ratio imaging system (Zeiss, Germany; for details see 57). Changes in $[Ca^{2+}]_i$ of single cells are expressed in a semiquantitative manner as 334 nm/380 nm emission ratios. Analyzed cells were identified as pinealocytes by means of S-antigen immunoreaction (38,57).

2.4. Determination of pCREB by Immunocytochemistry

After one to three days in culture, isolated pinealocytes were stimulated as indicated, fixed with 4% paraformaldehyde in phosphate-buffered saline (PBS) for 10 min and then washed in PBS. To block endogenous peroxidase, the preparations were treated with methanol containing 0.45% hydrogen peroxide for 10 min. After washing and preincubation with 10% normal goat serum (30 min), the specimens were incubated with a polyclonal antibody against the serine 133-phosphorylated form of CREB (pCREB; 1:500). The primary antibody was dissolved in PBS containing 0.3% Triton X-100 and 1% bovine serum albumin. Binding of the antibody was visualized using 1) biotin-conjugated anti-rabbit IgG as second antibody, 2) HRP-conjugated streptavidin as the third antibody, and 3) diaminobenzidine (DAB) as the coloring reagent. Some preparations were extensively washed and subsequently double-labeled with an S-antigen antibody (diluted 1:1000) to identify pCREB-immunoreactive cells as pinealocytes (37,38). Binding of the S-antigen antibody was visualized with a FITC-conjugated anti-rabbit IgG (Dako, Copenhagen, Denmark, diluted 1:200) (38,72).

2.5. Identification of Protein Kinases Involved in CREB Phosphorylation

Cultured organs or isolated pinealocytes were stimulated with NE with or without protein kinase inhibitors as indicated or left untreated as control. The PKA antagonist *Rp*-8-CPT-cAMPS was purchased from BioLog Lifescience Inst. (Bremen, Germany). The action mode of this class of antagonists has been described in detail (13,28,20). Chelerythrin and KN-62 were obtained from Calbiochem (Bad Soden, Germany), PD98059 from New England Biolabs (Schwalbach, Germany) and H-89 from BioMol (Hamburg, Germany).

2.6. Determination of PKA Subunit mRNA by Northern Blotting and Reverse Transcriptase Polymerase Chain Reaction (RT-PCR)

To analyze expression of the PKA subunit transcripts, RNA extraction was performed according to a modification of a previously described procedure (42,16) or by use of the Qiagen Mini-RNA kit. A single pineal gland or 100,000 cells were extracted in 300 µl Qiagen lysis buffer, processed according to the supplier's protocol and finally eluted into 50 µl DEPC water. Northern blots of total RNA samples obtained from rat pineal glands six hours after light on (midday; 12:00) and six hours after light off (midnight; 24:00) were hybridized with cDNA probes against human RIα, RIβ, RIIα and RIIβ. The use of human cDNA probes for the detection of rat mRNA is possible since the interspecies homology for PKA subunit isoforms is more than 90% at the nucleotide level (26). Thus, human probes were expected to detect rat mRNA as has also been shown for bovine mRNAs (42). One to 3 µl of the above described RNA preparations were used for a single reverse-transcriptase reaction performed with 4 µl $MgCl_2$ (25 mM), 2 µl PCR buffer (10×) (Perkin-Elmer, Konstanz, Germany), 8 µl dNTP-mix (2.5 mM each) (Roth, Germany), 1 µl oligo-dT primer (50 µM), 1 µl RNAse inhibitor (optional) and 1 µl MuLV-reverse transcriptase (Perkin-Elmer, Konstanz, Germany). The reaction conditions were 25°C for 10 min, 42°C for 15 min, 99°C for 5 min and cooling down to 4°C until use or long term storage. Twenty µl of the reversely transcribed DNA were directly diluted into 80 µl PCR-mix and amplified. Primers and amplification protocols for the PKA regulatory subunits were used as described (71). The Taq-polymerase was from Bio Labs Ltd., Israel. The primers for GAPDH were 5′-TGA-TGA-CAT-CAA-GAA-GGT-GG-3′ (forward) and 5′-TTT-CTT-ACT-CCT-TGG-AGG-CC-3′ (reverse). All primers were purchased from MWG Biotech (Munich, Germany).

2.7. Determination of PKA Subunits by Immunoblotting

Cultured organs or isolated pinealocytes were stimulated with NE as indicated or left untreated as control. The medium was then removed and preparations were transferred into sample buffer. Electrophoresis and blotting was done as described (42). The membranes were incubated with polyclonal antibodies against the following subunits of PKA: 1) catalytic subunit (C; 40kDa; 1:50,000; 59), 2) regulatory subunit type Iα (RIα; 49kDa; 1:15,000; 11), 3) regulatory subunit type Iβ (RIβ; 54kDa; 1:1,000; 69), 4) regulatory subunit type IIα (RIIα; 54kDa; 1:15,000; 11), 5) regulatory subunit type IIβ (RIIβ; 52kDa; 1:15,000; 69). The membranes were subsequently incubated with a horseradish peroxidase (HRP)- conjugated secondary antibody (New England BioLabs, Beverly, MA, USA) diluted 1:100,000. The signals were obtained by chemiluminescence detection on autoradiographic film (UltraSignal reagent and Clxposure film, Pierce, Rockford, Il., USA). To confirm equal loading of the lanes, immunoblots were stained with India Ink (Pelikan, Hannover, Germany) after chemiluminescence detection. Both, autoradiograms and stained blots were analyzed densitometrically (see below).

2.8. Detection of PKA Catalytic Subunit Translocation

Pinealocytes were treated for 1 hour with NE or left untreated. Nuclear extracts were prepared as previously described (23). In brief, pineal preparations were harvested in AT buffer (60mM KCl, 15mM NaCl, 14mM β-mercaptoethanol, 2mM EDTA, 15mM HEPES pH 7.9, 0.3M sucrose, 5μg/ml aprotinin, 10μg/ml leupeptin, 2μg/ml pepstatin, 0.1mM phenylmethylsulfonyl-chloride, 1mM NaF, 1mM Na_3VO_3 and 1mM Na_3MnO_4) containing 0.1% Triton X-100 and homogenized. The homogenate was layered onto AT-buffer containing 1M sucrose and centrifuged for 5min at 8.000g in an Eppendorf centrifuge to collect the nuclear fraction in the pellet. The supernatant contains the cytosolic fraction. The protein content of each fraction was measured and adjusted to 1μg/μl with 2× electrophoresis sample buffer.

Visualization of the PKA catalytic subunit (1:1000) was done by immunofluorescence (see above) using a CY3-conjugated anti-rabbit secondary antibody (1:1000; Dianova, Hamburg, Germany). Confocal images were obtained using the software package of a LSM510 connected to an Axiovert 100 microscope (Zeiss, Jena, Germany).

2.9. Computer-Assisted Semiquantitative Analysis of PKA-Subunit Immunoblots and pCREB Immunocyto-Chemistry

Computer-based image analysis systems (VIDAS and KS 300 systems, Kontron, Eching, FRG) were used to carry out a combined plani- and densitometric semiquantitative analysis of the data of the PKA subunit immunoreaction from immunoblots and the pCREB signals from immunocytochemical preparations (72; Wicht et al., in press). The immunoblots were digitized with a scanner (constant settings for all measurements: grey value resolution 8 bits, spatial resolution 200 pixel per inch, fixed settings of the A/D-converter). In order to achieve higher numerical values for darker (more intense) signals, the images were inverted (light areas were represented by low grey-values, dark areas by high values). The immunoreactive signals were segmented from the background; their size and intensity were quantified simultaneously by summing up the grey values of all pixels belonging to an individual signal. The results are given as "SUMDENS" (sumdensity) values. A plot of the amount of antigen against the SUMDENS values shows a linear relation between the SUMDENS values and the logarithm of the antigen concentration (Wicht et al., unpublished observation). A very similar method was used to quantify the immunocytochemical results of the pCREB reaction in dispersed pinealocytes. A light microscope equipped with a constant light source and black/white CCD-camera was used to record digital images from various regions in these preparations (constant settings for all recordings: 20×-objective, grey value resolution 8 bits, spatial resolution 512 × 512 pixel/image, fixed settings of the A/D-converter of the CCD-camera). Typically, four regions per culture were analyzed and care was taken to select only regions that contained an equally and comparably dense monolayer of cells. The unstained background between the cells was used as an intensity reference and was adjusted to a grey value of 255 by slight variations of the intensity of the light source. The resulting digital images required only minimal image processing and allowed the discrimination and segmentation of the immunoreactive nuclei using the same discrimination value for all measurements. After the segmentation, the sum of all (inverted, see above) grey values of the pixels belonging to the immunoreactive nuclei was determined; the results are also given as "SUMDENS" values. The SUMDENS values need to be corrected for the total area covered by cells. The resulting parameter is called SUMDENSCORR.

2.10. Radioimmunoassay

Medium was collected from each well of the in vitro preparations before termination of the experiment. Concentration of melatonin in the cell culture medium was measured by means of a commercial radioimmunoassay using ^{125}I-melatonin (ELIAS, Osceola, Wisconsin) as the tracer. The detection limit for melatonin in this assay is 1.5 pg/ml (41,44).

2.11. Transfection of Pinealocytes with ICER Constructs

A partial cDNA of the ICER 5′-end (0.2 kb) was generated by RT-PCR from RNA isolated from rat pineal glands as described earlier (65). The two primers used (forward: 5′-TTT-TGG-ACT-GTG-GTA-CGG-CC-3′; reverse: 5′-GGG GAC TGT GCA GGC TTC CT-3′) encompass the start site of translation of the ICER gene (63,45) and the ICER coding sequence upstream the first DNA binding domain. PCR fragments were size-selected, electroeluted, subcloned into the vector pBS SK$^-$ and sequenced before further use. Prior to transfection of pinealocytes, the ICER cDNA fragment was subcloned into the eukaryotic expression vector pRc/CMV (Invitrogen) in either antisense (pICERas) or sense (pICERs) orientation. As an additional control, cells were transfected with the vector alone (pRC/CMV). To monitor transfection efficiency within a given experiment, cells were cotransfected with 10 μg of the vector pCH110 (Pharmacia) that carries the β-galactosidase gene. Primary pinealocyte cultures were prepared, maintained as described above and transfected by the calcium phosphate co-precipitation technique (51). The culture medium was supplemented with 20 μg of pRC/CMV-DNA, pICERs or pICERas, respectively, and 10 μg pCH110 for 5 h. Subsequently, cultures were washed and after a recovery period of 16 h stimulated with 10^{-6} M NE, 10^{-6} M ISO and 10^{-7} M PE for 5 h, and melatonin was measured in the medium.

2.12. Statistical Analysis

All figures show the means (± SEM) of at least three experiments representing one of 2 to 5 replications. Treatment effects were statistically assessed using either Student's t-test or one-way ANOVA with subsequent Bonferroni tests for multiple comparisons using the program PRISM II™ (GraphPad Inc., CA, USA) for Windows 3.1 (Microsoft, Seattle, USA) with $p \leq 0.05$ as the criterion of significance.

3. RESULTS AND DISCUSSION

3.1. Adrenergic Receptors in the Membrane of the Rodent Pinealocyte

In rodents, norepinephrine (NE) is released nocturnally from the intrapineal nerve endings of sympathetic neurons originating from the superior cervical ganglion (21) and binds to α- and β-adrenergic receptors on the pinealocyte membrane (32). The α- and β-adrenergic receptors are the predominant neurotransmitter receptors in the rodent pineal organ. Of all rat tissues tested the pineal organ exerts the highest expression of β_1-adrenergic receptor mRNA (40). Beside these adrenergic receptors, which are in the focus of this contribution, many other, e.g. peptidergic-, monoaminergic-, and cholinergic receptors are present in the rodent pineal gland as well (61,39). However, the β_1-adrenergic receptor appears to be most important for AANAT mRNA upregulation and stimulation of melatonin biosynthesis (32,61,35).

The transcription of the β_1-adrenergic receptor (β_1-ADR) is developmentally regulated, fluctuates diurnally and is gated by a cAMP-dependent mechanism in rodent pinealocytes (see 14,46,51). β_1-ADR mRNA levels are low during the light phase and start to increase immediately after the onset of darkness (Figure 1). The maximum level is reached in the middle of the dark period and falls thereafter. This rhythm

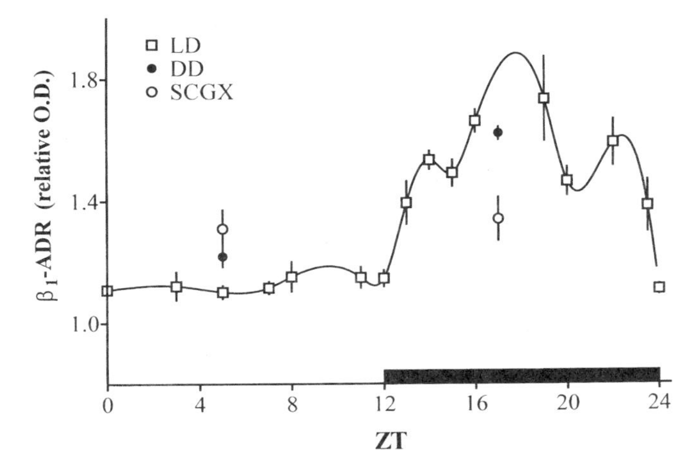

Figure 1. β_1-ADR mRNA is regulated in a diurnal and circadian manner. The β_1-ADR mRNA level (in O.D.) is low during the light phase and starts to increase immediately after onset of darkness (open squares). The maximum is reached in the middle of the dark period and falls thereafter. This rhythm persists in constant darkness (black circles), but is ablated in ganglionectomized animals (open circles).

persists in constant darkness, is ablated in ganglionectomized animals and is therefore truly circadian (51; Figure 1). Interestingly the expression of the α_{1B}-adrenergic receptor is regulated in a very similar way (19) and both the α- and the β-adrenergic receptors can be modulated posttranscriptionally (17). β_1-ADR transcription is not directly correlated with receptor abundancy, since the rhythmic fluctuation in ligand binding sites in the rat pineal gland reaches elevated levels at the earliest 7h after the nocturnal peak in β_1-ADR mRNA levels (52,51). From this time lag, it can be reasoned that the receptor protein rests in a sequestered state and that its incorporation into the pinealocyte membranes is suspended until the lack of ligand during daytime triggers this event. For the β_2-ADR it has been demonstrated that continuous exposure to agonists leads to a homologous receptor desensitization within minutes. This event is followed by a receptor down regulation which is maintained for hours on a steady-state receptor level by a cyclic AMP-dependent lowering of receptor transcription (17). However, these classical ligand-dependent regulations, common for many other G-protein coupled receptors, apply for pineal β_1-ADR fluctuations only to a limited extent: rhythms in β_1-ADR mRNA and protein are notably out of frame and driven either by the presence (β_1-ADR mRNA) or the absence (β-ADR protein) of the ligand (51). The findings suggest that the rhythm in biochemical activities of the rat pineal gland, which is induced by the nocturnally elevated release of NE, seems to be stabilized in two ways already at the β_1-ADR level: NE-induced β_1-ADR transcription provides the basis to react to daytime ligand depriviation with an upregulation in receptor abundancy. At the same time NE rapidly sequesters ligand binding sites in the rat pineal gland as a protective mechanism, diminishing cAMP-mediated physiological responses despite the presence of a constant NE stimulus. The constantly elevated level in β_1-ADR mRNA in the pineal gland of SCGX animals can be explained by the the absence of a protective function provided by the sympathetic nerve endings against circulating catecholamines (49). Similarly, the amount of β_1-ADR mRNA in the rat pineal gland is elevated until postnatal day 8 and is not yet regulated. It is only when the sympathetic innervation of the pineal gland gains function at the beginning of the second postnatal week (66,33) that β_1-ADR mRNA levels decrease during daytime and become rhythmic. Interestingly, this developmental pattern in β_1-ADR mRNA closely matches the dynamics in adrenergic binding sites (4) and is similar to the ontogeny of pineal NAT activity rhythm (33,22). Thus, the maturation in NE/β_1-ADR interaction (4) coincides with a complete maturation of the cyclic AMP signaling pathway (22,33,61,64).

3.2. Second Messengers I: Inositol-1,4,5-Trisphosphate (IP_3), Diacylglycerol and Calcium

The pineal α-adrenergic receptors appear to be coupled to their effectors by G-proteins, which, however, have not yet been identified. Possible candidates are the G_q or $G_{12/13}$-proteins. Via α-adrenergic receptors, NE induced phosphatidyl-inositol-bisphosphate (PIP_2) phosphodiesterase (also known as phospholipase C; PLC) (30) and phospholipase A2 (31). As a final consequence calcium is released from intracellular stores and activates, together with diacylglycerol (DAG), protein kinase C (PKC) and probably also calcium-calmodulin-dependent protein kinase (CaMK) and mitogen-activated protein kinase-kinase (MEK1). NE evoked a dose-dependent, biphasic rise in the intracellular concentration of free calcium ions ($[Ca^{2+}]_i$) in more than 95% of the

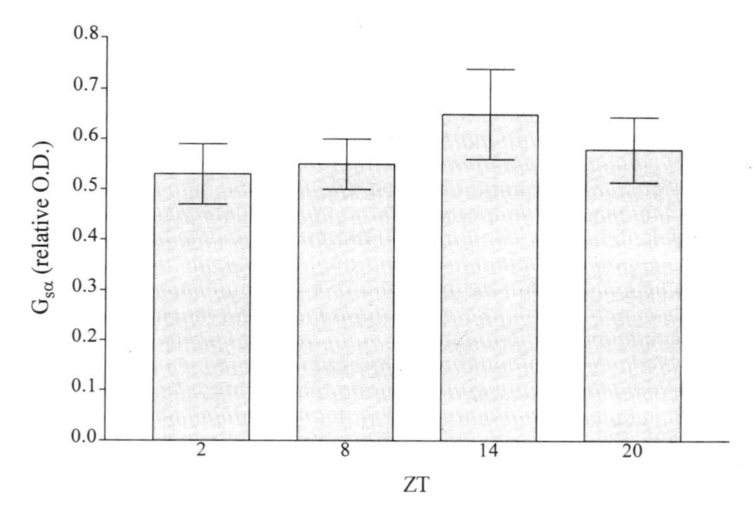

Figure 2. The $G_{s\alpha}$ mRNA level does not differ between day and night. In situ hybridization was performed with a $G_{s\alpha}$—specific probe and pineal organs taken at Zeitgeber time (ZT) 2, 8, 14 and 20. The autoradiographic film was analyzed densitometrically, the abscissa shows the mean OD-values ± SEM (N = 3).

S-antigen-positive pinealocytes (57,38). The basal $[Ca^{2+}]_i$ and the responses of the cells were rather uniform and showed a maximum at a concentration of 1 μM. The primary phase of the NE-response is characterized by a rapid tenfold increase in $[Ca^{2+}]_i$, which decreases rapidly to a secondary phase that persists as long as the cells are exposed to NE. When NE is removed the $[Ca^{2+}]_i$ rapidly returns to basal values and the pinealocytes regain NE-responsiveness immediately thereafter. This response is mediated by α-adrenergic receptors, as it is mimicked by phenylephrine, a pure α-adrenergic agonist, is blocked by the α-adrenergic antagonist prazosin, and cannot be elicited by the β-adrenergic agonist isoproterenol (38; 1998). The primary phase involves release of calcium from intracellular calcium stores, whereas the secondary phase depends on calcium influx. The mechanism of the calcium influx has not been elucidated yet, but it is not depending on L-type-voltage-gated calcium channels (57,38; 1998). All the characteristics of the NE-induced $[Ca^{2+}]_i$ elevation described above were observed in both rat and mouse pinealocytes (Figure 2).

3.3. Second Messengers II: Cyclic AMP (cAMP)

Rat pinealocytes were shown to contain the alpha-subunits of the G-proteins G_s, G_o and G_i (5,6). Their amount did not vary between day and night, but was upregulated under stimulus-deprivation, suggesting that these G-proteins may be involved in denervation supersensitivity (5,6). Also the mRNA levels of $G_{s\alpha}$ in the rat pineal gland remain constant at day and night (39; Figure 3). $G_{s\alpha}$ links the activation of the β-adrenergic receptors to the activation of adenylate cyclase which causes a drastic increase in intracellular cAMP-concentrations mirrored by a solid increase in extracellular cAMP levels (Co: mean 73 ± 6 fMol/ml; NE for 1 h: mean 1429 ± 226 fMol/ml; SEM; N = 4). Information about the types of adenylate cyclase (AC) expressed in the rodent pineal gland is sparse (70).

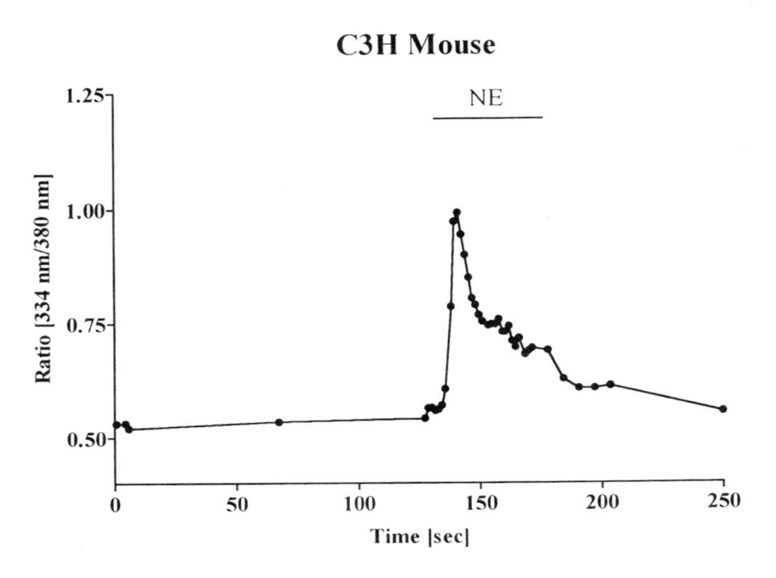

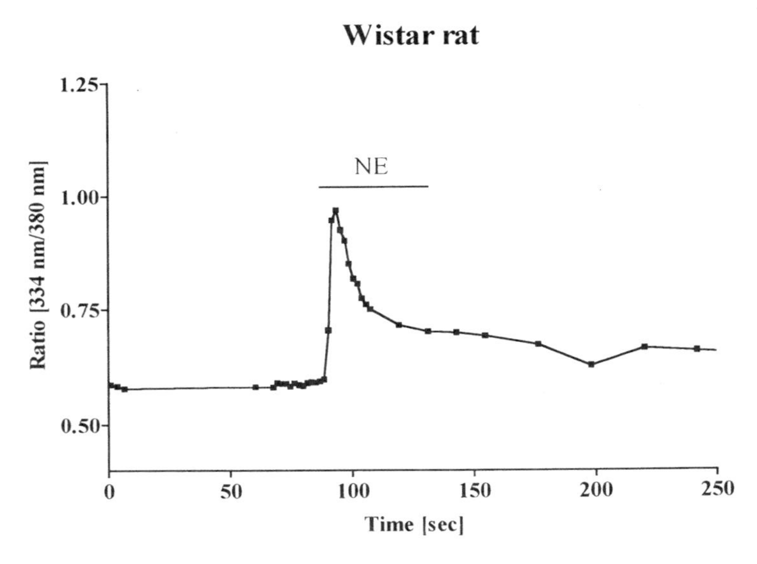

Figure 3. Norepinephrine (1 μM) causes elevation of the intracellular concentration of free calcium ions in both rat and mice pinealocytes. The response is characterized by a rapid spike followed by a sustained plateau. The abcissa shows the ratio of the intensities of the emitted light at 515 nm upon excitation at 334 nm and 380 nm in single FURA-2 loaded pinealocytes.

3.4. NE-Induced Phosphorylation of CREB

Phosphorylated CREB is known to stimulate transcription of genes with cyclic AMP responsive elements (CREs) in their promoter regions (29,47). Several studies have pointed out that NE-dependent activation of the cAMP pathway causes the phosphorylation of the activating transcription factor CREB which appears as an important link between activation of second messenger systems and stimulation of pineal gene transcription (3,53,67,58,38,43,39). Immunocytochemical demonstration of pCREB in isolated rat pinealocytes revealed that stimulation with NE or the β-adrenergic agonist isoproterenol results in a similar, time-dependent induction of pCREB immunoreactivity exclusively located in cell nuclei. The response is observed in more than 95% of the pinealocytes identified by the S-antigen immunoreaction. It reaches its maximum after 30 to 60 min and persists for up to 300 min, provided the cells are constantly exposed to NE (67). Upon removal of NE, the pCREB immunoreaction rapidly declines to background levels within 60 min. Stimulation of α_1-adrenergic receptors does not cause pCREB induction. These immunocytochemical results obtained with isolated rat pinealocytes are in full agreement with the gel mobility-shift analyses of homogenates from whole pineal glands (53) and suggest that the NE-induced phosphorylation of CREB in the pineal is pivotally linked to the β-adrenergic cascade. The observation that NE-induced phosphorylation of CREB occurs in virtually all pinealocytes underlines the fundamental role of this transcription factor for the regulation of pineal cAMP-inducible genes.

3.5. Identification of the Protein Kinases Involved in CREB Phosphorylation

An important question is which of the potentially NE-inducible protein kinase(s) mediate(s) CREB phosphorylation in rat pinealocytes. To investigate this problem we analyzed the NE-induced CREB-phosphorylation after inhibition of several protein kinases. We used KN-62 for the inhibition of calcium-calmodulin-dependent protein kinase (CaMK; 60), chelerythrin for protein kinase C (PKC; 73), PD98059 for the inhibition of mitogen-activated-protein-kinase-kinase type 1 (MEK1; 56) and the antagonists Rp-8-CPT-cAMPS (28,42) and H-89 (15) for the inhibition of the cyclic AMP-dependent protein kinase (PKA; 29). Immunocytochemistry and densitometric analyses of unstimulated controls and preparations treated with NE alone or NE plus the protein kinase antagonist revealed that only the PKA antagonist Rp-8-CPT-cAMPS reduced the NE-induced CREB phosphorylation in a dose-dependent manner (Figure 4). The PKA isozyme responsible for CREB phosphorylation in the rat pineal organ was analyzed using different pairs of cAMP analogs (26) that complement each other for the activation of either PKA type I (8-AHA-cAMP and 8-PIP-cAMP) (9,42,48) or type II (Sp-cDBIMPS and N6-PHE-cAMP) (10,26,42). Only cAMP analog combinations which selectively activate PKA type II elevated pCREB immunoreactivity in rat pinealocytes, whereas PKA type I-directed analog pairs were ineffective. This implies that PKA type II mediates the phosphorylation of CREB in the rat pineal organ (Maronde et al., unpublished observations).

Northern blot and reverse transcriptase polymerase chain reaction (RT-PCR) were employed to assess for the possible transcriptional regulation of PKA subunits in the rat pineal organ. The Northern blots showed a 3.2 kb signal for RIα with two additional signals at 3.0 kb and 1.7 kb as well as a 3.2 kb signal for RIIβ (Figure 5A/B; Maronde et al., unpublished results). These signals were similar to those described in the bovine pineal gland (42). Transcript levels of both RIα and RIIβ did not vary between midday (12:00) and midnight (24:00). RIβ and RIIα transcripts could not be detected by Northern blotting. The expression of RIIα, however, was revealed by RT-PCR as was the expression of RIα and RIIβ. In order to elucidate whether NE affects

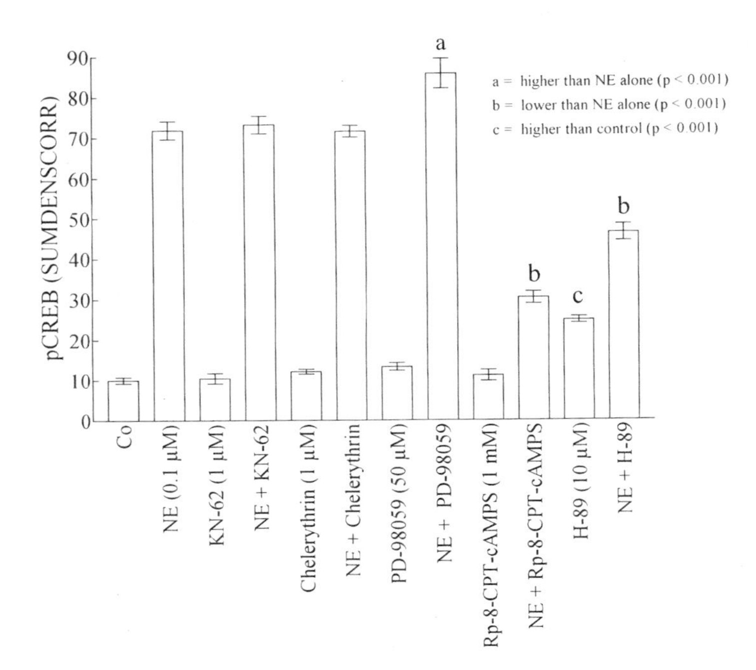

Figure 4. Semi-quantification of the amount of nuclear pCREB in rat pinealocytes treated with NE (0.1 μM) and different protein kinase inhibitors for 30 minutes. Among the different protein kinase inhibitors tested (CaMK, KN-62, 1 μM; PKC, chelerythrin 1 μM; MEK1, PD98059, 50 μM; PKA, H-89, 10 μM and Rp-8-CPT-cAMPS, 1 mM) only the PKA-antagonists inhibit CREB phosphorylation. Shown are the mean SUMDENSCORR-values (see Exp. Procedures) ± SEM of 4 replicates (Co = control).

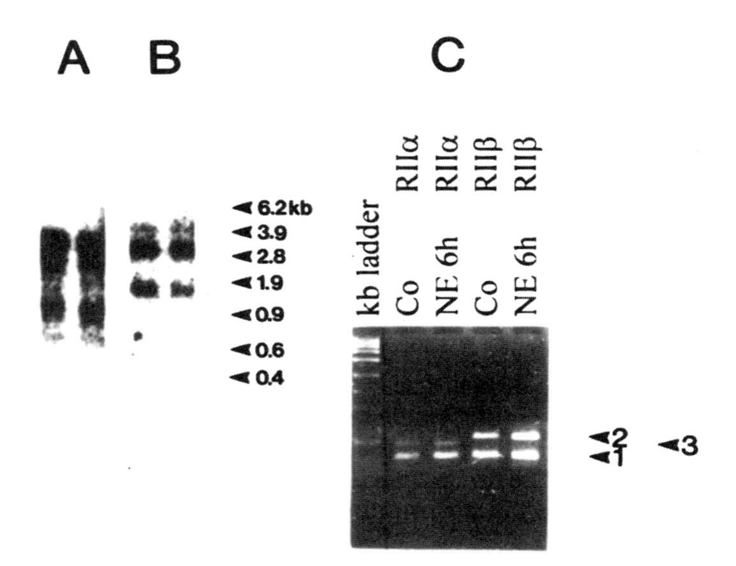

Figure 5. Expression of the cyclic AMP-dependent protein kinase regulatory subunits RIα, RIIα and RIIβ in rat pinealocytes. Rats were kept under 12L:12D. **A** and **B** show Northern blots of pineal RNA extracts obtained six hours after lights on (left lanes, respectively) and six hours after lights off (right lanes, respectively) probed with a RIα- (A) or a RIIβ- (B) specific probe. **C** Cultured rat pineal organs were left untreated (Co) or treated for six hours with NE (NE 6h). RT-PCR-reaction products obtained from pineal RNA extracts were amplified with either RIIα- (arrowhead 3) or RIIβ- (arrowhead 2)-specific primers and a GAPDH (arrowhead 1)-specific primer as a loading control.

PKA subunit transcription, RIα-, RIIα-, and RIIβ-mRNA levels were investigated in pinealocytes stimulated with NE for 6 hours. Transcript levels of RIα and RIIα did not change, but a slight upregulation of the RIIβ transcript was detected after 6h of NE treatment (Figure 5C) conforming to observations in other/ systems (68).

By use of specific antibodies protein levels of PKA subunits were analyzed in rat pineal glands or pinealocytes kept *in vitro* (43) and in pineal glands directly dissected out of the animal (*ex vivo*; Figure 6). All types of preparations revealed the presence of a signal at 40kDa corresponding to the catalytic subunit (C), detected a major 49kDa signal corresponding to the regulatory subunit RIα, a double 53/55kDa signal corresponding to the regulatory subunit RIIα, and a single 52kDa signal corresponding to the regulatory subunit RIIβ. No specific signal was found with the antiserum against the regulatory subunit RIβ (data not shown). None of the detectable subunits showed significant variation in the course of an LD cycle of 12:12 (Figure 6). Similar results were obtained with the pineal organ of C3H mice (von Gall et al., unpublished observations).

To examine possible effects of NE on the various PKA subunit proteins immunoblots were prepared from isolated rat pinealocytes that were either left untreated or stimulated for 30min or 6 hours with 1μM NE (43). The 40kDa signal of the catalytic subunit (C) and the 52kDa signal of RIIβ slightly decreased upon NE treatment after 30min and 6 hours. The RIα signal was reduced after 30min of NE-treatment, but reached control levels after 6 hours. These small changes may represent degradation of the subunits occurring after dissociation, since both regulatory and catalytic subunits are more sensitive to proteolysis in the monomeric state (26). The 55/53kDa double band of RIIα increased slightly after 30min and 6h as compared to controls. In corresponding immunoblots pCREB was undetectable in the controls, strongly induced after 30min and still elevated after 6h, whereas total CREB did not vary considerably (43). This comparison indicates that CREB phosphorylation in the rodent pineal organ is unrelated to the protein levels of PKA subunits.

Recent findings with rat cerebellum and cortex (71) as well as with FRTL-5 thyroid cell lines (23) suggested that PKA type I and soluble PKA type II are mainly involved in phosphorylation of cytosolic proteins, whereas particulate (anchored) PKA type II mediates phosphorylation of nuclear proteins like CREB. In the rat pineal gland RIα was found mostly in the soluble fraction whereas both RIIα and RIIβ were preferentially found in the particulate fraction, suggesting that the anchored form of PKA type II (which mediates phosphorylation of nuclear proteins) predominates in the rat pineal organ (Maronde et al., unpublished results). The functional role of PKA I in the pineal gland is not clear. It might keep cytosolic target proteins, like AANAT, phosphorylated (18,35) or mediate the cAMP-induced inhibition of proteasomal degradation of AANAT (27).

3.6. Nuclear Translocation of PKA Catalytic Subunit

Immunocytochemical investigations showed that in untreated pinealocytes the regulatory and catalytic subunits are exclusively located in the cytosol. Under NE-treatment a certain amount of the catalytic subunits entered the nucleus (Figure 7a) whereas the regulatory subunits remained in the cytosol (data not shown). NE-induced nuclear translocation of the PKA catalytic subunit could also be shown by immunoblotting of nuclear extracts (Figure 7b). The nuclear entry of the PKA catalytic subunit is regulated by a β-adrenergic cAMP pathway, since NE, as well as isoproterenol and

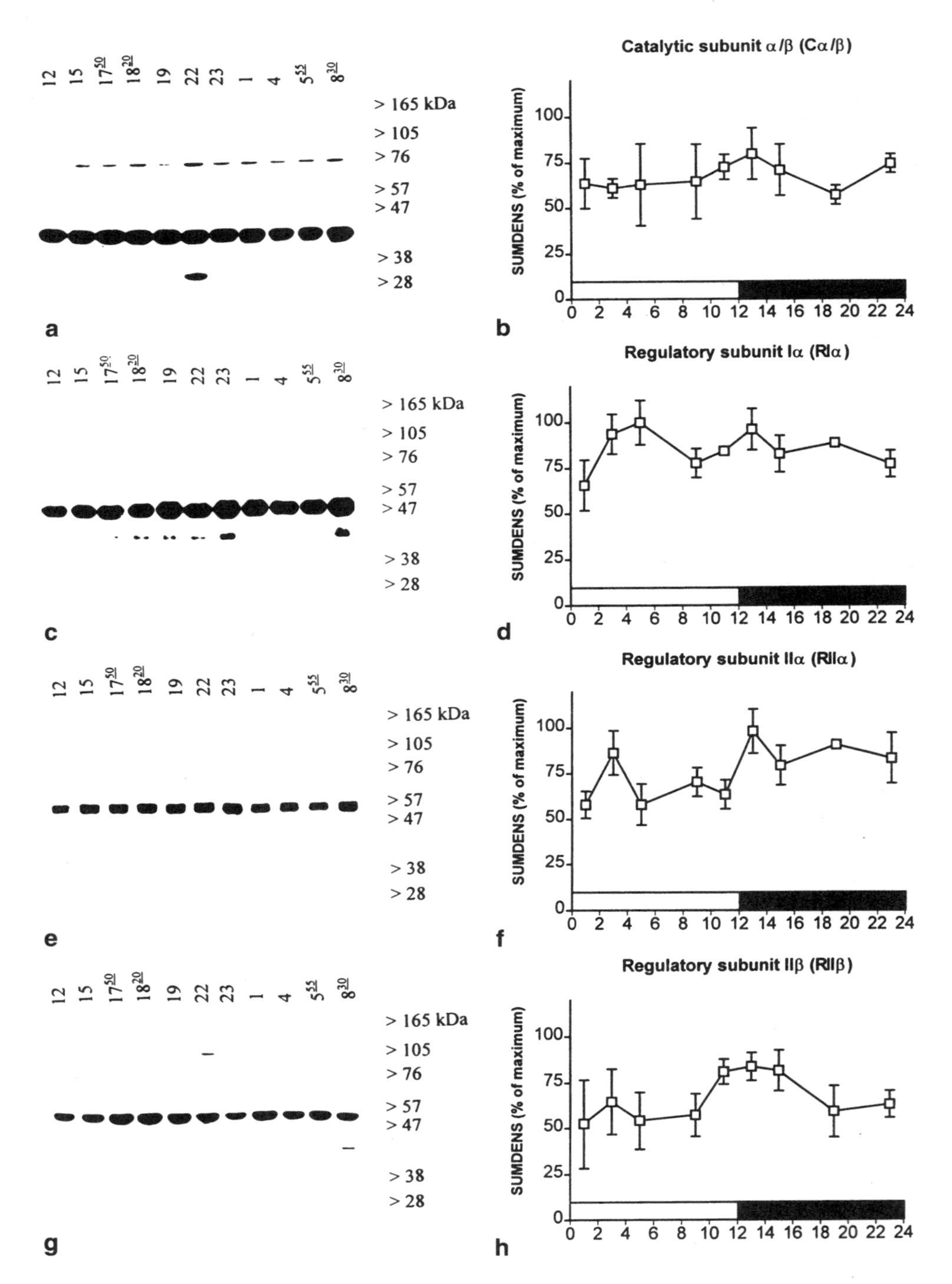

Figure 6. Immunochemical detection of the cyclic AMP-dependent protein kinase subunits C (a/b), RIα (c/d), RIIα (e/f) and RIIβ (g/h) in rat pineal glands obtained ex vivo at different times under 12 h L:12 h D. **a**, **c**, **e** and **g** show representative original immunoblots with the timepoints indicated on top of each blot and the size markers on the right side. **b**, **d**, **f** and **h** show semi-quantifications of a total of 3 to 4 such experiments. The open bars under each graph represents the light phase, the black bars the dark phase. The error bars represent means ± SEM.

A

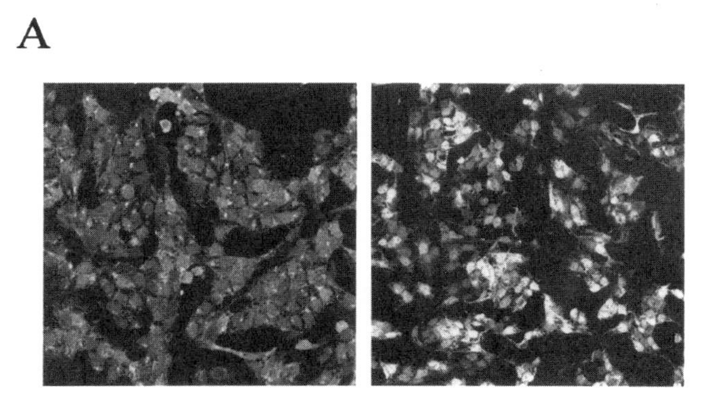

B

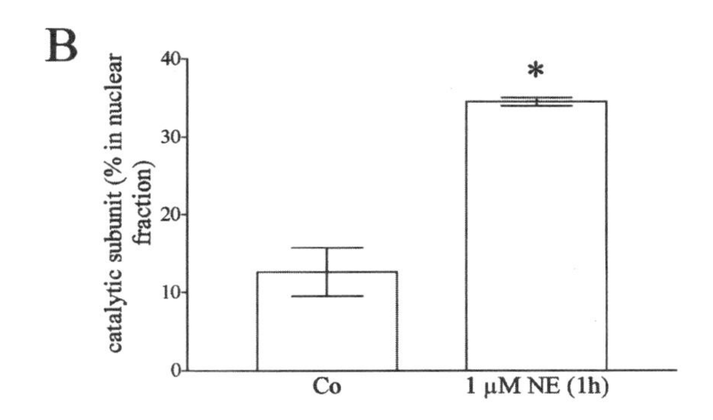

Figure 7. Translocation of the catalytic subunit of PKA (cPKA) into the nucleus after norepinephrine-stimulation for one hour. **A** The image on the left side shows cPKA-like-immunoreactivity in untreated rat pinealocytes, the image on the right side shows cPKA-like-immunoreactivity in rat pinealocytes treated with NE for one hour. The treated pinealocytes contain more immunoreactive material in their nuclei than the untreated controls. **B** Percentage of nuclear immunoreactivity for cPKA in extracts of pineal organs treated with NE for one hour or left untreated (Co). Shown are means ± SEM (t-test, $p \leq 0.05$; N = 4).

cAMP analogs induced nuclear translocation of cPKA to a very similar extent (Maronde et al., unpublished observations). These results suggest that nuclear translocation of catalytic subunits is an essential step for the induction of CREB phosphorylation in rat pinealocytes.

3.7. CREs as Targets of the Cyclic AMP Pathway

The phosphorylation of CREB at serine 133 leads to homodimerization and binding of the pCREB-homodimers to cyclic AMP-regulated elements (CRE; 29), present in the promoter of many cAMP-regulated genes. Binding of the pCREB homodimer to these CREs leads to an enhanced gene transcription. CREs are found in the promotor of several genes encoding important pineal proteins like the β_1-adrenergic receptor (17), inducible cyclic AMP early repressor (ICER) (63), FRA-2 (7) and AANAT (8). The CRE in the promoter region of the AANAT gene has been shown to belong to the class of non-classical non-palindromic elements consisting of a CRE halfsite and a CCAAT enhancer element (8). The promoter of the inhibitory transcription factor ICER contains four CREs (45). As shown for the rat ICER is raised from an alternative intronic promoter of the cyclic AMP-regulated element modulator (CREM) gene (63,61) and its expression is regulated by the NE/β_1-ADR/cAMP-signaling pathway. ICER is able to bind to CREs and leads to repressed transcription. One of the most rapidly and strongly repressed gene is probably ICER itself, since the ICER promoter region consists of four CRE elements (45). ICER also binds to CRE elements in the promoter of AANAT (24; Maronde et al., unpublished observations) and the β_1-adrenergic receptor gene (51).

3.8. Possible Roles For pCREB and ICER in Regulating AANAT Transcription

In the rat, the time course of NAT transcription closely matches with the time course of NAT activity (12,54), suggesting that, in the rat, the regulation of NAT activity comprises an important transcriptional component. Several lines of evidence point to a substantial role of pCREB in the transcriptional activation of AANAT: 1) the AANAT promotor region includes two CREs, 2) the kinetics of CREB phosphorylation falls into the temporal gap between NE-induced elevation of cAMP and AANAT transcription, 3) activators of PKA induce CREB phosphorylation and AANAT transcription, 4) inhibitors of PKA lead to a reduction in NE-induced CREB phosphorylation, binding of nuclear extracts to CREs, AANAT activity and melatonin biosynthesis with very similar kinetics (Figures 8 and 5) in vivo pCREB levels rise always before AANAT transcript rises. All findings suggest that pCREB is involved in the upregulation of AANAT transcription. This hypothesis could be substantiated in transfection experiments with newly developed CREB antagonists (A-CREB; 1). Such

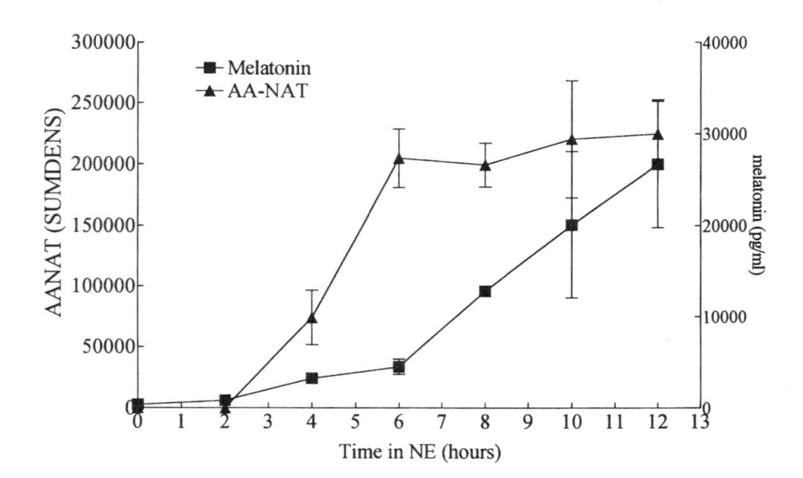

Figure 8. Timecourse of the amount of melatonin (squares) in the medium and the appearance of AANAT protein (triangles) in dispersed rat pinealocytes treated with NE (1 μM) for two to twelve hours. The ordinate shows time in hours, the left abscissa AANAT protein in SUMDENS-values and the right abscissa melatonin in pg/ml. The error bars represent mean ± SEM of 3–4 replicates.

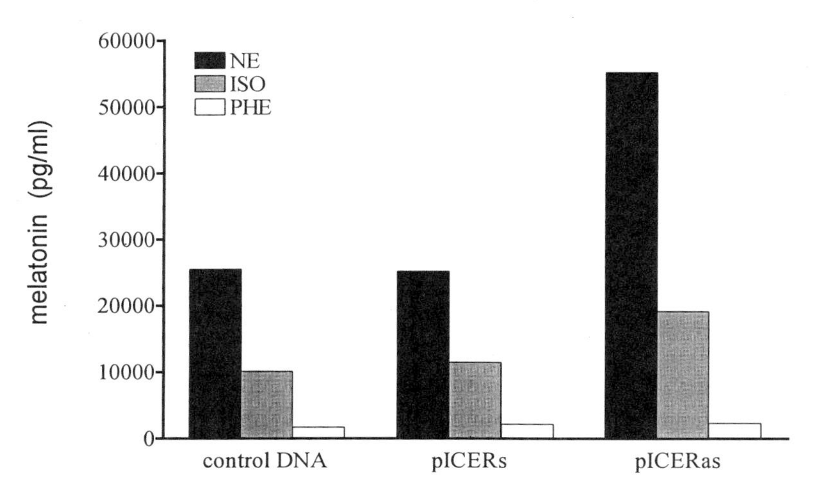

Figure 9. Rat pinealocytes transfected with either control DNA (empty vector) or an ICER-sense construct (pICERs) showed stimulation of melatonin biosynthesis under NE- (black columns) and ISO- (gray columns), but not under phenylephrine (PHE)- stimulation (white columns). Cells transfected with an ICER antisense construct (pICERas) showed disinhibition of melatonin biosynthesis upon NE- and ISO-treatment. NE, ISO and PHE were applied for six hours. Bars show a representative experiment (of three replicates).

experiments are currently underway in our laboratory (Pfeffer et al., unpublished results). Also, several lines of evidence point to a role of ICER in the transcriptional inhibition of AANAT: 1) the AANAT promotor includes two CREs, to one of which ICER binds better than to a perfect CRE (Maronde et al., unpublished observations), 2) the kinetics of ICER protein induction correlates temporally with the decline in AANAT transcription, 3) activators of PKA cause ICER induction and 4) ICER protein plateaus in vivo when AANAT transcript declines. A role of ICER in the regulation of melatonin biosynthesis also gains support from ICER antisense experiments. Rat pinealocytes transfected with either control DNA or an ICER-sense construct showed normal stimulation of melatonin biosynthesis upon stimulation with NE and ISO. However, those cells transfected with an ICER antisense construct (pICERas) inhibiting ICER upregulation showed an elevated ("disinhibited") melatonin biosynthesis upon NE-and ISO-treatment (Figure 9). It can be concluded from these data that ICER is involved in repressing AANAT transcription. Like pCREB (see above) also ICER fluctuates diurnally in both rats (Maronde et al., unpublished observations) and C3H mice (Von Gall et al., unpublished observations). Thus, the cAMP-induced and CRE-dependent transcriptional activity in the rodent pineal organ may be regulated by a changing ratio in the amount of activating (e.g. pCREB) and inhibiting (ICER) transcription factors (see also Figure 10).

3.9. Possible Role of Post-Transcriptional Regulation of AANAT and Melatonin Biosynthesis

Beside the transcriptional up- and downregulation of AANAT, posttranslational processes are also involved in the regulation of rat AANAT activity (27). In the latter investigation the authors showed that AANAT protein is protected by a

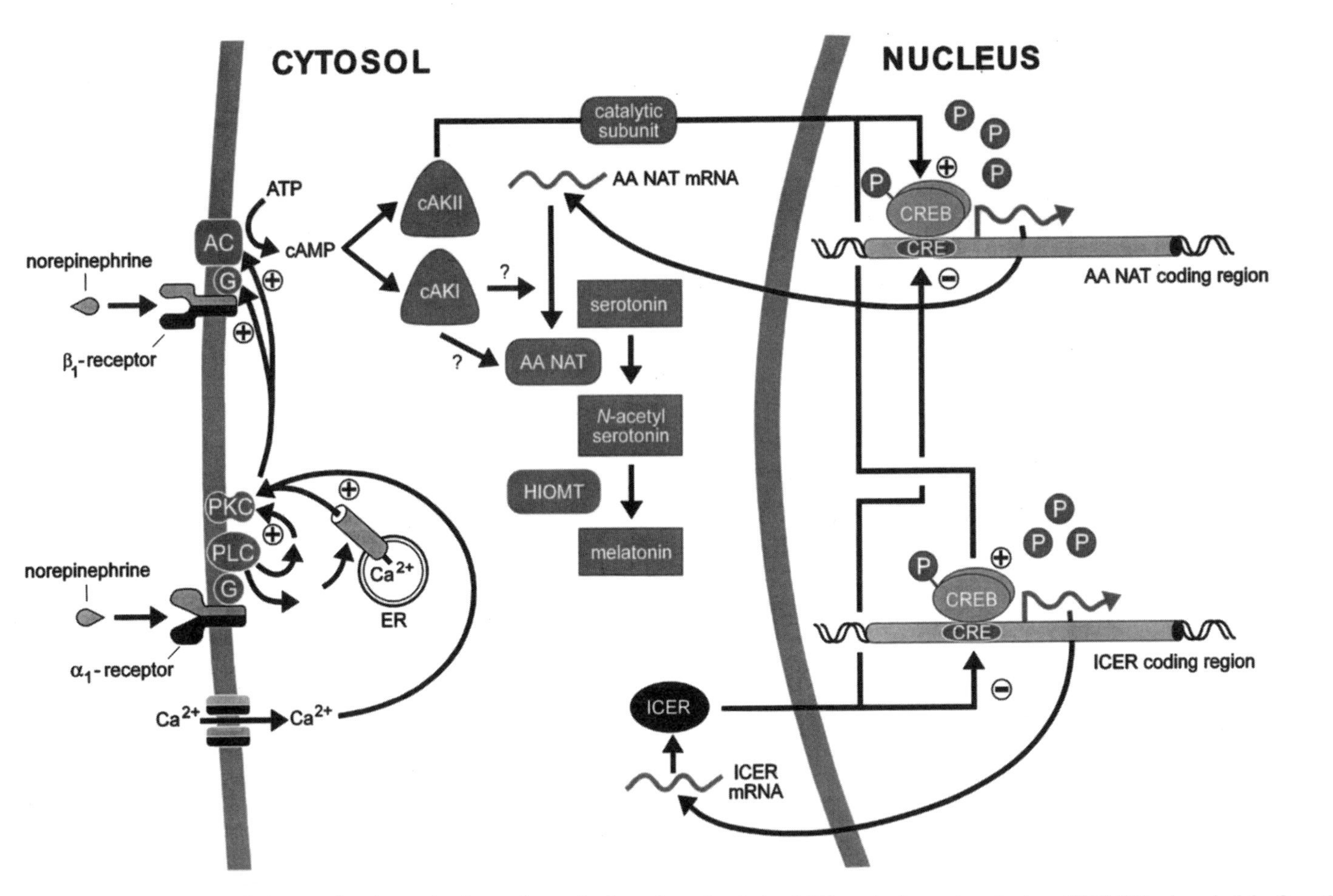

Figure 10. Scheme of some currently known signal transduction pathways in the rodent pineal gland. ER, endoplasmatic reticulum; HIOMT, hydroxyindole-*O*-methyl-transferase. The other abbreviations are explained in the text. Modified after 61,39).

cAMP-mediated inhibition of proteasomal proteolysis and that intact protein is indispensable for activity. Consequently the amount of AANAT protein determines the level of AANAT activity. In dispersed rat pinealocytes NE-treatment leads to a time-dependent upregulation of AANAT protein and melatonin biosynthesis (Figure 8). AANAT protein was not detectable in untreated cells and those treated for 2h with NE, but rose linearly between 2 and 6 hours after stimulation and reached a plateau which lasted for at least another six hours. The amount of melatonin in the medium showed a different kinetic; it rose slowly but steadily during the first six hours after stimulation and then showed a steeper and linear increase during the following six hours (Figure 8). Interestingly, we found that in C3H mice the melatonin increase in the medium after NE-stimulation is even more delayed than in rats (Von Gall et al., unpublished observations; see above).

4. COMPARATIVE ASPECTS OF SIGNAL TRANSDUCTION CASCADES REGULATING MELATONIN BIOSYNTHESIS

As outlined above the transcriptional regulation of the AANAT plays an important role in up- and downregulation of melatonin synthesis in the rodent pineal organ. This scenario, however, does not apply for all vertebrates, not even for all mammals. In the sheep, posttranscriptional mechanisms are more relevant for control the rhythmic melatonin biosynthesis, as the mRNA for AANAT fluctuates only marginally (18). Accordingly, plasma melatonin concentrations rise after the onset of darkness much faster in sheep than in rodents (2).

Chicken, trout and most other non-mammalian vertebrate species possess photoreceptive pineal organs. Thus light rather than adrenergic-stimuli appear as primary regulators of the melatonin synthesis in these species. In the chicken pineal organ AANAT mRNA fluctuates, but with a much smaller amplitude than in the rat (36). Thus, chicken AANAT seems to be regulated at both, the transcriptional and the post-transcriptional levels. Accordingly, ICER has been reported to fluctuate diurnally in the chicken pineal organ (25). In the trout pineal organ where melatonin biosynthesis is regulated by light/dark stimuli directly perceived in the organ and is not affected by NE, AANAT mRNA does not fluctuate (10). Darkness, the natural stimulus of the trout pineal melatonin biosynthesis, has virtually no effect on CREB phosphorylation (Kroeber et al., unpublished results). Furthermore cAMP and pCREB showed virtually no variation over a 12h light 12h dark cycle, whereas melatonin displayed its characteristic rise in darkness which depends on calcium influx. The data show that in the trout pineal organ phosphorylation of CREB is not essential for the induction of the melatonin biosynthesis and imply that calcium is more essential for stimulation of melatonin biosynthesis than cAMP. These comparative considerations suggest that pCREB and ICER levels are tightly regulated and play an important role in those species in which the regulation of the melatonin biosynthesis bears a substantial transcriptional component.

5. CONCLUSION AND PERSPECTIVE

In the rodent pineal organ stimulation of melatonin biosynthesis is tightly linked to activation of β-adrenergic receptors by norepinephrine which elevates intracellular cAMP levels, activates PKA, causes phosphorylation of the transcription factor CREB, and stimulates ICER expression (see above). Our findings that NE-induced melatonin biosynthesis is inhibited by PKA-antagonists, but not by CaMK-, PKC-, or MEK1-antagonists suggest an important role of the β-adrenergic-/cAMP-/PKA-/pCREB-/ICER-cascade for both transcriptional regulation of AANAT and melatonin biosynthesis in the rodent pineal organ. Our in vivo investigations with the rat (Maronde et al. unpublished observations) and the C3H mouse (Von Gall et al., unpublished observations) revealed conspicuous diurnal changes in the amount of pCREB and ICER protein related to AANAT expression in both rodent species. Thus, our in vitro observations apparently mirror physiologically relevant changes occuring under in vivo conditions.

Stimulus-coupled activation of transcription factors in the cAMP-signaling pathway is a widespread phenomenon in nervous tissue and can alter cellular function in a time frame ranging from seconds to lifespan. In particular, the transcription factors CREB and ICER have been shown to operate selectively in time and space and often in an antagonistic way on cAMP-linked cellular phenomena (55).

The rodent pineal organ will continue to serve as an excellent model system to further elucidate transcription factor-directed up- and downregulation of cAMP-inducible gene expression and to clarify the significance of the interaction between transcriptional and posttranscriptional mechanisms for the control of neuroendocrine and neuronal activity.

ACKNOWLEDGMENTS

We would like to thank Dr. D.C. Klein, NIH, Bethesda, USA for the AANAT-antibody, Prof. Dr. G. Schwoch, Göttingen, Germany for the PKA catalytic subunit antibody, Dr. K. Taskén, Oslo, Norway for regulatory subunit antisera and Dr. C.A. Molina, Newark, NJ, USA for the ICER-CREM antibody. These studies were supported by grants from the Deutsche Forschunggemeinschaft.

REFERENCES

1. Ahn, S., Olive, M., Aggarwal, S., Krylov, D., Ginty, D.D., and Vinson, C. A dominant-negative inhibitor of CREB reveals that it is a general mediator of stimulus-dependent transcription of c-fos. Mol Cell Biol 18:967–977, 1998.
2. Arendt, J. In: Melatonin and the Mammalian Pineal Gland, p. 201. London: Chapman and Hall, 1995.
3. Armstrong, R., Wen, W., Meinkoth, J., Taylor, S., and Montminy, M. A refractory phase in cyclic AMP-responsive transcription requires down regulation of protein kinase A. Mol Cell Biol 15:1826–1832, 1995.
4. Auerbach, D.A. β-Adrenergic receptors during development. In *Melatonin Rhythm Generating System*, (D.C. Klein ed.) Karger, Basel, pp. 97–107, 1981.
5. Babila, T. and Klein, D.C. Stimulus deprivation increases pineal $G_{s\alpha}$ and G_{β}. J Neurochem 59:1356–1362, 1992.
6. Babila, T. and Klein, D.C. Cholera toxin-induced G_s alpha down-regulation in neural tissue: studies on the pineal gland. Brain Res 638:151–156, 1994.
7. Baler, R. and Klein, D.C. Circadian expression of transcription factor Fra-2 in the rat pineal gland. J Biol Chem 270:27319–27325, 1995.
8. Baler, R., Covington, S., and Klein, D.C. The rat arylalkylamine N-acetyltransferase gene promoter cAMP activation via a cAMP-responsive element-CCAAT complex. J Biol Chem 272:6979–6985, 1997.

9. Beebe, S.J., Holloway, R., Rannels, S.R., and Corbin, J.D. Two classes of cAMP analogs which are selective for the two different cAMP-binding sites of type II protein kinase demonstrate synergism when added together in intact adipocytes. J Biol Chem 259:3539–3547, 1984.
10. Begay, V., Falcon, J., Cahill, G.M., Klein, D.C., and Coon, S.L. Transcripts encoding two melatonin synthesis enzymes in teleost pineal organ: circadian regulation in pike and zebrafish, but not trout. Endocrinology 139:905–912, 1998.
11. Berg, J.P., Ree, A.H., Sandvik, J.A., Taskén, K., Landmark, B.F., Torjesen, P.A., and Haug, E. 1,25-Dihydroxyvitamin D3 alters the effect of cAMP in thyroid cells by increasing the regulatory subunit type IIβ of the cAMP-dependent protein kinase. J Biol Chem 269:32233–32238, 1994.
12. Borjigin, J., Wang, M.M., and Snyder, S.H. Diurnal variation in mRNA encoding serotonin N-acetyltransferase in the pineal gland. Nature 378:783–785, 1995.
13. Bothelo, L.H.P., Rothermel, J.D., Coombs, R.V., and Jastorff, B. Phosphorothioate antagonists of cAMP action. In: *Initiation and Termination of Cyclic Nucleotide Action* (Corbin J.D. and Johnson R.A., eds.) Methods in Enzymology Vol. 159, pp. 159–172, Academic Press, San Diego, 1988.
14. Carter, D.A. Up-regulation of β-adrenoceptor messenger ribonucleic acid in the rat pineal gland: nocturnally, through a β-adrenoceptor-linked mechanism, and in vitro, through a novel posttranscriptional mechanism activated by specific protein synthesis inhibitors. Endocrinology 133:2263–2268, 1993.
15. Chijiwa, T., Mishima, A., Hagiwara, M., Sano, M., Hayashi, K., Inoue, T., Naito, K., Toshioka, T., and Hidaka, H. Inhibition of forskolin-induced neurite outgrowth and protein phosphorylation by a newly synthesized selective inhibition of cyclic AMP-dependent protein kinase, N-[2-(p-bromocinnamylamino)ethyl]-5 isoquinoline-sulfonamide (H89), of pheochromocytoma cells. J Biol Chem 265:5267–5272, 1990.
16. Chomczynski, P. and Sacchi, N. Single-step method of RNA isolation by acid guanidinium thiocyanate-phenol-chloroform extraction. Anal Biochem 162:156–159, 1987.
17. Collins, S., Caron, M.G., and Lefkowitz, R.J. From ligand binding to gene expression: New insights into the regulation of G-protein-coupled-receptors. Trends Biol Sci 17:37–39, 1992.
18. Coon, S.L., Roseboom, P.H., Baler, R., Weller, J.L., Namboodiri, M.A.A., Koonin, E.V., and Klein, D.C. Pineal serotonin N-acetyltransferase: expression cloning and molecular analysis. Science 270: 1681–1683, 1995.
19. Coon, S.L., McCune, S.K., Sugden, D., and Klein, D.C. Regulation of pineal α_{1B}-adrenergic receptor mRNA: day/night rhythm and β-adrenergic receptor/cyclic AMP control. Mol Pharmacol 51:551–557, 1997.
20. Dostmann, W.R.G. (*Rp*)-cAMPS inhibits cAMP-dependent protein kinase by blocking the cAMP-induced conformational transition. FEBS Lett 375:231–234, 1995.
21. Drijfhout, W., van der Linde, A., Kooi, S., Grol, C., and Westerink, B.H.C. Norepinephrine release in the rat pineal gland: The input from the biological clock measured by in vivo microdialysis. J Neurochem 66:748–755, 1996.
22. Ellison, N., Weller, J.L., and Klein, D.C. Development of a circadian rhythm in the activity of pineal serotonin N-acetyltransferase. J Neurochem. 19:1335–1341, 1972.
23. Feliciello, A., Giuliano, P., Porcellini, A., Garbi, C., Obici, S., Mele, E., Angotti, E., Grieco, D., Amabile, G., Cassano, S., Li, Y., Musti, A.M., Rubin, C.S., Gottesman, M.E., and Avvedimento, E.V. The v-Ki-ras oncogene alters cAMP nuclear signaling by regulating the location and the expression of cAMP-dependent protein kinase IIβ. J Biol Chem 271:25350–25359, 1996.
24. Foulkes, N.S., Duval, G., and Sassone-Corsi, P. Adaptive inducibility of CREM as transcriptional memory of circadian rhythms. Nature 381:83–85, 1996.
25. Foulkes, N.S., Whitmore, D., and Sassone-Corsi, P. Rhythmic transcription: The molecular basis of circadian melatonin synthesis. Biol. Cell 89:487–494, 1997.
26. Francis, S.H. and Corbin, J.D. Structure and function of cyclic nucleotide-dependent protein kinases. Ann Rev Physiol 56:237–272, 1994.
27. Gastel, J.A., Roseboom, P.H., Rinaldi, P.A., Weller, J., and Klein, D.C. Melatonin production: Proteasomal proteolysis in serotonin N-acetyltransferase regulation. Science 279:1358–1360, 1998.
28. Gjertsen, B.T., Mellgren, G., Otten, A., Maronde, E., Genieser, H.-G., Jastorff, B., Vintermyr, O.K., McKnight, G.S., and Døskeland, S.O. Novel Rp-cAMPS analogs as tools for inhibition of cAMP-kinase in cell culture. J Biol Chem 270:20599–20607, 1995.
29. Gonzalez, G.A. and Montminy, M.R. Cyclic AMP stimulates somatostatin gene transcription by phosphorylation of CREB at serine 133. Cell 59:675–680, 1989.
30. Ho, A.K. and Klein, D.C. Phosphatidylinositol phosphodiesterase (phospholipase C) activity in the pineal gland: characterization and photoneural regulation. J Neurochem 48:1033–1038, 1987a.

31. Ho, A.K. and Klein, D.C. Activation of α_1-adrenoceptors, protein kinase C, or treatment with intracellular free Ca^{2+}-elevating agents increases pineal phospholipase A_2 activity. Evidence that protein kinase C may participate in Ca^{2+}-dependent alpha$_1$-adrenergic stimulation of pineal phospholipase A_2 activity. J Biol Chem 262:11764–11770, 1987b.
32. Klein, D.C. Photoneural regulation of the mammalian pineal gland. In *Photoperiodism, Melatonin and the Pineal Gland*. Ciba Foundation Symposium 117, Pitman, London, pp. 38–56, 1985.
33. Klein, D.C., Namboodiri, M.A.A., and Auerbach, D.A. The melatonin rhythm generating system: developmental aspects. Life Sci 28:1975–1986, 1981.
34. Klein, D.C., Moore, R.Y., and Reppert, S.M. Suprachiasmatic Nucleus—the Mind's Clock. New York, Oxford: Oxford University Press, pp. 1–467, 1991.
35. Klein, D.C., Roseboom, P.H., and Coon, S.L. New light is shining on the melatonin rhythm enzyme—the first postcloning view. Trends Endocrinol Metabol 7:106–112, 1996.
36. Klein, D.C., Coon, S.L., Roseboom, P.H., Weller, J.L., Bernard, M., Gastel, J.A., Zatz, M., Iuvone, M., Rodriguez, I.R., Begay, V., Falcon, J., Cahill, G.M., Cassone, V.M., and Baler, R. The melatonin rhythm-generating enzyme: molecular regulation of serotonin N-acetyltransferase in the pineal gland. Rec Progr Horm Res 52:307–358, 1997.
37. Korf, H.-W., Oksche, A., Ekström, P., Zigler, J.S., Gery, I., and Klein, D.C. Pinealocyte projections into the mammalian brain revealed with S-antigen antiserum. Science 231:735–737, 1986.
38. Korf, H.-W., Schomerus, C., Maronde, E., and Stehle, J.H. Signal transduction in the rat pineal organ: Ca^{2+}, pCREB, and ICER. Naturwissenschaften 83:535–543, 1996.
39. Korf, H.-W., Schomerus, C., and Stehle, J.H. The pineal organ, its hormone melatonin, and the photoneuroendocrine system. Adv Anat Embryol Cell Biol 146:1–100, 1998.
40. Machida, C.A., Bunzow, J.R., Searles, J.R., Van Tol, H., Tester, B., Neve, K.A., Teal, P., Nipper, V., and Civelli, O. Molecular cloning and expression of the rat β_1-adrenergic receptor gene. J Biol Chem 265:12960–12965, 1990.
41. Maronde, E., Middendorff, R., Mayer, B., and Olcese, J. The effect of NO-donors in bovine and rat pineal cells: stimulation of cGMP and cGMP-independent inhibition of melatonin synthesis. J Neuroendocrinol 7:207–214, 1995.
42. Maronde, E., Middendorff, R., Telgmann, R., Taskén, K., Hemmings, B., Müller, D., and Olcese, J. Melatonin synthesis in the bovine pineal gland is regulated by cyclic AMP-dependent protein kinase type II. J. Neurochem. 68:770–777, 1997a.
43. Maronde, E., Stehle, J.H., Schomerus, C., and Korf, H.-W. Control of CREB phosphorylation and its role for induction of melatonin synthesis in rat pinealocytes. Biol Cell 89:505–511, 1997b.
44. Middendorff, R., Maronde, E., Paust, H.-J., Müller, D., Davidoff, M., and Olcese, J. Expression of C-type natriuretic peptide (CNP) in the bovine pineal gland. J Neurochem 67:517–524, 1996.
45. Molina, C.A., Foulkes, N.S., Lalli, E., and Sassone-Corsi, P. Inducibility and negative autoregulation of CREM: an alternative promotor directs the expression of ICER, an early response repressor. Cell 75:1–20, 1993.
46. Møller, M., Phansuwan-Pujito, P., Morgan, K.C., and Badiu, C. Localization and diurnal expression of mRNA encoding the β_1-adrenoceptor in the rat pineal gland: an in situ hybridization study. Cell Tissue Res 288:279–284, 1997.
47. Montminy, M.R., Gonzales, G.A., and Yamamoto, K.K. Regulation of cAMP-inducible genes by CREB. Trends Neurosci 13:184–188, 1990.
48. Øgreid, D., Ekanger, R., Suva, R.H., Miller, J.P., Sturm, P., Corbin, J.D., and Døskeland, S.O. Activation of protein kinase isozymes by cyclic nucleotide analogs used singly or in combination—principles for optimizing the isozyme specificity of analog combinations. Eur J Biochem 150:219–227, 1985.
49. Parfitt, A.G. and Klein, D.C. Sympathetic nerve endings in the pineal gland protect against acute stress-induced increase in *N*-acetyltransferase (E.C. 2.3.1.5) activity. Endocrinology 99:840–851, 1976.
50. Pfeffer, M. and Stehle, J.H. Ontogeny of a diurnal rhythm in arylalkylamine-N-acetyltransferase mRNA in rat pineal gland. Neurosci Lett 248:163–166, 1998.
51. Pfeffer, M., Kühn, R., Krug, L., Korf, H.-W., and Stehle, J.H. Rhythmic variation in β_1-adrenergic receptor mRNA-levels in the rat pineal gland: circadian and developmental regulation. Eur J Neurosci 10:2896–2904.
52. Romero, J.A. and Axelrod, J. Pineal β-adrenergic receptor: diurnal variation in sensitivity. Science 184:1091–1093, 1974.
53. Roseboom, P.H. and Klein, D.C. Norepinephrine stimulation of pineal cyclic AMP response element-binding protein phosphorylation: primary role of a β-adrenergic receptor/cyclic AMP mechanism. Mol Pharmacol 47:439–449, 1995.

54. Roseboom, P.H., Coon, S.L., Baler, R., McCune, S.K., Weller, J.L., and Klein, D.C. Melatonin synthesis: analysis of the more than 150-fold nocturnal increase in serotonin N-acetyltransferase messenger ribonucleotide acid in the rat pineal gland. Endocrinology 137:3033–3044, 1996.
55. Sassone-Corsi, P. Transcription factors responsive to cAMP. Annu Rev Cell Dev Biol 11:355–377, 1995.
56. Sato, N., Kamino, K., Tateishi, K., Satoh, T., Nishiwaki, Y., Yoshiiwa, A., Miki, T., and Ogihara, T. Elevated amyloid β protein (1–40) level induces CREB phosphorylation at serine-133 via p44/42 MAP kinase (Erk1/2)-dependent pathway in rat pheochromocytoma PC12 cells. Biochem Biophys Res Comm 232:637–642, 1997.
57. Schomerus, C., Laedtke, E., and Korf, H.-W. Calcium responses of isolated, immunocytochemically identified rat pinealocytes to nordarenergic, cholinergic and vasopressinergic stimulations. Neurochem Int 27:163–175, 1995.
58. Schomerus, C., Maronde, E., Laedtke, E., and Korf, H.-W. Vasoactive intestinal peptide (VIP) and pituitary adenylate cyclase-activating polypeptide (PACAP) induce phosphorylation of the transcription factor CREB in subpopulations of rat pinealocytes: immunocytochemical and immunochemical evidence. Cell Tissue Res 286:305–313, 1996.
59. Schwoch, G., Hamann, A., and Hilz, H. Antiserum against the catalytic subunit of adenosine 3′,5′-cyclic monophosphate-dependent protein kinase. Biochem J 192:222–230, 1980.
60. Sheng, M., Thompson, M.A., and Greenberg, M.E. CREB: a Ca^{2+}-regulated transcription factor phosphorylated by calmodulin-dependent kinases. Science 252:1427–1430, 1991.
61. Stehle, J.H. Pineal gene expression: dawn in a dark matter. J Pineal Res 18:179–190, 1995.
62. Stehle, J.H., Rivkees, S., Lee, J., Weaver, D., Deeds, J., and Reppert, S.M. Molecular cloning of the cDNA for an A_2-like adenosine receptor. Mol Endocrinol 6:384–393, 1992.
63. Stehle, J.H., Foulkes, N., Molina, C., Simonneaux, V., Pévet, P., and Sassone-Corsi, P. Adrenergic signals direct rhythmic expression of transcriptional repressor CREM in the pineal gland. Nature 365:314–321, 1993.
64. Stehle, J.H., Foulkes, N.S., Pevet, P., and Sassone-Corsi, P. Developmental maturation of pineal gland function: synchronized CREM inducibility and adrenergic stimulation. Mol Endocrinol 9:706–716, 1995.
65. Stehle, J.H., Pfeffer, M., Kühn, R., and Korf, H.-W. Light-induced expression of transcription factor ICER (inducible cAMP early repressor) in rat suprachiasmatic nucleus is phase-restricted. Neurosci Lett 217:169–172, 1996.
66. Tamarkin, L., Reppert, S.M., Orloff, D.J., Klein, D.C., Yellon, S.M., and Goldmann, B.D. Ontogeny of the pineal melatonin rhythm in the Syrian (*Mesocricetus auratus*) and Siberian (*Phodopus sungorus*) hamsters and in the rat. Endocrinology 107:1061–1063, 1980.
67. Tamotsu, S., Schomerus, C., Stehle, J.H., Roseboom, P., and Korf, H.-W. Norepinephrine-induced phosphorylation of the transcription factor CREB in isolated rat pinealocytes: an immunocytochemical study. Cell Tissue Res 282:219–226, 1995.
68. Taskén, K.A., Knutsen, H.K., Attramadal, H., Taskén, K., Jahnsen, T., and Eskild, W. Different mechanisms are involved in cAMP-mediated induction of mRNAs for subunits of cAMP-dependent protein kinases. Mol Endocrinol 5:21–28, 1991.
69. Taskén, K.A., Skålhegg, B.S., Solberg, R., Andersson, K.B., Taylor, S.S., Lea, T., Blomhoff, H.K., Jahnsen, T., and Hansson, V. Novel isozymes of cAMP-dependent protein kinase exist in human cells due to formation of RI alpha-RI beta heterodimeric complexes. J Biol Chem 268:21276–21282, 1993.
70. Tsavara, E., Pouille, Y., Defer, N., and Honoune, J. Diurnal variation of adenylyl cyclase type 1 in the rat pineal. Proc Natl Acad Sci USA 93:11208–11212, 1996.
71. Ventra, C., Porcellini, A., Feliciello, A., Gallo, A., Poalillo, M., Mele, E., Avvedimento, V.E., and Schettini, G. The differential response of protein kinase A to cyclic AMP in discrete brain areas correlates with the abundance of regulatory subunit II. J Neurochem 66:1752–1761, 1996.
72. Wicht, H., Korf, H.-W., and Schaad, N.C. Morphological and immunocytochemical heterogeneity of cultured pinealocytes from one week and two-month old rats: planimetric and densitometric investigations. J Pineal Res 14:128–137, 1993.
73. Xie, H. and Rothstein, T.L. Protein kinase C mediates activation of nuclear cAMP response element-binding protein (CREB) in B lymphocytes stimulated through surface Ig. J. Immunol. 154:1717–1723, 1995.

15

REGULATION OF MELATONIN SYNTHESIS IN THE OVINE PINEAL GLAND

An *in Vivo* and *in Vitro* Study

Karen Privat, Michelle Fevre-Montange, Christine Brisson,
Didier Chesneau, and Jean-Paul Ravault

INSERM U433, Faculté Laënnec
Rue Guillaume Paradin
69372 Lyon Cédex 08, France
Station Physiologie de la Reproduction des Mammifères Domestiques
INRA, 37380 Nouzilly, France

In all vertebrates, melatonin synthesis displays diurnal variations, being high at night (1). This nocturnal increase in melatonin production in the rat pineal gland is under the control of the circadian release of norepinephrine (NE) from sympathetic nerves terminating in the gland (2). Melatonin is produced from circulating tryptophan after a four-step enzymatic pathway involving successively the tryptophan hydroxylase (L-tryptophan tetrahydropteridin oxydoreductase, TPOH), the aromatic aminoacid decarboxylase, the serotonin N-acetyltransferase (arylalkylamine N-acetyltransferase, NAT) and the hydroxyindole-O-methyltransferase (HIOMT).

Many studies on melatonin regulation have focused on the enzyme NAT which converts seroronin to N-acetylserotonin in the penultimate step of the melatonin synthetic pathway (3,4). Little is known about the regulation of TPOH, the first enzyme in the melatonin synthesis and the rate limiting enzyme in serotonin synthesis. This enzyme is present at high levels in the pineal gland of several species and it has been suggested that NE may influence the expression of the TPOH gene in the gland. In the rat and chicken pineal gland, TPOH and NAT enzymatic activity and mRNA levels have been shown to follow a day/night rhythm (5–9).

In the present study, we have measured TPOH mRNA expression in the sheep pineal gland throughout the light/dark cycle by a semi-quantitative reverse transcription polymerase chain reaction (RT-PCR) analysis and compared it with the expression of NAT mRNAs. We showed a significant ($p < 0.05$, Mann-Whitney) nocturnal increase of 28.5 and 38% in TPOH and NAT mRNA expression, respectively (Figure

Melatonin after Four Decades, edited by James Olcese.
Kluwer Academic / Plenum Publishers, New York, 2000.

1), which supports the existence of a transcriptional regulation of the melatonin synthesis in the ovine pineal gland.

Nevertheless, these transcriptional mechanisms can not entirely explain the 10-fold increase in plasma melatonin level measured during the dark period (data not shown).

In addition, we developed the culture of ovine pineal cells, in order to study the possible adrenergic modulation of melatonin synthesis and release. Incubation with isoproterenol, a β-adrenergic agonist, elevates to about 900% the production of melatonin (Table 1).

This stimulation is almost totally blocked by propranolol, a β-adrenergic antagonist, and p-chlorophenylalanine, which inhibits TPOH enzymatic activity. Thus, TPOH could be the rate-limiting enzyme in melatonin synthesis in the ovine pineal gland. The adrenergic stimulation of melatonin production is also partially inhibited by the presence of an inhibitor of transcription (actinomycin D) or translation (cycloheximide) in the culture medium. Transcriptional and post-transcriptional mechanisms seem to be implicated in the regulation of melatonin synthesis in the ovine pineal gland.

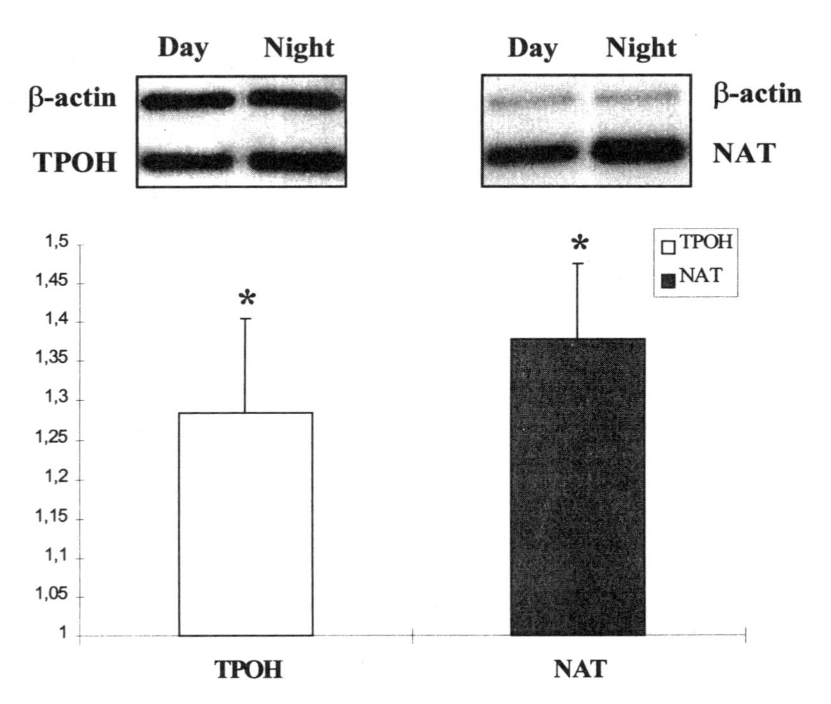

Figure 1. Day/night expression of TPOH and NAT mRNAs from ovine pineal gland. Quantitative RT-PCR of TPOH (day: n = 7; night: = 8) and NAT (day: n = 7; night: = 9) mRNAs from «day» and «night» ovine pineal glands. (*: $p < 0.05$)(bars: SEM). Male lambs (Ile de-France, 6-months old, from breeding of INRA (Nouzilly, France)), reared under defined lighting conditions (8L:16D, lights on: 02:00 pm, off: 10:00 pm), were sacrificed by decapitation during the light (10 animals at 05:30 pm) or dark (10 animals at 08:30 am) period. The RT-PCR reactions were performed with 0.1 μg of total RNA extracted from pineal gland using 4 μl of RT products. After electrophoresis, DNAs were transferred and hybridized with appropriate internal ^{32}P-labelled probes. The membrane sections were counted in a liquid scintillation counter. Experiments were repeated three times.

Table 1. Pharmacological treatments of ovine pinealocytes in culture

Pharmacological treatment (6 hours)	Melatonin (pg/ml plasma) (± SEM)	Inhibition of isoproterenol stimulation (in % ± SEM)
Control	203 ± 10	—
Iso 10^{-6} M	1909 ± 34	(100 ± 2)
Iso 10^{-6} M – Propr 10^{-5} M	251 ± 24	−87 ± 1
Iso 10^{-6} M – Chx 10^{-5} M	836 ± 28	−56 ± 2
Iso 10^{-6} M – ActD 10^{-5} M	1323 ± 65	−31 ± 4
Iso 10^{-6} M – pCPA 5.10^{-4} M	358 ± 12	−81 ± 1

Male lambs (Ile de-France, 6-months old, from breeding of INRA (Nouzilly, France)), reared under natural lighting conditions, were sacrificed by decapitation early in the morning. Ovine pinealocytes were prepared by enzymatic and mechanical dissociation. Cells (10^5/well in 24 wells plates) were maintained in culture in DMEM/F12 medium supplemented with glutamine (2 mM), penicillin (100 u/ml), steptomycin (100 µg/ml), amphotericin B (0.25 µg/ml) and FCS (10%), under constant conditions of incubation (37°C, 5% CO_2 in air). Pharmacological treatments were performed 6 days later. After 6 hours of incubation with appropriate dilution of different drugs in complete culture medium minus FCS, culture media were collected and frozen before radio-immunoassay for melatonin content.
(Iso: isoproterenol; Propr: propranolol; Chx: cycloheximide; ActD: actinomycin D; pCPA: p-chlorophenylalanine).

REFERENCES

1. Axelrod, J. The pineal gland: a neurochemical transducer. Science 184:1341–1348, 1974.
2. Wurtman, R.J., Axelrod, J., Sedvall, G., and Moore, R.Y. Photic and neural control of the 24-hour norepinephrine rhythm in the rat pineal gland. J. Pharmacol. Exp. Ther. 157:487–492, 1967.
3. Klein, D.C., Schaad, N.L., Namboordiri, M.A., Yu, L., and Weller, J.L. Regulation of pineal serotonin N-acetyltransferase activity. Biochem. Soc. Trans. 20:299–304, 1992.
4. Coon, S.L., Roseboom, P.H., Baler, R., Weller, J.L., Namboodiri, M.A., Koonin, E.V., and Klein, D.C. Pineal serotonin N-acetyltransferase: expression cloning and molecular analysis. Science 270:1681–1683, 1995.
5. Ehret, M., Pevet, P., and Maitre, M. Tryptophan hydroxylase synthesis is induced by 3′,5′-cyclic adenosine monophosphate during circadian rhythm in the rat pineal gland. J. Neurochem. 57:1516–1521, 1991.
6. Besancon, R., Simonneaux, V., Jouvet, A., Belin, M.F., and Fèvre-Montange, M. Nycthemeral expression of tryptophan hydroxylase mRNAs in the rat pineal gland. Mol. Brain Res. 40:136–138, 1996.
7. Florez, J.C., Seidenman, K.J., Barrett, R.K., Sangoram, A.M., and Takahashi, J.S. Molecular cloning of chick pineal tryptophan hydroxylase and circadian oscillation of its mRNA levels. Mol. Brain Res. 42:25–30, 1996.
8. Roseboom, P.H., Coon, S.L., Baler, R., McCune, S.K., Weller, J.L., and Klein, D.C. Melatonin synthesis: analysis of the more than 150-fold nocturnal increase in serotonin N-acetyltransferase messenger ribonucleic acid in the rat pineal gland. Endocrinol. 137:3033–3045, 1996.
9. Bernard, M., Klein, D.C., and Zatz, M. Chick pineal clock regulates serotonin N-acetyltransferase mRNA rhythm in culture. Proc. Natl. Acad. Sci. U.S.A. 94:304–309, 1997.

16

MELATONIN MODULATION OF PROLACTIN AND GONADOTROPHIN SECRETION

Systems Ancient and Modern

Gerald Lincoln

MRC Reproductive Biology Unit
37 Chalmers Street
Edinburgh EH3 9EW, United Kingdom

1. ABSTRACT

Recent studies in sheep indicate that the pineal melatonin signal which transduces effects of photoperiod acts at separate sites in the pituitary gland and brain to regulate seasonality in prolactin (PRL) and gonadotrophin secretion. The pituitary gland is the proposed site for control of PRL based on the observation that hypothalamo-pituitary disconnected (HPD) rams continue to show normal patterns of PRL secretion in response to changes in photoperiod or treatment with melatonin. Lactotrophs do not express melatonin receptors, thus this pituitary effect is assumed to be mediated by cells in the pars tuberalis via "tuberalin". The mediobasal hypothalamus (MBH) is the putative target for gonadotrophin control since: i) gonadotrophin secretion is dependent on pulsatile GnRH secretion from the MBH, ii) local administration of melatonin in the MBH, but not in other areas of the brain and pituitary gland, readily reactivates GnRH-induced LH and FSH secretion in photo-inhibited rams; and iii) treatment of HPD rams with a chronic pulsatile infusion of GnRH stimulates gonadotrophin secretion irrespective of photoperiod. Complementary studies conducted by others in the Syrian hamster, have shown that lesions in the MBH block the action of melatonin on gonadotrophin but not on prolactin secretion; this supports the "dual-site hypothesis". Since all photoperiodic mammals are essentially similar in hyper-secreting PRL under long days, the pituitary control mechanism for PRL is regarded as

Tel: 0131 229 2575; Fax: 0131 228 5571; e-mail: g.lincoln@ed-rbu.mrc.ac.uk

Melatonin after Four Decades, edited by James Olcese.
Kluwer Academic / Plenum Publishers, New York, 2000.

conserved (ancient) with the pleiotrophic actions of PRL inducing a summer physiology (e.g. growth of summer pelage). In contrast, the variation between species in the timing of the gonadal cycle indicates that evolution has independently modified the melatonin-sensitive neural circuits in the MBH to permit the species-specific timing of the mating season.

2. MELATONIN AS A SEASONAL TIME CUE

Melatonin transduces the effect of changes in daylength in photoperiodic mammals and thus times seasonal cycles in reproduction, moulting, somatic growth, fattening and other seasonal characteristics. This is the best characterised biological role(s) of melatonin (1,60,61). The photoperiodic transduction mechanism involves light perception by the retina, entrainment of the circadian rhythm generating system in the SCN and the inhibitory effect of light on the secretion of melatonin by the pineal gland (see other chapters). The result is that melatonin is secreted only at night, and the duration of secretion varies with the annual cycle in nightlength and thus daylength. The *duration* of melatonin secretion appears to be the critical parameter affecting the melatonin target tissues since daily infusions of melatonin of appropriate duration and maintained for many weeks, mimics the effect of long day (short period infusion) and short days (long period infusion) on seasonal physiology (4,10). Over the past 3 decades, the importance of melatonin in the photoperiod-transduction-relay has been estalished in a wide range of species (rodents, carnivores, ungulates, primates and some marsupials; 3,21), and may apply to man (68). The current challenge is to unravel where and how melatonin acts to induce these seasonal effects.

The aim of this chapter is to provide a brief summary of recent studies carried out in sheep which have investigated the potential sites of action of melatonin in the photoperiodic regulation of gonadotrophin and PRL secretion. Most of the illustrated data derives from the Soay ram model.

3. PHOTOPERIODIC INDUCTION AND ENTRAINMENT OF SEASONAL CYCLES

3.1. Photoperiod

In rams, exposure to an artificial lighting regimen of alternating 16-weekly periods of long and short days induces cyclical changes in gonadotrophin and PRL secretion (Figure 1). An abrupt switch from long to short days promotes an increase in gonadotrophin secretion and a decrease in PRL secretion, while the switch back to long days has the reverse effects. Since FSH acts along with testosterone on the Sertoli cells in the seminiferous tubules of the testis to regulate the efficiency of spermatogenesis (13), the testes become fully enlarged during short days as normally occurs in autumn in animals living outdoors. Circulating FSH concentrations decline at the peak of the tesicular cycle due to the negative feedback effect of testosterone and inhibin acting at the level of the hypothalamus and pituitary gland (45,64). The high PRL concentrations induced by long days have potentially many effects since PRL receptors are expressed in liver, kidney, pancreas, muscle, skin, adipose tissue, gonads and selected sites in brain (7). One conspicuous effect is on the pelage cycle with PRL acting in the

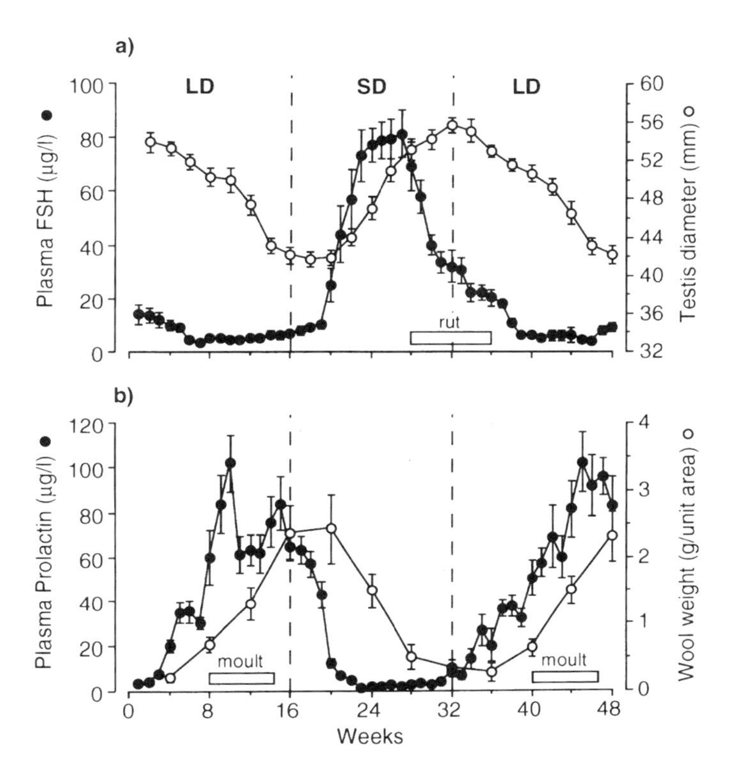

Figure 1. Cyclical changes in: a) FSH secretion and testicular size, and b) PRL secretion and wool growth in Soay rams exposed to alternating 16 weekly periods of long (16L:8D) and short days (8L:16D). The timing of the phases of aggressive behaviour (rut) and the pelage moult (horizontal bar) are also shown (32).

dermal papilla of the hair follicle to trigger reactivation of hair growth and the associated spring moult (11,14).

Regular changes between long and short days has the effect of accelerating the seasonal cyclicity and two complete reproductive cycles can be induced within 12 months (27). If the light changes are more frequent, however, rams fail to express a decline in gonadotrophin secretion during brief exposure to long days and may remain permanently sexually active state (59). While photoperiod has a clear inductive effect on seasonality, sheep do not need to experience changes in photoperiod to express long-term cyclicity. Under constant long or short days, both rams and ewes continue to show long-term variations in both gonadotrophin and PRL secretion, although the cycles amonst individuals become asynchronous and damp out with time (15,24,25,27). The long-term cycles are thus generated endogenously as circannual rhythms, but under natural conditions the annual cycle in photoperiod acts to induce and entrain the cycles such that the period of the endogenous rhythm is adjusted to match the sidereal year, and phase of each component system is timed to the appropriate season.

3.2. Melatonin

The first demonstration in sheep that melatonin secretion from the pineal gland relays effects of photoperiod to time seasonal characteristics was achieved by housing

Soay rams under a driving 32-week lighting regimen (28). Under these conditions, sheep without a functional pineal gland (pinealectomised or superior cervical ganglionectomised) were shown to be unable to adjust their long-term physiological rhythms to the changing photoperiod, although the intrinsic cyclicity persisted (44). Definitive evidence that the daily rhythm in melatonin secretion from the pineal gland mediates effects of photoperiod was provided from studies in which pinealectomised ewes were treated with programmed infusions of melatonin designed to mimic the normal daily patterns of melatonin secretion (4,6). Short-duration infusions (8-h daily) were found to suppress gonadotrophin (LH) secretion, as normally occurs in pineal intact ewes under long days, while conversely long-duration infusions promoted gonadotrophin secretion. The responses were unaffected by ambient photoperiod (71), or the phase of the light-dark cycle in which the treatments were delivered (67). Subsequent studies demonstrated that the circannual rhythm in gonadotrophin secretion can be entrained in pinealectomised ewes by exposure to long or short-day melatonin signals given for just 70 days at one specific time each year. A bout of long day signals was most effective (70), thus indicating the importance of the summer photoperiod in maintaining seasonal entrainment under natural conditions.

Many studies in sheep have shown that treatments with exogenous melatonin which effectively increase the duration of exposure to melatonin when given in spring and summer, or under long days, cause a premature increase in gonadotrophin secretion and an early onset of the mating season, and a corresponding decrease in PRL secretion. This can be achieved by the daily oral administration of melatonin given in the afternoon to expand the daily melatonin phase (1,26), or by simply by introducing a subcutaneous implant containing melatonin which permanently increases the circulating concentrations of melatonin (39). Although a constant-release implant produces a non-physiological, continuous melatonin signal with high daytime concentrations, the effect when initiated under long days is similar to that provoked by transfer to short days (58). This includes the effect of reversing refractoriness to long days, and the development of refractoriness to the short-day signal for both gonadotrophin and PRL secretion (37,39). Implantation of melatonin has subsequently been exploited commercially as a convenient way to manipulate the timing of the breeding season and the pelage cycle in several domesticated species (sheep, deer, mink).

4. SITES OF ACTION OF MELATONIN

4.1. Melatonin Receptors

In situ autoradiography using 2-(^{125}I)iodomelatonin as a ligand has been used to map the distibution of putative melatonin receptors in the ovine brain and pituitary gland with the aim of identifying target sites for the action of melatonin (Figure 2). Specific, high affinity binding of melatonin has been demonstrated in many sites in the hypothalamus, limbic system and brain, with highest levels of binding in the pars tuberalis of the pituitary gland (5,20). Using in *situ hybridization* with a specific probe for the melatonin receptor (mt1 receptor) melatonin receptor gene expression has been demonstrated in the pars tuberalis/zona tuberalis of the ovine pituitary (62,69), but not in the brain sites indicated by binding studies. This may reflect a lower of expression in the neural sites, or the presence of another form of receptor. The latter possibility is indicated by failure to find a transcript corresponding to the mt1 receptor in

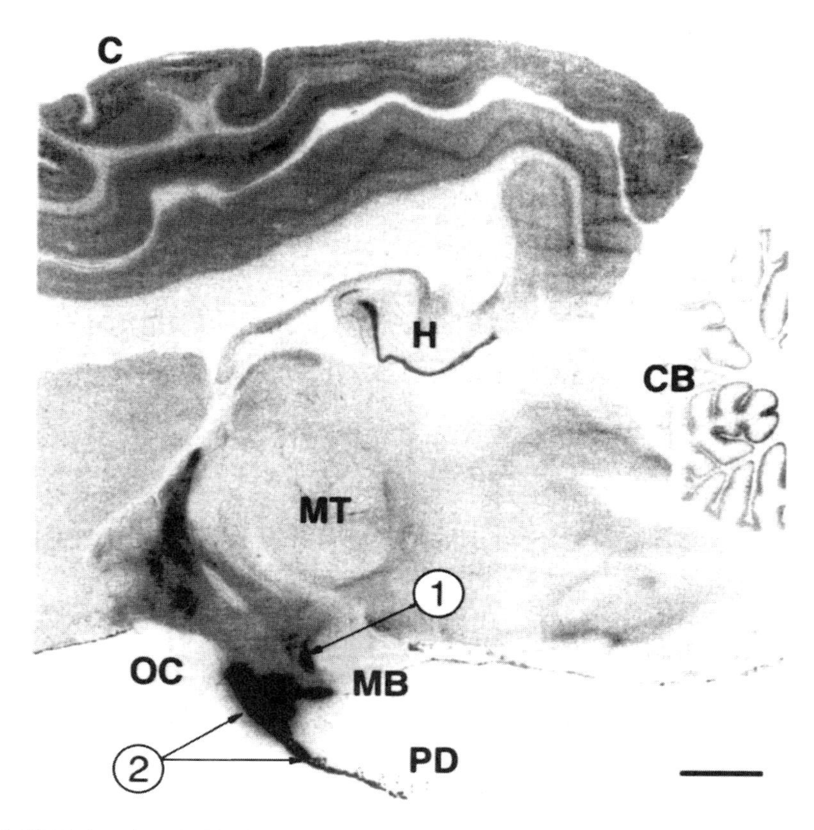

Figure 2. 2-(^{125}I)Iodomelatonin binding in the ovine brain and pituitary gland as revealed by in situ autoradiography. The putative areas involved in the photoperiodic regulation of gonadotrophin secretion (1) and PRL secretion (2) are indicated (see text). Abbreviations: H—hippocampus. OC—optic chiasma, MB mammillary body, MT—medial thalamus, PD—pars distalis. Data from Helliwell & Williams (20) are reproduced by permission.

hypothalamic RNA extracts amplified by RT PCR (P. Morgan, personal communication). The affinity of G-protein coupled receptors, including the melatonin receptor, is altered in the presence of guanine nucleotides resulting in an apparent drop in receptor number (55). This response can be demonstrated for 2-(^{125}I)iodomelatonin binding the pars tuberalis but not for binding in the hippocampus and other areas of the brain (57,69) consistent with the view that there is a intrinsic difference between the G-protein coupled melatonin receptor in the pars tuberalis and the putative melatonin receptor(s) in the areas of the ovine brain.

4.2. Melatonin Micro-Implants

Melatonin has been administered locally in different sites in the brain and pituitary gland in sheep to elucidate where the hormone acts to mediate the different effects of photoperiod (34,40,41,47). This involved chronic placement of bilateral micro-implants of melatonin (contained at the tip of a 22 g needle) placed in possible target sites in animals preconditioned to long days (Figure 3). This design tested whether the

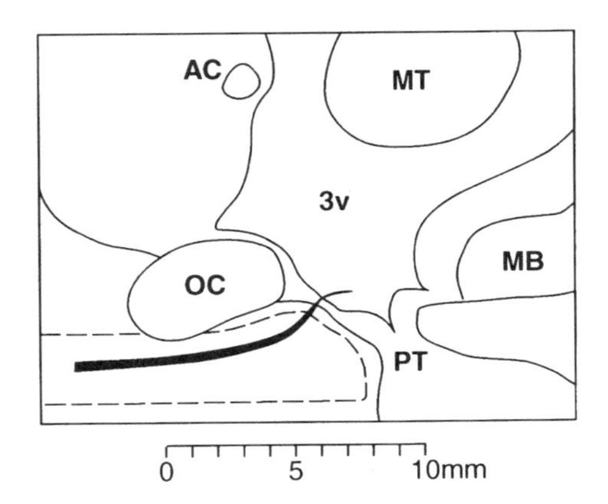

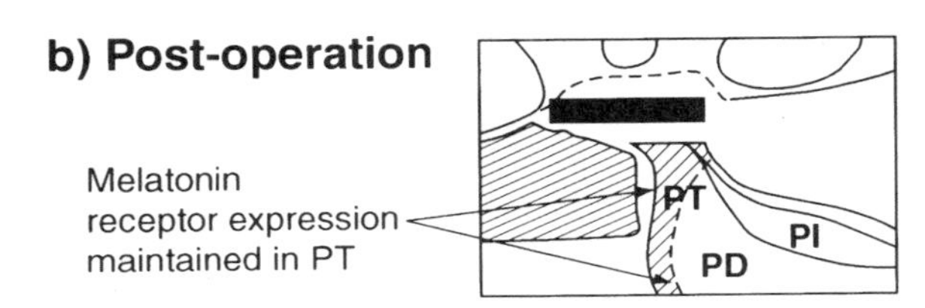

Figure 3. Localisation and effect of placing bilateral microimplants of melatonin in: (A) septum of forebrain, (B) mediobasal hypothalamus (MBH), and (C) pars distalis on FSH and PRL secretion in Soay rams housed under long days (•, experimental group; ○, sham control group). The implants consisted of 22 g needles with melatonin fused inside the tip (see insert), or empty implants as controls, and were left in place for 14 weeks (horizontal bar). Only treatment in the MBH consistently reversed the effect of long days on FSH and PRL secretion (34). Bottom panel shows HPD ram—12 months post-op.

exogenous melatonin from the implant would block the effect of the endogenous melatonin signal, and activate a short-day response. In Soay rams, only microimplants of melatonin placed in the mediobasal hypothalamus (MBH), and to a lesser extent, in the adjacent pars tuberalis of the pituitary gland were consistently effective at blocking the effect of long days. This response included premature activation of FSH secretion and growth of the testes, and a coincident decrease in PRL secretion and growth of the winter pelage (40,41; Figure 3). These effects were local since there was no detected change in melatonin concentrations in the peripheral blood, and the response was reversed after removal of the micro-implants. The latter effect also demonstrated that the treatment had reactivated responsivness to long days (reversed long-day refractoriness). The delivery of a more physiological intermittent signal using reverse-dialysis was also effective at inducing short-day responses (33).

Complementary studies carried out in ovariectomised, oestradiol-supplemented Ile-de-France ewes have also shown that melatonin microimplants placed in the MBH, and not in other regions of the hypothalamus, provoke an increase in LH secretion in a proportion of animals, with a variable PRL response (47). Notably, however, microimplants positioned directly in or close to the pars tuberalis, the site expressing the highest concentration of melatonin receptors, did not modify LH secretion (48,50). Most recently it has been shown that melatonin microimplants placed in the premammillary region of the MBH consistently activates LH secretion in ewes, indicating that this may be the specific site for the action of melatonin in relaying the photoperiodic effect on the reproductive axis (49).

4.3. Hypothalamo-Pituitary Disconnection and Surgical Ablation

The hypothalamo-pituitary disconnection (HPD) Soay sheep model in which the pituitary gland is surgically separated from the hypothalamus has been used to further investigate the potential sites of action of melatonin in the regulation of seasonality (35). The surgical procedure involves visualising the stalk region of the pituitary gland via a paramedial, transnasal, transphenoidal approach, entering the median eminence above the pituitary portal vasculature, and evacuating the neural tissue from the tuberal cinereum (12; Figure 4). This permanently destroys the terminal fields of the median eminence and much of the arcuate nucleus region of the hypothalamus, but spares the lateral and more rostral regions of the hypothalamus and the preoptic area. It also permits the survival of the majority of the pituitary gland including the pars tuberalis which continues to express a high concentration of melatonin receptors (69). To limit reconnection of neural tissue a small piece of aluminium foil is placed in the infundibulum to form a barrier between the hypothalamus and the pituitary gland (Fig. 4). Following HPD surgery, sheep show clinical signs of pituitary disconnection including polyuria (loss of posterior pituitary arginine vasopressin/AVP secretion), gonadal regression (loss of gonadotrophin releasing hormone/GnRH support to gonadotrophin secretion) and gradual weight gain (loss of hypothalamic control of energy balance), but remain in good health and require no hormonal replacement therapy to maintain viability.

In HPD Soay rams PRL secretion was shown to vary in response to changes in photoperiod and the administration of melatonin despite the absence of regulation by the hypothalamus (Figure 5). The circulating PRL concentrations were slightly higher that normal presumably due to the loss of the inhibitory influence of dopamine from the hypothalamus, but treatment of HPD rams with the specific dopamine D2

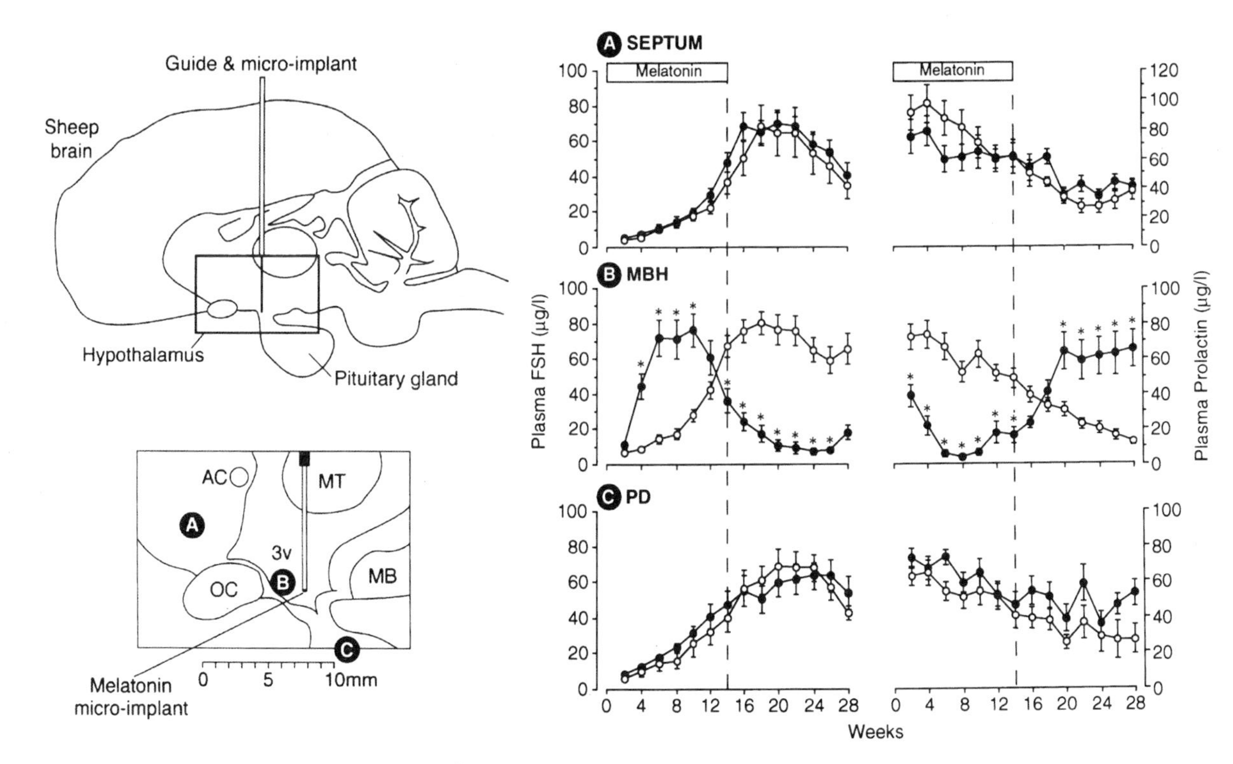

Figure 4. Hypothalamo-pituitary disconnection (HPD) Soay sheep model used to investigate the sites of action of melatonin. The HPD operation removes the median eminence and destroys communication between the hypothalamus and pituitary gland, but maintains the viability of the pituitary gland including the cells expressing melatonin receptors in the pars tuberalis (PT). Abbreviations: AC—anterior commissure, MB mammillary body, MT—medial thalamus, OC—optic chiasma, PD—pars distalis, PI—pars intermdia, 3V—third cerebral ventricle.

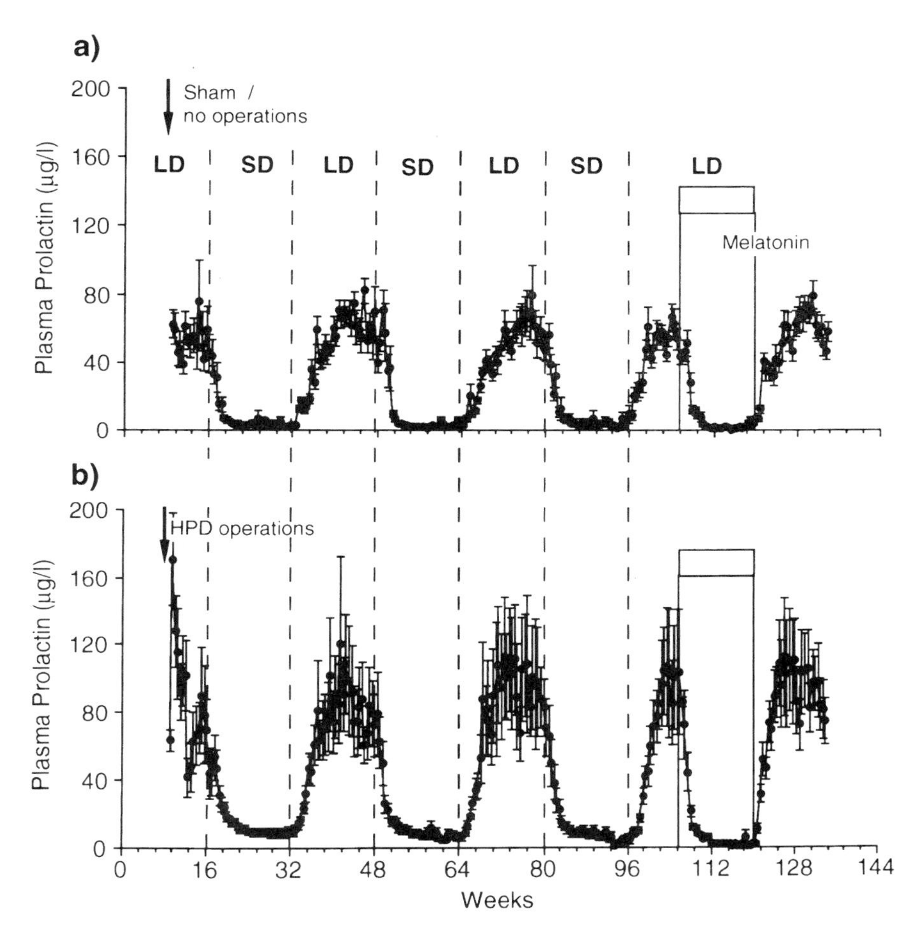

Figure 5. Long-term cyclical changes in blood plasma concentrations of PRL in a) control and b) HPD Soay rams exposed to alternating 16-weekly periods of long (16L:8D, LD) and short days (8L:16D, SD), and treated with a subcutaneous implant on melatonin (horizontal bar) under long days. The sham and HPD operations occurred at week 8. Note, the persistence of normal photoperiod- and melatonin-induced PRL cycles in HPD rams (35,36).

receptor antagonist, sulpiride, failed to elicit PRL secretion under either long or short days, providing evidence that a dopaminergic system does not mediate the effect of photoperiod/melatonin on PRL secretion (36). Similar results have been obtained in foetal sheep where the more radical HPD operation still spares the ability of the foetus to modulate PRL secretion in relation to the ambient photoperiod (22). In this case the melatonin signal is generated by the mother and transferred to the foetus via the placenta.

These results provide clear support for the view that melatonin acts directly in the pituitary gland to mediate effects of photoperiod on PRL secretion. Since melatonin receptors are not expressed by lactotrophs in sheep (69), this effect is presumed to be relayed via the cells in the pars tuberalis which express the receptor in abundance. This is consistent with the recent observation that cultured pars tuberalis cells secrete a unidentified factor called "tuberalin" which upregulates gene expression in lactotrophs (and other cells in the pars distalis) and stimulates PRL secretion (19,56).

The pars tuberalis appears to be under inhibitory regulation by melatonin, thus short-duration exposure to melatonin daily (long day signal) may favour high tuberalin secretion thus stimulating PRL secretion, while long or continuous exposure to melatonin (short day signal) may inhibit tuberalin secretion to cause the observed decline in PRL secretion.

HPD sheep show acute changes in PRL secretion in response to manipulations of photoperiod and melatonin, but also show chronic changes indicative of the development of refractoriness. This has been observed in HPD Soay rams treated continuously with a subcutaneous implant of melatonin for 48 weeks while under long days (37). This induced an initial decline in the blood concentrations of PRL, followed by a partial recovery after 8–12 weeks, with a similar profile to intact controls. The treatment with a second melatonin implant after some 20 weeks failed to inhibit PRL secretion demonstrating that the system was now unresponsive. In the studies in HPD foetal sheep, it has been observed that the PRL response varies with photoperiodic history (23). Taken together, these results support the view that the pars tuberalis-lactotroph-cellular relay within the pituitary gland has the capacity to generate temporal changes in PRL secretion and to carry a "memory" of the previous photoperiod.

4.4. HPD Animals Treated with GnRH

The HPD Soay ram model has also been used to assess whether melatonin acts in the pituitary gland to affect gonadotrophin secretion (38). This has involved treating HPD rams with a standard pulsatile infusion of GnRH to reactivate the gonadotrophin/testicular axis in animals exposed to long and short days (different melatonin signal, Figure 6). The replacement of GnRH is necessary since the HPD surgery destroys the terminals of the GnRH neurones and blocks the endogenous GnRH drive to the gonadotrophs which become quiescent (12,43). The chronic treatment with pulsatile GnRH for 10 weeks induced an increase in LH and FSH secretion, and full maturation of the testes in HPD rams, but there was no effect of the prevailing photoperiod on these reproductive responses. This was despite differences in prolactin secretion between the groups. These results support the conclusion that melatonin does not act in the pituitary gland to affect GnRH-induced gonadotrophin secretion, and the changes in PRL secretion which result from the local action of melatonin in the pituitary gland do not affect gonadotrophin secretion in the adult ram.

5. DIFFERENTIAL CONTROL OF GONADOTROPHIN AND PRL SECRETION

5.1. Implant and Lesion Experiments

The simplest interpretation of the experimental studies using the cerebral microimplants and the HPD model is that the melatonin signal that transduces the effects of photoperiod acts at different sites to regulate gonadotrophin and PRL secretion. The caudal region of the MBH appears to be the principle site for the control of gonadotrophin secretion, while the pars tuberalis/pituitary gland appears to be the principle site for the control of PRL secretion. The initial observation that microimplants of melatonin placed in the MBH simultaneously affects both FSH and PRL

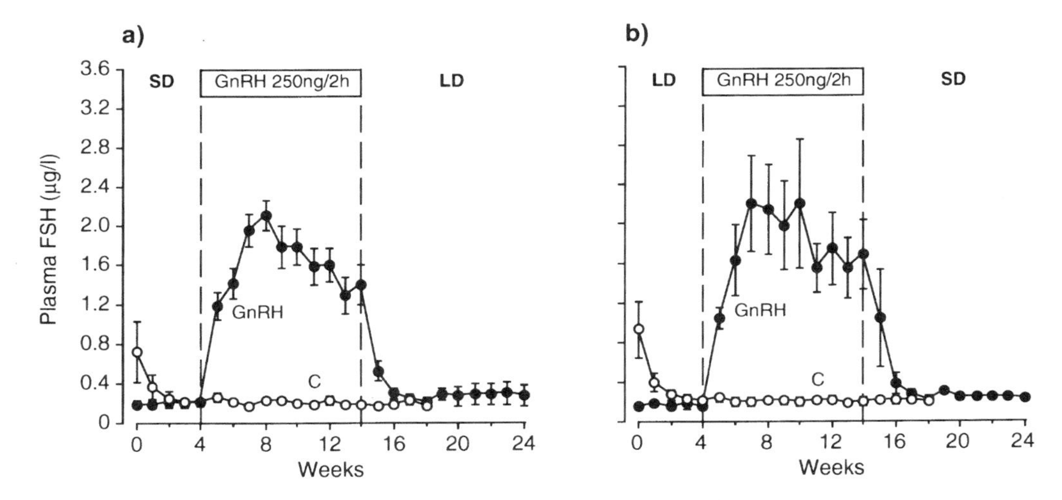

Figure 6. Long-term changes in blood plasma concentrations of FSH in HPD Soay rams treated with a pulsatile infusion of GnRH for 10 weeks (or no treatment—c) while exposed to: a) short days (8L:16D, SD) and b) long days (16L:8D, LD). Note, a similar response irrespective of the ambient photoperiod (38).

secretion (Figure 3), is therefore explained by the diffusion of melatonin from the implant to the two closely adjacent control sites. In the MBH, the increased exposure to melatonin activates gonadotrophin secretion, while in the pituitary gland the same change supresses PRL secretion. This produces the inverse relationship between the two endocrine systems which is such a notable feature in sheep seasonality (e.g. Figure 1).

In the Syrian hamster, selective electrolytic lesions of the dorso-medial region of the MBH which shows enhanced binding of 2-(^{125}I)iodomelatonin, blocks the inhibitory effect of short days, or appropriately programmed daily infusions of melatonin on gonadotrophin secretion and testicular activity (51). This surgical intervention, however, spares the inhibitory PRL response. Since the GnRH neurosecretory cells, that regulate gonadotrophin secretion are not located close to the putative target site in the MBH, it is presumed that melatonin acts via the neural networks which modulate pulsatile GnRH secretion. Dopaminergic, serotonergic and opioidergic pathways are particularly implicated (52,66). For the control of PRL in the pituitary gland, melatonin is thought to act via the pars tuberalis, thus in both the hypothalamus and pituitary gland the target cells expressing the melatonin receptor are one component of a cellular relay.

5.2. Two Different Systems

The concept that melatonin acts via different transduction pathways to affect gonadotrophin and PRL secretion is also consistent with the differences in timing of the endocrine events in relation to manipulations of photoperiod or the administration of melatonin. In sheep, a decrease in blood PRL concentrations can occur within 3 days following an abrupt switch from long to short days (46), while it takes 4 weeks for LH pulse frequency and FSH concentrations to begin to increase (42). In the longer term, the refractory response to a fixed photoperiod develops differently for the two endocrine systems (63) and the cyclical changes in the blood concentrations of PRL and gonadotrophins become dissociated (25,30). In addition, there is evidence from studies in Syrian hamsters that photoperiodic history has minimal effect on the PRL response to a photoperiodic challenge but has a marked effect on the gonadotrophin/gonadal response (18). In the Siberian hamster the critical daylength for the long-day activation of the PRL/pelage axis is different from that for the activation of gonadotrophin/gonadal axis (16).

Furthermore, there is no consistent temporal association across species. In paticular, most, if not all photoperiodic species show an increase in PRL secretion in summer in response to long days irrespective of the characteristics of the gonadotrophin response to photoperiod (long vs short day breeding species). This lack of a relationship between PRL and gonadotrophin secretion applies for closely related species within one taxonomic group (e.g. deer; 29,53) and even within a single species (Figure 7). In sheep, artificial selection aimed at extending the duration of the lambing season has produced breeds of sheep which show a seasonal maximum in FSH secretion earlier in the year compared to the wild-type. This genetic change, however, has not been accompanied by a corresponding change in the timing of the seasonal PRL cycle which peaks in mid-summer for all the domesticated breeds (9,31,45). Thus, artificial selection and natural selection can apparently independently influence the neuroendocrine mechanisms controlling PRL and gonadotrophin secretion indicative of separate control.

6. SITES ANCIENT AND MODERN

Overall, the data support a simple "dual site" hypothesis which proposes that melatonin acts independently in mediobasal hypothalamus to affect gonadotrophin secretion but acts in the pituitary gland to affect PRL secretion (Figure 8). Since mammalian species are remarkably similar in the photoperiodic control of PRL secretion it is likely that the melatonin-PRL control mechanism in the pituitary gland has been conserved during the evolution. This may be because all photoperiodic species share common requirement for an endocrine signal of summer. PRL is a pleiotrophic hormone inducing multiple effects on growth and metabolism and may thus consitute such a signal. PRL induces the development a summer pelage, for example, which is a common feature in mammals, and is clearly adaptive. The presence of high concentrations of melatonin receptors in the pars tuberalis in all photoperiodic mammals so far studied (54), is consistent with a generalised role of the pars tuberalis in the regulation of PRL secretion. In addition, the pars tuberalis is a conspicuous feature in the anatomy of the pituitary gland in reptiles and amphibians (17), and undergoes marked seasonal activation in spring in some species (e.g. frog: 65), thus the melatonin regulation of this tissue may have developed at an early phase of vertebrate evolution and constitutes an ancestral mechanism.

In contrast, the marked variation between species in the photoperiodic regulation of GnRH/gonadotrophin secretion indicates that there has been divergence in the neural mechanisms mediating the effect of melatonin on the gonadal cycle. The selective pressure to produce different responses to the melatonin signal must relate to the requirement to time conception and birth to optimise reproductive success, and this varies between species related to differences in body size, gestation length and ecology

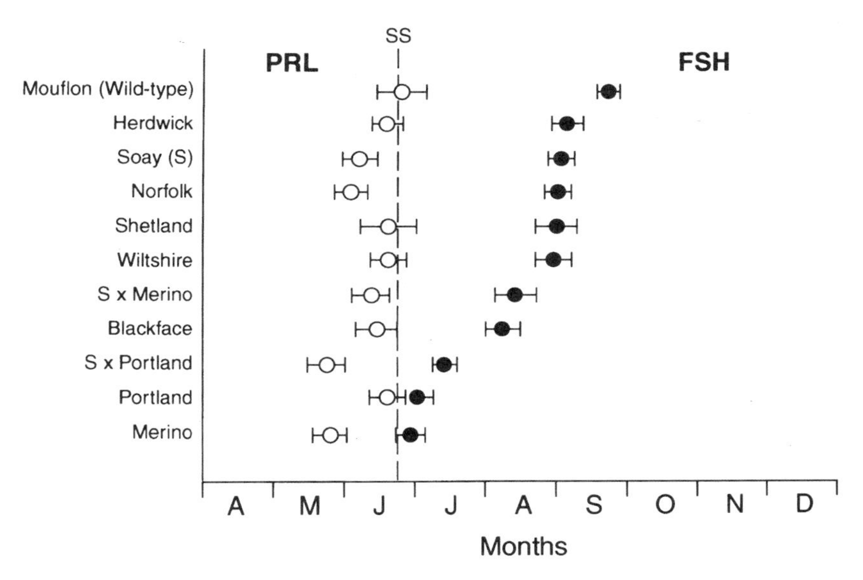

Figure 7. Timing of the seasonal peak in the blood concentrations of PRL and FSH in wild (mouflon), feral (Soay) and domesticated breeds of sheep selected for a prolongation of the lambing seaon (31,45).

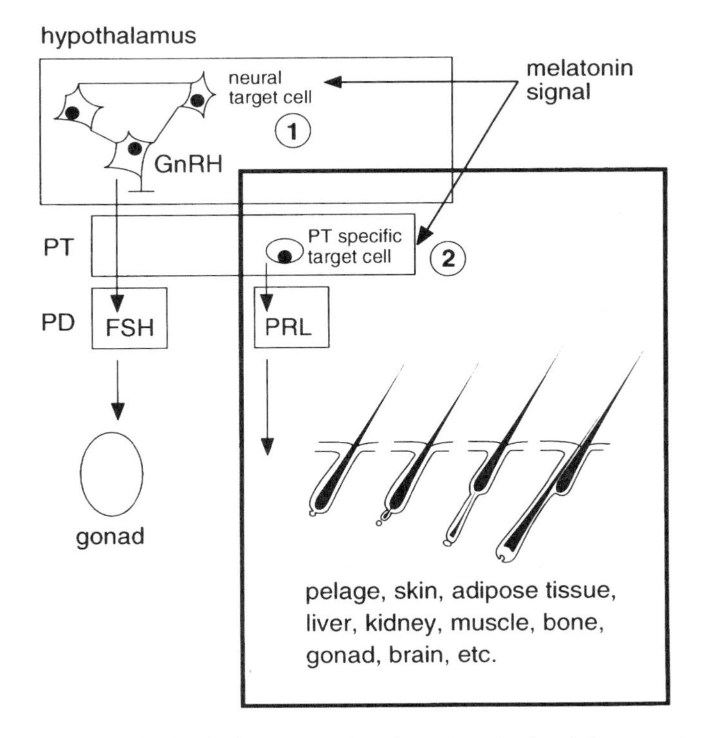

Figure 8. The "dual site hypothesis which proposes that the melatonin signal that transduces the effects of photoperiod acts: 1) in the mediobasal hypothalamus (MBH) via neural network to affect GnRH/gonadotrophin secretion and the gonadal axis, and (2) in the pituitary gland via the pars tuberalis (PT) to affect PRL secretion and the associated pleiotrophic responses.

(8). Currently the phenotype of the melatonin responsive cells in the MBH is unknown, but the prediction is that different inhibitory/stimulatory neural pathways will be targeted by melatonin in the different species to mediate the species-specific reproductive response.

REFERENCES

1. Arendt J. Role of the pineal gland and melatonin in seasonal reproductive function in mammals. Oxford Rev Rep Biol 1986;8:266–320.
2. Arendt J. In "Melatonin and the mammalian pineal gland" Chapman & Hall: London 1995 pp 110–160.
3. Arendt J. Melatonin and the pineal gland: influence on mammalian seasonal and circadian physiology. Rev Reprod 1998;3:13–22.
4. Bittman E.L. and Karsch F.J. Nightly duration of pineal melatonin secretion determines the reproductive response to inhibitory day length in the ewe. Biol Reprod 1984;585–593.
5. Bittman E.L. and Weaver D.R. The distribution of melatonin binding sites in neuroendocrine tissues of the ewe. Biol Reprod 1990;43:986–993.
6. Bittman E.L., Dampsey R.J., and Karsch F.J. Pineal melatonin secretion drives the reproductive response to daylength in the ewe. Endocrinology 1983;113:2276–2283.
7. Bole-Feysot C., Goffin V., Edery M., Binart N., and Kelly P.A. Prolactin (PRL) and its receptor: actions, signal transduction pathways and phenotypes observed in PRL receptor knockout mice. Endocr Rev 1998;19:225–268.
8. Bronson F.H. Mammalian Reproductive Biology. Chicago Press: Chicago 1993 pp 28–59.
9. Carr W.R. and Land R.B. Seasonal variation in plasma concentrations of prolactin in castrated rams of breeds of sheep with different seasonality of reproduction. J Reprod Fert 1982;66:231–235.
10. Carter D.S. and Goldman B.D. Antigonadal effects of timed melatonin infusion in pinealectomized male Djungarian hamster (*Phodopus sungorus*): Duration is the critical parameter. Endocrinology 1983;113:1261–1267.
11. Choy V.J., Nixon A.J., and Pearson A.J. Distribution of prolactin receptor immunoreactivity in ovine skin and changes during the wool follicle cycle. J Endocrinol 1997;155:265–275.
12. Clarke I.J., Cummins J.T., and de Kretser D.M. Pituitary gland function after disconnection from direct hypothalamic influences in sheep. Neuroendocrinol 1983;36:376–384.
13. Courot M., Hochereau-de-Reviers M.T., and Monet-Kuntz D. Endocrinology of spermatogenesis in the hypophysectomized ram. J Reprod Fertil 1979;26:165–173.
14. Dicks P., Russel A.J.F., and Lincoln G.A. The role of prolactin in the reactivation of hair follicles in relation to moulting in cashmere goats. J Endocrinol 1994;143:441–448.
15. Ducker M.J., Bowman J.C., and Temple A. The effects of constant photoperiod on the expression of oestrus in the ewe. J Reprod Fertil 1973;19:143–150.
16. Duncan M.J., Goldman B.D., Dipinto M.N., and Stetson M.H. Testicular function and pelage colour have different critical day-lengths in the Djungarian hamster (*Phodopus sungorus sungorus*). Endocrinology 1985;116:424–430.
17. Fitzgerald K.T. The structure and function of the pars tuberalis of the vertebrate adenohypophysis. Gen Comp Endocrinol 1979;37:383–399.
18. Hastings M.H., Walker A.P., and Powers J.B. Differential effects of photoperiodic history on the responses of gonadotrophin and prolactin in intermediate daylengths in the male Syrian hamster. J Biol Rhythms 1989;4:335–350.
19. Hazelrigg D.G., Hastings M.H., and Morgan P.J. Production of a prolactin releasing factor by the ovine *pars tuberalis*. J Neuroendocrinol 1996;8:489–492.
20. Helliwell R.J.A. and Williams L.M. Melatonin binding sites in the ovine brain and pituitary: characterisation during the oestrus cycle. J Neuroendocrinol 1992;4:287–294.
21. Hinds L.A. and Loudon A.S. Mechanisms of seasonality in marsupials: a comparative view In "Marsupial Biology" Eds N. Saunders and L. Hinds, UNSW Press: Sydney 1995 pp 41–49.
22. Houghton D.C., Young I.R., and McMillen I.C. Response of prolactin to different photoperiods after surgical disconnection of the hypothalamus and pituitary in sheep foetuses. J Reprod Fertil 1995;104:199–206.
23. Houghton D.C., Young I.R., and McMillen I.C. Photoperiodic history and hypothalamic control of prolactin secretion before birth. Endocrinology 1997;138:1506–1511.

24. Jansen H.T. and Jackson G.L. Circannual rhythms in the ewe: patterns of ovarian cycles and prolactin under two different constant photoperiods. Biol Reprod 1993;49:627–634.
25. Karsch F.J., Robinson J.E., Woodfill C.J.I., and Brown M.B. Circannual cycles in luteinizing hormone and prolactin secretion in ewes during prolonged exposure to a fixed photoperiod: evidence for an endogenous reproductive rhythm. Biol Reprod 1989;41:1034–1046.
26. Kennaway D.J., Gilmore T.A., and Seamark R.F. Effect of melatonin feeding on serum prolactin and gonadotrophin levels and the onset of seasonal estrous cyclicity in sheep. Endocrinology 1982;1982:1766–1772.
27. Langford G.A., Ainsworth L., Marcus G.J., and Shrestha J.N.B. Photoperiod entrainment of testosterone, luteinizing hormone, follicle-stimulating hormone, and prolactin cycles in rams in relation to testis size and sperm quality. Biol Reprod 1987;37:489–499.
28. Lincoln G.A. Photoperiodic control of seasonal breeding in the ram: participation of the cranial sympathetic nervous system. J Endocrinol 1979;82:135–147.
29. Lincoln G. Seasonal breeding in deer. In "Biology of Deer Reproduction" Eds P. Fennessy and K. Drew The Royal Society of New Zealand Bull: 1985 pp 165–180.
30. Lincoln G.A. Significance of seasonal cycles in prolactin secretion in male mammals In "Perspectives in Andrology" Eds M. Serio Raven Press: New York 1989 pp 299–306.
31. Lincoln G.A. Correlation with changes in horns and pelage, but not reproduction, of seasonal cycles in the secretion of prolactin in rams of wild, feral and domesticated breeds of sheep. J Reprod Fertil 1990;90:285–296.
32. Lincoln G.A. Photoperiod, pineal and seasonality in large mammals. In "Pineal Research" Eds J. Arendt and P. Pevet John Libbey: London 1991 pp 211–218.
33. Lincoln G.A. Administration of melatonin into the mediobasal hypothalamus as a continuous or intermittent signal affects the secretion of follicle stimulating hormone and prolactin in the ram. J Pineal Res 1992;12:135–144.
34. Lincoln G.A. Effects of placing micro-implant of melatonin in the pars tuberalis, pars distalis and the lateral septum of the forebrain on the secretion of follicle stimulating hormone and prolactin, and testicular size in rams. J Endocrinol 1994;142:267–276.
35. Lincoln G.A. and Clarke I.J. Photoperiodically-induced cycles in the secretion of prolactin in hypothalamo-pituitary disconnected rams: evidence for translation of the melatonin signal in the pituitary gland. Neuroendocrinol 1994;6:251–260.
36. Lincoln G.A. and Clarke I.J. Evidence that melatonin acts in the pituitary gland through a dopamine-independent mechanism to mediate effects of daylength on the secretion of prolactin in the ram. J Neuroendocrinol 1995;7:637–643.
37. Lincoln G.A. and Clarke I.J. Refractoriness to a static melatonin signal develops in the pituitary gland for the control of prolactin secretion in the ram. Biol Reprod 1997;57:460–467.
38. Lincoln G.A. and Clarke I.J. Absence of photoperiodic modulation of gonadotrophin secretion in HPD rams following chronc pulsatile infusions of GnRH. J Neuroendocrinol 1998;10:461–471.
39. Lincoln G.A. and Ebling F.J.P. Effect of constant-release implants of melatonin on seasonal cycles in reproduction, prolactin secretion and moulting in rams. J Reprod Fertil 1985;73:241–253.
40. Lincoln G.A. and Maeda K.-I. Effects of placing micro-implants of melatonin in the mediobasal hypothalamus and preoptic area on the secretion of prolactin and b-endorphin in rams. J Endocrinol 1992;134:437–448.
41. Lincoln G.A. and Maeda K.-I. Reproductive effects of placing micro-implants of melatonin in the mediobasal hypothalamus and preoptic area in rams. J Endocrinol 1992;132:210–215.
42. Lincoln G.A. and Peet M.J. Photoperiodic control of gonadotrophin secretion in the ram: a detailed study of the temporal changes in plasma levels of follicle-stimulating hormone, luteinizing hormone and testosterone following an abrupt switch from long to short days. J Endocrinol 1977;74:355–367.
43. Lincoln G.A. and Richardson M. Photo-neuroendocrine control of seasonal cycles in body weight, pelage growth and reproduction: lessons from the HPD sheep model. Comp Biochem Physiol C pharmacol Toxicol Endocrinol 1998;119:283–294.
44. Lincoln G.A., Libre E.A., and Merriam G.R. Long-term reproductive cycles in rams after pinealectomy or superior cervical ganglionectomy. J Reprod Fertil 1989;85:687–704.
45. Lincoln G.A., Lincoln C.E., and McNeilly A.S. Seasonal cycles in the blood plasma concentration of FSH, inhibin and testosterone, and testicular size in rams of wild, feral and domesticated breeds of sheep. J Reprod Fertil 1990a;88:623–633.

46. Lincoln G.A., McNeilly A.S., and Cameron C.L. The effects of a sudden decrease or increase in daylength on prolactin secretion in the ram. J Reprod Fertil 1978;52:305–311.
47. Malpaux B., Daveau A., Maurice F., Gayrard V., and Thiery J.C. Short-day effects of melatonin on luteinizing hormone secretion in the ewe: evidence for central sites of action on the mediobasal hypothalamus. Biol Reprod 1993;48:752–760.
48. Malpaux B., Daveau F., Maurice F., Locatelli A., and Thiery J.-C. Evidence that melatonin binding sites in the pars tuberalis do not mediate the photoperiodic actions of melatonin on LH and prolactin secretion in ewes. J Reprod Fertil 1994;101:625–632.
49. Malpaux B., Daveau A., Maurice-Mandon F., Duarte G., and Chemineau P. Evidence that melatonin acts in the premammillary hypothalamic area to control reproduction in the ewe: presence of binding sites and stimulation of luteinizing hormone secretion by *in situ* microimplant delivery. Endocrinology 1998;139:1508–1516.
50. Malpaux B., Skinner D.C., and Maurice F. The ovine pars tuberalis does not appear to be targeted by melatonin to modulate luteinising hormone secretion but may be important for prolactin release. J Neuroendocrinol 1995;7:199–206.
51. Maywood E.S. and Hastings M.H. Lesions of the iodomelatonin-binding sites of the mediobasal hypothalamus spare the lactotroph but block the gonadotroph response of male Syrian hamsters to short photoperiod and to melatonin. Endocrinology 1995;136:144–153.
52. Meyer S.L. and Goodman R.L. Neurotransmitters involved in mediating the steroid-dependent suppression of pulsatile luteinizing hormone secretion in anestrous ewes: effects of receptor antagonists. Endocrinology 1985;116:2054–2061.
53. Montford S.L., Brown J.L., Bush M., Wood T.C., Wemmer C., Vargas A., Williamson L.R., Montali R.J., and Wildt D.E. Circannual inter-relationships among reproductive hormones, gross morphometry, behaviour, ejaculate characteristics and testicular histology in Eld's deer stags (Cervus eldi thamin). J Reprod Fertil 1993;98:471–480.
54. Morgan P.J. and Williams L.M. The pars tuberalis of the pituitary: a gateway for neuroendocrine output. Rev Reprod 1996;1:153–161.
55. Morgan P.J., Lawson W., Davidson G., and Howell H.E. Guanine nucleotides regulate the affinity of melatonin receptors on the ovine pars tuberalis. Neuroendocrinol 1989;50:359–362.
56. Morgan P.J., Webster C.A., Mercer J.G., Ross A.W., Hazelrigg D.G., Maclean A., and Barrett P. The ovine pars tuberalis secretes a factor(s) which regulates gene expression in both lactotrophic and non-lactotrophic pituitary cells. Endocrinology 1996;9:4018–4026.
57. Morgan P.J., Williams L.M., Barrett P., Lawson W., Davidson G., Hannah L., and MacLean A. Differential regulation of melatonin receptors in sheep, chicken and lizard brains by cholera and pertussis toxins and guanine nucleotides. Neurochem Int 1996;28:259–269.
58. O'Callaghan D., Roche J.F., Boland M.P., and Karsch F.J. Does a melatonin implant mimic a short-day photoperiodic effect in ewes. J Anim Sci 1989;Suppl 1:(Abstract 879).
59. Pelletier J. and Almeida G. Short light cycles induce persistent reproductive activity of Ile-de-France rams. J Reprod Fertil 1987;34:215–226.
60. Reiter R.J. Pineal melatonin: cell biology of its synthesis and its physiological interactions. Endocrinology Reviews 1991;12:151–180.
61. Reppert S.M. and Weaver D.R. Melatonin madness. Cell 1995;83:1059–1062.
62. Reppert S.M., Weaver D.R., and Ebisawa T. Cloning and characterization of a mammalian melatonin receptor that mediates reproductive and circadian responses. Neuron 1994;13:1–20.
63. Sweeney T., Kelly G., and O'Callaghan D. Seasonal variation in long-day stimulation of prolactin secretion in ewes. Biol Reprod 1999;60:128–133.
64. Tillbrook A.J., de Kretser D.M., and Clarke I.J. Human recombinant inhibin A and testosterone act directly at the pituitary to suppress plasma concentrations of FSH in castrated rams. J Endocrinol 1993;138:181–189.
65. Vandenberghe M.P., Dierickx K., and Goosens N. Seasonal changes of the ultrastructure of the pars tuberalis of the hypophysis in *Rana temporaria*. Z Zellforsch 1973;145:459–469.
66. Viguie C., Thibault J., Thiery J.-C., Tillet Y., and Maulpaux B. Photoperiodic modulation of monoamines and amino-acids involved in the control of prolactin and LH secretion in the ewe: evidence for a regulation of tyrosine hydorxylase activity. J Neuroendocrinol 1996;8:465–474.
67. Wayne N.L., Malpaux B., and Karsch F.J. How does melatonin code for day length in the ewe: duration of nocturnal melatonin release or coincidence of melatonin with a light-entrained sensitive period. Biol Reprod 1988;39:66–75.

68. Wehr T.A. The duration of human melatonin secretion and sleep respond to changes in daylength. J Clin Endocrinol Metab 1991;73:1276–1280.
69. William L., Lincoln G.A., Mercer J.G., Barrett P., Morgan P.J., and Clarke I.J. Melatonin receptors in the brain and pituitary gland of hypothalamo-pituitary disconnected Soay rams. J Neuroendocrinol 1997;9:639–643.
70. Woodfill C.J.I., Wayne N.L., Moenter S.M., and Karsch F.J. Photoperiodic synchronization of a circannual reproductive rhythm in sheep: Identification of seasonal specific time cues. 71 duration of nocturnal melatonin secretion in determining the reproductive response to inductive photoperiod in the ewe. Biol Reprod 1985;32:523–529.
71. Yellon S.M., Bittman E.L., Lehman M.N., Olster D.H., Robinson J.E., and Karsch F.J. Importance of duration of nocturnal melatonin secretion in determining the reproductive response to inductive photoperiod in the ewe. Biol Repord 1985;32:523–529.

17

THE SIBERIAN HAMSTER AS A MODEL FOR STUDY OF THE MAMMALIAN PHOTOPERIODIC MECHANISM

Bruce D. Goldman

Department of Physiology and Neurobiology
University of Connecticut
U-154, Storrs, Connecticut 06269

Early studies in mammals (12) and other organisms established the role of the circadian system in photoperiodism. Specifically, several types of experiments demonstrated that the circadian system is used in the process of photoperiodic time measurement (PTM)—i.e., the measurement of day length. The results of these studies were generally compatible with the external coincidence hypothesis: Light falling during a particular phase of the circadian cycle has a photoinductive action; that is, the organism exhibits long day-type responses only if light is present during this photoinductive phase. Thus, light has two crucial roles in this model of the photoperiodic mechanism—(a) to entrain the circadian system and (b) to induce long day responses when coincident with the photoinductive phase (17).

SIBERIAN HAMSTER AS MODEL FOR STUDY OF PHOTOPERIODISM

The Siberian hamster (*Phodopus sungorus*) has served as a useful subject for investigation into the neuroendocrine basis of photoperiodism. Observations made in this species have contributed very significantly to current concepts regarding the importance of the circadian system and the pineal gland in the photoperiodic mechanism. Siberian hamsters exhibit several robust seasonal variations that are regulated primarily by day length. These include seasonal changes in reproduction, body weight and fat stores, thermoregulation, and pelage. During the summer, most Siberian hamsters exhibit reproductive activity, maximum body lipid stores, and a gray pelage. In winter, most individuals have regressed gonads and low circulating concentrations

Melatonin after Four Decades, edited by James Olcese.
Kluwer Academic / Plenum Publishers, New York, 2000.

of reproductive hormones, reduced body weight and fat stores, and a mostly white pelage (22). Winter hamsters frequently show daily torpor, in which body temperature falls as low as 15°C during the early hours of the morning while the animals are torpid and returns to 36–37°C during the remainder of the circadian cycle when animals are awake and may be active (21).

Early work clearly established the importance of the pineal gland for photoperiodic responses in Siberian hamsters (22), and continuous release implants of melatonin (MEL) were shown to exert marked effects on the reproductive system, resembling the changes evoked by certain photoperiod manipulations (23,25). Photoperiodic responses are apparent in Siberian hamsters by 15–20 days of age. In fact, juvenile male hamsters show more rapid reproductive responses to changes in day length as compared to adults (5,24,49). This rapid response was exploited in experiments that investigated the precise role of the pineal MEL rhythm (PMR) in photoperiodism in this species.

These early studies in Siberian hamsters were paralleled by numerous studies in Syrian hamsters that employed continuous release implants (34,45) or daily injections of MEL (42). These studies revealed that exogenous MEL could exert actions that either mimicked or antagonized the effects of short photoperiods. In both Syrian (43) and Siberian (25) hamsters, MEL was found to exert striking effects on the reproductive system in pinealectomized as well as pineal-intact animals. The results in pinealectomized animals were particularly important in that they indicated that MEL was acting as a pineal hormone, rather than acting within the pineal to regulate the secretion of some other pineal product. However, the results obtained with daily MEL injections differed in pineal-intact vs pinealectomized Syrian hamsters: In intact animals the effectiveness of single daily injections varied depending on the time-of-day of injection. Injections evoked gonadal regression when they were given during the afternoon or late in the night (2 h prior to lights-on), but not when given earlier in the night or during the morning (42,44). In pinealectomized hamsters, single daily injections of MEL did not induce gonadal regression; however, regression did occur when animals were given 3 injections of MEL/day, administered at 3 h intervals. With this regimen, it did not appear that the time-of-day of injections was of importance in the pinealectomized hamsters. It was suggested that these results might be explained by additive effects of exogenous and endogenous MEL, with elevated levels of endogenous MEL occurring during the dark phase (44). However, in another study, pinealectomized Syrian hamsters exhibited partial testis regression when given daily MEL injections at 2 h after lights-off, but did not respond to injections administered at other times (40).

DURATION OF NOCTURNAL MEL SECRETION AND PHOTOPERIODIC RESPONSE

Later studies employed timed daily infusions of MEL to allow experimental manipulation of not only the phase, but also the duration, of the daily period of elevated circulating MEL. These experiments revealed that the duration of exposure to MEL was crucial. In pinealectomized juvenile male Siberian hamsters, daily MEL infusions of 8–12 h duration inhibited testicular growth (6); infusions of 4 h or 6 h duration actually stimulated testis development (7). In pinealectomized hamsters, the nature of the gonadal response (inhibition or stimulation) appeared to depend only on the duration of each daily MEL infusion, not on the time-of-day at which the infusions were administered. However, in pineal-intact hamsters, short duration (5 h) MEL infusions inhibited testis growth when the infusions were timed to overlap with the endogenous, nocturnal elevation of pineal MEL; infusions of the same duration but administered beginning 2 h after lights-on failed to prevent testis development. These observations supported the hypothesis that the time-of-day effect of exogenous MEL observed in experiments with pineal-intact animals was related to temporal summation of endogenous and exogenous MEL (16).

Similar timed daily infusion studies were performed in sheep (4,47) and in Syrian hamsters (30). In these species also, the results indicated that the duration of elevated MEL levels was the most important determinant of the effects of the hormone on the reproductive system.

CIRCADIAN REGULATION OF PMR

As discussed above, early studies of mammalian photoperiodism led to two major types of findings: (a) evidence that the circadian system is employed in the measurement of day length and (b) support for the involvement of the PMR, and more specifically the duration of the nocturnal increase in pineal MEL secretion, as an endocrine "signal" related to the photoperiod. The relation between these two types of conclusions can be understood by examining the influence of the circadian system on the PMR. It was clear from early studies that the circadian oscillator(s) of the suprachiasmatic nuclei (SCN) is essential for regulating the PMR in mammals; this contrasts with the situation for birds and reptiles, where the pineal gland has both photoreceptors and an intrinsic circadian rhythmicity (46). Later results in laboratory rats were interpreted as suggestive that circadian regulation of the PMR was best explained by a dual-oscillator model. Furthermore, these observations indicated that both the initiation and the termination—and therefore the duration—of nocturnal MEL secretion are tightly regulated by the circadian system (10,27).

The relation between the circadian system and the duration of the MEL peak as a photoperiod signal is further clarified by examining results obtained with experimental paradigms that were used to test for circadian involvement in PTM. Perhaps the most convincing demonstration of a circadian mechanism in PTM derived from experiments employing light pulses applied in non-24 h T-cycles (12). In a later study, male Siberian hamsters showed testis regression when exposed to 1 h light pulses given in T-cycles of 24.0 h and 24.33 h; hamsters receiving 1 h light pulses in a T-cycle of 24.78 h maintained large testes. The T-cycles that were associated with testis regression resulted in long duration peaks in the circadian PMR, whereas the T-cycle that resulted in testis stimulation was associated with a short duration peak in the PMR (10).

Another paradigm used to test for circadian involvement in PTM employs a brief light pulse administered in the middle of the long dark phase of a short day photoperiod. In mammals, this night-interruption paradigm evokes long day type responses, as contrasted to the short day type responses that are evident in controls not receiving the night interruption light pulse. Light pulses given around the middle of the night have been shown to result in the inhibition of pineal MEL for the remainder of the dark phase in rats (27), thereby shortening the duration of the nocturnal MEL signal. Thus, the photoperiodic action of this paradigm can also be explained in terms of its effect on the duration of nocturnal MEL secretion.

ROLE OF CIRCADIAN SYSTEM IN RESPONSIVENESS TO MEL

Though evidence from several timed infusion studies in hamsters and sheep suggests that the time-of-day of exposure to MEL does not influence the photoperiodic response, there is other evidence that suggests some type of involvement of the circadian system in responsiveness to MEL. This evidence stems from studies investigating the frequency at which MEL signals must be presented in order to evoke photoperiodic responses. Siberian hamsters failed to respond to short day-like MEL infusions when the infusions were administered at 36 h or 48 h intervals (13). Syrian hamsters exhibited testicular regression when given long duration MEL infusions at intervals of 24 h or 25 h, but regression did not occur in hamsters receiving MEL infusions at 28 h intervals (30,31). It may be mere coincidence that the "limits of frequency" for MEL signal effectiveness lies so close to 24 h. However, observations in tau mutant Syrian hamsters suggest that there is a strong relation between the length of the circadian cycle and the limits of frequency for MEL signals. Tau mutant hamsters have freerunning circadian rhythms with period lengths (tau) of approximately 20 h. These animals responded to MEL infusions presented at 20 h intervals, but not to signals presented at 24 h intervals (41). However, MEL infusions administered at random times on consecutive days did result in testicular regression in male wild-type Syrian hamsters (20). Thus, the precise nature of the relation between the period length of circadian oscillations and the required frequency of MEL signals remains to be clarified.

Whereas it is clear that the circadian system drives the mammalian PMR, it has also been suggested that MEL has effects on the circadian system; however, there is uncertainty regarding the biological significance of these actions. The major evidence in support of MEL effects on the circadian system are: (a) It has been possible to entrain rats (9,33) and Syrian and Siberian hamsters (28) to daily injections or infusions of MEL. (b) Several species exhibit uptake of radiolabeled MEL in the SCN (3). (c) MEL can induce phase resetting in SCN tissue in vitro (32). (d) MEL can phase shift the human circadian clock and has been employed as a chronobiotic (29,35). Despite these observed actions of MEL, pinealectomized mammals generally exhibit "normal" circadian rhythms, so it is not clear that the PMR has an important role in the regulation of circadian rhythms in this vertebrate class. In non-mammalian vertebrates, the PMR appears to function primarily as a component of a multioscillatory circadian system (46). Therefore, to better understand the evolution of vertebrate MEL functions it will be important to learn more about putative actions of MEL on the mammalian circadian system.

GENETIC INFLUENCES ON PHOTORESPONSIVENESS

Another type of link between the circadian system and the photoperiodic role of the PMR is illustrated by studies of inter-individual variability in photoperiodic responsiveness in Siberian hamsters. When hamsters from our general breeding colony (UNS; unselected) were raised in 16 L and transferred to 10 L in adulthood, about 25–35% of the animals failed to exhibit inhibition of the reproductive system or to show a molt to winter pelage. By artificial selection, it was possible to develop a breeding stock in which >90% of the hamsters failed to exhibit these species-typical short day responses. In DD, these photoperiod nonresponsive (PNR) hamsters exhibited freerunning period lengths (tau$_{DD}$) that were <24h; in comparison, tau$_{DD}$ for photoperiod responsive hamsters was almost always <24h. The longer taus of the PNR hamsters were associated with a very different phase angle of entrainment to short day photoperiod as compared to that seen in the UNS animals. In 10L, the UNS hamsters exhibited onset of wheel-running activity within 1h after lights-off, whereas the PNR animals initiated activity at about 5–6h after the beginning of the dark period. It was suggested that the altered phase relation of the circadian system to the L:D cycle may result in light falling on the photoinductive portion of the circadian cycle in the PNR hamsters, but not in the UNS animals. This could explain the failure of PNR hamsters to undergo reproductive inhibition in short photoperiods. The importance of the phase angle of entrainment in short days was supported by the observation that PNR hamsters did undergo gonadal regression when exposed to DD for 8 weeks (14).

ENVIRONMENTAL INFLUENCES ON PHOTORESPONSIVENESS

In addition to the genetic component to photoperiod responsiveness in Siberian hamsters, these animals also show strong effects of environmental history, as demonstrated in two ways:

1) When PNR hamsters were exposed to short photoperiod and allowed free access to running wheels beginning very soon after weaning (18–20 days postpartum), they uniformly exhibited the typical short day responses—gonad regression and molt to winter pelage. These animals also displayed a phase angle of entrainment to the L:D cycle that was indistinguishable from that shown by UNS hamsters. PNR hamsters exposed to short days at weaning but without running wheels failed to exhibit the short day reproductive response or the pelage molt. When PNR animals were exposed to short days and allowed wheel access beginning in adulthood, they did not show the short day responses (14,15). These observations indicate that running wheel activity expressed early in life can negate the photoperiod effects of the PNR genotype. There are a number of questions that remain to be answered vis-à-vis this paradigm: What aspect of running wheel activity (i.e., metabolic expenditure, circadian nature of the activity) is involved in restoring the short day responsiveness of PNR hamsters? Does the running wheel effect observed in the laboratory have any significance for the behavior of hamsters in the field? What neural pathway is responsible for the effects of wheel-running on phase angle of entrainment to short days and photoperiod responsiveness?
2) When three groups of UNS male hamsters were reared from birth until approximately 2 months of age in 18L, 16L, and 14L, respectively, and then transferred to 10L for 8 weeks, the number of animals showing testis regression varied in relation to the photoperiod of rearing. Animals exposed to the longer photoperiods during the first 2 months of life were less likely to exhibit reproductive inhibition in 10L as compared to those reared in 14L. A similar pattern was observed for PNR males, though these animals generally showed lower frequencies of gonadal regression than the UNS hamsters for each of the photoperiod treatment groups (Dhandapani and Goldman, unpublished data). These findings correspond closely to the results of an earlier study that employed an unselected breeding stock of Siberian hamsters (19). Our

observations appear to support the concept, proposed by Gorman and Zucker, that photoperiod history has a major impact on the probability of response to short day exposure in adulthood (18). This appears to be true not only for the general laboratory population of Siberian hamsters, but also for animals subjected to artificial selection for failure to respond to short days.

The decrease in short day responsiveness following exposure to very long day lengths might be expected to result in a higher probability of winter breeding for animals born early in the breeding season (since these individuals would experience the longest days of mid-summer) as compared to those born later in the summer. Most Siberian hamsters probably do not survive for more than one year in nature; therefore, the photoperiod mechanism described above would likely result in the oldest individuals comprising the pool of winter breeders, whereas younger animals would channel their energy toward a "gamble" on survival through the winter to breed the following spring. This could represent a successful reproductive strategy: The older hamsters would be fully grown and presumably might be more successful at coping with the environmental challenge of winter breeding. Also, because of the relatively short life-expectancy of Siberian hamsters, forgoing winter breeding in a gamble on survival until spring might represent a greater risk for the older individuals in the population.

DEVELOPMENTAL ASPECTS OF PHOTOPERIODISM

Studies in montane voles (*Microtus montanus*) first revealed that the photoperiod in effect during gestation can influence the photoperiodic responses of the developing animals (26). Similar studies were performed in Siberian hamsters, revealing that in this species as well the dam is able to transmit a photoperiodic message to her fetuses. Thus, when male Siberian hamsters were raised from birth in 14L they exhibited larger testes at 1 month of age if their mothers had been exposed to a shorter day length during gestation than if the mother had experienced only photoperiods >14L during pregnancy (36,39). This mechanism may enable juvenile hamsters to rapidly assess the significance of a moderately long photoperiod (e.g., 14L) that occurs in both early and late summer. Thus, if photoperiod has increased between the time of gestation and postpartum exposure to 14L, this would be indicative of the early part of the breeding season. Under these conditions, hamsters would rapidly achieve puberty and would reproduce during the latter part of the summer season. By contrast, hamsters born in late summer would experience a decrease in day length between gestation and postpartum life; this might serve as an accurate cue that would lead to withholding reproduction until the following spring.

The ability of the dam to influence the photoperiod responses of her offspring is dependent on some action of her pineal gland; pups born to pinealectomized dams failed to exhibit differential responses to a postpartum 14L photoperiod in relation to the day length in effect during gestation. Further, when pinealectomized dams were administered daily MEL infusions during late pregnancy, the photoperiodic responses of their male pups were related to the duration of the MEL infusions: Mothers receiving long duration MEL infusions (to simulate a short day gestational photoperiod) gave birth to pups that showed relatively rapid testis growth in 14L; testis growth was slower in males born to mothers that had received long duration MEL infusions (simulating a long day gestational photoperiod). These results may indicate that the fetuses are able to use the mother's MEL signal as an indicator of the gestational photoperiod (48). This hypothesis is supported by the appearance of MEL binding sites in the brain and pars tuberalis of late fetal hamsters (8).

Further studies were performed to explore how the photoperiodic mechanism of the pups is modified by the photoperiod experience of the dam during gestation. The results indicated that the responsiveness of juvenile male hamsters to daily MEL infusions, administered at 14–32 days after birth, was not altered by the gestational photoperiod (37). However, the PMR of the male pups was influenced by gestational photoperiod. When raised from birth in 14L and examined 18 days postpartum, males whose mothers had been exposed to shorter day lengths produced longer duration MEL peaks as compared to males whose mothers were exposed to long days during pregnancy. This effect of gestation photoperiod on the PMR of the juvenile males could explain the differences in rate of postnatal testis development between pups gestated in long and short days, respectively. It seems likely that the maternal influence could be exerted via an action of the dam's PMR on the circadian system of the fetus—perhaps involving MEL receptors in the fetal SCN (8). Interestingly, no comparable effect of gestation photoperiod was observed for the PMR of female hamster pups (38).

SITES OF MEL ACTION

Sites of uptake of radiolabeled MEL have been identified in numerous mammals. The pars tuberalis of the pituitary exhibits uptake in almost all mammals and may be the only brain-related binding site in some species. Other uptake sites have been identified, and these vary among species (3). Siberian hamsters exhibit uptake of MEL in the pars tuberalis and SCN, and also in the periventricular and reuniens nuclei of the thalamus (11). SCN lesions prevented the action of systemic MEL infusions to inhibit reproductive parameters in Siberian hamsters (2), but not in Syrian hamsters (30). Microinfusions of MEL into the brains of juvenile male Siberian hamsters inhibited testis growth when MEL was delivered directly to the SCN, periventricular, or reuniens nuclei; infusions in other brain regions did not have this effect (1). Further studies will be required to determine whether the photoperiod-related actions of MEL at the hypothalamic and thalamic binding sites are redundant, and also to establish the role of MEL-binding in the hamster pars tuberalis.

SUMMARY

The Siberian hamster has been a useful model for studies of mammalian photoperiodism for a number of reasons:

1) Siberian hamsters are hardy animals that are easily maintained and bred in the laboratory.
2) The species exhibits a large number of seasonal, photoperiod-driven, pineal-dependent responses. Thus, the Siberian hamster is an excellent species in which to examine whether several different types of photoperiod responses share similar mechanistic features with respect to their control by MEL. Are all the responses cued to the duration of the nocturnal MEL peak? Does

MEL act at a single site to influence all the types of responses, or are there separate MEL target sites for different responses?

3) Juvenile Siberian hamsters exhibit an unusually rapid (for mammals) response to photoperiod change or to MEL treatments, making them ideal subjects for certain types of photoperiod-related studies.
4) Populations of Siberian hamsters show individual variations in photoperiod responsiveness, and the differences are at least partly heritable. These hamsters also exhibit strong influences of environmental history on short day responsiveness. Thus, the species may be a valuable model for the investigation of both genetic and environmental influences on the photoperiodic mechanism.
5) Siberian hamsters have proved to be useful animals in which to study maternal influences on the developing photoperiodic mechanism of the fetus.

ACKNOWLEDGMENT

This paper is dedicated in memory of Klaus Hoffmann. His numerous studies of photoperiodism in Siberian hamsters were important in the development of this area of research and established a sound basis for the use of this species in further investigations of circadian and photoperiodic mechanisms.

REFERENCES

1. Badura, L.L. and Goldman, B.D. Central sites mediating reproductive responses to melatonin in juvenile male Siberian hamsters. *Brain Res* **598**:98–106, 1992.
2. Bartness, T.J., Goldman, B.D., and Bittman E.L. SCN lesions block the reception of melatonin daylength signals in Siberian hamsters. *Am J Physiol* **260**:R102–R112, 1990.
3. Bittman, E.L. The sites and consequences of melatonin binding in mammals. *Am Zool* **33**:200–211, 1993.
4. Bittman, E.L. and Karsch, F.J. Nightly duration of pineal melatonin determines the reproductive response to inhibitory day length in ewe. *Biol Reprod* **30**:585–593, 1984.
5. Brackmann, M. and Hoffmann, K. Pinealectomy and photoperiod influence testicular development in the Djungarian hamster. *Naturwissenschaften* **64**:341, 1977.
6. Carter, D.S. and Goldman, B.D. Antigonadal effects of timed melatonin infusion in pinealectomized male Djungarian hamsters (Phodopus sungorus sungorus): Duration is the critical parameter. *Endocrinology* **113**:1261–1267, 1983a.
7. Carter, D.S. and Goldman, B.D. Progonadal role of the pineal in the Djungarian hamster (Phodopus sungorus sungorus): Mediation by melatonin. *Endocrinology* **113**:1268–1273, 1983b.
8. Carlson, L.L., Weaver, D.R., and Reppert, S.M. Melatonin receptors and signal transduction during development in Siberian hamsters (*Phodopus sungorus*). *Dev Brain Res* **59**:83–88, 1991.
9. Cassone, V.M., Wade, W.S., Brooks, D.S., and Lu, J. Melatonin, the pineal gland, and circadian rhythms. *J Biol Rhythms* **8**:S73–S81, 1993.
10. Darrow, J.M. and Goldman, B.D. Circadian regulation of pineal melatonin and reproduction in the Djungarian hamster. *J Biol Rhythms* **1**:39–54, 1986.
11. Duncan, M.J., Takahashi, J.S., and Dubocovich, M.L. Characteristics and autoradiographic localization of 1-[^{125}] iodomelatonin binding sites in Djungarian hamster brain. *Endocrinology* **125**:1011–1018, 1989.
12. Elliott, J.A. Circadian rhythms and photoperiodic time measurement in mammals. *Fed Proc* **35**:2339–2346, 1976.
13. Elliott, J.A., Bartness, T.J., and Goldman, B.D. Effect of melatonin infusion duration and frequency on gonad, lipid, and body mass in pinealectomized male Siberian hamsters. *J Biol Rhythms* **4**:439–455, 1989.

14. Freeman, D.A. and Goldman, B.D. Evidence that the circadian system mediates photoperiodic nonresponsiveness in Siberian hamsters: The effect of running wheel access on photoperiodic responsiveness. *J Biol Rhythms* **12**:100–109, 1997a.
15. Freeman, D.A. and Goldman, B.D. Photoperiod nonresponsive Siberian hamsters: Effect of age on the probability of nonresponsiveness. *J Biol Rhythms* **12**:110–121, 1997b.
16. Goldman, B.D., Darrow, J.M., and Yogev, L. Effects of timed melatonin infusions on reproductive development in the Djungarian hamster (*Phodopus sungorus*). *Endocrinology* **114**:2074–2083, 1984.
17. Goldman, B.D. and Elliott, J.A. Photoperiodism and seasonality in hamsters: Role of the pineal gland. In: Processing of Environmental Information in Vertebrates, MH Stetson (ed.), Springer-Verlag, New York, pp. 203–218, 1988.
18. Gorman, M.R. and Zucker, I. Seasonal adaptations of Siberian hamsters. II. Pattern of change in day length controls annual testicular and body weight rhythms. *Biol Reprod* **53**:116–125, 1995.
19. Gorman, M.R. and Zucker, I. Environmental induction of photononresponsiveness in the Siberian hamster, *Phodopus sungorus*. *Am J Physiol* **272**:R887–R895, 1997.
20. Grosse, J., Maywood, E.S., Ebling, F.J.P., and Hastings, M.H. Testicular regression in pinealectomized Syrian hamsters following infusions of melatonin delivered on non-circadian schedules. *Biol Reprod* **49**:666–674, 1993.
21. Heldmaier, G. and Steinlechner, S. Seasonal pattern and energetics of short daily torpor in the Djungarian hamster, *Phodopus sungorus*. *Oecologia* **48**:265–270, 1981.
22. Hoffmann, K. The influence of photoperiod and melatonin on testis size, body weight, and pelage colour in the Djungarian hamster (*Phodopus sungorus*). *J Comp Physiol* **95**:267, 1973.
23. Hoffmann, K. Testicular involution in short photoperiods inhibited by melatonin. *Naturwissenschaften* **61**:364, 1974.
24. Hoffmann, K. Effect of short photoperiods on puberty, growth and moult in the Djungarian hamster (*Phodopus sungorus*). *J Reprod Fertil* **54**:29, 1978.
25. Hoffmann, K. and Kuderling, I. Antigonadal effects of melatonin in pinealectomized Djungarian hamsters. *Naturwissenschaften* **64**:339–340, 1977.
26. Horton, T.H. Growth and reproductive development of male *Microtus montanus* is affected by the prenatal photoperiod. *Biol Reprod* **31**:499–504, 1984.
27. Illnerova, H. and Vanecek, J. Two-oscillator structure of the pacemaker controlling the circadian rhythm of N-acetyltransferase in the rat pineal gland. *J Comp Physiol A* **145**:539–548, 1982.
28. Kirsch, R., Beignaoui, S., Gourmelen, S., and Pevet, P. Daily melatonin infusion entrains free-running activity in Syrian and Siberian hamsters. In: Light and Biological Rhythms in Man, L Wetterberg (ed.), Pergamon Press, pp. 107–120, 1993.
29. Lewy, A.J., Ahmed, S., and Sack, R.L. Phase shifting the human circadian clock using melatonin. *Behav Brain Res* **73**:131–134, 1995.
30. Maywood, E.S., Buttery, R.C., Vance, G.H.S., Herbert, J., and Hastings, M.H. Gonadal responses of the male Syrian hamster to programmed infusions of melatonin are sensitive to signal duration and frequency but not to signal phase nor to lesions of the suprachiasmatic nuclei. *Biol Reprod* **43**:174–182, 1990.
31. Maywood, E.S., Grosse, J., Lindsay, J., Karp, J.D., Powers, J.B., Ebling, F.J.P., Herbert, J., and Hastings, M.H. The effect of signal frequency on the gonadal response of male Syrian hamsters to programmed melatonin infusions. *J Neuroendocrinol* **4**:37–43, 1992.
32. McArthur, A.J., Gillette, M.U., and Prosser, R.A. Melatonin directly resets the rat suprachiasmatic clock *in vitro*. *Brain Res* **565**:158–161, 1991.
33. Redman, J., Armstrong, S., and Ng, K.T. Free-running activity rhythms in the rat: Entrainment by melatonin. *Science* **219**:1089–1091, 1982.
34. Reiter, R.J., Vaughan, M.K., Blask, D.E., and Johnson, L.Y. Melatonin: its inhibition of pineal antigonadotrophic activity in male hamsters. *Science* **185**:1169–1171, 1974.
35. Sack, R.L. and Lewy, A.J. Melatonin as a chronobiotic: Treatment of circadian desynchrony in night workers and the blind. *J Biol Rhythms* **112**(6):595–603, 1997.
36. Shaw, D. and Goldman, B.D. Influence of prenatal and postnatal photoperiods on postnatal testis development in the Siberian hamster (*Phodopus sungorus*). *Biol Reprod* **52**:833–838, 1995a.
37. Shaw, D. and Goldman, B.D. Influence of prenatal photoperiods on postnatal reproductive responses to daily infusions of melatonin in the Siberian hamster (*Phodopus sungorus*). *Endocrinology* **136**:4231–4236, 1995b.
38. Shaw, D. and Goldman, B.D. Gender differences in influence of prenatal photoperiods on postnatal pineal melatonin rhythms and serum prolactin and follicle-stimulating hormone in the Siberian hamster (*Phodopus sungorus*). *Endocrinology* **136**:4237–4246, 1995c.

39. Stetson, M.H., Elliott, J.A., and Goldman, B.D. Maternal transfer of photoperiodic information influences the photoperiodic response of prepubertal Djungarian hamsters (*Phodopus sungorus sungorus*). *Biol Reprod* **34**:664–669, 1986.
40. Stetson, M.H. and Watson-Whitmyre, M. Effects of exogenous and endogenous melatonin on gonadal function in hamsters. *J Neural Transmission* **21**:55–80, 1986.
41. Stirland, J.A., Hastings, M.H., Loudon, A.S.I., and Maywood, E.S. The *tau* mutation in the Syrian hamster alters the photoperiodic responsiveness of the gonadal axis to melatonin signal frequency. *Endocrinology* **137**:2183–2186, 1996.
42. Tamarkin, L., Westrom, W.K., Hamill, A.I., and Goldman, B.D. Effect of melatonin on the reproductive systems of male and female hamsters: a diurnal rhythm in sensitivity to melatonin. *Endocrinology* **99**:1534–1541, 1976.
43. Tamarkin, L., Hollister, C.W., Lefebvre, N.G., and Goldman, B.D. Melatonin induction of gonadal quiescence in pinealectomized hamsters. *Science* **198**:953–955, 1977a.
44. Tamarkin, L., Lefebvre, N.G., Hollister, C.W., and Goldman, B.D. Effect of melatonin administered during the night on reproductive function in the Syrian hamster. *Endocrinology* **101**:631–634, 1977b.
45. Turek, F.W., Desjardins, C., and Menaker, M. Melatonin antigonadal and progonadal effects in male golden hamsters. *Science* **190**:280–282, 1975.
46. Underwood, H. and Goldman, B.D. Vertebrate circadian and photoperiodic systems: Role of the pineal gland and melatonin. *J Biol Rhythms* **2**:279–315, 1987.
47. Wayne, N.L., Malpaux, B., and Karsch, F.J. How does melatonin code for day length in the ewe: Duration of nocturnal melatonin release or coincidence of melatonin with a light-entrained sensitive period. *Biol Reprod* **39**:66–75, 1988.
48. Weaver, D.R. and Reppert, S.M. Maternal melatonin communicates daylength to the fetus in Djungarian hamsters. *Endocrinology* **119**:2861–2863, 1986.
49. Yellon, S.M. and Goldman, B.D. Photoperiod control of reproductive development in the male Djungarian hamster (*Phodopus sungorus*) *Endocrinology* **114**:664–670, 1984.

18

HOW DOES THE MELATONIN RECEPTOR DECODE A PHOTOPERIODIC SIGNAL IN THE PARS TUBERALIS?

Peter J. Morgan, Sophie Messager, Catriona Webster, Perry Barrett, and Alexander Ross

Molecular Neuroendocrinology Unit
Rowett Research Institute
Greenburn Road, Bucksburn, Aberdeen
Scotland, United Kingdom, AB21 9SB

1. INTRODUCTION

The biological function of melatonin in mammals is to convey temporal information about photoperiod to the neuroendocrine system, which regulates many major physiological axes (19). This process allows mammals to make anticipatory adaptive changes in their physiology and behaviour that enables optimal survival throughout the annual climatic cycle. Most apparent of these adaptations is the seasonality of reproduction (9). This strategy limits reproductive competence and behaviour of the male and female of the species to a particular time of the year, allowing their offspring to be born during spring when temperatures are rising and food availability is increasing. There are also other important physiological changes that can be equally dramatic, yet are less well studied. These include changes in pelage (10) appetite and body weight (13). The mechanisms involved in these processes are unknown, but understanding their control in the seasonal mammal offers potentially unique insights into the fundamental neuroendocrine regulation of several major physiological axes.

The key to this understanding is the mode of action of the melatonin at the cellular level. Knowledge of the molecular pharmacology and function of the melatonin receptors therefore is an essential prerequisite. Arguably no target site has contributed more, to our understanding of melatonin function at its receptor, than the pars tuberalis of the pituitary.

Melatonin after Four Decades, edited by James Olcese.
Kluwer Academic / Plenum Publishers, New York, 2000.

2. THE PARS TUBERALIS AS A TARGET SITE FOR MELATONIN

Until a decade ago the pars tuberalis (PT) had been dismissed as an accessory endocrine gland to the pars distalis (PD) (18). A major re-appraisal of this view was demanded when melatonin receptors were localised to this gland in rodents and sheep, yet these receptors were apparently absent from the pars distalis, at least in the post-neonate (17). This suggested that the pars tuberalis has a function related to photoperiod and melatonin, and one distinct from the pars distalis. Considerable focus and interest in the pars tuberalis was generated also because of all the central target sites for melatonin identified through radioligand binding in mammals, it was the only site labelled by 2-^{125}I-iodomelatonin in each of the species examined (17), suggesting a common function. In some species such as the ferret, it is the only central site of action that has been identified (24). These data, together with its anatomical position, provided strong support for a major role of the PT in the photoperiodic actions of melatonin. This has provided the impetus for studying the pars tuberalis as a target site for melatonin, and it has thereby become an important model gland for understanding the cellular mode of action of melatonin.

3. A PHYSIOLOGICAL ROLE FOR THE PARS TUBERALIS

Physiological studies involving hypothalamic-pituitary disconnected (HPD) Soay rams have shown that photoperiodically regulated cycles in plasma prolactin can still be maintained in the absence of hypothalamic input to the pituitary (8). As melatonin receptors are not located on the pars distalis, a direct action of melatonin upon the lactotrophs cannot explain the photoperiodic effects of melatonin on plasma prolactin levels. The possibility that the loss of hypothalamic pituitary input to the pituitary in HPD animals results in the "up-regulation" of melatonin expression in lactotrophs has also been discounted (25). Therefore the most plausible hypothesis is for the pars tuberalis of the pituitary to act as an endocrine intermediate in the regulation of prolactin, decoding the effects of melatonin and translating them into a new intrapituitary endocrine signal (5,18,19).

Using culture medium conditioned by ovine pars tuberalis cells we have demonstrated that a factor is secreted by pars tuberalis cells to stimulate prolactin release by ovine pars distalis (5,20). We have called this factor "tuberalin" (20). Conditioned medium from ovine pars tuberalis cells also stimulates c-fos expression in identified lactotrophs (20), and on the basis of simple chromatographic separations, it seems likely that this is also a biological activity associated with tuberalin (unpublished observations). These effects of tuberalin are not mimicked by medium conditioned by pars distalis cells. On the other hand, the factor causing the induction of c-fos expression in GH3 cells in response to medium conditioned by pars tuberalis cells (20) is chromatographically distinct from tuberalin and is also present in medium conditioned by pars distalis cells (unpublished observations). Thus tuberalin is a pars tuberalis specific secretion that increases prolactin secretion and gene expression in ovine lactotrophs, and thereby may contribute to the indirect photoperiodic regulation of prolactin synthesis and secretion.

4. DECODING THE MELATONIN SIGNAL

The ovine pars tuberalis has become a valuable pharmacological model for studying the function of the melatonin receptor, not only as it is amenable to cell culture and biochemical experiment, but also as it is a photoperiodically relevant target tissue. Using primary cultures of this gland it has been possible to define the acute signal transduction characteristics of the melatonin receptor.

Reverse transcription PCR, using degenerate primers, has been used to amplify melatonin receptor sequences from mRNA extracted from ovine pars tuberalis (oPT), and only the Mel1a (mt_1) receptor sub-type has been amplified, although we have found that this exists in two allelic forms (GenBANK accession number AF045219) (1). We have also amplified, cloned and sequenced an ovine melatonin-related receptor, which does not bind melatonin, from the ovine pars tuberalis (3), yet we have found no evidence for a sheep Mel1b (mt_2) receptor in this tissue or from genomic DNA (1) (unpublished observations). Competitive binding studies on oPT cell membranes using Luzindole, which has a weak selectivity for the hMel1b receptor sub-type relative to hMel1a receptor expressed in HEK293 cells (2), confirms that the binding affinity in the oPT matches that of the Mel1a receptor (7) (Figure 1). This suggests that the functional melatonin receptor expressed in the sheep PT is solely the Mel1a receptor.

In oPT melatonin acts through a pertussis toxin-sensitive G-protein to prevent and reverse forskolin stimulated cyclic AMP levels (14,15). Notably however melatonin does not affect basal cyclic AMP levels. The recombinant sheep and human Mel1a receptors stably expressed in L-cells and HEK293 cells respectively have similar signal transduction characteristics (1,2). These data suggest that the melatonin receptor acts via a Gi-protein to inhibit cyclic AMP. We have also shown that melatonin can inhibit PMA induced c-fos expression in a pertussis toxin sensitive manner in oPT cells (22). As this response is not mediated through cyclic AMP, it indicates that second messenger pathways other than cyclic AMP are susceptible to acute inhibition through the

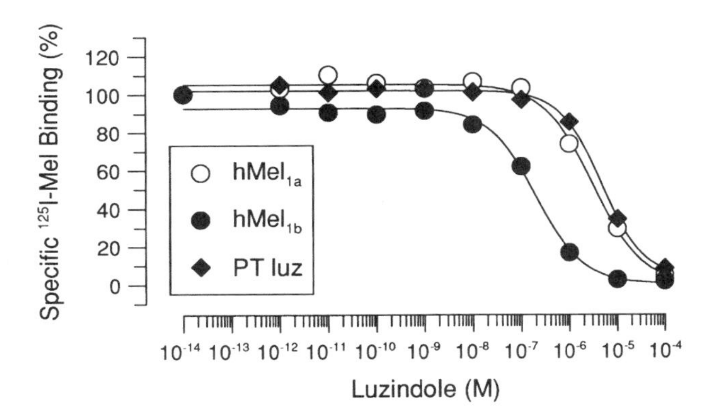

Figure 1. Competitive displacement of 2-^{125}I-iodomelatonin binding by Luzindole (2-benzyl-N-acetyltryptamine) from human Mel1a and human Mel1b receptors stably expressed in HEK293 cells, and comparison to displacement from melatonin receptors in oPT. Data show that Luzindole has a 10 fold selectivity for the Mel1b receptor, and receptor in oPT has an affinity like Mel1a for Luzindole. The data are re-drawn from Conway et al. 1997 (2) and Howell and Morgan 1991 (7). (NB. 10^{-14} defines the binding in the absence of Luzindole).

melatonin receptor in oPT cells (22). Melatonin has not been shown to have any independent acute effects through several signalling pathways in oPT cells, including activation of mitogen activated protein kinase (MAPK) (6), mobilisation of calcium (16), the turnover of phospholipases C and D (11,16), activation of protein kinase C (22) and the activation of c-fos (22). All these data suggest that the main function of melatonin is to prevent cellular activation by another external stimulus to the oPT cells, thereby rendering the cells functionally silent.

A prediction, if the above were true, would be that the activity of the PT would be quiescent during the hours of darkness co-incident with high levels of melatonin, and activated following light onset and the decline in melatonin. Therefore in the context of a photoperiodic role by melatonin, where the duration of the nocturnal melatonin signal determines the nature of the biological response, a simplistic interpretation would be that the PT is stimulated by another exogenous stimulus at all times when melatonin is not around. Thus under long days elevated nocturnal melatonin levels would be short, and the PT would be stimulated for an extended period. By contrast under short days the long duration of the nocturnal melatonin signal would allow the PT to be stimulated for a shorter period. This is a simplistic interpretation that takes no account of any periodicity in the availability of a humoral stimulus to the PT or of any change in receptor sensitivity to the stimulus during prolonged periods of stimulation.

In a recent paper by Sun et al., the first cloning of a mammalian homologue of Drosophila period gene, Per1 (RIGUI), was reported (23). Suprisingly, in addition to the expression of this gene in the SCN, where it was anticipated, Per expression was observed at ZT24 in the PT of the 129/SvEvBrd mice strain (23). Furthermore the expression in the PT was shown to be under circadian regulation as it continued under conditions of constant darkness (23). Although the secretion of melatonin in the 129/SvEvBrd strain of mice was not established, the authors showed that in C57 BL/6 mice, which have a genetic defect in melatonin production, there was no expression of Per1 in the PT (23). They concluded that melatonin present in the 129/SvEvBrd mouse strain was driving the rhythm of Per expression in the PT (23). If this conclusion is correct the data imply that melatonin may be driving Per expression in the PT during the hours of darkness, contrary to the prediction above. We have therefore investigated the expression of Per in sheep pars tuberalis as a marker of gene expression to define how nocturnal melatonin may be regulating its expression, and to test whether melatonin may be stimulating the PT gland.

A 404 bp cDNA fragment of the ovine Per1 gene was generated by reverse-transcription PCR from mRNA isolated from sheep PT (21). This was subsequently cloned in pGEM-T, and sequenced to verify that a *bone fide* part of the sheep Per1 gene had been obtained. Alignment of the amplified region, 287–690 of the human Per1 gene (GenBANK accession number AB002107) showed strong homology to the human sequence (Figure 2). The cDNA probe was used to define the regulation and temporal expression of the oPer1 gene in sheep PT cells by Northern blot. In the absence of any stimulation, oPT cells displayed only a minimal signal, a single band at 4.9 kb, which did not change in expression over 48 h of culture. Forskolin, a diterpene used to stimulate increased cyclic AMP, increased oPer1 mRNA expression dramatically after 2–4 h in culture, but this response returned to basal levels of expression within 6 h, indicating only a transient period of expression (Figure 3 and data not shown, see Morgan et al. 1998 (21)). Such transient expression is typical of early response genes, and consistent with this oPer1 was shown to increase dramatically in oPT cells

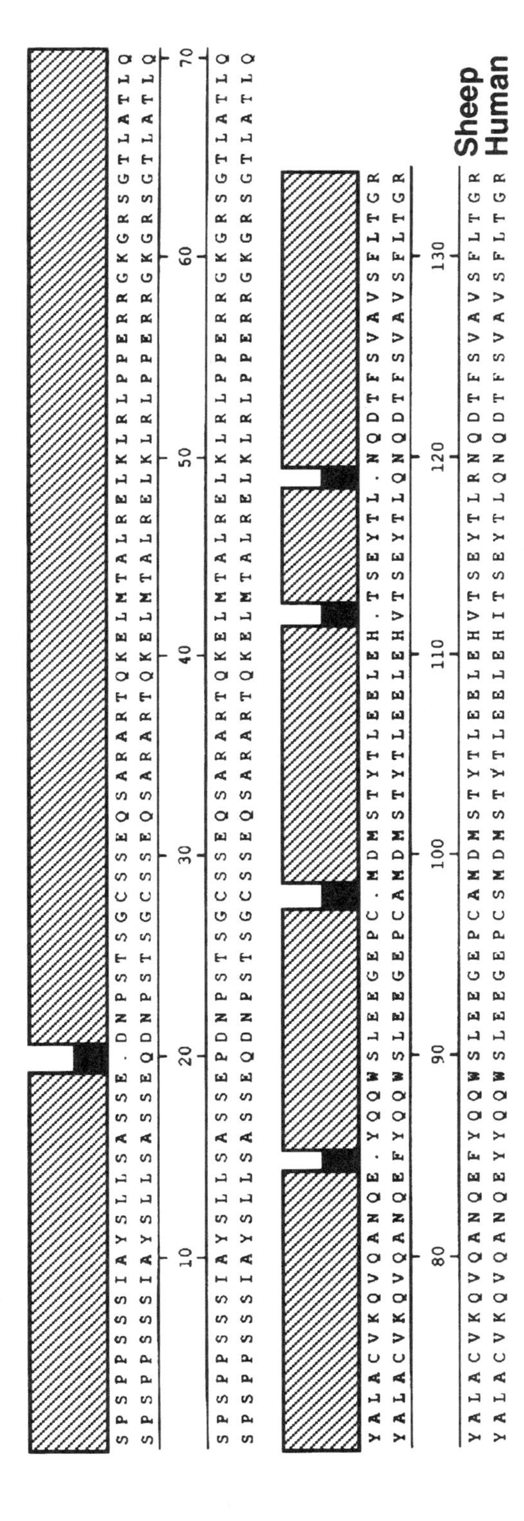

Figure 2. Alignment of predicted protein fragment of oPer1, translated from ovine PCR product, with its corresponding region in human Per1. It shows the close similarity of this region at the amino acid level, with only 5 changes occuring over 134 amino acid. This confirmed that a fragment of sheep Per1 had been amplified and cloned (GenBANK accession number AF044911).

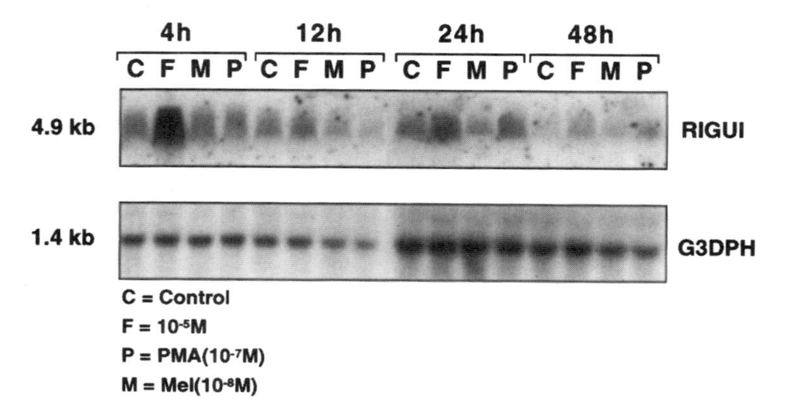

Figure 3. Northern blot showing the expression of oPer1 (4.9 kb band) in ovine PT cells over a 48 h period. G3PDH (1, 4 kb band) provides a reference for loading. The data show how the expression of oPer1 is increased after 4 h of forskolin stimulation, but is relatively unaffected by any treatments at any other of the time points. (From Morgan et al. 1998 (21)).

in response to forskolin stimulation after 2 h even in the presence of the 10 μg/ml cycloheximide. As cycloheximide is a protein synthesis inhibitor, the ability of oPer1 mRNA expression to be increased in its presence demonstrates that de novo protein synthesis is not required for enhanced gene expression of the oPer1 gene. These data confirm that oPer1 is an early response gene (21).

A parallel Northern blot showed that the expression of oPer1 in the pituitary is not restricted to the pars tuberalis. Expression within the pars distalis was also shown, although the response to stimulation by forskolin in the absence of cycloheximide was much weaker that in the PT. This probably reflects the weaker effect of forskolin of cyclic AMP levels in the PD that has been observed previously (12).

The phorbol ester, phorbol 12, 13, myristate acetate, stimulates protein kinase C, and we have shown that it will stimulate c-fos expression in oPT cells (22) (Fig. 3). This powerful stimulus had no effect on the induction of oPer1 expression, and therefore suggests that protein kinase C mediated pathways are not involved in the regulation of oPer1 expression.

As we have shown previously for the induction of c-fos, another early response gene, melatonin had no effect of the induction of oPer1 expression in oPT cells, but it strongly inhibited the forskolin-induced response, consistent with its known effect of preventing or reversing forskolin-induced cyclic AMP stimulation (data not shown). Therefore these data in ovine PT cell cultures are entirely consistent with all other data concerning the acute effects of melatonin, showing its ability to prevent cellular activation, yet having no independent effect over a 48 h period (see Fig. 3). Therefore these data do not support the notion that melatonin stimulates Per1 expression response directly.

We next tested how oPer1 expression changes *in vivo*. Six Suffolk cross-bred ewes were housed in a 12L:12D photoperiod. Subsequently three were killed 3 h after lights off, and three 3 h after lights on. The expression of oPer1 in both the PT and PD was measured by *in situ* hybridisation using an antisense ^{35}S-riboprobe for oPer1 followed by quantification by computing densitometry. Consistent with the Northern

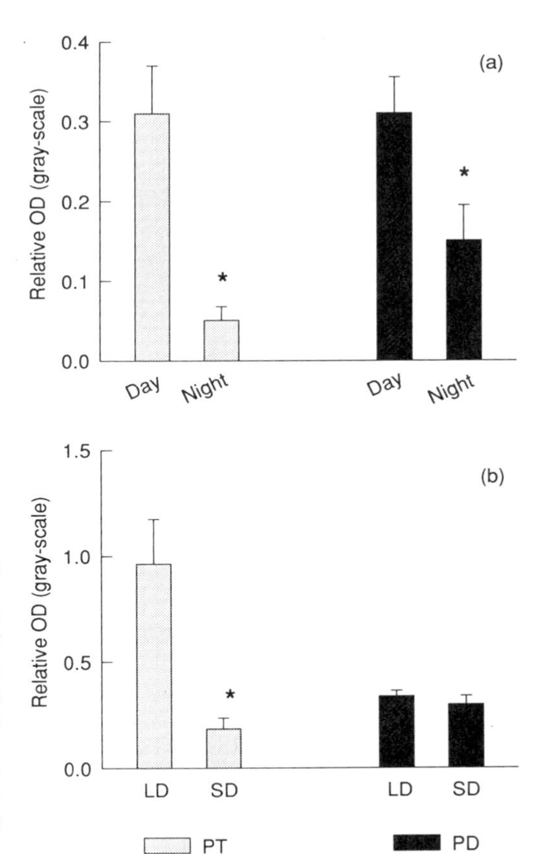

Figure 4. Densitometric analysis of oPer1 gene expression in sheep PT and PD, measured following *in situ* hybridisation using an ^{35}S-antisense riboprobe to oPer1. (a) shows diurnal expression of oPer1 in sheep pituitary measured 3 h after lights on (day) and 3 h after lights off (night). The sheep were maintained on a 12L:12D photoperiod. (b) shows effect of photoperiodic background on oPer1 expression in pituitary, where sheep were maintained on 16L:8D (LD) or 8L:16D (SD). * indicates significant day-night or LD-SD difference ($p < 0.001$).

blot findings oPer1 expression was detected in both the PT and PD. However the level of expression was affected by the time of day they were killed. In both the PT and PD a significantly higher level of expression was observed in the sheep killed 3 h after lights on, relative to those sheep killed 3 h into the dark period ($p < 0.001$) (Figure 4). This indicates that there is a diurnal pattern of expression of oPer1 in both the PT and PD, and this infers a diurnal pattern of stimulation to both glands. To assess the effect of long and short days on the expression of oPer1 in the pituitary, Soay rams which had been entrained to the following photoperiodic cycles were used: 8 week long days (LD; 16L:8D) or 8 weeks short days (SD; 8L:16D) (see Morgan et al. 1998 for details (21)). Animals from each photoperiod were killed 2 h after lights on, and then oPer1 expression assessed by *in situ* hybridisation and computing densitometry as above. A marked difference ($p < 0.001$) in the expression of oPer1 in the PT of LD and SD was observed, whereas no difference in expression was observed between the PD of LD and SD animals (Figure 4). This indicates that the difference in the level of oPer1 expression in the PT is due to the photoperiodic background, and hence melatonin. The lack of effect on the PD is explained, as there are no melatonin receptors on the PD.

These data allow several inferences to be drawn about how melatonin acts on the PT. Firstly, the increased level of oPer1 gene expression 3 h after lights on relative to 3h after lights off, in conjunction with the transient increase in expression of oPer1 mRNA in PT cells within 2h following stimulation with forskolin, infers that a humoral stimulus drives the expression of oPer1 in the pituitary (PT and PD) following lights on (and melatonin decline). From experiments using primary PT cell cultures we predict that when melatonin levels are high the PT will not be stimulated. Secondly the reduced amplitude in oPer1 mRNA expression following a short relative to a long photoperiodic signal (long vs short duration melatonin) infers that the duration of the nocturnal melatonin signal influences either the amplitude of the response to the humoral stimulus or alters the timing of the peak. Recent work in Syrian hamsters measuring Per1 expression at multiple time points shows that it is the amplitude of the peak in gene expression that is affected (unpublished observations). Thus these data provide the first evidence to show how a durational melatonin signal is interpreted by its target gland through gene expression. Namely duration is decoded into a signal of amplitude. Therefore while melatonin may be important in preventing cellular activation, these data suggest that an important aspect of melatonin signalling is a distinct "programming" effect related to its duration, and this influences subsequent cellular sensitivity to stimulation.

We have shown previously, using primary cell cultures, that melatonin alters the sensitivity of PT cells to produce cyclic AMP in response to stimulation by forskolin in a duration-dependent manner (4). This sensitisation process is pertussis-toxin sensitive (unpublished observations) and requires only physiological concentrations of melatonin (i.e. 100pM) (4), suggesting that the Mel1a receptor mediates this response. The mechanism involved is unknown, but it does not require protein synthesis, and is unlikely to involve any known signal transduction pathway (4). Nevertheless the results together with those above infer that melatonin, through its receptor, alters cellular sensitivity to stimulation.

There is however, a major discrepancy between the results produced *in vivo* and *in vitro*. *In vivo* the effect of short photoperiod (hence long duration melatonin) is to reduce the amplitude of oPer1 expression relative to the long photoperiod (short duration melatonin) animals. In PT cells long duration melatonin signals (8–16h) increase the amplitude of cyclic AMP production relative to short duration melatonin signals (0–4h). Such an amplified response in cyclic AMP would be expected to amplify oPer1 expression in short day animals, contrary to the abrogated response obtained. The reason for this apparent discrepancy is not immediately clear, and at the present time we cannot make all the parts of the jig-saw fit together. Nevertheless, both the *in vivo* and *in vitro* data strongly suggest that the mechanism through which the duration of the melatonin signal is relayed to the PT involves altered sensitivity of intracellular signalling pathways leading to a change in the amplitude of gene expression. Thus in terms of its biological function, the PT would be less biochemically active during short days than during long days.

ACKNOWLEDGMENTS

The authors would like to thank the Scottish Office Agriculture, Environment and Fisheries Department for financial support in this work. SM would like to acknowledge the support of the Fondation Singer-Polignac, France.

REFERENCES

1. Barrett, P., Conway, S., Jockers, R., Strosberg, A.D., Guardiola, B., Delagrange, P., and Morgan, P.J. (1997). Cloning and functional analysis of a polymorphic variant of the ovine Mel 1a melatonin receptor. Biochim. Biophys. Acta *1356*, 299–307.
2. Conway, S., Drew, J.E., Canning, S.J., Barrett, P., Jockers, R., Strosberg, A.D., Guardiola-lemaitre, B., Delagrange, P., and Morgan, P.J. (1997). Identification of Mel(1a) melatonin receptors in the human embryonic kidney cell line HEK293: Evidence of G protein-coupled melatonin receptors which do not mediate the inhibition of stimulated cyclic AMP levels. FEBS lett *407*, 121–126.
3. Drew, J.E., Barrett, P., Williams, L.M., Conway, S., and Morgan, P.J. (1998). The ovine melatonin-related receptor: cloning and preliminary distribution and binding studies. J Neuroendocrinol *10*, 651–661.
4. Hazlerigg, D.G., Gonzalez-Brito, A., Lawson, W., Hastings, M.H., and Morgan, P.J. (1993). Prolonged exposure to melatonin leads to time-dependent sensitization of adenylate cyclase and down-regulates melatonin receptors in pars tuberalis cells of ovine pituitary. Endocrinol *132*, 285–292.
5. Hazlerigg, D.G., Hastings, M.H., and Morgan, P.J. (1996). Production of a prolactin releasing factor by the ovine pars tuberalis. Journal of Neuroendocrinology *8*, 489–492.
6. Hazlerigg, D.G., Thompson, M., Hastings, M.H., and Morgan, P.J. (1996). Regulation of mitogen-activated protein kinase in the pars tuberalis of the ovine pituitary: interactions between insulin like growth factor-1 and forskolin. Endocrinol *137*, 210–218.
7. Howell, H.E. and Morgan, P.J. (1991). Luzindole (2-benzyl-N-acetyltryptamine), 5-methoxytryptamine, N-acetyltryptamine and 6-methoxy-2-benzoxazoline activity in ovine pars tuberalis cells. Adv Pineal Res *5*, 205–207.
8. Lincoln, G.A. and Clarke, I.J. (1994). Photoperiodically-induced cycles in secretion of prolactin in hypothalamo-pituitary disconnected rams. Evidence for translation of the melatonin signal in the pituitary gland. J Neuroendocrinol *6*, 251–260.
9. Lincoln, G.A. and Short, R.V. (1980). Seasonal breeding: nature's contraceptive. Rec Prog Hormone Res *36*, 1–52.
10. Lincoln, G.A. (1990). Correlation with changes in horns and pelage, but not reproduction, of seasonal cycles in the secretion of prolactin in rams of wild, feral and domesticated breeds of sheep. J Reprod Fert. 90:285–296.
11. McNulty, S., Morgan, P.J., Thompson, M., Davidson, G., Lawson, W., and Hastings, M.H. (1994). Phospholipases and melatonin signal transduction in the ovine pars tuberalis. Mol Cell Endcrinol *99*, 73–79.
12. McNulty, S., Ross, A., Barrett, P., Hastings, M.H., and Morgan, P.J. (1994). Melatonin regulates the phosphorylation of CREB in ovine pars tuberalis cells. J Neuroendocrinol *6*:523–532.
13. Mercer, J.G. (1998). Regulation of appetite and body weight in seasonal mammals. Comp Biochem Physiol. C Pharmacol Toxicol Endocrinol *119*, 295–303.
14. Morgan, P.J., Davidson, G., Lawson, W., and Barrett, P. (1990). Both pertussis toxin-sensitive and insensitive G-proteins link melatonin receptor to inhibition of adenylate cyclase in the ovine pars tuberalis. J Neuroendocrinol *2*, 773–776.
15. Morgan, P.J., Lawson, W., and Davidson, G. (1991). Interaction of forskolin and melatonin on cyclic AMP generation in pars tuberalis cells of ovine pituitary. J Neuroendocrinol *3*, 497–501.
16. Morgan, P.J., Hastings, M.H., Thompson, M., Barrett, P., Lawson, W., and Davidson, G. (1991). Intracellular signalling in the ovine pars tuberalis: an investigation using aluminium fluoride and melatonin. J Mol Endocrinol *7*, 137–144.
17. Morgan, P.J., Barrett, P., Howell, H.E., and Helliwell, R. (1994). Melatonin receptors: localization, molecular pharmacology and physiological significance. Neurochem Int *24*, 101–146.
18. Morgan, P.J. and Williams, L.M. (1996). The pars tuberalis of the pituitary: a gateway for neuroendocrine output. Reviews in Reproduction *1*, 153–161.
19. Morgan, P.J. and Mercer, J.G. (1994). Control of seasonality by melatonin. Proceedings of the Nutrition Society *53*, 483–493.
20. Morgan, P.J., Webster, C.A., Mercer, J.G., Ross, A.W., Hazlerigg, D.G., Maclean, A., and Barrett, P. (1996). The ovine pars tuberalis secretes a factor(s) that regulates gene expression in both lactotrophic and nonlactotrophic pituitary cells. Endocrinol *137*, 4018–4026.
21. Morgan, P.J., Ross, A.W., Graham, E.S., Adam, C., Messager, S., and Barrett, P. (1998). oPer1 is an early response gene under photoperiodic regulation in the ovine pars tuberalis. J Neuroendocrinol *10*: 319–323

22. Ross, A.W., Webster, C.A., Thompson, M., Barrett, P., and Morgan, P.J. (1998). A novel interaction between inhibitory melatonin receptors and protein kinase C-dependent signal transduction in ovine pars tuberalis cells. Endocrinol *139*, 1723–1730.
23. Sun, Z.S., Albrecht, U., Zhuchenko, O., Bailey, J., Eichele, G., and Lee, C.C. (1997). RIGUI, a putative mammalian ortholog of the Drosophila period gene. Cell *90*, 1003–1011.
24. Weaver, D.R. and Reppert, S.M. (1990). Melatonin receptors are present in the ferret pars tuberalis and pars distalis, but not in brain. Endocrinol *127*, 2607–2609.
25. Williams, L.M., Lincoln, G.A., Mercer, J.G., Barrett, P., Morgan, P.J., and Clarke, I.J. (1997). Melatonin receptors in the brain and pituitary gland of the hypothalamo-disconnected sheep. J Neuroendocrinol *9*, 639–643.

19

DAILY AND CIRCADIAN EXPRESSION PATTERNS OF mt_1 MELATONIN RECEPTOR mRNA IN THE RAT PARS TUBERALIS

H. Y. Guerrero,[1,2] F. Gauer,[1] P. Pevet,[1] and M. Masson-Pevet[1]

[1] Neurobiologie des Fonctions Rythmiques et Saisonnières
UMR-CNRS 7518
Université Louis Pasteur
Strasbourg, France
[2] Cátedra de Fisiología
Escuela de Medicina "José María Vargas"
Universidad Central de Venezuela, Caracas
Venezuela

The mammalian pineal gland transduces lighting conditions into an endocrine message, the nocturnal synthesis and secretion of melatonin (17,20). Photoperiod is then coded hormonally by the duration of the nocturnally elevated plasma melatonin levels (7,14,23). Herewith, melatonin is involved in the regulation of seasonal biological rhythms, like reproduction (14,15,23). Among the large number of structures containing melatonin receptors, the pars tuberalis (PT) of the pituitary is the only one presenting a high number of 2-^{125}I-melatonin binding sites in all the mammalian species studied so far (14). To date, in mammals, two different melatonin receptor cDNAs have been cloned, called Mel_{1a} and Mel_{1b} (mt_1 and MT_2, respectively) (18,19). These authors showed also that the distribution of mt_1 mRNA is coincident with that of 2-^{125}I-melatonin binding. In response to the photoperiodic changes, seasonal variations of 2-^{125}I-melatonin binding in the PT of seasonal breeders have been described (2,4,13,16,21,22,23). At present time, these receptors are thought to be involved in mediating the photoperiodic effects of melatonin in the seasonal control of prolactin secretion (8,9,10,12).

Address correspondence to: Hilda GUERRERO, Neurobiologie des Fonctions Rythmiques et Saisonnières, UMR-CNRS 7518, Université Louis Pasteur, 12, rue de l'Université, 67000 Strasbourg, France. Telephone number: + (33) 3 88 35 85 16, Fax number: + (33) 3 88 24 04 61, E-Mail: guerrero@neurochem.u-strasbg.fr

Melatonin after Four Decades, edited by James Olcese.
Kluwer Academic / Plenum Publishers, New York, 2000.

In the rat, generally considered as a "non seasonal" species, both diurnal and circadian variations of the 2-^{125}I-melatonin receptor density have been described in the PT (3,5).

The aim of the present work was to study the daily and circadian expression patterns of mt_1 melatonin receptor mRNA in the rat PT, and to see whether it correlates with the daily and circadian changes in 2-^{125}I-melatonin binding. This study was performed in animals kept both in light/dark conditions and in constant darkness.

MATERIAL AND METHODS

Young adult male Wistar rats (Iffa Credo, France) were bred in a 12h light/12h dark regime (lights on at 07.00h), with free access to commercial chow and water. Two groups of animals were used: one kept in this light/dark regime (L/D group), and the other one transferred 3 days before the sacrifice in constant darkness (D/D group). For the two groups, a dim red light was permanently on, and the temperature was set at 20° ± 2°C. Animals from both groups were sacrificed by decapitation at different time-points along the 24 hours. Brains were rapidly removed and frozen in isopentane maintained at −30°C. Serial coronal sections (20 μm thickness) of the hypothalamic region containing the PT were cut on a cryostat, thaw mounted onto gelatin-coated slides and processed for the detection of the 2-^{125}I-melatonin binding or mt_1 mRNA by quantitative autoradiography and in situ hybridization, respectively.

Autoradiographic binding and in situ hybridization procedures as well as quantitative analysis of the autoradiograms were performed as previously described (6,16).

For melatonin assay, trunk blood of every animal was collected. Plasma melatonin concentrations were determined by radioimmunoassay (RIA) using a specific rabbit antiserum (R 19540) provided by INRA (Nouzilly, France) at a final dilution of 1/250,000, and 2-^{125}I-iodo-melatonin.

RESULTS

The present analysis of mt_1 receptor mRNA levels in the PT reveals the presence of daily and circadian rhythms of the expression of mt_1 melatonin receptors. Daily variations of mt_1 receptor mRNA show an increase during the second half of the dark period (Figure 1).

In animals kept in constant darkness, the mt_1 mRNA expression also increased during the second half of the subjective night (Figure 2). The increase in mt_1 mRNA expression in both cases preceded the increase in 2-^{125}I-melatonin binding by 4–6 hours. The 2-^{125}I-melatonin binding showed daily and circadian variations with high values during the day and the lowest value during the night (Figures 1, and 2). This pattern confirms the results already obtained in this species (3). The 2-^{125}I-melatonin binding variations were negatively correlated with plasma melatonin concentration in both groups.

DISCUSSION

These results show that the expression of mt_1 melatonin receptor mRNA is rhythmic in L/D and in D/D conditions, with a quite similar 24h time course pattern in both conditions. The daily variations in mt_1 melatonin receptor expression are thus circadian in the rat PT.

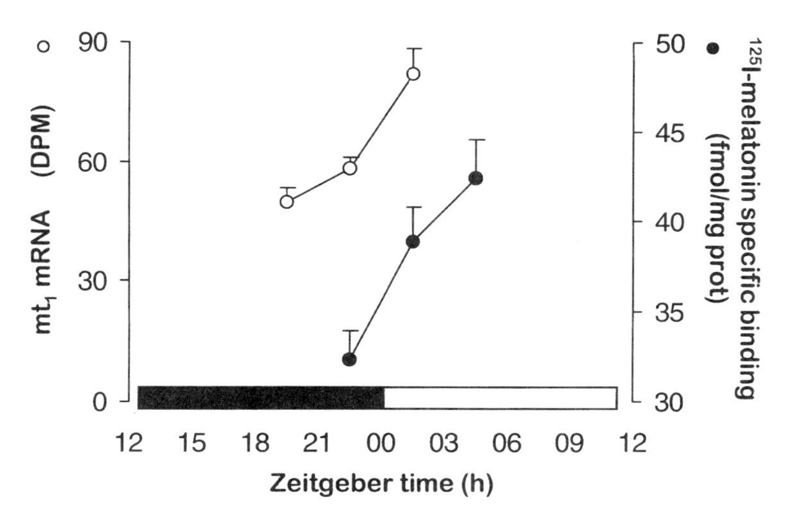

Figure 1. Diurnal variations in mt_1 mRNA levels (○) and in 2-^{125}I-melatonin specific binding (●) in the pars tuberalis of rats kept in a 12 h light/12 h dark regime (lights on at 07.00 h). Each point represents the mean ± SEM of five animals.

In a previous study, we showed in this species the presence of an endogenous rhythm in the density of PT melatonin receptors, which was not driven by the external light/dark cycle (3). The present results confirm and extend these findings. We show here that the 2-^{125}I-melatonin binding increase at the light/dark transition or at the late subjective night in L/D and D/D, respectively, follows by approximately 5 hours a parallel increase in mt_1 mRNA expression. At a first glance, it could indicate that the melatonin receptor density might correlate, with a phase delay, to the mt_1 mRNA synthesis. However, the significance of this time lag must be carefully analyzed since

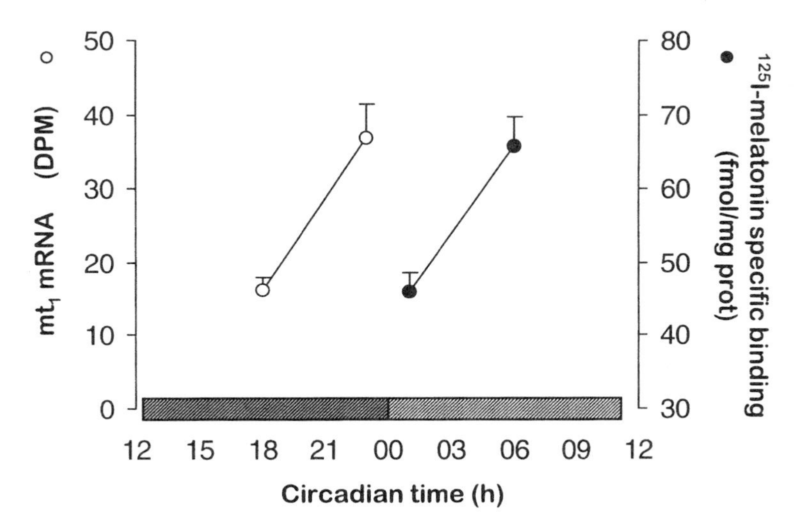

Figure 2. Circadian variations in mt_1 mRNA levels (○) and in 2-^{125}I-melatonin specific binding (●) in the pars tuberalis of rats placed for three days in continuous darkness. Each point represents the mean ± SEM of five animals.

many different factors could be interfering. For example, since the mRNA levels, as well as the protein levels, are depending on the balance between synthesis and degradation, it would be necessary to consider the turnover rates of both the mRNA and the protein. Moreover, it was described in the ovine PT in vitro, that forskolin which induces an important stimulation of the mt_1 mRNA levels, affects only moderately the melatonin receptor density (1). For these authors posttranscriptional mechanisms could exist which would regulate the receptor protein translation. Then, although our results demonstrate the circadian rhythmicity of the mt_1 mRNA in the rat PT, the physiological significance of this fluctuation on the receptor protein remains to be investigated.

A second arising question concerns the role of the nocturnal melatonin peak in the regulation of the mt_1 mRNA transcription. The mt_1 mRNA levels in L/D and D/D conditions starts to increase during the night, at about the same time when the nocturnal plasma melatonin peak starts to decrease. This observation could suggest that a decrease in plasma melatonin concentration could remove an inhibition of the mt_1 receptor transcription. There is some evidence that melatonin could inhibit the mt_1 mRNA expression in the ovine PT. Indeed, Barrett and colleagues (1), showed that melatonin suppresses partially the spontaneous increase in mt_1 gene expression and reverses the stimulatory effect of forskolin induced increase in mt_1 mRNA levels in cultured cells of sheep PT. Thus, it seems possible that melatonin could regulate negatively the synthesis of mt_1 mRNA and thus indirectly the synthesis of mt_1 receptors. Measurement of mt_1 mRNA throughout the L/D cycle after suppression of the melatonin peak, e.g., after pinealectomy, will allow us to answer this question. Nevertheless, the possibility of a direct effect of melatonin on the regulation of the receptor protein must not be underestimated.

In conclusion, this study provides the first evidence of a circadian regulation of the mt_1 mRNA subtype in the pars tuberalis in vivo. Further studies are in progress to provide more clues about the physiological mechanism of melatonin receptors.

ACKNOWLEDGMENTS

The authors wish to thank Professor S.M. Reppert (Massachusetts General Hospital, Boston, MA, USA) for kindly donating the pCRII vector containing the Mel_{1a} melatonin receptor partial cDNA.

REFERENCES

1. Barrett, P., MacLean, A., Davidson, G., and Morgan, P.J. Regulation of the Mel_{1a} melatonin receptor mRNA and protein levels in the ovine pars tuberalis: evidence for a cyclic adenosine 3′,5′-monophosphatase-independent Mel_{1a} receptor coupling and an autoregulatory mechanism of expression. Mol. Endocrinol. 10:892–902, 1996.
2. Gauer, F., Masson-Pévet, M., Saboureau, M., George, D., and Pévet, P. Differential seasonal regulation of melatonin receptor density in the pars tuberalis and suprachiasmatic nuclei: A study in the hedgehog (*Erinaceus europeus*). J. Neuroendocrinol. 5:685–690, 1993a.
3. Gauer, F., Masson-Pévet, M., Skene, D.J., Vivien-Roels, B., and Pévet, P. Daily rhythms of melatonin binding sites in the rat pars tuberalis and suprachiasmatic nuclei; evidence for a regulation of melatonin receptors by melatonin itself. Neuroendocrinology. 57:120–126, 1993b.

4. Gauer, F., Masson-Pévet, M., and Pévet, P. Seasonal regulation of melatonin receptors in rodent pars tuberalis: correlation with reproductive state. J. Neural. Transm. 96:187–195, 1994a.
5. Gauer, F., Masson-Pévet, M., Stehle, J., and Pévet, P. Daily variations in melatonin receptor density of rat pars tuberalis and suprachiasmatic nuclei are distinctly regulated. Brain Res. 641:92–98, 1994b.
6. Gauer, F., Schuster, C., Pévet, P., and Masson-Pévet, M. Effect of a light pulse on melatonin receptor density and mRNA expression in Siberian hamster suprachiasmatic nuclei. Neurosc. Lett. 233:49–52, 1997.
7. Goldman, B.D. and Darrow, J.M. The pineal gland and mammalian photoperiodism. Neuroendocrinology. 37:386–396, 1983.
8. Hazlerigg, D.G. and Morgan, P.J. The pars tuberalis: a pituitary target for melatonin. In Therapeutic potential of melatonin. G.J.M. Maestroni, A. Conti, R.J. Reiter, eds, Frontiers in Hormone Research, Vol. 23, Karger, Basel, pp. 3–13, 1997.
9. Lincoln, G.A. and Clarke, M.J. Photoperiodically-induced cycles in the secretion of prolactin in hypothalamo-pituitary disconnected rams: evidence for translation of the melatonin signal in the pituitary gland. J. Neuroendocrinol. 6:251–260, 1994.
10. Lincoln, G.A., Clarke, M.J., and Sweeney, T. "Hamster-like" cycles in testicular size in the absence of gonadotropin secretion in HPD rams exposed to long-term changes in photoperiod and treatment with melatonin. J. Neuroendocrinol. 8:855–866, 1996.
11. Masson-Pévet, M., George, D., Kalsbeek, A., Saboureau, M., Lakhdar-Ghazal, N., and Pévet,. P. An attempt to correlate brain areas containing melatonin-binding sites with rhythmic functions: a study in five hibernator species. Cell Tissue Res. 278:97–106, 1994.
12. Maywood, E.S., Bittman, E.L., and Hastings, M.H. Lesions of the melatonin- and androgen-responsive tissue of the dorsomedial nucleus of the hypothalamus block the gonadal response of male Syrian hamsters to programmed infusions of melatonin. Biol. Reprod. 54:470–477, 1996.
13. Messager, S., Caillol, M., George, D., and Martinet, L. Seasonal variation of melatonin binding sites in the pars tuberalis of the male mink (*Mustela vison*). J. Neuroendocrinol. 9:523–528, 1997.
14. Pévet, P. The role of the pineal gland in the photoperiodic control of reproduction in different hamster species. Reprod. Nutr. Develop. 28:443–458, 1988.
15. Ralph, M.R., Foster, R.G., Davis, F.C., and Menaker, M. Transplanted suprachiasmatic nucleus determines circadian period. Science 247:975–978, 1990.
16. Recio, J., Pévet, P., Vivien-Roels, B., Míguez, J.M., and Masson-Pévet, M. Daily and photoperiodic melatonin binding changes in the suprachiasmatic nuclei, paraventricular thalamic nuclei and pars tuberalis of the female siberian hamster (*Phodopus sungorus*). J. Biol. Rhythms 11:325–332, 1996.
17. Reiter, R.J. Pineal melatonin: Cell biology of its synthesis and of its physiological interactions. Endocrine Rev. 12:151–180, 1991.
18. Reppert, S.M., Weaver, D.R., and Ebisawa, T. Cloning and characterisation of a mammalian melatonin receptor that mediates reproductive and circadian responses. Neuron. 13:1–20, 1994.
19. Reppert, S.M., Godson, C.G., Mahle, C.D., Weaver, D.R., Slaugenhaupt, S.A., and Gusella, J.F. Molecular characterization of a second melatonin receptor expressed in human retina and brain: the Mel_{1b} melatonin receptor. Proc. Natl. Acad. Sci. USA. 92:8734–8738, 1995.
20. Rollag, M.D. and Niswender, G.D. Radioimmunoassay of serum concentrations of melatonin in sheep exposed to different lighting regimens. Endocrinology. 98:482–489, 1976.
21. Skene, D.J., Masson-Pévet, M., and Pévet, P. Seasonal changes in melatonin binding sites in the pars tuberalis of male Eurpean hamster and the effect of testosterone manipulation. Endocrinology. 132:1682–1686, 1993.
22. Stanton, T.L., Siuciak, J.A., Dubocovich, M.L., and Krause, D.N. The area of 2-^{125}I-melatonin binding in the pars tuberalis of the ground squirrel is decreasing during hibernation. Brain Res. 557:285–288, 1991.
23. Tamarkin, L., Baird, C.J., and Almeida, O.F.X. Melatonin: a coordinating signal for mammalian reproduction? Science. 227:714–720, 1985.
24. Vanecek, J. and Jansky, L. Short days induce changes in specific melatonin binding in hamster median eminence and anterior pituitary. Brain Res. 477:387–390, 1989.

20

MOLECULAR PHARMACOLOGY AND FUNCTION OF MELATONIN RECEPTOR SUBTYPES

Margarita L. Dubocovich, Monica I. Masana, and Susan Benloucif

Department of Molecular Pharmacology and Biological Chemistry
Northwestern Drug Discovery Program
Northwestern University Institute for Neuroscience
Northwestern University Medical School
303 East Chicago Avenue
Chicago, Illinois 60611

1. INTRODUCTION

1.1. Discovery of Melatonin and Its Receptors

The secretion of melatonin primarily from the vertebrate retina and pineal gland with high levels at night is controlled by circadian clocks locally within the retina and the hypothalamic suprachiasmatic nucleus and is synchronized by environmental light (1,2). The first biological activity of melatonin can be traced back to 1917 when McCord and Allan (3) discovered that extracts of bovine pineal gland caused blanching of *Rana pipiens* tadpole skin. This bioassay was used to isolate melatonin from pineal extracts which led to the elucidation of its chemical structure (4). The property of melatonin to aggregate pigment granules (melanosomes) of amphibian dermal melanophores was used 1) to postulate the presence of melatonin receptors and to propose that N-acetyltryptamine is a melatonin receptor antagonist (5), 2) to establish the first structure-activity relationships of melatonin analogues (5), and 3) to demonstrate, in cultured *Xenopus laevis* melanophores, that activation of melatonin receptors inhibits cAMP formation through coupling to a pertussis toxin-sensitive G-protein (6). Subsequently, expression cloning led to the isolation of the first cDNA encoding a melatonin receptor from the *Xenopus laevis* melanophore (7). This landmark discovery facilitated the cloning and characterization of mammalian melatonin receptor subtypes, belonging to a novel subfamily of seven transmembrane domain G-protein coupled receptors

Melatonin after Four Decades, edited by James Olcese.
Kluwer Academic / Plenum Publishers, New York, 2000.

(8). The identification of the molecular structure of this new family of G-protein coupled mammalian melatonin receptors is leading the search into the discovery of melatonin receptor subtypes in native tissues as potential therapeutic targets.

1.2. Melatonin Functions in Mammals

Functional melatonin receptors in a mammalian tissue was first described in the rabbit retina, where melatonin inhibits dopamine release (9,10). The introduction of the competitive melatonin receptor antagonist luzindole, allowed the unequivocal characterization of this receptor (9–14). These findings, together with the development of 2-[^{125}I]-iodomelatonin as a high affinity iodinated radioligand (15–16) allowed the localization and pharmacological characterization of putative melatonin receptors in a number of neuronal and non-neuronal tissues from a variety of vertebrate species including human (1,8,14,17–19).

Melatonin receptors in the neural retina and superior colliculus are targets for the regulation of visual function (9–10,13–14) and receptors in the suprachiasmatic nucleus of the hypothalamus are responsible for melatonin's action on circadian rhythms and neuroendocrine function most notably regulation of reproduction in seasonal breeding mammals (2,8,14,18,20–21). Non-neuronal melatonin receptors in cerebral and caudal arteries may regulate cardiovascular function and temperature (22) and in the pars tuberalis of seasonally breading species are likely to regulate neuroendocrine function (8,17,21). Of interest is to determine the melatonin receptor subtypes involved in mediating these various functions of melatonin, potential therapeutic targets for drug discovery.

2. MAMMALIAN MELATONIN RECEPTORS

2.1. Melatonin Receptor Nomenclature and Classification

This review will use the nomenclature and classification for melatonin receptors adopted by the International Union of Pharmacology (23) in 1998 (Figure 1). Accordingly, the melatonin receptors are referred to with the letters MT (*M*ela*T*onin). This nomenclature uses lower case "mt" followed by its number to describe receptors for which only the molecular structure is known (e.g., mt_1 former Mel_{1a}) (8). Melatonin receptors with a well defined functional pharmacology in a native tissue, as well as known molecular structure are referred in upper case followed by a number subscript (e.g. MT_2 former Mel_{1b}) (8,24–25). Upper case in italics (*MT*) followed by the corresponding number is reserved for receptors pharmacologically characterized in native tissues for which the molecular structure is not known (e.g., MT_3 former ML_2) (14).

2.2. Melatonin Receptor Subtypes

The original classification of melatonin receptors differentiated the ML_1 and ML_2 subtypes based on kinetics and pharmacological characteristics (12) (Figure 1). High (30–300 pM) affinity (ML_1) 2-[^{125}I]-iodomelatonin binding sites found in mammalian retina, suprachiasmatic nucleus, and pars tuberalis, show a pharmacology profile (2-iodomelatonin ≥ melatonin >> N-acetylserotonin) characterized by a low affinity for the precursor of melatonin, N-acetylserotonin. This pharmacological

	mt_1	mt_2	MT_3
Nomenclature	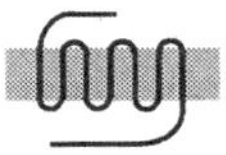	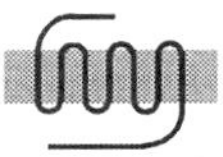	
Other Names	Mel_{1a}, ML_{1A}	Mel_{1b}, ML_{1B}	ML_2
Affinity (K_D)	45 pM	140 pM	0.3-2 nM
Pharmacology	MLT >> NAS	MLT>>NAS	MLT=NAS
Selective Agonists	---	---	N-Acetyl 5HT 5-MCA-NAT
Selective Antagonists	--- ---	4P-ADOT 4P-PDOT	PRAZOSIN

Figure 1. Nomenclature for mammalian melatonin receptors. The nomenclature described here was adopted by the Nomenclature Committee of International Union of Pharmacology in 1998. For further details please see reference 25. Luzindole: MLT: melatonin (N-acetyl-5-methoxytryptamine); NA 5-HT: N-acetyl-5-hydroxytryptamine; 5-MCA-NAT: 5-methoxy-carbonylamino- N-acetyltryptamine; 4-P-PDOT: 4-phenyl 2-propionamidotetraline; 4-P-ADOT: 4-phenyl 2-acetamidotetraline.

profile corresponds closely to that of the functional melatonin receptor of rabbit retina (14).

Cloning studies revealed two mammalian melatonin receptors (Mel_{1a} and Mel_{1b}, now termed mt_1 and MT_2) encoding 2-[125I]-iodomelatonin binding sites showing the general pharmacology of the high affinity (ML_1) melatonin receptor. These two melatonin receptors were defined as unique subtypes on the basis of their molecular structure and chromosomal localization (8,26). The human melatonin receptor subtypes show 60% homology at the amino acid level and distinct pharmacological profiles of partial agonists and antagonists (24). The use of cell lines expressing the human mt_1 and MT_2 melatonin receptors has led to the discovery of subtype selective analogues (24) (Figure 1).

Activation of high affinity melatonin receptors in both neuronal and non-neuronal native tissues and transfected cell lines inhibits cAMP formation through a pertussis toxin sensitive inhibitory G proteins (17,27–28). However, coupling of the high affinity melatonin receptors to modulation of cGMP formation (29), synthesis of diacylglycerol and release of arachidonic acid, changes in calcium influx (30) and potassium conductances (31) were also reported. Melatonin receptor mediated vasoconstriction appears to involve closing of calcium-activated potassium channels (BK_{Ca}) (32). In cell lines expressing the h mt_1 and h MT_2 melatonin receptors, melatonin inhibits forskolin stimulated cAMP accumulation (27–28). In this preparation a parallel signal transduction mechanism has been proposed for mt_1 receptors whereby $G_i\beta\gamma$ subunits may activate phospholipase Cβ subsequent to PGF2α stimulation (27).

The putative MT_3 melatonin receptor (former ML_2) binds 2-[^{125}I]-iodomelatonin with nanomolar affinity and shows a pharmacological profile (2-iodomelatonin > melatonin = N-acetylserotonin) distinct from that of the high affinity site where melatonin has higher affinity than N-acetylserotonin (12,33–34) (Figure 1). 2-[^{125}I]-Iodomelatonin and the novel MT_3 subtype selective radioligand 2-[^{125}I]-5-MCA-NAT bind with low nanomolar affinity (K_D: 0.9–10 nM) and show fast kinetics of association and dissociation to membranes of hamster brain, kidney and testes, to mouse brain and to RPMI 1846 melanoma cells (34–35). Melatonin receptors with pharmacology and kinetics similar to those in hamster brain have been described in brown fat (36). Activation of the MT_3 receptor appears to signal through increases in phosphoinositide turnover (37). In guinea pig colon melatonin induced contraction through activation of a receptor with a pharmacological profile comparable to that of the MT_3 receptor (38). In summary, MT_3 melatonin binding sites show distinct pharmacological profiles and distribution than high affinity melatonin receptors (i.e., mt_1 and MT_2).

2.3. Subtype Selective Melatonin Analogues

The structural molecular differences between the human recombinant mt_1 and MT_2 melatonin receptors are reflected in distinct pharmacological profiles of melatonin analogues to compete for either 3H-melatonin or 2-[^{125}I]-iodomelatonin binding (24,39). A number of synthetic melatonin analogues (e.g., S20098, 6-chloromelatonin, GR 196429) that mimic the effect of melatonin in functional responses (e.g., inhibition of dopamine release) do show small differences in affinities for the mammalian subtypes (24–25). For example 6-chloromelatonin competes with 57 times higher affinity for 2-[^{125}I]-iodomelatonin binding to the MT_2 than the mt_1 human melatonin receptors expressed in CHO cells.

Subtype selective melatonin receptor antagonists are essential to identify melatonin receptor subtypes in native tissues. In this review we describe the specificity and selectivity of three competitive melatonin receptor antagonists, luzindole, 4P-ADOT and 4P-PDOT that are used to identify melatonin receptor subtypes in native tissues (14,24) (Figure 2). These antagonists show melatonin receptor specificity as they did not compete for binding of forty-nine radioligands to receptors, channels, transporters and second messengers and various degree of selectivity for the h MT_2 melatonin receptor subtype (24–25). Figure 2 shows the chemical structures of luzindole, 4P-ADOT and 4P-PDOT and their affinity constants to compete for 2-[^{125}I]-iodomelatonin binding to the recombinant h mt_1 and h MT_2 melatonin receptors stably expressed in CHO cells. Luzindole (Ki = 7.3 ± 2.0 nM, n = 4), 4P-ADOT (Ki = 0.4 ± 0.02 nM, n = 3) and 4P-PDOT (Ki = 0.41 ± 0.04 nM, n = 3) showed higher affinity for competition with 2-[^{125}I]-iodomelatonin binding to the h MT_2 melatonin receptor. The calculated affinity ratios (Ki_{MT2}/Ki_{mt1}) showed that luzindole has 25 fold higher affinity for the h MT_2 than the h mt_1 melatonin receptor subtype, while 4P-ADOT and 4P-PDOT are 951 and 1560 fold MT_2 subtype selective (Figure 2).

2.4. Melatonin Receptor Subtype Function

Specific functions of melatonin believed to be mediated through activation of melatonin receptors include: inhibition of dopamine release in retina (9–11,13), acute inhibition of electrical activity (40–41), phase shifts of circadian rhythms in the suprachiasmatic nucleus slice (42–43) and potentiation by melatonin of endogenous and exogenous norepinephrine mediated vasoconstriction (22,32). This hypothesis is based on the observation that melatonin responses were blocked by either the melatonin receptor antagonists, luzindole or S20928. The use of subtype selective melatonin analogues is essential to identify functional melatonin receptor subtypes in

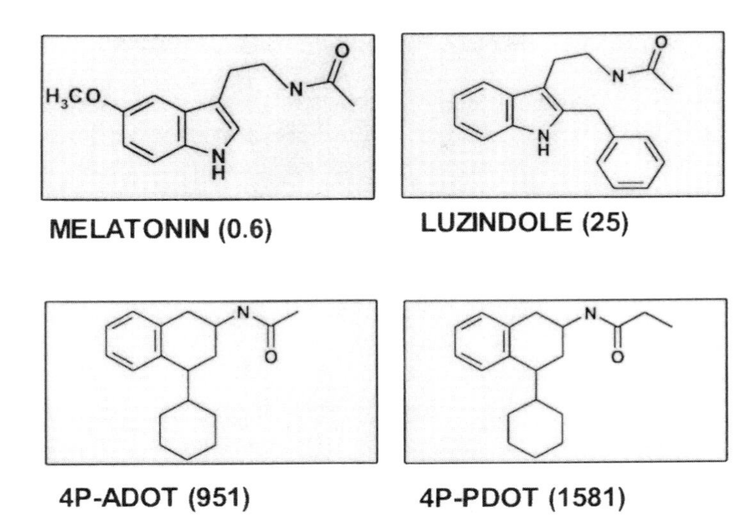

Figure 2. Chemical structures of melatonin and competitive melatonin receptor antagonists. Melatonin, luzindole (2-benzyl-N-acetyltryptamine), 4P-ADOT and 4P-PDOT competed for 2-[125I]-iodomelatonin binding to the human mt_1 and MT_2 melatonin receptor subtypes expressed in CHO cells (24–25). The number in parenthesis shows the affinity ratios (Ki mt_1/ Ki MT_2) with represent the fold differences in affinity for the subtypes. The higher the number the higher the affinity for the MT_2 subtype. Analogues with affinity ratios of equal or higher than 100 are considered subtype selective (25).

native tissues. Although targeted gene disruption or the use of antisense provide evidence for the function of a given receptor protein, the presence of a receptor subtype is better characterized using subtype selective competitive melatonin receptor agonists and antagonists. Here, we describe the use of subtype selective MT_2 melatonin receptor antagonists to characterize functional melatonin receptor subtypes in mammals.

2.4.1. Retina Receptors. In rabbit retina melatonin at picomolar concentrations inhibits the calcium dependent release of 3H-dopamine through activation of a presynaptic heteroreceptor. The release of 3H-dopamine is antagonized by luzindole (K_B = 20 nM) as well as by the MT_2 subtype selective melatonin receptor antagonists, 4P-ADOT and 4P-PDOT (K_B = 1.6 nM and 0.3 nM, respectively) (24). The excellent correlation found between the affinity of a number of melatonin receptor antagonists to block the inhibition of dopamine release by melatonin (K_B) with the affinity (Ki values) of these analogues to compete for 2-[^{125}I]-iodomelatonin binding to the MT_2 recombinant melatonin receptor strongly supports the classification of the presynaptic melatonin heteroreceptor of rabbit retina as an MT_2 subtype (24).

2.4.2. Melatonin Receptors in the Circadian Timing System. The mt_1 melatonin receptor was initially suggested to mediate circadian functions in mammals due to the high levels of mRNA expression within the mammalian SCN, which correlated with the specific 2-[^{125}I]-iodomelatonin binding (8). However, recently Liu et al. (41) reported that targeted disruption of the mt_1 melatonin receptor in the C57BL/6 mouse blocks the melatonin-induced inhibition of neuronal firing in SCN slices without impairing the melatonin-mediated phase shifts of circadian firing rhythms. This phase shifting effect of melatonin is pertussis toxin sensitive, suggesting the involvement of a G-protein coupled receptor (41–42).

In order to assess the melatonin receptor subtype mediating phase shifts of circadian rhythms in mammals we investigated the effect of subtype selective melatonin receptor antagonists on the wheel running activity in the C3H/HeN mouse. We used the C3H/HeN mouse in these studies because it produces melatonin in the pineal gland (44) and in this strain melatonin phase-shifts circadian activity rhythms with periods of sensitivity identical to those found in humans (45–46).

The suprachiasmatic nucleus of the C3H/HeN mouse shows high density of 2-[^{125}I]-iodomelatonin binding sites (18) as well as expression of both the mt_1 and MT_2 melatonin receptor mRNA (detected by *in situ* hybridization histochemistry with digoxigenin labeled oligonucleotide probes) (25). However, the selective melatonin receptor antagonist, 4P-ADOT did not compete for 2-[^{125}I]-iodomelatonin binding to the mouse SCN. These findings suggest that either the density of the MT_2 melatonin receptor protein in the C3H/HeN mouse SCN is below the limits of detection or 2-[^{125}I]-iodomelatonin is not able to recognize the native MT_2 melatonin receptor protein in brain frozen sections. It is noteworthy that undetectable levels of 2-[^{125}I]-iodomelatonin binding were also reported in the SCN of the mt_1 knockout C57BL/6 mouse (41).

In C3H/HeN mice, as in humans, melatonin administration for three consecutive days phase advances circadian activity rhythms when given at the end of the subjective day (CT 10) (18,45–46). Treatment with vehicle followed by saline administration for three consecutive days at CT 10 did not affect the rhythm of wheel running activity (45). Melatonin administration at CT 10 induced advances in the phase of the circadian activity rhythms in a dose dependent manner (0.3 to 30 μg/mouse) (Figure 3). The dose of melatonin inducing a half-maximal phase advance (EC_{50}) was 0.72 μg/mouse with a maximal advance of 0.98 ± 0.08 h ($n = 15$) at 9 μg/mouse. The selective MT_2 melatonin receptor antagonists, 4P-ADOT and 4P-PDOT (90 μg/mouse, sc) did not affect the phase of circadian activity rhythms when given alone at CT 10 (Figure 3). Both antagonists, however, shifted to the right the dose response curve to melatonin, as they significantly reduced the phase shifting effects of 0.9 and 3 μg melatonin (25) (Figure 3). The melatonin receptor antagonist luzindole, which shows 25 fold higher affinity for the MT_2 than the mt_1 melatonin receptor also antagonized the phase advance by melatonin at CT 10 (25). Together this study suggests that activation of the MT_2 melatonin receptor subtype within the circadian timing system mediates the phase advances of circadian activity rhythms.

2.5. Summary

The MT_2 melatonin receptor appears to be emerging as the subtype involved in mediating important physiological functions by melatonin in the visual, circadian and vascular systems. Pharmacological studies using subtype selective antagonists have demonstrated that the melatonin receptor involved in mediating inhibition of dopamine release from rabbit retina is the MT_2 subtype (24). In the mammalian SCN, activation of G-protein coupled receptors by melatonin appear to mediate two distinct functional responses, i.e., acute inhibition of neuronal firing through the mt_1 subtype and time-dependent phase shifts of circadian rhythms through the MT_2 subtype (40–41). The dual effect of melatonin on phenylephrine-mediated vasoconstriction in rat caudal artery appears to be mediated by activation of two distinct receptor

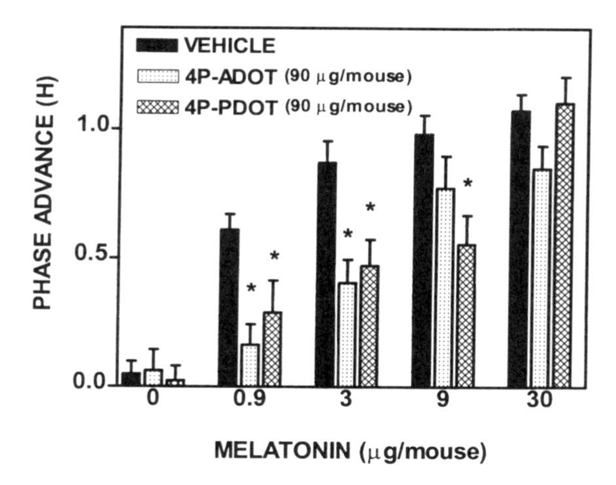

Figure 3. The MT_2 selective melatonin receptor antagonists blocked the melatonin-induced phase advances of circadian rhythms in the C3H/HeN mouse. Circadian wheel running activity rhythms were recorded from C3H/HeN mice held in constant dark. Mice received two treatments a day on three consecutive days. Each day the first treatment was at CT 10 and the second treatment 10 min later. The ordinate represents the phase advance of circadian activity rhythms at CT 10. C3H/HeN mice were first treated with vehicle and ten minutes later with saline or melatonin (0.9–30 μg/mouse). Mice treated with 4P-ADOT (90 μg/mouse) or 4P-PDOT (90 μg /mouse) did not show change in phase, however, they antagonized the phase advance induced by various doses of melatonin (0.9–30 μg/mouse). The effect of the antagonists for each dose of melatonin was assessed by one way ANOVA ($p < 0.01$ for 0.9 μg; $p < 0.001$ for 3 μg and $p < 0.05$ for 9 μg) (Data from reference 5).

subtypes, i.e., vasoconstriction by the mt_1 subtype and vasodilation by the MT_2 subtype (47).

3. MELATONIN RECEPTORS AS THERAPEUTIC TARGETS

Molecular and pharmacological studies suggest the presence of melatonin receptors in the human central nervous system. Melatonin receptors were localized to the human suprachiasmatic nucleus and to the molecular layer of the cerebellum by receptor autoradiography with 2-[^{125}I]-iodomelatonin (19,21). In postmortem human cerebellar membranes 2-[^{125}I]-iodomelatonin binds to a site showing pharmacological characteristics similar to those of the human mt_1 melatonin receptor (24,39). Both mt_1 and MT_2 mRNAs have been amplified from human tissues using reverse transcription polymerase chain reaction. The mt_1 mRNA was localized to the suprachiasmatic nucleus and retina and the MT_2 to the retina, hippocampus and whole brain (8,21,23). Using *in situ* hybridization histochemistry, the mt_1 melatonin receptor was localized to the suprachiasmatic nucleus and to the granule cell layer of the cerebellum of human postmortem brain (21,48). Recently, we demonstrated the localization of MT2 mRNA to human cerebellar Bergmann glia and astrocytes (W. Al-Ghoul and M.L. Dubocovich, unpublished). Together, these data suggest that both melatonin receptor subtypes are present in human brain and retina, which represent potential therapeutic targets for drug action.

Melatonin-mediated activation of the mt_1 subtype in the SCN and/or other limbic system areas may mediate somnogenic effects (41,49), while activation of the MT_2

subtype may be involved in the regulation of circadian rhythms (25). Thus, it follows that MT_2 selective melatonin receptor agonists and/or antagonists may be used to treat disorders involving alterations in the phase of the circadian clock as observed in depression (50), blindness (20), delayed sleep phase syndrome (20), or following a rapid change in the light dark/cycle such as jet travel and shift work (20). We conclude that the use of specific and subtype selective mt_1 and MT_2 melatonin receptor antagonists will elucidate the functional role of melatonin in mammals and may prompt the development of subtype selective analogues for the treatment of insomnia and circadian sleep and mood disorders.

ACKNOWLEDGMENTS

The work reported in this paper was supported by MH 42922 and MH 52685.

REFERENCES

1. Krause, D.N. and Dubocovich, M.L. Regulatory sites in the melatonin system of mammals. Trends Neurosci 13:464–470, 1990.
2. Reiter, R.J. Pineal melatonin: cell biology of its synthesis and its physiological interactions. Endocr Rev 12:151–180, 1991.
3. McCord, C.P. and Allen, F.P. Evidence associating pineal gland function with alterations in pigmentation. J Exp Zool 23:207–224, 1917.
4. Lerner, A.B., Case, J.D., and Heizelman, R.V. Structure of melatonin. J Am Chem Soc 81:6084–6085, 1959.
5. Heward, C.B. and Hadley, M.E. Structure-activity relationships of melatonin and related indoleamines. Life Sci 17:1167–1178, 1975.
6. White, B.H., Sekura, R.D., and Rollag, M.D. Pertussis toxin blocks melatonin induced pigment aggregation in *Xenopus* dermal melanophores. J Comp Physiol 157:153–159, 1987.
7. Ebisawa, T., Karne, S., Lerner, M.R., and Reppert, S.M. Expression cloning of a high affinity melatonin receptor from Xenopus dermal melanophores. Proc Natl Acad Sci USA 91:6133–6137, 1994.
8. Reppert, S.M., Weaver, D.R., and Godson, C. Melatonin receptors step into the light: cloning and classification of subtypes. Trends Pharmacol Sci 17:100–102, 1996.
9. Dubocovich, M.L. Melatonin is a potent modulator of dopamine release in the retina. Nature 306:782–784, 1983.
10. Dubocovich, M.L. Characterization of a retinal melatonin receptor. J Pharmacol Exp Ther 234:395–401, 1985.
11. Dubocovich, M.L. Luzindole (N-0774): A novel melatonin receptor antagonist. J Pharmac Exp Ther 246:902–910, 1988.
12. Dubocovich, M.L. Pharmacology and function of melatonin receptors. FASEB J 2:2765–2773, 1988.
13. Dubocovich, M.L. Pharmacology and function of melatonin in retina. Osborne, N.N. and Chader, G. Progress in Retinal Research. Oxford, Pergamon Press 8:129–151, 1988.
14. Dubocovich, M.L. Melatonin receptors: Are There Multiple Subtypes? Trends Pharmacol Sci 16:50–56, 1995.
15. Vakkuri, O., Leppaluoto, J., and Vuolteenaho, O. Development and validation of a melatonin radioimmunoassay using radioiodinated melatonin as tracer. Acta Endocrinol (Copenh) 106:152–157, 1984.
16. Dubocovich, M.L. and Takahashi, J.S. Use of 2-[125I]-iodomelatonin to characterize melatonin binding sites in chicken retina. Proc Natl Acad Sci USA 84:3916–3920, 1987.
17. Morgan, P.J., Barrett, P., Howell, H.E., and Helliwell, R. Melatonin receptors: localization, molecular pharmacology and physiological significance. Neurochem Int 24:101–146, 1994.
18. Dubocovich, M.L., Benloucif, S., and Masana, M.I. Melatonin receptors in the mammalian suprachiasmatic nucleus. Behav Brain Res 73:141–147, 1996.

19. Reppert, S.M., Weaver, D.R., Rivkees, S.A., and Stopa, E.G. Putative melatonin receptors in a human biological clock. Science 242:78–81, 1988.
20. Armstrong, S.M. and Redman, J.R. Melatonin and circadian rhythmicity. Yu, H.S., and Reiter, R.J. In Melatonin: Biosynthesis, Physiological Effects, and Clinical Applications. CRC Press 187–224, 1993.
21. Weaver, D.R. and Reppert, S.M. The Mel_{1a} melatonin receptor gene is expressed in human suprachiasmatic nuclei. NeuroReport 8:109–112, 1996.
22. Mahle, C.D., Goggins, G.D., Agarwal, P., Ryan, E., and Watson, A.J. Melatonin modulates vascular smooth muscle tone. J Biol Rhythms 12:690–696, 1997.
23. Dubocovich, M.L., Cardinali, D.P., Guardiola-Lemaitre, B., Hagan, R.M., Krause, D.N., Sugden, D., Vanhoutte, P.M., and Yocca, F.D. Melatonin receptors. The IUPHAR Compendium of Receptor Characterization and Classification. IUPHAR Media, London, pp 187–193, 1998.
24. Dubocovich, M.L., Masana, M.I., Iacob, S., and Sauri, D.M. Melatonin receptor antagonists that differentiate between the human Mel_{1a} and Mel_{1b} recombinant subtypes are used to assess the pharmacological profile of the rabbit retina ML_1 presynaptic heteroreceptor. Naunyn Schmiederbergs' Arch Pharmacol 355:365–375, 1997.
25. Dubocovich, M.L., Yun, K., Al-Ghoul, W., Benloucif, S., and Masana, M.I. Selective MT2 melatonin receptor antagonists block melatonin-mediated phase advances of circadian rhythms. FASEB 12:1211–1220, 1998.
26. Slaugenhaupt, S.A., Roca, A.L., Liebert, C.B., Altherr, M.R., Gusella, J.F., and Reppert, S.M. Mapping of the gene for the Mel_{1a} melatonin receptor to human chromosome 4 *(MTNR1A)* and mouse chromosome 8 (*MTNR1a*). Genomics 27:355–357, 1995.
27. Godson, C. and Reppert, S.M. The Mel_{1a} melatonin receptor is coupled to parallel signal transduction pathways. Endocrinology 138:397–404, 1997.
28. Witt-Enderby, P.A. and Dubocovich, M.L. Characterization and regulation of the human ML_{1A} melatonin receptor stably expressed in chinese hamster ovary cells. Mol Pharmacol 50:166–174, 1996.
29. Vacas, M.I., Keller-Sarmiento, M.I., and Cardinali, D.P. Melatonin increases cGMP and decreases cAMP levels in the medial basal hypothalamus in vitro. Brain Res 225:207–211, 1982.
30. Vanecek, J. and Klein, D.C. Mechanism of melatonin signal transduction in the neonatal rat pituitary. Neurochem Int 27:273–278, 1995.
31. Jiang, Z.-G., Nelson, C.S., and Allen, C.N. Melatonin activates an outward current and inhibits I_h in rat suprachiasmatic nucleus neurons. Brain Research 687:125–132, 1995.
32. Geary, G.G., Krause, D.N., and Duckles, S.P. Melatonin directly constricts rat cerebral arteries through modulation of potassium channels. Am J Physiol 273:H1530–H1536, 1997.
33. Duncan, M.J., Takahashi, J.S., and Dubocovich, M.L. Characteristics and autoradiographic localization of 2-[125I]-iodomelatonin binding sites in Djungarian hamster brain. Endocrinology 125:1011–1018, 1989.
34. Molinari, E.J., North, P.C., and Dubocovich, M.L. Characterization of ML_2 melatonin binding sites using the selective radioligand 2-[125I]-iodo-5-methoxycarbonylamino-N-acetyltryptamine. Eur J Pharmacol 301:159–168, 1996.
35. Pickering, D.S. and Niles, L.P. Expression of nanomolar-affinity binding sites for melatonin in Syrian hamster RPMI 1846 melanoma cells. Cell Sig 4:201–207, 1992.
36. Le Gouic, S., Atgie, C., Viguerie-Bascands, N., Hanoun, N., Larrouy, D., Ambid, L., Raimbault, S., Ricquier, D., Delagrange, P., Guardiola-Lemaitre, B., Penicaud, L., and Casteilla, L. Characterization of a melatonin binding site in Siberian hamster brown adipose tissue. Eur J Pharmacol 339:271–278, 1997.
37. Popova, J.S. and Dubocovich, M.L. Melatonin receptor-mediated stimulation of phosphoinositide breakdown in chick brain slices. J Neurochem 64:130–138, 1995.
38. Lucchelli, A., Santagostino-Barbone, and M.G., Tonini, M. Investigation into the contractile response of melatonin in the guinea-pig isolated proximal colon: the role of 5-HT4 and melatonin receptors. Br J Pharmacol 121:1775–1781, 1997.
39. Beresford, I.J.M., North, P.C., Oakley, N.R., Starkey, S., Brown, J., Foord, S.M., Andrews, J., Coughlan, J., Stratton, S., Dubocovich, M.L., and Hagan, R.M. GR 196429: A nonindolic agonist at high-affinity melatonin receptors. J Pharmacol Exp Ther 285:1239–1245, 1998.
40. Ying, S.-H., Rusak, B., Delagrange, P., Mocaer, E., Renard, P., and Guardiola-Lemaitre, B. Melatonin analogues as agonists and antagonists in the circadian system and other brain areas. Eur J Pharmacol 296:33–42, 1996.
41. Liu, C., Weaver, D.R., Jin, X., Shearman, L.P., Pieschl, R.L., Gribkoff, V.K., and Reppert, S.M. Molecular dissection of two distinct actions of melatonin on the suprachiasmatic circadian clock. Neuron 19:91–102, 1997.

42. McArthur, J.J., Hunt, A.E., and Gillette, M.U. Melatonin action and signal transduction in the rat suprachiasmatic circadian clock: activation of protein kinase C at dusk and dawn. Endocrinology 138:627–634, 1997.
43. Starkey, S.J., Walker, M.P., Beresford, I.J.M., and Hagan, R.M. Modulation of the rat suprachiasmatic circadian clock by melatonin *in vitro.* Neuroreport 6:1947–1951, 1995.
44. Goto, M., Oshima, I., Tomita, T., and Ebihara, S. Melatonin content of the pineal gland in different mouse strains. J Pineal Res 7:195–204, 1989.
45. Benloucif, S. and Dubocovich, M.L. Melatonin and light induce phase shifts of circadian rhythms in the C3H/HeN mouse. J Biol Rhythms 11:113–125, 1996.
46. Lewy, A.J., Ahmed, S., Latham-Jackson, J.M., and Sack, R.L. Melatonin shifts human circadian rhythms according to a phase-response curve. Chronobiol Int 9, 380–392, 1992.
47. Doolen, S., Krause, D., Dubocovich, M., and Duckles, S. Melatonin mediates two distinct responses in vascular smooth muscle. Eur J Pharmacol 345:67–69, 1998.
48. Mazzucchelli, C., Pannacci, M., Nonno, R., Lucini, V., Fraschini, F., and Stankov, B.M. The melatonin receptor in the human brain: cloning experiments and distribution studies. Mol Brain Res 39:117–126, 1996.
49. Zhdanova, I.V., Wurtman, R.J., Lynch, H.J., Ives, J.R., Dollins, A.B., C, M., Matheson, J.K., and Schomer, D.L. Pharmacodynamics and drug action: Sleep-inducing effects of low doses of melatonin ingested in the evening. Clin Pharmacol Ther 57:552–558, 1995.
50. Wirz-Justice, A. Biological rhythms in mood disorders. In Pharmacology: The Fourth Generation of Progress (Bloom, F.E. and Kupfer, D.J., eds) pp. 999–1017, Raven Press, Ltd., New York, 1995.

21

MECHANISMS OF MELATONIN ACTION IN THE PITUITARY AND SCN

Jiri Vanecek[1] and Kazuto Watanabe[2]

[1] Institute of Physiology
Academy of Sciences
Prague, Czech Republic
[2] Department of Physiology
Dokkyo University School of Medicine
Mibu, Tochigi, Japan

1. SUMMARY

We have compared melatonin effects in two different cell types in order to determine general intracellular mechanisms of its action. In neonatal rat pituitary, melatonin acts *via* the specific membrane receptors to inhibit GnRH-induced LH release. The melatonin effect disappears in adulthood due to the disappearance of the receptors. The mechanism of the melatonin action involves inhibition of the GnRH induced increase of intracellular calcium ($[Ca^{2+}])_i$. Our observations indicate that melatonin has dual inhibitory effect on GnRH-induced $[Ca^{2+}]_i$: it inhibits mobilisation of Ca^{2+} from endoplasmic reticulum as well as Ca^{2+} influx through voltage sensitive channels. Besides, melatonin also decreases basal and GnRH- or forskolin-induced increase of cAMP concentration in the pituitary. Although cAMP is not of primary importance for regulation of LH release, the cAMP decrease may participate in the mechanism of inhibitory melatonin action on LH release.

Rat suprachiasmatic nuclei (SCN) have a high density of the melatonin receptors throughout the postnatal life. Cultures of dispersed SCN cells show circadian rhythm of vasopressin (AVP) release, with several fold increase in the middle of the day and decrease during night. Melatonin inhibits the spontaneous AVP release. Melatonin also inhibits the AVP release induced by vasoactive intestinal peptide (VIP). Intracellular mechanisms of the melatonin effect may involve cAMP, because melatonin inhibits the VIP-induced increase of cAMP and increase of cAMP formation by forskolin stimulates AVP release from the cultures. On the other hand, involvement of intracellular calcium in the regulation of AVP release may not be excluded. VIP induces $[Ca^{2+}]_i$ increase in 14% of the SCN cells and AVP release is stimulated by Ca^{2+} ionophore

Melatonin after Four Decades, edited by James Olcese.
Kluwer Academic / Plenum Publishers, New York, 2000.

ionomycin. Our observations indicate that some of the mechanisms of melatonin action are similar in the pituitary and SCN.

2. MELATONIN EFFECTS

The pineal hormone melatonin is involved in photoperiodic regulations of reproductive functions and in entrainment of daily rhythms. In order to determine general intracellular mechanisms of melatonin action, we have compared melatonin effects in two different cell types. Melatonin acts through the specific high-affinity membrane receptors ($K_d \sim 10^{-11}$ M) (22,24,36,38). Distribution of the melatonin receptors has been determined by *in vitro* autoradiography using ^{125}I-melatonin. Melatonin receptors are present in discrete areas of the rat brain. High density of the receptors has been found in suprachiasmatic nuclei of the hypothalamus, in *area postrema* and in *pars tuberalis* of the pituitary (23,37). The receptor density in the rat *pars distalis* is age-dependent: it is about 30 fmol per mg of protein on embryonic day 20 and postnatal day 1, but within 30 postnatal days decreases 10 times, i.e. below 3 fmol/mg protein (23).

We have studied melatonin effects and their intracellular mechanisms in neonatal rat anterior pituitary and suprachiasmatic nuclei (SCN). In both cases, the primary cell cultures were used in our experiments. There are several similar aspects of melatonin action in both cell types. Melatonin has generally inhibitory effects: it inhibits release of luteinizing hormone (LH) from the pituitary gonadotrophs and release of vasopressin (AVP) from the SCN neurones (12,27,35). Moreover, melatonin inhibits cAMP accumulation in both systems. On the other hand, there are also specific effects of melatonin in each of the systems.

2.1. Anterior Pituitary

The release of luteinizing hormone from gonadotrophs is low under basal conditions and is markedly stimulated by GnRH. Melatonin has no effect on basal LH release, but it inhibits secretion of LH induced by GnRH (9; Figure 1). The melatonin effect on LH release was first described with cultured neonatal rat pituitaries and later confirmed using dispersed cells in culture (11,27). Inhibitory effect of melatonin on GnRH-induced LH release has been also observed *in vivo* (10).

Melatonin inhibits the effects of GnRH in a dose-dependent manner. A minimal inhibition of LH-release is seen with 10^{-10} M and the maximal effect is attained with 10^{-8} to 10^{-7} M melatonin (9). The melatonin effect is specific, the order of potency of various indoles is 2-iodomelatonin > melatonin > 6-hydroxymelatonin > N-acetylserotonin > 5-methoxytryptamine >> 5-hydroxytryptamine.

Melatonin inhibits not only release of LH but also of FSH from gonadotrophs (14). Melatonin acts probably directly on gonadotrophs, because it inhibits GnRH-induced LH release from enriched gonadotroph fraction (11). A melatonin effect on other cell types in the pituitary may not be completely excluded although it has not been possible to demonstrate any effect of melatonin on release of other pituitary hormones (14).

The inhibitory effect of melatonin on LH- and FSH-release is age dependent (9,13). While in 4 to 8-day-old rats, melatonin inhibits GnRH-induced LH release by about 60%, the melatonin effect gradually decreases starting day 10 and disappears almost completely after day 15 of age. The developmental changes of melatonin

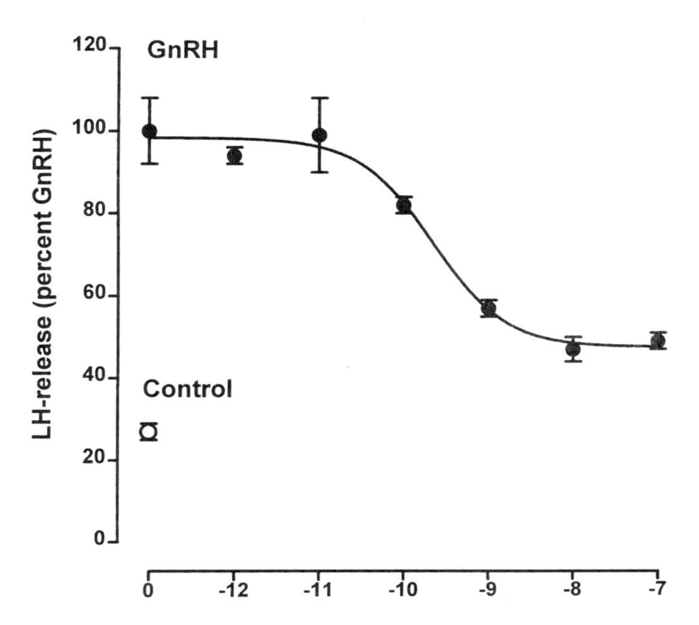

Figure 1. Melatonin effect on GnRH-induced LH release from cultured pituitary cells of neonatal rat. The primary cell cultures were incubated for 3 hr in the presence of various concentration of melatonin and 2 nM GnRH. LH release into the incubation medium was determined by radioimmunoassay. Each point represents the mean (±SEM) from at least 3 independent cultures.

potency correlate with the postnatal changes of the melatonin receptor density in the rat pituitary (23).

2.2. Suprachiasmatic Nuclei

Cultured hypothalamic slices containing SCN have been shown to release vasopressin in a circadian fashion (5). Later it has been found that dispersed suprachiasmatic neurones also show a circadian rhythm in AVP release (16,39). The release is low during subjective night and increases during subjective day peaking around circadian time (CT) (6). Melatonin has inhibitory effects on the spontaneous AVP release (35; Figure 2). Melatonin administered early in the morning delays the spontaneous increase of AVP release and decreases its amplitude. Melatonin added at the time of the peak, advances and accelerates the decrease of AVP release.

The inhibitory effect of melatonin is time dependent. When applied during subjective day, melatonin decreases the spontaneous AVP release by about 50%. However, markedly smaller inhibition is seen after melatonin administration at night and melatonin has no effect when applied around midnight.

Vasopressin release may be increased by vasoactive intestinal peptide (VIP). Melatonin also inhibits the VIP-induced AVP release induced from cultured SCN neurones (35). The melatonin effect is dose-dependent, inhibition starts at 10^{-10} M and maximal effect is reached at 10^{-8} to 10^{-7} M melatonin. EC_{50} is about 0.4 nM.

The effects on AVP release correlate with melatonin effects on spontaneous electric activity of the SCN neurones. The electric activity of SCN neurones in cultured hypothalamic slices shows a circadian rhythm peaking at noon and decreasing during

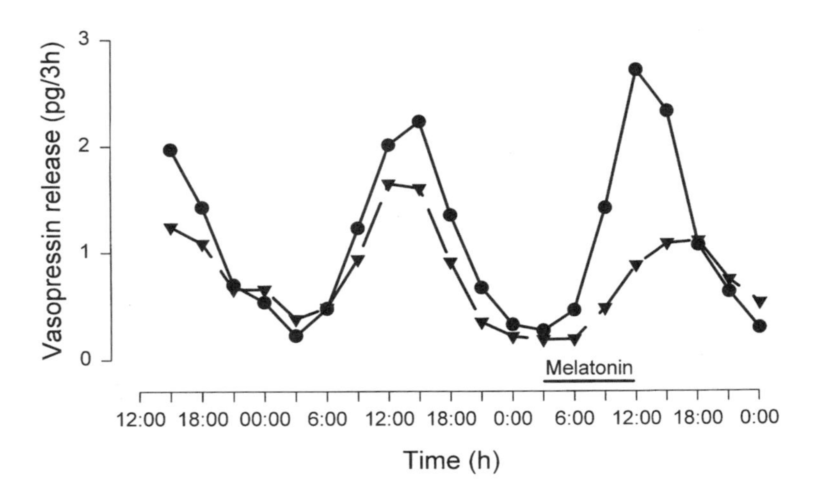

Figure 2. Melatonin effect on spontaneous release of vasopressin from cultured suprachiasmatic neurones. Two primary cultures of the neurones were incubated for 60 hr. The incubation medium was changed every 3 hr and vasopressin concentration determined by radioimmunoassay. Melatonin (100 nM) was added during the period indicated by the horizontal line to the culture No. 2 (▼). ● control, untreated culture.

late subjective day and night (6). Addition of melatonin decreases the frequency of the spontaneous firing (18,20). The effect is rapid, reversible and time-dependent. The maximal inhibitory effect is seen between CT 9 and CT 15, when most of the cells are inhibited by melatonin. At other times, the majority of the cells is unresponsive to melatonin.

3. INTRACELLULAR MECHANISMS OF MELATONIN ACTION

3.1. G-Proteins

Melatonin receptors are coupled to the intracellular effectors *via* GTP-binding proteins. In the presence of GTP or non-hydrolysable GTP derivatives affinity of the melatonin receptors for ^{125}I-melatonin is decreased (15). The effects of melatonin in the pituitary, including inhibition of LH release and the effects on 2nd messengers are abolished after preincubation with pertussis toxin (PTX; 27). This indicates that the receptors are coupled to G-protein(s) belonging to the G_i family. Also the effects of melatonin in SCN are mediated by G_i: preincubation with PTX blocks the effect of melatonin on spontaneous electric activity (8). G_i family consists of 5 subspecies: 3 G_i and 2 G_o proteins (2,17). It is not clear which of these G proteins is involved in transduction of the melatonin signal.

3.2. Cyclic AMP

Melatonin has inhibitory effects on cAMP in both systems. In the cultured hemipituitaries of neonatal rats, melatonin inhibits cAMP accumulation induced by GnRH or forskolin (30,31). Melatonin also decreases the basal concentration of cAMP in the cultured hemipituitaries. Dispersed pituitary cell cultures were later used to study inhibitory effects of melatonin and its derivatives on forskolin-induced cAMP accumulation (25; Figure 3). The effect of melatonin on cAMP is dose-dependent (EC_{50} = 0.15 nM) and mediated by PTX-sensitive G-protein.

In cultured SCN cells, melatonin inhibits the VIP-induced increase of cAMP (33; Figure 3). The melatonin effect is dose-dependent, EC_{50} is 0.21 nM. Melatonin has no effect on basal cAMP, however.

The molecular mechanism of the melatonin effect is not clear. Because melatonin inhibits cAMP accumulation in the pituitary even in the presence of phosphodiesterase inhibitor 3-isobutyl-1-methylxanthine it has been concluded that the indole inhibits adenylyl cyclase (25,30). However, no direct evidence showing this effect is available.

The melatonin-induced decrease of cAMP is not of primary importance for inhibition of LH-release. Melatonin inhibits the GnRH-induced LH release even in the presence of permeable cAMP derivative 8-bromo-cAMP. Melatonin induced decrease of intracellular free calcium ($[Ca^{2+}]_i$) is the most important signal for inhibition of LH release: when the decrease of $[Ca^{2+}]_i$ is prevented by Ca^{2+} ionophore, melatonin does not inhibit LH release. Nevertheless, cAMP may be involved in transduction of the melatonin signal, because it stimulates Ca^{2+} influx in subpopulations of gonadotrophs. Melatonin-induced decrease of cAMP may thus inhibit Ca^{2+} influx

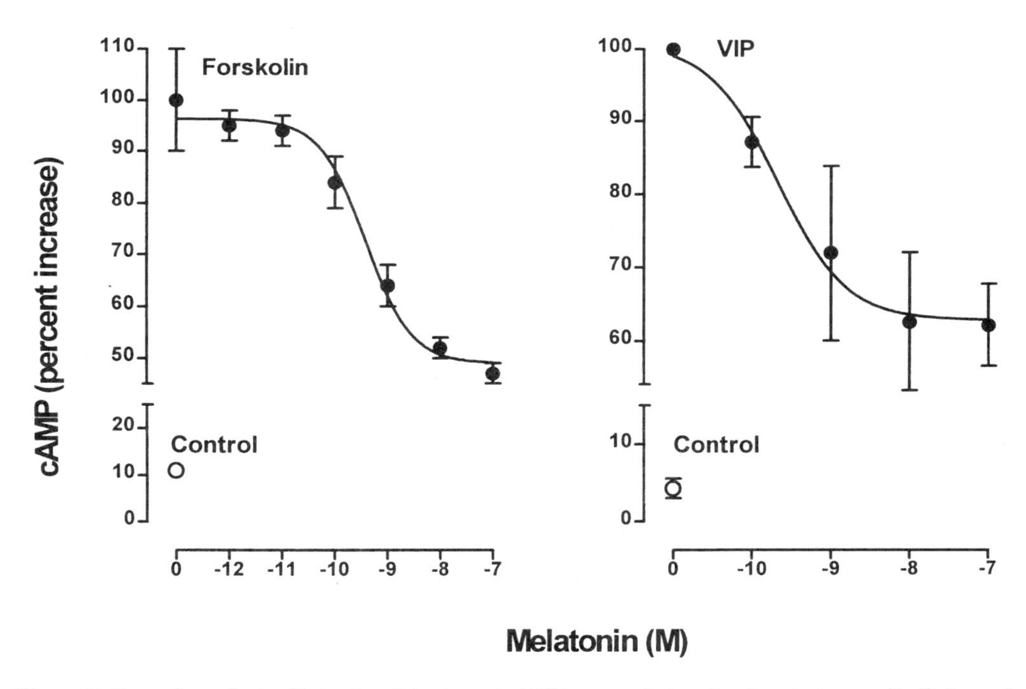

Figure 3. Dose-dependent effect of melatonin on cAMP accumulation. In the pituitary cells (left part) forskolin (10 µM) was used to increase cAMP accumulation in the presence of various concentration of melatonin and cAMP accumulation in the cells was determined after 30 min incubation. Suprachiasmatic cells (right part) were incubated in the presence of vasoactive intestinal peptide (VIP, 100 nM) together with various concentration of melatonin for 90 min and cAMP concentration in the incubation medium was measured. Each point represents the mean (±SEM) from 4 to 6 independent cultures.

resulting in inhibition of LH release (26). The involvement of cAMP in the inhibitory effect of melatonin on AVP release from SCN neurones is currently being studied. Our preliminary data indicate that an increase of cAMP induced by forskolin stimulates AVP release from the neurones. Melatonin may thus inhibit AVP release from SCN neurones *via* the decrease of cAMP.

3.3. Intracellular Free Ca^{2+}

Because the GnRH-induced increase of $[Ca^{2+}]_i$ is the primary signal for release of LH from gonadotrophs (3,4,21), melatonin effects on $[Ca^{2+}]_i$ have been studied. GnRH induces $[Ca^{2+}]_i$ increase by two mechanisms. Initially Ca^{2+} is released from intracellular IP_3-sensitive stores which is followed by Ca^{2+} influx through voltage-sensitive channels (7,21). Our data indicate that melatonin may inhibit both pathways.

When gonadotrophs are incubated in Ca^{2+}-free medium, the GnRH-induced $[Ca^{2+}]_i$ increase is mediated by mobilisation from intracellular stores. Because in the neonatal rat gonadotrophs the Ca^{2+} stores are quite limited, the Ca^{2+} spike is transient and followed by rapid decrease to basal levels in most of the cells. In the presence of melatonin, the GnRH-induced Ca^{2+} spike is inhibited in 30% of gonadotrophs (19). This finding indicates, that melatonin inhibits GnRH-induced Ca^{2+} mobilisation from intracellular stores in about 1/3 of the gonadotrophs.

How melatonin inhibits the Ca^{2+} mobilisation is not clear. Three possible mechanisms could be involved: 1) inhibition of phospholipase C and IP_3 formation; 2) inhibition of IP_3 binding on its receptor in endoplasmic reticulum or inhibition of the channel opening; 3) increased Ca^{2+} clearance from cytosol. Based on indirect evidence, the most likely mechanism is inhibition of phospholipase C. This enzyme metabolises phosphatidylinositol bisphosphate into diacylglycerol and inositol trisphosphate (IP_3), which in turn induces the release of Ca^{2+} from endoplasmic reticulum (1). Melatonin has been shown to inhibit the GnRH-induced formation of diacylglycerol in the gonadotrophs (32) and may also inhibit the IP_3 accumulation (J. Vanecek, unpublished data).

Apart from inhibiting Ca^{2+} mobilisation, melatonin also blocks Ca^{2+} influx through voltage-sensitive channels (27,28). In the neonatal rat gonadotrophs cultured in Ca^{2+}-supplemented medium, the GnRH administration induces Ca^{2+} spike followed either by a sustained plateau or by calcium oscillations (19). Melatonin added after the GnRH-induced spike decreases $[Ca^{2+}]_i$ in about 50% of the gonadotrophs. The melatonin effect has been mimicked by inhibitor of voltage-sensitive Ca^{2+} channels verapamil and in Ca^{2+}-free medium the melatonin effect is abolished. These observations indicate that melatonin inhibits Ca^{2+} influx through voltage-sensitive channels. Mechanism of the effect may involve hyperpolarization of plasma membrane, because melatonin has been shown to increase membrane potential in neonatal rat gonadotrophs (28,29). Hyperpolarization closes the voltage-sensitive channels and inhibits the Ca^{2+} influx. This pathway may represent the most important mechanism for inhibition of LH release by melatonin. In the presence of the L-channel agonist Bay K, the melatonin inhibitory effect on LH release is markedly reduced while the inhibitor of voltage-sensitive Ca^{2+} channels nifedipine inhibits GnRH-induced LH release to a similar degree as melatonin. Moreover, the effects of melatonin and nifedipine are not additive. This conclusion correlates with the known important role of intracellular Ca^{2+} in the regulation of LH release.

Melatonin may affect $[Ca^{2+}]_i$ also in the SCN neurones. Our preliminary data indicate that VIP induces $[Ca^{2+}]_i$ increase in about 14% of the SCN cells. In some of them, the $[Ca^{2+}]_i$ increase is blocked in the presence of melatonin. However, these studies are complicated, because the VIP response desensitises rapidly and only a few of the cells are sensitive to melatonin. Nevertheless, taken all together our observations indicate that in gonadotrophs and in SCN neurones melatonin has similar effects and may act through similar mechanisms.

ACKNOWLEDGMENTS

This work has been supported by grants 309/97/0513 and 309/96/0682 from GA CR and grant No. A5011705 from IGA AVCR. Authors would like to thank to Dr. D.C. Klein, Laboratory of Developmental Neurobiology, NICHD, MD for generous gift of cAMP antiserum, to J.J. van Heerikhuize from Netherlands Institute for Brain Research, Amsterdam for AVP antiserum and to A.F. Parlow and to National Hormone and Pituitary Program, NIDDK, for LH antiserum and antigen for iodination.

REFERENCES

1. Berridge M.J. 1993 Inositol trisphosphate and calcium signalling. Nature 361:315–325.
2. Carty D.J., Padrell E., Codina J., Birnbaumer L., Hildebrandt J.D., and Iyengar R. 1990 Distinct guanine nucleotide binding and release properties of the three Gi proteins. J Biol Chem 265:6268–6273.
3. Chang J.P., Stojilkovic S.S., Graeter J.S., and Catt K.J. 1988 Gonadotropin-releasing hormone stimulates luteinizing hormone secretion by extracellular calcium-dependent and -independent mechanisms. Endocrinology 123:87–97.
4. Conn P.M., Huckle W.R., Andrews W.V., and McArdle C.A. 1987 The molecular mechanism of action of gonadotropin releasing hormone (GnRH) in the pituitary. Recent Prog Horm Res 43:29–68.
5. Earnest D.J. and Sladek C.D. 1986 Circadian rhythms of vasopressin release from individual rat suprachiasmatic explants in vitro. Brain Res 382:129–133.
6. Green D.J. and Gillette R. 1982 Circadian rhythm of firing rate recorded from single cells in the rat suprachiasmatic brain slice. Brain Res 245:198–200.
7. Izumi S., Stojilkovic S.S., and Catt K.J. 1989 Calcium mobilization and influx during the biphasic cytosolic calcium and secretory responses in agonist-stimulated pituitary gonadotrophs. Arch Biochem Biophys 275:410–428.
8. Liu C., Weaver D.R., Jin X.W., Shearman L.P., Pieschl R.L., Gribkoff V.K., and Reppert S.M. 1997 Molecular dissection of two distinct actions of melatonin on the suprachiasmatic circadian clock. Neuron 19:91–102.
9. Martin J.E., Engel J.N., and Klein D.C. 1977 Inhibition of the in vitro pituitary response to luteinizing hormone-releasing hormone by melatonin, serotonin and 5-methoxytryptamine. Endocrinology 100:675–680.
10. Martin J.E., McKellar S., and Klein D.C. 1980 Melatonin inhibition of the in vivo pituitary response to luteinizing hormone-releasing hormone in the neonatal rat. Neoroendocrinology 31:13–17.
11. Martin J.E., McKeel D.W., and Jr., Sattler C. 1982 Melatonin directly inhibits rat gonadotroph cells. Endocrinology 110:1079–1084.
12. Martin J.E. and Klein D.C. 1976 Melatonin inhibition of the neonatal pituitary response to luteinizing hormone-releasing factor. Science 191:301–302.
13. Martin J.E. and Sattler C. 1979 Developmental loss of the acute inhibitory effect of melatonin on the in vitro pituitary luteinizing hormone and folliclestimulating hormone responses to luteinizing hormone releasing hormone. Endocrinology 105:107–112.
14. Martin J.E. and Sattler C. 1982 Selectivity of melatonin pituitary inhibition for luteinizing hormone-releasing hormone. Neuroendocrinology 34:112–116.
15. Morgan P.J., Lawson W., Davidson G., and Howell H.E. 1989 Guanine nucleotides regulate the affinity of melatonin receptors on the ovine pars tuberalis. Neuroendocrinology 50:359–362.

16. Murakami N., Takamure M., Takahashi K., Utunomiya K., Kuroda H., and Etoh T. 1991 Long-term cultured neurons from rat suprachiasmatic nucleus retain the capacity for circadian oscillation of vasopressin release. Brain Res 545:347–350.
17. Padrell E., Carty D.J., Moriarty T.M., Hildebrandt J.D., Landau E.M., and Iyengar R. 1991 Two forms of the bovine brain Go that stimulate the inositol trisphosphate-mediated Cl- currents in Xenopus oocytes. Distinct guanine nucleotide binding properties. J Biol Chem 266:9771–9777.
18. Shibata S., Cassone V.M., and Moore R.Y. 1989 Effects of melatonin on neuronal activity in the rat suprachiasmatic nucleus in vitro. Neurosci Lett 97:140–144.
19. Slanar O., Zemkova H., and Vanecek J. 1997 Melatonin inhibits GnRH-induced Ca2+ mobilization and influx through voltage-regulated channels. Biol Signals 6:284–290.
20. Stehle J., Vanecek J., and Vollrath L. 1989 Effects of melatonin on spontaneous electrical activity of neurons in rat suprachiasmatic nuclei: an in vitro iontophoretic study. J Neural Transm 78:173–177.
21. Tasaka K., Stojilkovic S.S., Izumi S., and Catt K.J. 1988 Biphasic activation of cytosolic free calcium and LH responses by gonadotropin-releasing hormone. Biochem Biophys Res Commun 154:398–403.
22. Vanecek J., Pavlik A., and Illnerova H. 1987 Hypothalamic melatonin receptor sites revealed by autoradiography. Brain Res 435:359–362.
23. Vanecek J. 1988 The melatonin receptors in rat ontogenesis. Neuroendocrinology 48:201–203.
24. Vanecek J. 1988 Melatonin binding sites. J Neurochem 51:1436–1440.
25. Vanecek J. 1995 Melatonin inhibits increase of intracellular calcium and cyclic AMP in neonatal rat pituitary via independent pathways. Mol Cell Endocrinol 107:149–153.
26. Vanecek J. 1998 Melatonin inhibits release of luteinizing hormone via decrease of [Ca2+]i and cyclic AMP. Physiol Res 47:329–335.
27. Vanecek J. and Klein D.C. 1992 Melatonin inhibits gonadotropin-releasing hormone-induced elevation of intracellular Ca2+ in neonatal rat pituitary cells. Endocrinology 130:701–707.
28. Vanecek J. and Klein D.C. 1992 Sodium-dependent effects of melatonin on membrane potential of neonatal rat pituitary cells. Endocrinology 131:939–946.
29. Vanecek J. and Klein D.C. 1993 A subpopulation of neonatal gonadotropin-releasing hormone-sensitive pituitary cells is responsive to melatonin. Endocrinology 133:360–367.
30. Vanecek J. and Vollrath L. 1989 Melatonin inhibits cyclic AMP and cyclic GMP accumulation in the rat pituitary. Brain Res 505:157–159.
31. Vanecek J. and Vollrath L. 1990 Developmental changes and daily rhythm in melatonin-induced inhibition of 3′,5′-cyclic AMP accumulation in the rat pituitary. Endocrinology 126:1509–1513.
32. Vanecek J. and Vollrath L. 1990 Melatonin modulates diacylglycerol and arachidonic acid metabolism in the anterior pituitary of immature rats. Neurosci Lett 110:199–203.
33. Vanecek J. and Watanabe K. 1998 Melatonin inhibits increase of cyclic AMP in rat suprachiasmatic neurons induced by vasoactive intestinal peptide. Neurosci Lett 252:21–24.
34. Watanabe K., Koibuchi N., Ohtake H., and Yamaoka S. 1993 Circadian rhythms of vasopressin release in primary cultures of rat suprachiasmatic nucleus. Brain Res 624:115–120.
35. Watanabe K., Yamaoka S., and Vanecek J. 1998 Melatonin inhibits spontaneous and VIP-induced vasopressin release from suprachiasmatic neurons. Brain Res 801:216–219.
36. Weaver D.R., Namboodiri M.A., and Reppert S.M. 1988 Iodinated melatonin mimics melatonin action and reveals discrete binding sites in fetal brain. FEBS Lett 228:123–127.
37. Williams L.M., Martinoli M.G., Titchener L.T., and Pelletier G. 1991 The ontogeny of central melatonin binding sites in the rat. Endocrinology 128:2083–2090.
38. Williams L.M. and Morgan P.J. 1988 Demonstration of melatonin-binding sites on the pars tuberalis of the rat. J Endocrinol 119:R1–3.

22

THE ROLES OF MELATONIN IN DEVELOPMENT

David R. Weaver

Laboratory of Developmental Chronobiology
Pediatric Service, GRJ 1226
Massachusetts General Hospital
32 Fruit Street
and
Department of Pediatrics
Harvard Medical School
Boston, Massachusetts, 02115

1. MELATONIN IN PRENATAL DEVELOPMENT: HORMONE OR PHEROMONE?

Briefly consider the perceptual world of the fetus. Consider the role the mother plays in sheltering the fetus from the external environment. In this context, it seems remarkable that the mother is also actively involved in transferring environmental information to the fetus. During gestation, the mother generates signals which allow the fetus to perceive the length of the light portion of the lighting cycle (day length) as well as the phase (timing) of the light-dark cycle. We refer to these two different forms of prenatal communication as maternal-fetal communication of daylength and maternal-fetal communication of circadian phase, respectively (Figure 1). The maternal melatonin rhythm appears to play a role in transmitting both attributes of the lighting cycle from mother to fetus. Melatonin is thus an important component of the perceptual world of the fetus. In this context, it may be worth considering melatonin as a pheromone (a chemical substance from one member of a species which communicates information to another) rather than as a hormone.

Telephone: (617) 726-8450; Fax: (617) 726–1694; e-mail: weaver@helix.mgh.harvard.edu

Melatonin after Four Decades, edited by James Olcese.
Kluwer Academic / Plenum Publishers, New York, 2000.

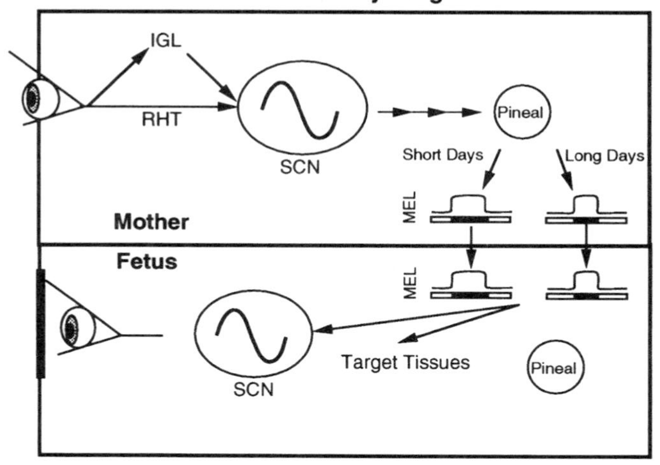

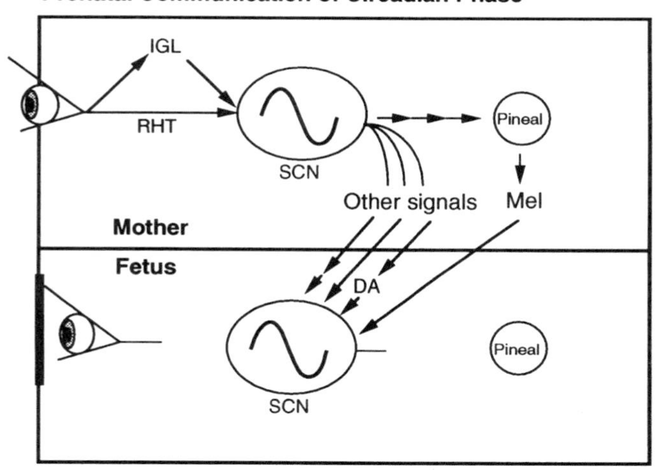

Figure 1. Schematic models illustrating the roles of melatonin in perception of the prenatal lighting cycle. Top panel: Prenatal Communication of Day Length. The mother (top box) detects light via the eye at a time when the fetus is sheltered from environmental lighting and the anatomical pathways for detection of light are not developed. Rhythms in melatonin (MEL) released from the maternal pineal are accurately reflected in the fetal circulation. Melatonin interacts with target tissues which may include the suprachiasmatic nucleus (SCN) to allow fetal perception of day length, which influences postnatal reproductive development (see section 3). Lower Panel: Prenatal Communication of Circadian Phase. Melatonin is one of several temporal signals whose generation is dependent upon the maternal SCN. Melatonin and dopamine (DA) interact with receptors in the fetal SCN to impart time-of-day information (see section 4). Abbreviations: IGL, intergeniculate leaflet; RHT, retinohypothalamic tract. Adapted from Ref. 53.

The two most widely recognized physiological roles of melatonin are in the seasonal regulation of reproduction and in the regulation of circadian rhythms. In its "pheromonal" role, melatonin influences these important physiological systems even while the developing animal is incapable of detecting light directly or of producing melatonin. This chapter will review these and related roles of melatonin in development. First, a brief discussion of the sources of melatonin reaching the developing animal is necessary.

2. SOURCES OF MELATONIN DURING DEVELOPMENT

The primary source of circulating melatonin in adult mammals is the pineal gland (38). Melatonin levels are elevated at night. As reviewed in detail elsewhere in this volume, melatonin production from the pineal is regulated by the circadian clock in the suprachiasmatic nuclei. Other tissues appear capable of producing melatonin. Melatonin is produced rhythmically in the retina, but the retina does not appear to contribute significantly to circulating melatonin levels in rodents. N-acetyltransferase activity and NAT gene expression occur in other tissues, including brainstem and pituitary (12). With precursor loading it is possible to detect significant production of melatonin in the gastrointestinal tract, although melatonin synthesized in the gastrointestinal tract does not appear to contribute to circulating levels of the hormone (24). It is nevertheless possible the local production of melatonin within the abdomen may provide a pool capable of reaching the fetus.

The primary source of melatonin for the developing mammal is the maternal pineal gland. The melatonin rhythm in maternal circulation is accurately reflected in the fetus, allowing melatonin to serve as a "chrono-pheromone." Melatonin levels in fetal plasma closely parallel those in maternal plasma, and rapid transfer of melatonin across the placenta has been demonstrated in several species including rodents, sheep, and non-human primates (34,59,69,86,87,89). Furthermore, maternal pinealectomy abolishes the rhythm of melatonin in fetal circulation (46,87; see Figure 2).

Maternal melatonin also reaches the developing animal after birth. Transfer of melatonin from mother to offspring in milk has been demonstrated (50), and rhythmicity of serum melatonin levels in suckling rats appears due to transfer of maternal melatonin via the milk (69).

Rhythmic melatonin production from the developing pineal is first significant during the second to third week of postnatal life in rodents (67). In sheep and humans, rhythmicity of melatonin levels in serum, originating in the developing pineal gland, also begins during the postnatal period (2,11,46).

3. MELATONIN AND REPRODUCTION

A major physiological role of melatonin in adult mammals is the regulation of seasonal reproduction (for reviews see 3,21,33,68 and this volume). The pineal gland is necessary for appropriate perception of seasonal changes in day length and thus for the proper timing of reproduction in species which breed seasonally. The duration of the nocturnal melatonin elevation is regulated by the photoperiod. Melatonin ultimately affects reproductive activity by modulating the activity of hypothalamic neuroendocrine circuits whose activity is necessary for gonadal function. In the broadest

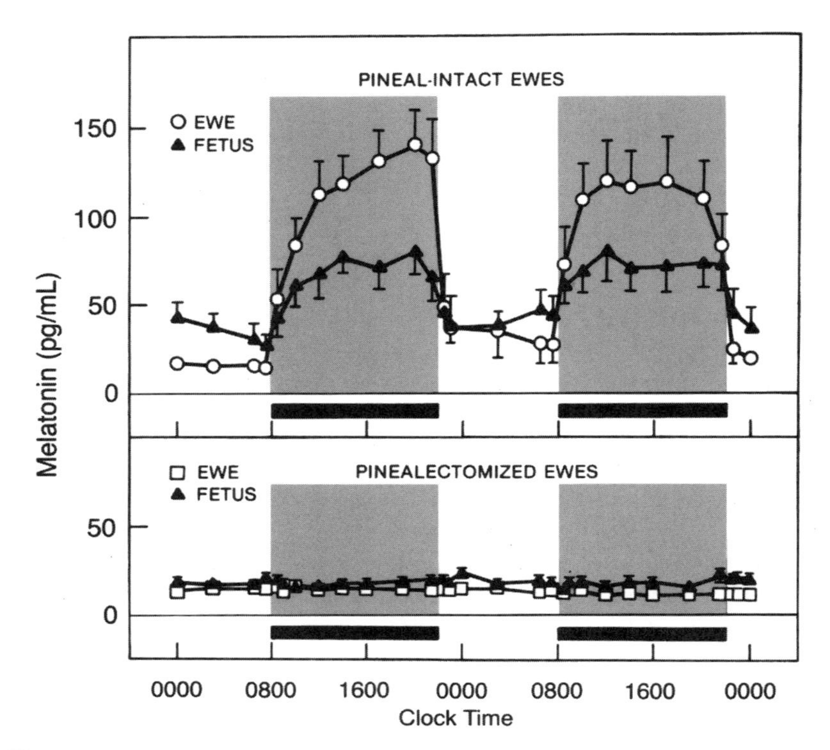

Figure 2. The maternal pineal gland is the source of melatonin rhythms in fetal sheep. Top panel: Melatonin rhythms in the mother and fetus are similar in phase and amplitude. Lower panel: Removal of the maternal pineal abolishes maternal and fetal melatonin rhythms. Modified from Ref. 87.

sense, then, melatonin broadly influences development by restricting the season of conception, and thus the season of birth, such that the offspring are born at a season with the most favorable environmental conditions possible.

The influence of melatonin on reproductive development begins during the prenatal period and extends into postnatal life. The following sections will detail the influence of melatonin on reproduction during development.

3.1. Postnatal Photoperiod and the Timing of Puberty

In many seasonally breeding species, the timing of initial reproductive development (puberty) is strongly influenced by daylengths experienced during the postnatal period. The most well-studied species include Siberian hamsters and sheep (for reviews see 3,21). Administration of melatonin in the appropriate temporal (durational) patterns can influence the timing of puberty in these species. In Siberian hamsters, for example, exposure to long days stimulated postnatal reproductive development. Infusion of long-day patterns of melatonin similarly stimulates gonadal growth, while short-day patterns of melatonin suppress puberty (8; Figure 3). While the details are species-specific, the same general phenomenon exists in other photoperiodic species: specific photoperiodic requirements must be met to allow rapid pubertal development, and these photoperiodic requirements are transduced to the neuroendocrine axis by melatonin (21,29,63,83,85,87). Melatonin implants alter the timing of pubertal development in many photoperiodic species, likely by preventing accurate perception of the environmental daylength normally imparted by the melatonin rhythm.

In Syrian hamsters, initial reproductive development is not influenced by daylength or melatonin. Syrian hamsters detect daylengths and melatonin prior to puberty, but the response is delayed until after puberty due to a long latency to respond (57).

Correlation between a developmental decline in melatonin levels with the timing of puberty in humans led to speculation that melatonin regulates the timing of puberty (72). Subsequent investigation indicates that this developmental decline in melatonin levels is due at least in part to developmental changes in body mass (and thus volume

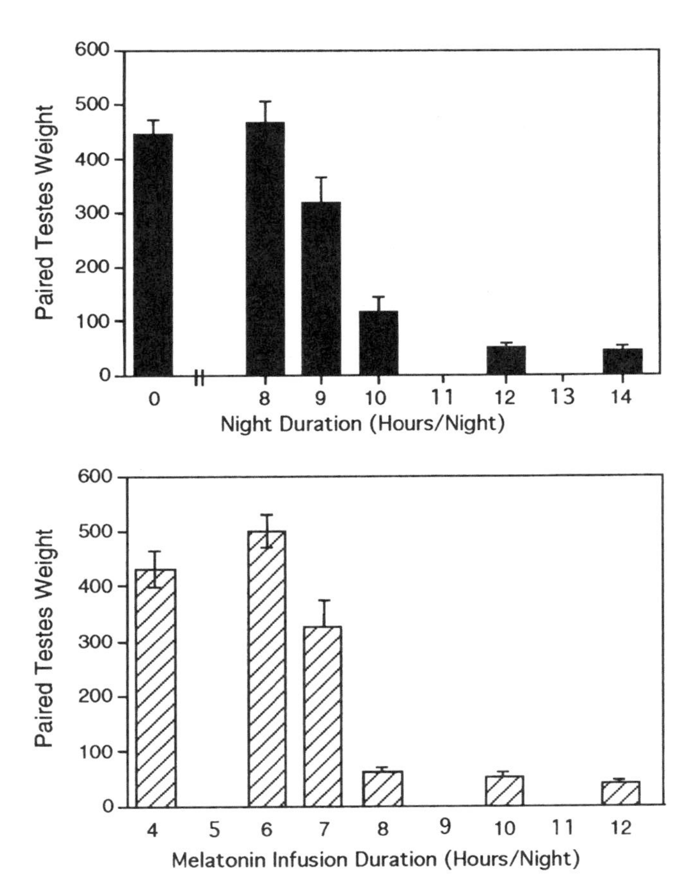

Figure 3. Postnatal photoperiod and melatonin influence the rate of reproductive development in Siberian hamsters. Top panel: Pineal-intact juvenile males were housed in photoperiods ranging from constant light to 10 hours light:14 hours darkness beginning at 18 days of age. Lower panel: Juvenile male hamsters were pinealectomized at 18 days of age and received infusions of melatonin (10 ng per night) each night from day 18 to day 30. All animals were born to pineal-intact dams housed in 16 hours of light per day. Testicular weights were assessed at 30 days of age. Long nightly melatonin infusions mimicked the effect of exposing males to short days, and the duration-response curves for melatonin and darkness are parallel. Data from Ref. 8.

of distribution), and is without a strict relationship to pubertal development (9,48,88). While endogenous melatonin does not appear to play a role in timing human puberty, no data are available to draw a conclusion with respect to the effects of exogenous melatonin on puberty in humans.

3.2. Prenatal Programming of Postnatal Reproductive Development

Photoperiodic information reaches the fetus during prenatal life, and can have a dramatic impact on reproductive development (For reviews see 14,77). The initial observation that day lengths experienced during the prenatal period affect pubertal development began with studies of montane voles (*Microtus montanus*) conducted by Teresa Horton. Field work had shown that male voles born after the summer solstice were unlikely to undergo rapid reproductive development, and these late-season males would often delay puberty until the following Spring. This seemed contrary to photoperiodic mechanisms, as the day length just after the summer solstice is among the longest of the year. Horton's laboratory studies revealed that the photoperiod experienced during prenatal life affected the timing of postnatal reproductive development (25,26). Cross-fostering studies revealed that the effect of daylength was perceived during the prenatal period (27).

Subsequent studies have primarily focused on Siberian hamsters, with several groups demonstrating that prenatal photoperiod influences postnatal reproductive development (Figure 1a; Figure 4 top; see 77 for review). Pups reared in an

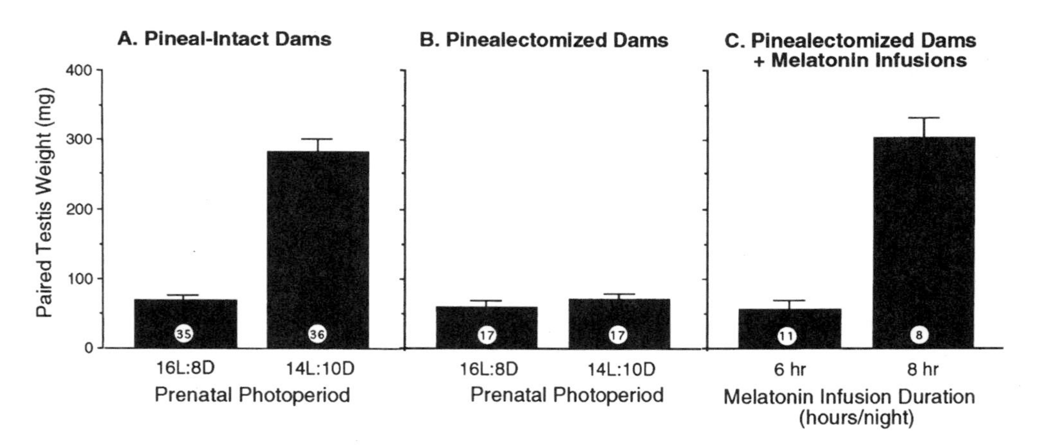

Figure 4. The prenatal photoperiod and prenatal exposure to melatonin influence the rate of reproductive development in male Siberian hamsters. All offspring were reared in a 14 hour light: 10 hour dark (14L:10D) lighting cycle after birth, and paired testis weights were determined at euthanasia on postnatal day 34. Values represent mean ± SEM. Sample sizes are indicated within each bar. A. Prenatal perception of day length by offspring of pineal intact-dams. Animals experiencing a decrease in daylength from 16L:8D to 14L:10D at birth have slower reproductive development than hamsters maintained in the intermediate 14L:10D photoperiod throughout pre- and postnatal development. B. Maternal pinealectomy prior to breeding prevents fetal perception of daylength. C. Melatonin communicates daylength information to the fetus. The nightly duration of infusion of melatonin was varied (50 ng in 0.2 cc delivered over 6 or 8 hours per night, for 3–7 nights at the end of gestation) to reproduce melatonin pattern that would occur in pineal-intact hamsters maintained in photoperiods of 16L:8D or 14L:10D, respectively. The effects of varying melatonin infusion duration mimicked the effects of varying prenatal photoperiod. Modified from Ref. 77.

"intermediate" postnatal photoperiod differ in their rate of reproductive development if the prenatal photoperiods experienced by their mothers differed (63,64,75). At more extreme photoperiods, the postnatal photoperiod "overrides" the influence of the prenatal photoperiod (63,65). The influence of prenatal daylength occurs prenatally (65). Considering the role of the pineal gland in photoperiodic regulation, it seemed likely that removal of the maternal pineal gland would prevent prenatal perception of daylength. Several groups have shown that the maternal pineal gland is necessary for prenatal communication of daylength information (29,75; Figure 4 center). To specifically assess the role of melatonin, we delivered infusions of melatonin into pinealectomized dams during pregnancy. Timed infusions of melatonin were delivered for the last several nights of gestation, with melatonin infusion duration (e.g., the hours of infusion per night) being varied to mimic the physiological pattern of the hormone that would occur under various lighting schedules. The nightly duration of melatonin exposure during gestation determined the rate of pubertal development (75,79; Figure 4 lower). Infusions delivered during the day and during the night appeared equally effective; the important variable was the infusion duration. In a subsequent study, we restricted the number of infusions to determine the time of peak sensitivity to melatonin, and also to define the minimum stimulus. A single gestational melatonin infusion did not influence postnatal reproductive development, but two infusions on consecutive nights did provide a sufficient signal (79). A period of maximum sensitivity to melatonin infusions occurs during late gestation. Studies using timed injection of melatonin (rather than infusions) support the conclusion that melatonin is a critical cue for transfer of day length information to the fetus (28,65). In keeping with the historical theme of this conference, it is worth noting that our desire to identify sites of melatonin action within fetal hamster brain prompted our first studies of melatonin receptor localization (80). Our melatonin receptor work recently led to our identification of a family of melatonin receptor subtypes (see 54 for review).

Maternal transfer of photoperiodic information has been referred to as "prenatal programming" of postnatal reproductive development, as the influence of the prenatal photoperiod can be observed in animal reared in constant light, in the absence of postnatal melatonin exposure (28,65).

One way to interpret these data, and the field data from voles, is that the developing animal takes a reading of daylength (melatonin duration) during late fetal life and compares this with the melatonin signal derived from the developing pineal around 15–20 days of age (67). This model suggests that a postnatal melatonin pattern is interpreted differently depending on an animal's prenatal photoperiodic history. Shaw and Goldman (62) have reported that the response to postnatal melatonin infusions is in fact *not* dependent on the prenatal photoperiod; reproductive response to postnatal melatonin infusions was dictated solely by infusion duration, without an influence of the prenatal photoperiod. Assessment of the melatonin rhythms in populations of juvenile hamsters with different prenatal photoperiodic histories revealed a difference in the duration of the nocturnal melatonin elevation (61). These results suggest a model in which the effect of prenatal photoperiod is to alter the SCN-generated pattern of postnatal melatonin production, rather than having an effect on programming neuroendocrine function per se. This model is unable to account for the impact of prenatal photoperiodic history on animals reared in constant light, in the absence of postnatal photoperiodic information (cf. 28,65).

Finally, Horton et al. (30) have proposed that there are other signals, besides melatonin, involved in the prenatal communication of daylength information in Siberian hamsters. Pregnant hamsters treated with melatonin-containing implants that "swamped out" the endogenous melatonin profile were still able to communicate day length information to their fetuses. Thus, while the pineal gland is necessary for prenatal communication of daylength information (75), and melatonin is sufficient (65,75,79), additional pineal-derived signals may be involved.

The prenatal influence of melatonin on development is not limited to Siberian hamsters and montane voles. Lee, Zucker and colleagues have shown that perinatal development of meadow vole pups (*Microtus pennsylvanicus*) is influenced by prenatal photoperiod and prenatal melatonin treatment (35–37). Remarkably, there is also an influence of the duration the mother has been exposed to short day lengths, e.g., time since exposure to short days (35,36).

Melatonin influences other neuroendocrine parameters in developing animals prior to puberty, For example, prolactin levels in adult and neonatal sheep are regulated by daylength (20,21). Prolactin levels of fetal sheep are influenced by the photoperiod the mother experienced during gestation, with long days resulting in higher prolactin levels (20). Melatonin implants reduce prolactin levels in fetal sheep (4,59). Melatonin acts via the pars tuberalis to influence prolactin secretion in adult sheep (31,44). The same mechanisms appears to be operative in the fetal lamb (46). There is also an influence of photoperiodic history on prolactin: prolactin levels in neonatal lambs exposed to 12 hours of light per day following birth are strongly influenced by the prenatal daylength (20). Photoperiodic history influences prolactin secretion even in fetal sheep (31).

3.3. Seasonal Embryonic Diapause

A fascinating, ecologically important role for melatonin in regulating reproduction during early fetal development is in the initiation and maintenance of seasonal embryonic diapause (for review, see 58). Following fertilization, the blastocyst develops for several days and then "arrests" for a variable period (Figure 5). Implantation and reactivation of development occur such that offspring are born in Spring, with the most favorable conditions for survival. The period of delayed implantation (embryonic diapause) is regulated by the environmental lighting cycle, via melatonin. Season embryonic diapause occurs in many different mammalian species including mustelids (e.g., skunks, ferrets, badgers, weasels, mink), pinipeds (e.g., Australian sea lions and Antarctic fur seals, harbor seals), insectivores (several bat species), canids (wolves and coyotes) bears, and marsupials.

Western spotted skunks, tammar wallabies, and mink have been studied most extensively with respect to the role of the pineal and melatonin. In each of these three species, pinealectomy or denervation of the pineal prevents seasonal embryonic diapause, and melatonin treatment influences the length of diapause (6,7,40,41,45,49). The data from all species are consistent with the interpretation that melatonin acts in the pregnant mother to influence neuroendocrine function, particularly prolactin secretion, and that diapause is caused by alterations in the uterine environment. Melatonin does not appear to directly affect the embryo.

In skunks, melatonin treatment delays implantation by several months, and causes a long-term suppression of prolactin levels. Destruction of the SCN does not block the effect of exogenous melatonin (6). In contrast, anterior hypothalamic lesions disinhibit prolactin production and result in precocious termination of diapause, regardless of the

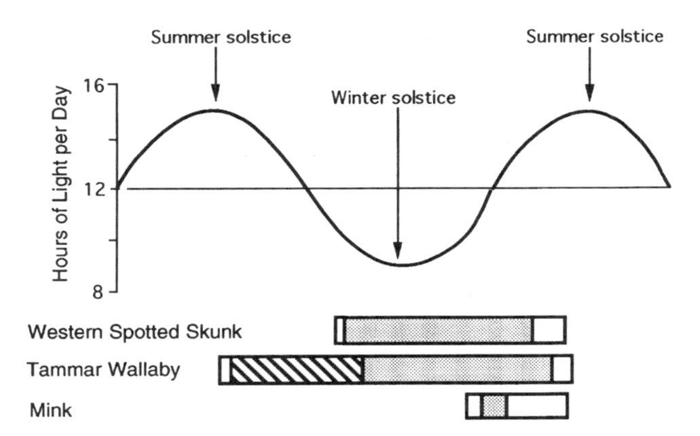

Figure 5. Schematic representation of seasonal embryonic diapause. A period of embryonic diapause (stippled bar) interrupts fetal development (open bar) for each species. In wallabies, the initial period of diapause is induced by suckling stimuli from the pouch young and is termed lactational diapause (cross-hatched bar) (cf. Ref. 49). The upper portion indicates the annual change in day length at a mid-temperate (ca. 45°) latitude; note that annual day length variation at more polar latitudes would be more extreme than shown. Data are presented relative to solstices to allow presentation of species from both Northern and Southern hemispheres.

melatonin milieu (7,32). It is not clear whether melatonin acts within or through the anterior hypothalamus; the anterior hypothalamus may be a leg in the final common pathway in prolactin regulation downstream of the melatonin target sites. Considering that high-affinity melatonin receptor binding is restricted to the pars tuberalis in skunks (18), these data suggest a model in which melatonin regulates production of a factor from the pars tuberalis that influences prolactin secretion. A similar proposal has been made to explain the influence of melatonin, mediated by the pars tuberalis, on prolactin secretion in sheep (39). For further discussion of the pars tuberalis as a site of melatonin action, see the chapter by P.J. Morgan (this volume).

3.4. Melatonin and Embryonic Survival

Melatonin is generally viewed as remarkably non-toxic (see 66,73), and this also appears to be the case during fetal life. Melatonin is without effect on development of mouse embryos in vitro (42). A second study reports that melatonin does have toxicity to embryos (10), but the concentrations of melatonin used were 100–200 ug/ml (ca. 0.5–1.0 mM), astronomically high when compared to endogenous melatonin levels (<1 nM) or with the micromolar levels which result following pharmacological doses.

An influence of melatonin injection on embryonic survival has been reported in meadow voles (22). Melatonin injections (10 ug, 2 hr before lights off in 14L:10D) reduced survival rates of female (but not male) pups when given prior to blastocyst implantation. Injections later in gestation were without effect. This effect of melatonin may underlie seasonal changes in litter size, and perhaps seasonal changes in sex ratios. The mechanism by which melatonin modulates prenatal mortality in a sex-specific manner is not known.

4. MELATONIN AND PRENATAL COMMUNICATION OF CIRCADIAN PHASE

In mature rodents, light is the most potent signal for synchronization of endogenous circadian rhythmicity to environmental cyclicity. Direct and indirect pathways from the retina to the suprachiasmatic nucleus (SCN) are involved in the process of photic entrainment. During development, entrainment of the biological clock begins prior to the development of the retinohypothalamic tract. Indeed, studies in rats, hamsters, mice, and monkeys indicate that the biological clock in the SCN is functional (rhythmic) and entrained prior to birth (Figure 1, lower panel; for reviews see 13,14,53,56). Prenatal entrainment requires the presence of the maternal SCN, indicating that the mother plays an active role in the entrainment of her fetuses (15,51). The potency of the mother as an entraining cue (Zeitgeber) decreases during the first week after birth, coincident with the appearance of circadian sensitivity to light. Maternal communication of time-of-day information is thus thought to reflect a transient mechanism for entrainment which precedes the anatomical and functional development of photic entrainment mechanisms (for further discussion, see 78).

Reppert and Schwartz (52) removed various endocrine organs in an attempt to identify a hormonal signal necessary for entrainment of the rat fetus. Removal of the pituitary, ovaries, adrenals, thyroid/parathyroid or pineal (each surgical procedure performed in separate groups of animals) did not disrupt prenatal synchronization (although results with pinealectomy were somewhat unclear). An alternative strategy is to identify stimuli which can entrain the fetal circadian clock when given periodically during gestation to SCN-lesioned dams. Three treatments which can entrain the fetal clock have been identified in this way: restricted feeding, timed administration of D1-dopamine receptor agonist, and timed administration of melatonin.

In rats, periodic feeding of SCN-lesioned dams during gestation leads to fetal entrainment (76). Periodic feeding (4 hours food access per 24 hour day on gestational days 8–19) induces rhythmicity in maternal activity, body temperature, and likely in metabolic parameters and substrate availability. Precursor loading can alter fetal brain neurotransmitter levels (1), so periodic feeding may act in part through stimulating monoamine levels in the fetal brain.

A second manipulation that can entrain fetal rodents is periodic stimulation of D1-dopamine receptors. D1-dopamine receptor gene expression is high in the fetal SCN and activation of SCN D1-dopamine receptors induces c-fos gene expression (5,70,71,78,81,82). Other studies have shown that the dopaminergic activation of c-fos gene expression is brief, developmentally restricted to the prenatal and early neonatal period, and is dependent upon the D1-dopamine receptor (5,78,82). In view of the correlation between photic activation of c-fos gene expression and phase shifting of the adult SCN, these data suggested that activation of D1-dopamine receptors in the fetal SCN may lead to entrainment of the fetus. Indeed, treatment of pregnant, SCN-lesioned hamsters with the D1-dopamine receptor agonist SKF 38393 (8 mg/kg once per day on GD 12–15) entrains fetal hamsters and induces c-fos gene expression in the fetal hamster SCN (71). More recently, Viswanathan and Davis (70) have shown that a single injection of SKF 38393 on gestational day 15 can entrain fetal hamsters. Several groups have described a transient dopaminergic input to the developing SCN, which may participate in fetal entrainment under physiological conditions (47,60, and references therein).

Timed administration of melatonin is the third "signal" capable of entraining the fetal circadian clock. Melatonin plays an important role in circadian organization of several non-mammalian vertebrates, and exogenous melatonin can influence circadian rhythmicity in a variety of mammals (see 74 and this volume for reviews). Fred Davis and colleagues have shown that melatonin is a powerful entraining stimulus for the fetus: as with SKF 38393, a single injection late in gestation can synchronize the fetal circadian clocks (14). Sensitivity to melatonin begins late in gestation and persists into the early neonatal period (13,14,16,23; Figure 6). The postnatal decline in sensitivity to melatonin parallels the postnatal decline in melatonin receptor binding in the Syrian hamster SCN (19).

Receptors for both dopamine and melatonin are expressed in the developing SCN. These receptors generally have antagonistic effects on signal transduction

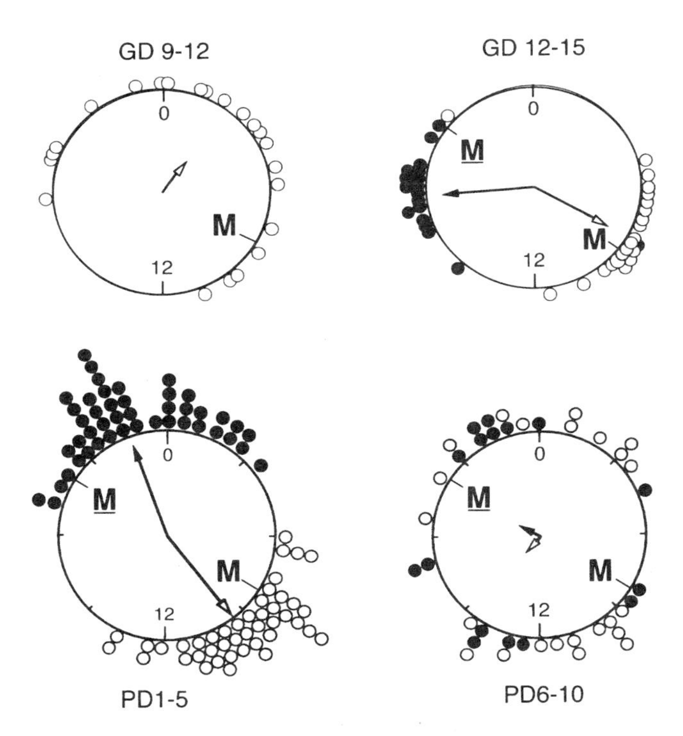

Figure 6. Definition of a sensitive period for entrainment by melatonin. Each large circle 24 hours on the day of weaning, and each smaller circle around the perimeter represents the time of activity onset of one pup. Synchrony among pups receiving a treatment is indicated by a clustering of pup phases. Filled and open circles represent pups whose dams received melatonin at night and in the morning, respectively. Timing of melatonin administration in night groups is indicated by underlined *M*, while morning groups are indicated by M (not underlined). All pups were gestated and reared in dim constant light by dams receiving SCN lesions on gestational day 7. The length of the arrow within each circle represents the degree of synchrony within the population as assessed by the Rayleigh test. Significant synchrony among pups was established by 4–5 melatonin injections delivered late in gestation (GD 12–15) and in the early postnatal period (postnatal day [PD] 1–5). Injections earlier in gestation (GD 9–12) and later in postnatal life (PD 6–10) did not produce the same synchronizing effect. Saline injections at comparable periods were without effect (not shown). Note that injections of melatonin at opposite phases established average phases that were roughly opposite. melatonin doses were 10–100 ug per animal in pregnancy, and an equivalent amount to neonates (0.167 ug/g). Data from Refs. 13,16,23.

mechanisms: D1-dopamine receptors are coupled to the stimulation of cAMP accumulation, while melatonin receptors are negatively coupled to adenylyl cyclase (54). This antagonism extends also to prenatal entrainment: the phase set by prenatal injection of SKF 38393 is opposite the phase established by prenatal injections of melatonin (70,71). It is possible that the three identified mechanisms for prenatal entrainment are actually interrelated, with dopamine representing a final common pathway. Periodic feeding may cause rhythmic stimulation of dopamine levels in the fetal SCN, while melatonin may inhibit synaptic release of dopamine. Melatonin inhibits dopamine release from rabbit retina and rodent hypothalamus (17,90), so an effect of melatonin on dopamine release within the fetal SCN is very plausible. This model predicts that both melatonin and dopamine will be ineffective in entraining animals that lack functional D1-dopamine receptors. Dopamine, but not melatonin, is expected to entrain mice lacking melatonin receptors. Unfortunately, direct assessment of circadian phase in fetal mice has not been possible to date (5). Postnatal assessment of circadian phase in mice promises to be a useful approach provided that the phase recorded after weaning is an accurate reflection of prenatal, rather than postnatal, events.

5. CONCLUSIONS

The pattern of melatonin in the maternal bloodstream is accurately reflected in the fetal circulation. Transplacental melatonin exposure has at least two important functions. Melatonin influences neuroendocrine development of seasonally breeding mammals. Melatonin is also one of several signals capable of entraining the fetal circadian clock. In view of its ability to communicate temporal information from mother to fetus, this enigmatic molecule should be viewed as a (chrono) pheromone as well as being a hormone.

REFERENCES

1. Aravelo R., Afonso D., Castro R., and Rodriguez M. (1991). Fetal brain serotonin synthesis and catabolism is under control by mother intake of tryptophan. Life Sci 49:53–66.
2. Attanasio A., Rager K., and Gupta D. (1986). Ontogeny of circadian rhythmicity for melatonin, serotonin, and N-acetylserotonin in humans. J Pineal Res 3:251–256.
3. Bartness T.J., Powers J.B., Hastings M.H., Bittman E.L., and Goldman B.D. (1993). The timed infusion paradigm: What has it taught us about the melatonin signal, its reception, and the photoperiodic control of seasonal responses? J Pineal Res 15:161–190.
4. Bassett J.M., Curtis N., Hanson C., and Weeding C.M. (1989). Effects of altered photoperiod or maternal melatonin administration on plasma prolactin concentrations in fetal lambs. J Endocrinol 122:633–643.
5. Bender M., Drago J., and Rivkees S.A. (1997). D1 receptors mediate dopamine action in the fetal suprachiasmatic nuclei: studies of mice with targeted deletion of the D1 dopamine receptor gene. Mol Brain Res 49:271–277.
6. Berria M., DeSantis M., and Mead R.A. (1988). Effects of suprachiasmatic nuclear ablation and melatonin on delayed implantation in the spotted skunk. Neuroendocrinology 48:371–375.
7. Berria M., DeSantis M., and Mead R.A. (1989). Lesions to the anterior hypothalamus prevent the melatonin-induced lengthening of delayed implantation. Endocrinology 125:2897–2904.
8. Carter D.S. and Goldman B.D. (1983). Antigonadal effects of timed melatonin infusion in pinealectomized male Djungarian hamsters (*Phodopus sungorus sungorus*): duration is the critical parameter. Endocrinology 113:1261–1267.
9. Cavallo A. and Ritschel W.A. (1996). Pharmacokinetics of melatonin in human sexual maturation. J Clin Endocrinol Metab 81:1882–1886.

10. Chan W.Y. and Ng T.B. (1995). Changes induced by pineal indoles in post-implantation mouse embryos. Gen Pharmacol 26:1113–1118.
11. Claypool L.E., Wood R.I., Yellon S.M., and Foster D.L. (1989). The ontogeny of melatonin secretion in the lamb. Endocrinology 124:2135–2143.
12. Coon S.L., Roseboom P.H., Baler R., Weller J.L., Namboodiri M.A.A., Koonin E.V., and Klein D.C. (1995). Pineal serotonin N-acetyltransferase: expression cloning and molecular analysis. Science 270:1681–1683.
13. Davis F.C. (1989). Use of postnatal behavioral rhythms to monitor prenatal circadian function. In: S.M. Reppert (Ed.) Development of Circadian Rhythmicity and Photoperiodism in Mammals, Perinatology Press, Ithaca NY, pp. 45–65.
14. Davis F.C. (1997). Melatonin: role in development. J Biol Rhythms 12:498–508.
15. Davis F.C. and Gorski R.A. (1988). Development of hamster circadian rhythms: role of the maternal suprachiasmatic nucleus. J Comp Physiol [A] 162:601–610.
16. Davis F.C. and Mannion J. (1988). Entrainment of hamster pup circadian rhythms by prenatal melatonin injections to the mother. Am J Physiol 255:R439–448.
17. Dubocovich M.L. (1983). Melatonin is a potent modulator of dopamine release in the retina. Nature 306:782–784.
18. Duncan M.J. and Mead R.A. (1992). Autoradiographic localization of binding sites for 2-[125I]iodomelatonin in the pars tuberalis of the western spotted skunk (*Spilogale putorius latifrons*). Brain Res 569:152–155.
19. Duncan M.J. and Davis F.C. (1993). Developmental appearance and age related changes in specific 2-[125I]iodomelatonin binding sites in the suprachiasmatic nuclei of female Syrian hamsters. Dev Brain Res 73:205–212.
20. Ebling F.J., Wood R.I., Suttie J.M., Adel T.E., and Foster D.L. (1989). Prenatal photoperiod influences neonatal prolactin secretion in the sheep. Endocrinology 125:384–391.
21. Foster D.L., Ebling F.J.P., Claypool L.E., and Wood R.I. (1989). Photoperiodic timing of puberty in sheep. In: S.M. Reppert (Ed.) Development of Circadian Rhythmicity and Photoperiodism in Mammals, Perinatology Press, Ithaca NY, pp. 103–153.
22. Gorman M.R., Ferkin M.H., and Dark J. (1994). Melatonin influences sex-specific prenatal mortality in meadow voles. Biol Reprod 51:873–878.
23. Grosse J., Velickovic A., and Davis F.C. (1996). Entrainment of Syrian hamster circadian activity rhythms by neonatal melatonin injections. Am J Physiol 270:R533–540.
24. Heuther G., Poeggeler B., Reimer A., and George A. (1992). Effects of tryptophan administration on circulating melatonin levels in chicks and rats: evidence for stimulation of melatonin synthesis and release in the gastrointestinal tract. Life Sci 51:945–953.
25. Horton T.H. (1984a). Growth and maturation of *Microtus montanus*: effects of photoperiods before and after weaning. Can J Zool 62:1741–1746.
26. Horton T.H. (1984b). Growth and reproductive development of male *Microtus montanus* is affected by the prenatal photoperiod. Biol Reprod 31:499–504.
27. Horton T.H. (1985). Cross-fostering of voles demonstrates in utero effect of photoperiod. Biol Reprod 33:934–939.
28. Horton T.H., Ray S.L., and Stetson M.H. (1989). Maternal transfer of photoperiodic information in Siberian hamsters. III. Melatonin injections program postnatal reproductive development expressed in constant light. Biol Reprod 41:34–39.
29. Horton T.H., Stachecki S.A., and Stetson M.H. (1990). Maternal transfer of photoperiodic information in Siberian hamsters. IV. Peripubertal reproductive development in the absence of maternal photoperiodic signals during gestation. Biol Reprod 42:441–449.
30. Horton T.H., Ray S.L., Rollag M.D., Yellon S.M., and Stetson M.H. (1992). Maternal transfer of photoperiodic information in Siberian hamsters. V. Effects of melatonin implants are dependent on photoperiod. Biol Reprod 47:291–296.
31. Houghton D.C., Young I.R., and McMillen I.C. (1997). Photoperiodic history and hypothalamic control of prolactin secretion before birth. Endocrinology 138:1506–1511.
32. Kaplan J.B., Berria M., and Mead R.A. (1991). Prolactin levels in the western spotted skunk: changes during pre- and periimplantation and effects of melatonin and lesions to the anterior hypothalamus. Biol Reprod 44:991–997.
33. Karsch F.J., Bittman E.L., Foster D.L., Goodman R.L., Legan S.J., and Robinson J.E. (1984). Neuroendocrine basis of seasonal reproduction. Recent Prog Horm Res 40:185–232.
34. Klein D.C. (1972). Evidence for the placental transfer of 3 H-acetyl-melatonin. Nat New Biol 237:117–118.

35. Lee T.M. and Zucker I. (1988). Vole infant development is influenced perinatally by maternal photoperiodic history. Am J Physiol 255:R831–838.
36. Lee T.M., Smale L., Zucker I., and Dark J. (1987). Influence of daylength experienced by dams on post-natal development of young meadow voles (*Microtus pennsylvanicus*). J Reprod Fertil 81:337–342.
37. Lee T.M., Spears N., Tuthill C.R., and Zucker I. (1989). Maternal melatonin treatment influences rates of neonatal development of meadow vole pups. Biol Reprod 40:495–502.
38. Lewy A.J., Tetsuo M., Markey S.P., Goodwin F.K., and Kopin I.J. (1980). Pinealectomy abolishes plasma melatonin in the rat. J Clin Endocrinol Metab 50:204–205.
39. Lincoln G.A. and Clarke I.J. (1994). Photoperiodically-induced cycles in the secretion of prolactin in hypothalamo-pituitary disconnected rams: evidence for translation of the melatonin signal in the pituitary gland. J Neuroendocrinol 6:251–260.
40. May R. and Mead R.A. (1986). Evidence for pineal involvement in timing implantation in the western spotted skunk. J Pineal Res 3:1–8.
41. McConnell S.J., Tyndale-Biscoe C.H., and Hinds L.A. (1986). Change in duration of elevated melatonin concentration is the major factor in photoperiod response in the tammer, *Macropus eugenii*. J Reprod Fertil 77:623–632.
42. McElhinny A.S., Davis F.C., and Warner C.M. (1996). The effects of melatonin on cleavage rate on C57BL/6 and CBA/Ca preimplantation embryos cultured in vitro. J Pineal Res 21:44–48.
43. McMillen I.C. and Nowak R. (1989). Maternal pinealectomy abolishes the diurnal rhythm in plasma melatonin concentrations in the fetal sheep and pregnant ewe during late gestation. J Endocrinol 120:459–464.
44. McMillen I.C., Houghton D.C., and Young I.R. (1995). Melatonin and the development of circadian and seasonal rhythmicity. J Reprod Fertil Suppl 49:137–146.
45. Murphy B.D., DiGregorio G.B., Douglas D.A., and Gonzalez-Reyna A. (1990). Interactions between melatonin and prolactin during gestation in mink (*Mustela vison*). J Reprod Fertil 89:423–429.
46. Nowak R., Young I.R., and McMillen I.C. (1990). Emergence of the diurnal rhythm in plasma melatonin concentrations in newborn lambs delivered to intact or pinealectomized ewes. J Endocrinol 125:97–102.
47. Novak C.M. and Nunez A.A. (1998). Tyrosine hydroxylase- and/or aromatic L-amino acid decarboxylase- containing cells in the suprachiasmatic nucleus of the Syrian hamster (*Mesocricetus auratus*). J Chem Neuroanat 14:87–94.
48. Reiter R.J. (1998). Melatonin and human reproduction. Ann Med 30:103–108.
49. Renfree M.B., Lincoln D.W., Almeida O.F., and Short R.V. (1981). Abolition of seasonal embryonic diapause in a wallaby by pineal denervation. Nature 293:138–139.
50. Reppert S.M. and Klein D.C. (1978). Transport of maternal [^{3}H]-melatonin to suckling rats and the fate of [^{3}H]-melatonin in the neonatal rat. Endocrinology 102:582–586.
51. Reppert S.M. and Schwartz W.J. (1986a). Maternal suprachiasmatic nuclei are necessary for maternal coordination of the developing circadian system. J Neurosci 6:2724–2729.
52. Reppert S.M. and Schwartz W.J. (1986b). Maternal endocrine extirpations do not abolish maternal coordination of the fetal circadian clock. Endocrinology 119:1763–1767.
53. Reppert S.M. and Weaver D.R. (1989). Maternal transduction of light-dark information for the fetus. In: W.P. Smotherman and S.R. Robinson (Eds.) Behavior of the Fetus., Telford Press, Caldwell NJ, pp. 119–139.
54. Reppert S.M. and Weaver D.R. (1995). Melatonin madness. Cell 83:1059–1062.
55. Reppert S.M., Chez R.A., Anderson A., and Klein D.C. (1979). Maternal-fetal transfer of melatonin in the non-human primate. Pediatr Res 13:788–791.
56. Reppert S.M., Weaver D.R., and Rivkees S.A. (1989). Prenatal function and entrainment of a circadian clock. In: S.M. Reppert (Ed.) Development of Circadian Rhythmicity and Photoperiodism in Mammals, Perinaology Press, Ithaca NY, pp. 25–44.
57. Rollag M.D., Dipinto M.N., and Stetson M.H. (1982). Ontogeny of the gonadal response of golden hamsters to short photoperiod, blinding, and melatonin. Biol Reprod 27:898–902.
58. Sandell M. (1990). The evolution of seasonal delayed implantation. Q Rev Biol 65:23–42.
59. Seron-Ferre M., Vergara M., Parraguez V.H., Riquelme R., and Llanos A.J. (1989). Fetal prolactin levels respond to a maternal melatonin implant. Endocrinology 125:400–403.
60. Strother W.N., Norman A.B., and Lehman M.N. (1998). D1-dopamine receptor binding and tyrosine hydroxylase- immunoreactivity in the fetal and neonatal hamster suprachiasmatic nucleus. Dev Brain Res 106:137–144.

61. Shaw D. and Goldman B.D. (1995a). Gender differences in influence of prenatal photoperiods on postnatal pineal melatonin rhythms and serum prolactin and follicle-stimulating hormone in the Siberian hamster (*Phodopus sungorus*). Endocrinology 136:4237–4246.
62. Shaw D. and Goldman B.D. (1995b). Influence of prenatal photoperiods on postnatal reproductive responses to daily infusions of melatonin in the Siberian hamster (*Phodopus sungorus*). Endocrinology 136:4231–4236.
63. Shaw D. and Goldman B.D. (1995c). Influence of prenatal and postnatal photoperiods on postnatal testis development in the Siberian hamster (*Phodopus sungorus*). Biol Reprod 52:833–838.
64. Stetson M.H., Elliott J.A., and Goldman B.D. (1986). Maternal transfer of photoperiodic information influences the photoperiodic response of prepubertal Djungarian hamsters. Biol Reprod 34:664–669.
65. Stetson M.H., Ray S.L., Creyaufmiller N., and Horton T.H. (1989). Maternal transfer of photoperiodic information in Siberian hamsters. II. The nature of the maternal signal, time of signal transfer, and the effect of the maternal signal on peripubertal reproductive development in the absence of photoperiodic input. Biol Reprod 40:458–465.
66. Sugden D. (1983). Psychopharmacological effects of melatonin in mouse and rat. J Pharmacol Exp Ther 227:587–591.
67. Tamarkin L., Reppert S.M., Orloff D.J., Klein D.C., Yellon S.M., and Goldman B.D. (1980). Ontogeny of the pineal melatonin rhythm in the Syrian (*Mesocricetus auratus*) and Siberian (*Phodopus sungorus*) hamsters and in the rat. Endocrinology 107:1061–1064.
68. Underwood H. and Goldman B.D. (1987). Vertebrate circadian and photoperiodic systems: role of the pineal gland and melatonin. J Biol Rhythms 2:279–315.
69. Velazquez E., Esquifino A.I., Zueco J.A., Ruiz Albusac J.M., and Blazquez E. (1992). Evidence that circadian variations of circulating melatonin levels in fetal and suckling rats are dependent on maternal melatonin transfer. Neuroendocrinology 55:321–326.
70. Viswanathan N. and Davis F.C. (1997). Single prenatal injections of melatonin or the D1-dopamine receptor agonist SKF 38393 to pregnant hamsters sets the offsprings' circadian rhythms to phases 180 degrees apart. J Comp Physiol [A] 180:339–346.
71. Viswanathan N., Weaver D.R., Reppert S.M., and Davis F.C. (1994). Entrainment of the fetal hamster circadian pacemaker by prenatal injections of the dopamine agonist SKF 38393. J Neurosci 14:5393–5398.
72. Waldhauser F., Weiszenbacher G., Frisch H., Zeitlhuber U., Waldhauser M., and Wurtman R.J. (1984). Fall in nocturnal serum melatonin during prepuberty and pubescence. Lancet 1:362–365.
73. Weaver D.R. (1997). Melatonin: A "wonder drug" to wonder about. J Biol Rhythms 12:682–689.
74. Weaver D.R. (1999). Melatonin and circadian rhythms in vertebrates: Physiological roles and pharmacological effects. In: F.W. Turek and P.C. Zee (Eds.) Neurobiology of Sleep and Circadian Rhythms, Marcel Dekker, 197–262.
75. Weaver D.R. and Reppert S.M. (1986). Maternal melatonin communicates daylength to the fetus in Djungarian hamsters. Endocrinology 119:2861–2863.
76. Weaver D.R. and Reppert S.M. (1989a). Periodic feeding of SCN-lesioned pregnant rats entrains the fetal biological clock. Dev Brain Res 46:291–296.
77. Weaver D.R. and Reppert S.M. (1989b). Maternal communication of daylength information to the fetus. In: S.M. Reppert (Ed.) Development of Circadian Rhythmicity and Photoperiodism in Mammals, Perinatology Press, Ithaca NY, pp. 209–219.
78. Weaver D.R. and Reppert S.M. (1995). Definition of the developmental transition from dopaminergic to photic regulation of c-fos gene expression in the rat suprachiasmatic nucleus. Mol Brain Res 33:136–148.
79. Weaver D.R., Keohan J.T., and Reppert S.M. (1987). Definition of a prenatal sensitive period for maternal-fetal communication of day length. Am J Physiol 253:E701–704.
80. Weaver D.R., Namboodiri M.A.A., and Reppert S.M. (1988). Iodinated melatonin mimics melatonin action and reveals discrete binding sites in fetal brain. FEBS Lett 228:123–127.
81. Weaver D.R., Rivkees S.A., and Reppert S.M. (1992). D1-dopamine receptors activate c-fos expression in the fetal suprachiasmatic nuclei. Proc Natl Acad Sci USA 89:9201–9204.
82. Weaver D.R., Roca A.L., and Reppert S.M. (1995). c-fos and jun-B mRNAs are transiently expressed in fetal rodent suprachiasmatic nucleus following dopaminergic stimulation. Dev Brain Res 85:293–297.
83. Wilson M.E. and Gordon T.P. (1989). Short-day melatonin pattern advances puberty in seasonally breeding rhesus monkeys (*Macaca mulatta*). J Reprod Fertil 86:435–444.

84. Yellon S.M. and Goldman B.D. (1984). Photoperiodic control of reproductive development in the male Djungarian hamster (*Phodopus sungorus*). Endocrinolgy 114:664–670.
85. Yellon S.M. and Foster D.L. (1986). Melatonin rhythms time photoperiod-induced puberty in the female lamb. Endocrinology 119:44–49.
86. Yellon S.M. and Longo L.D. (1987). Melatonin rhythms in fetal and maternal circulation during pregnancy in sheep. Am J Physiol 252:E799–802.
87. Yellon S.M. and Longo L.D. (1988). Effect of maternal pinealectomy and reverse photoperiod on the circadian melatonin rhythm in the sheep and fetus during the last trimester of pregnancy. Biol Reprod 39:1093–1099.
88. Young I.M., Francis P.L., Leone A.M., Stovell P., and Silman R.E. (1988). Constant pineal output and increasing body mass account for declining melatonin levels during human growth and sexual maturation. J Pineal Res 5:71–85.
89. Zemdegs I.Z., McMillen I.C., Walker D.W., Thorburn G.D., and Nowak R. (1988). Diurnal rhythms in plasma melatonin concentrations in the fetal sheep and pregnant ewe during late gestation. Endocrinology 123:284–289.
90. Zisapel N., Egozi Y., and Laudon M. (1982). Inhibition of dopamine release by melatonin: regional distribution in the rat brain. Brain Res 246:161–163.

23

INVESTIGATION OF THE HUMAN Mel 1a MELATONIN RECEPTOR USING ANTI-RECEPTOR ANTIBODIES

Lena Brydon,[1,2] Perry Barrett,[2] Peter J. Morgan,[2] A. Donny Strosberg,[1] and Ralf Jockers[1]

[1] CNRS-UPR 0415 and Université Paris VII
Institut Cochin de Génétique Moléculaire
22 rue Méchain, F-75014 Paris
France

[2] Rowett Research Institute
Greenburn Road, Bucksburn
Aberdeen AB2 9SB, United Kingdom

1. INTRODUCTION

1.1. Anti-Receptor Antibodies

Few reports describe the successful production of antibodies specific for G protein-coupled receptors (GPCRs). Antibodies have been prepared against partially or highly purified receptors (1–4). However, the difficulty in preparing sufficient quantities of these hydrophobic, low abundance proteins for multiple injections in animals hampered the large scale production of anti-receptor antibodies by this method and alternative strategies were sought.

The recent application of molecular cloning techniques has greatly expanded our knowledge of the primary sequence of membrane receptors. This knowledge permits the design of synthetic peptides corresponding to selected receptor regions, which may then be used as immunogens for antibody production. This approach has permitted successful production of antibodies against a vast range of GPCRs including β-adrenergic receptors (5), dopamine receptors (6), muscarinic acetylcholine receptors (7), and serotonin receptors (8), and is now the method of choice for production of anti-receptor antibodies.

Anti-receptor antibodies are invaluable tools for investigation of the structure and function of GPCRs. Immunoblotting with antibodies permits analysis of receptor

Melatonin after Four Decades, edited by James Olcese.
Kluwer Academic / Plenum Publishers, New York, 2000.

molecular mass and structural modifications such as receptor glycosylation and disulphide bridge formation (9,10). Antibodies directed against specific receptor domains have also provided crucial information about the structural topography of seven transmembrane receptors, through selective permeabilisation of receptor-bearing cells and immunofluorescence (11). Furthermore, purification of receptors by immunoaffinity chromatography using anti-receptor antibodies, can yield large quantities of receptor sufficient for structural analysis (12).

Anti-receptor antibodies may also be used to probe functional receptor domains. For example, antibodies against B2 bradykinin receptors were used to localise receptor-ligand binding sites by competing for bradykinin binding to receptors (13) and sites of covalent insertion of β-adrenergic receptor photoaffinity ligands were defined using site-directed antibodies to immunoprecipitate receptor fragments in tandem with microsequencing (14).

Immunochemistry using anti-receptor antibodies has permitted high-resolution localization of specific receptor subtypes in different tissues and cells (15,16). Antibodies detect receptors at an expressed protein level unlike other detection methods such as *in situ* hybridisation and polymerase chain reaction (17,18) and can directly distinguish receptor isoforms, in contrast to detection by ligand autoradiography (18).

A final illustration of the utility of anti-receptor antibodies is in the investigation of receptor signal transduction. Signalling pathways associated with a specific receptor may be investigated by immunoprecipitation of receptor complexes with anti-receptor antibodies and subsequent analysis of receptor-associated proteins. For example, G proteins coupled to specific receptors in a given tissue have been identified by co-precipitation using anti-receptor antibodies (8,19,20). We decided to use the same approach to investigate G protein coupling to Mel 1a melatonin receptors.

1.2. Signal Transduction by the Mel 1a Melatonin Receptor

The hormone melatonin plays an important role in a variety of physiological responses including circadian rhythm regulation, seasonal reproduction in mammals, sleep and vision (21). It is thought to mediate its effects through high affinity GPCRs. The Mel 1a melatonin receptor subtype (now also referred to as the mt_1 receptor) is located in the suprachiasmatic nuclei (SCN) of the hypothalamus and the *pars tuberalis* of the pituitary and is believed to mediate melatonin's actions at these sites (22). However, the biochemical mechanisms underlying the effects of melatonin are not yet fully understood.

The multiplicity of pharmacological effects associated with Mel 1a receptor activation suggests that receptors may couple to more than one G protein. Activated Mel 1a receptors have been shown to inhibit cAMP accumulation through activation of pertussis toxin-sensitive G ($G_{i/o}$) proteins (23). Recombinant Mel 1a receptors were also shown to activate Kir3.1/3.2 potassium ion channels, to potentiate PGF2α-promoted stimulation of phospholipase C and to modulate PKC and PLA_2 via $G_{\beta\gamma}$ subunits liberated during $G_{i/o}$ protein activation (24–26). However, Mel 1a receptor signalling may not be restricted to proteins of the G_i family. We have developed antibodies specific for the human Mel 1a melatonin receptor and used these antibodies to investigate the G protein coupling profile of this receptor.

2. RESULTS AND DISCUSSION

Polyclonal antibodies were raised against a peptide corresponding to the C-terminal region of the Mel 1a receptor (YKWKPSPLMTNNNVVKVDSV-COO^-, peptide 536) (Figure 1). This region was chosen since the C-terminal of proteins is often considered as more likely to generate antibodies against synthetic peptides than other parts often involved in the folding of proteins (27). Indeed, the selection of a C-terminal peptide immunogen has previously led to the successful production of a number of anti-receptor antibodies including anti-β3AR, anti-β2AR, and antibodies against muscarinic acetylcholine receptor subtypes (5,7,28). Furthermore, this sequence shows little or no homology with the corresponding regions of other high affinity melatonin receptors, suggesting little potential cross reactivity with the human Mel 1b subtype and high selectivity for the Mel 1a subtype (Figure 1).

Results demonstrated that 536-antibodies specifically recognise the recombinant human Mel 1a receptor in both its native and denatured form. In preliminary immunoblot experiments on Mel 1a receptor-expressing cells, antibodies detected a predominant protein with an apparent molecular weight of 60 kDa. To confirm that this 60 kDa protein corresponds to the Mel 1a receptor, the receptor was labelled with ^{125}I-MLT and partially purified prior to analysis with 536-antibodies in immunoblots. Partial purification was carried out by solubilising ^{125}I-MLT-labelled receptors from transfected cell membranes, followed by separation of receptors by native gel electrophoresis and extraction of the radiolabelled region of the gel. Antibodies specifically recognised a protein of approximately 60 kDa in the extracted Mel 1a receptor fraction. This result confirmed that 536-antibodies specifically recognize the recombinant, ^{125}I-MLT-labelled Mel 1a receptor. According to its amino acid sequence, the predicted molecular weight of the Mel 1a receptor is 39 kDa. This indicates that the expressed receptor is subject to post-translational modifications. The broad migration pattern of the receptor protein

Peptide 536	**KWKPSPLMTNNNVVKVDSV**
h Mel 1a	....DRV**KWKPSPLMTNNNVVKVDSV***
m Mel 1a	....DKI**K**C**KPSPL**IP**NNN**LI**KVDSV***
o Mel 1a	....DRI**K**R**KPSPL**IA**N**H**N**LI**KVDSV***
ch Mel 1a	....DRIRS**KPSPL**I**TNNN**Q**VKVDSV***
h Mel 1b	....AEGLQS**P**A**P**PIIGVQHQADAL*
ch Mel 1c	....EGL**K**S**KPSP**AV**TNNN**QAEIHL*
xe Mel 1c(α)	....EGL**K**S**KPSP**AV**TNNN**QADMYV*

Figure 1. Alignment of peptide 536 with carboxy-terminal sequences of melatonin receptors. Single-letter code is used to denote amino acids. Letters in **bold** indicate conserved amino acid residues and * indicates carboxy-termini. Nucleotide sequences for melatonin receptors were obtained from Genbank using the following accession numbers: P48039 (h Mel 1a), Q61184 (m Mel 1a), P48040 (o Mel 1a), P49285 (ch Mel 1a), P49286 (h Mel 1b), P49288 (ch Mel 1c) and U67879 (xe Mel 1c(α)).

in SDS-PAGE suggests that the receptor is glycosylated. In support of this hypothesis, the Mel 1a receptor contains 2 potential accessible glycosylation sites in its N terminal domain.

In order to establish whether 536-antibodies specifically recognised the receptor in its native form, antibodies were tested for their ability to precipitate Mel 1a receptors from transfected COS M6 and HEK 293 cells. 536-antibodies successfully immunoprecipitated 40% of solubilised ^{125}I-MLT-labelled receptors from both cell types. Detectable precipitation was negligible when antibodies were pre-incubated with peptide 536 or replaced with pre-immune serum. Furthermore, no precipitation was detected when the same experiments were repeated with cells transfected with the Mel 1b receptor, confirming specificity for the Mel 1a subtype. The quantity of immunoprecipitated receptor compares favourably with results observed for other anti-receptor antibodies e.g. antibodies specific for $5HT_{1A}$ serotonin receptors (29) and D_2 dopamine receptors (30). Specific interaction between antibodies and native receptors was further confirmed by visualization of the ^{125}I-MLT-labelled receptor-antibody complex after separation by native gel electrophoresis and detection by autoradiogaphy. The Mel 1a receptor complex alone was detected at approximately 550–700 kDa. The presence of 536-antibodies induced a mobility shift in a portion of complexes (35–40%) to a higher molecular weight representing the formation of a new (antibody-receptor-ligand) complex.

Melatonin receptors are known to signal through G proteins. In order to determine whether G proteins were functionally associated with the solubilized, ^{125}I-MLT-labelled Mel 1a receptor, the complex was incubated with GppNHp, a non-hydrolysable GTP analogue which activates G proteins and thus induces dissociation of the ternary complex (ligand-receptor-G protein). Dissociation can be determined by loss of high-affinity agonist (^{125}I-MLT) binding. Immunoprecipitation and native gel electrophoresis experiments showed that high affinity ^{125}I-MLT binding to Mel 1a receptor complexes was vastly reduced by treatment with GppNHp. Detectable immunoprecipitation was reduced by approximately 50% and there was a marked displacement of receptor-bound ligand to free ligand on the native gel. The sensitivity of Mel 1a receptor complexes to guanine nucleotides indicated that G proteins remained functionally associated with isolated Mel 1a receptors, and thus co-precipitate with receptors.

Indeed, by treating precipitated Mel 1a receptor complexes with GppHNp, co-precipitated G proteins were displaced from the complex and subsequently identified on immunoblots with anti-G protein antibodies. Mel 1a receptors were found to specifically couple to $G_{i\alpha2}$, $G_{i\alpha3}$ and $G_{q/11\alpha}$ proteins in HEK 293 cells. G protein-coupling was strictly receptor dependent since these proteins were not detected when experiments were performed using non-transfected HEK 293 cells. Coupling was also dependent on ligand occupancy indicating that hormone binding stabilises the receptor-G_α complex and little of the receptor is pre-coupled to G_α in absence of agonist activation. The lack of association between Mel 1a receptors and other G_α proteins present in HEK, $G_{i\alpha1}$, $G_{o\alpha}$, $G_{12\alpha}$, $G_{z\alpha}$ and $G_{s\alpha}$, further demonstrates the selectivity of receptor-G protein coupling.

G_β subunits were also found coupled to receptors, further demonstrating the presence of functional G proteins in the immunoprecipitated complex. Since β subunits are usually found tightly complexed with γ subunits, it is likely that the latter are also present in the receptor complex. In contrast to receptor-G_α coupling, physical association between G_β subunits and the receptor was independent of ligand occupancy. Mel 1a receptors appear to be pre-coupled to $G_{\beta\gamma}$ since G_β remains associated with receptors in the presence or absence of agonist. Direct interaction of $G_{\beta\gamma}$ subunits with GPCRs has been demonstrated for several other members of this family (31–33) but the functional significance of this interaction is unknown.

Previous reports have demonstrated that Mel 1a receptors inhibit adenylyl cyclase in a PTX-sensitive manner, indicating coupling with $G_{i/o}$ proteins (22,34). Detection of G_{i2} and G_{i3} proteins in Mel 1a receptor complexes, directly confirms coupling between Mel 1a receptors and G_i proteins. Furthermore, the absence of G_{i1} and G_o from complexes suggests selective coupling among different members of the $G_{i/o}$ family. Detection of $G_{q/11\alpha}$ proteins in Mel 1a receptor complexes provides new evidence for coupling between Mel 1a melatonin receptors and PTX-insensitive G proteins. Studies on functional implications of $G_{q/11}$ coupling to Mel 1a receptors in HEK 293 cells are currently underway in our laboratory.

3. CONCLUSIONS/PERSPECTIVES

In conclusion, we have shown that Mel 1a receptors selectively couple to G_{i2}, and G_{i3} proteins which likely mediate the inhibitory effect of these receptors on cAMP accumulation. Furthermore, the association between Mel 1a receptors and $G_{q/11}$ demonstrates that these receptors are capable of coupling to both PTX-sensitive and PTX-insensitive G proteins. Signalling through $G_{q/11}$ proteins either via phospholipase C or other effector systems constitutes promising new candidates for melatonin receptor signalling. In future work, antibodies may also be used to answer questions regarding other properties of Mel 1a receptors. For example, immunoprecipitation and immunochemistry with anti-receptor antibodies may be used to localise Mel 1a receptors in tissues and cells from structures known to bind ^{125}I-MLT.

ACKNOWLEDGMENTS

We are grateful to Dr. L. Camoin (ICGM, Paris) for his expert advice on choice of peptide antigen and antibody purification. We thank Dr. G. Milligan (University of Glasgow, Scotland) and Drs. V. Homburger and G. Guillon (University of Montpellier, France) for providing polyclonal anti-G protein antisera. This work was supported by grants from CNRS, the Université de Paris, and Institut de Recherches Internationales Servier. During this work RJ was supported by the Société de Secours des Amis des Sciences and the Fondation pour la Recherche Médicale.

REFERENCES

1. Chapot, M.P. et al. (1990) *Eur. J. Biochem.* 12, 137–144.
2. Couraud, P.O., Delavier, K.C., Durieu, T.O., and Strosberg, A.D. (1981) *Biochem. Biophys. Res. Commun.* 30, 1295–1302.
3. Kaveri, S.V., Cervantes-Olivier, P., Delavier-Klutchko, C., and Strosberg, A.D. (1987) *Eur. J. Biochem.* 167, 449–456.
4. Raposo, G. et al. (1989) *Eur. J. Cell. Biol.* 50, 340–352.

5. Guillaume, J.L. et al. (1994) *Eur. J. Biochem.* 224, 761–770.
6. Boundy, V.A. et al. (1993) *Mol. Pharmacol.* 43, 666–676.
7. Wang, P., Luthin, G.R., and Ruggieri, M.R. (1995) *J. Pharmacol. Exp. Ther.* 273, 959–966.
8. Raymond, J.R., Olsen, C.L., and Gettys, T.W. (1993) *Biochemistry* 32, 11064–11073.
9. Emrich, T., Forster, R., and Lipp, M. (1994) *Cell Mol. Biol.* 40, 413–419.
10. Moxham, C.P., Ross, E.M., George, S.T. and, Malbon, C.C. (1988) *Mol. Pharmacol.* 33, 486–492.
11. Wang, H., Lipfert, L., Malbon, C.C., and Bahout, S. (1989) *J. Biol. Chem.* 264, 14424–14431.
12. Kwatra, M.M. et al. (1995) *Protein Expression and Purification* 6, 717–721.
13. Abd Alla, S. et al. (1996) *J Biol Chem* 271, 1748–55.
14. Wong, S.K. F., Slaughter, C., Ruoho, A.E., and Ross, E.M. (1988) *J. Biol. Chem.* 263, 7925–7928.
15. Wang, H.Y., Berrios, M., and Malbon, C.C. (1989) *Biochem. J.* 263, 519–532.
16. Ruggieri, M.R. et al. (1995) *J. Pharmacol. Exp. Ther.* 274, 976–982.
17. Giros, B. et al. (1989) *Nature* 342, 923–926.
18. Mansour, A. et al. (1990) *J. Neurosci.* 10, 2587–2600.
19. Gu, Y.-Z. and Schonbrunn, A. (1997) *Mol. Endocrinol.* 11, 527–537.
20. Matesic, D.F., Manning, D.R., and Luthin, G.R. (1991) *Mol. Pharmacol.* 40, 347–353.
21. Morgan, P.J., Barrett, P., Howell, H.E., and Helliwell, R. (1994) *Neurochem. Int.* 24, 101–146.
22. Reppert, S.M., Weaver, D.R., and Ebisawa, T. (1994) *Neuron* 13, 1177–1185.
23. Reppert, S.M., Weaver, D.R., and Godson, C. (1996) *Trends Pharmacol. Sci.* 17, 100–102.
24. Nelson, C.S., Marino, J.L., and Allen, C.N. (1996) *Neuroreport* 7, 717–720.
25. McArthur, A.J., Hunt, A.E., and Gillette, M.U. (1997) *Endocrinology* 138, 627–634.
26. Godson, C. and Reppert, S.M. (1997) *Endocrinology* 138, 397–404.
27. Van Regenmortel, M.H. and Daney de Marcillac, G. (1988) *Immunol. Lett.* 17, 95–107.
28. Dixon, R. et al. (1987) *Nature* 326, 73–77.
29. El Mestikawy, S. et al. (1990) *Neurosci. Lett.* 118, 189–192.
30. Chazot, P.L., Doherty, A.J., and Strange, P.G. (1993) *Biochem. J.* 289, 789–794.
31. Phillips, W.J. and Cerione, R.A. (1992) *J. Biol. Chem.* 267, 17032–17039.
32. Kurstjens, N.P. et al. (1991) *Eur. J. Biochem.* 197, 167–176.
33. Taylor, J.M. et al. (1996) *J. Biol. Chem.* 271, 3336–3339.
34. Morgan, P.J., Barrett, P., Howell, H.E., and Helliwell, R. (1994) *Neurochem. Int.* 24, 101–146.

24

A PHARMACOLOGICAL INTERACTION BETWEEN MELATONIN AND THE α_2-ADRENOCEPTOR IN CUCKOO WRASSE MELANOPHORES

Lena G. E. Mårtensson and Rolf G. G. Andersson

Department of Pharmacology
Faculty of Health Sciences
S-581 85 Linköping
Sweden

1. INTRODUCTION

Melanophores from fish scales can aggregate or disperse their pigment granules (melanosomes) within seconds as a means of background adaptation or social signalling. The scales from the teleost fish cuckoo wrasse (*Labrus ossifagus L.*) can be isolated and the melanophores can be used as a simple model to answer various pharmacological and physiological questions. The melanophores are sympathetically innervated, and pigment aggregation occurs when postsynaptic α_2-adrenoceptors are activated by electrical stimulation of the nerve endings surrounding the melanophores, which entails the concomitant release of noradrenaline, or by adding different pharmacological substances to the surrounding medium (1).

For each experiment, scales are removed from the dark area of the dermis and transferred to a saline buffer solution and then subjected to different treatments. The effects are investigated under a light microscope and evaluated by using a slight modification of the melanophore index of Hogben and Slome (2); 1 stands for maximal pigment aggregation and 5 stands for a state of complete pigment dispersion (Figure 1).

Denervation of the melanophores can be accomplished either by chemical treatment or by allowing degeneration, occurring naturally when the scales are isolated in cell culture medium (3,4). However, the melanophores remain able to aggregate and disperse their pigment granules for weeks. After denervation, the

Melatonin after Four Decades, edited by James Olcese.
Kluwer Academic / Plenum Publishers, New York, 2000.

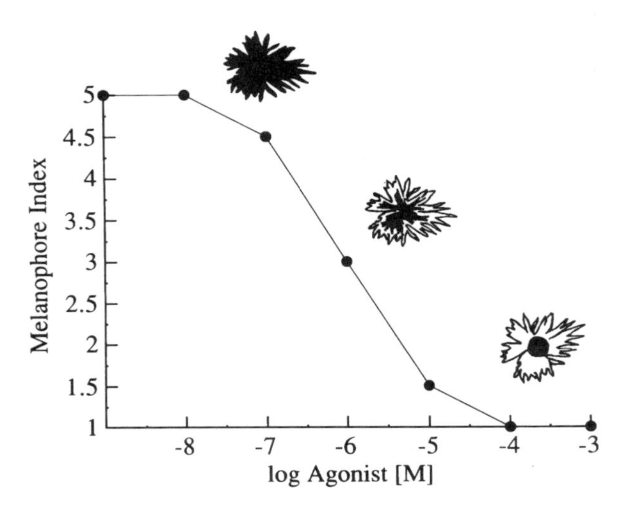

Figure 1. A schematic illustration of the Hogben and Slome melanophore index. An index of 5 stands for a fully dispersed melanophore, an index of 3 for partly aggregated melanophores and an index of 1 for fully aggregated melanophores.

melanophores become supersensitive and a contemporaneous change occurs in the pharmacology.

2. A TWO-SITE RECEPTOR MODEL

Experimental data from earlier pharmacological studies on cuckoo wrasse melanophores suggest a two-site receptor model: the two agonists noradrenaline and B-HT 920 have different pharmacological properties and seem to bind to two different sites. For example, yohimbine does not interact with noradrenaline and B-HT 920 in the same way, and the equivalent is true for amiloride (5) and UK 14,304 (6). Furthermore, signal transduction mechanisms are not the same for B-HT 920 and noradrenaline. Neither of these studies implicated melatonin as a substance of interest. As a result of these studies, a new receptor model was suggested (7).

A receptor model consists in most cases of a receptor with an active site, which binds the endogenous ligand, and, in doing so, activates a transducer, usually a G-protein. The G-protein mediates the signal to a second messenger, which in turn activates the intracellular response. The sensitivity of the system is dependent on occupancy, efficacy and the number of spare receptors. Occupancy refers to the number of receptors engaged by an agonist; efficacy reflects how effective the agonist is in inducing a cellular response; and there are spare receptors if it is not necessary that all receptors be occupied to produce a maximal response. In an in vitro system, such as the prepared cuckoo wrasse melanophores, different agonists react differently to the indicated parameters. This may be explained by the suggested model comprising local and non-local receptor signaling (Figure 2). An agonist with less efficacy may induce only a local physiological response, and an agonist with greater efficacy could cause a global response. However, the model does not reveal whether the local or non-local response originates from the same receptor with different binding sites or if two receptors are present.

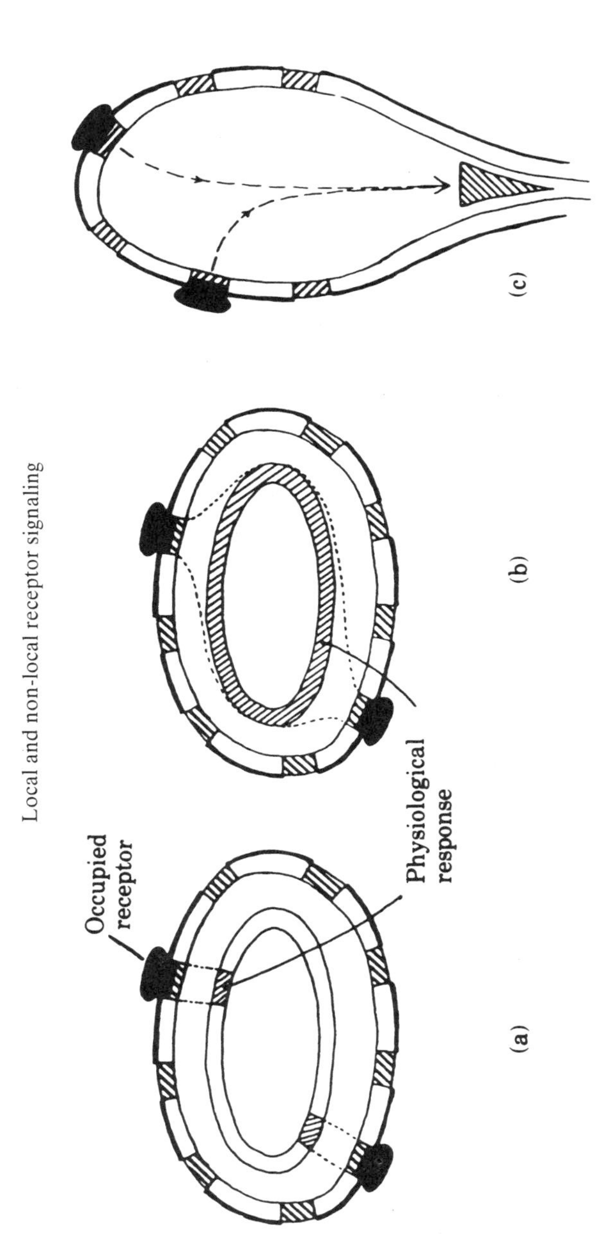

Figure 2. Schematic illustration of receptor-triggered signalling in the cell. a) A receptor-agonist complex causes a local physiological response in the cell. b) A receptor-agonist complex causes a physiological response in a large part of or in the whole cell. c) Receptor-agonist complexes cause physiological response, which are added at a specific location of the cell. The hatched areas inside the cell indicate the spread of the physiological response. (With kind permission from Academic Press; ref 7.)

2. MELATONIN MODULATION IN INNERVATED MELANOPHORES

With the two-site receptor model in mind, we suggest a new possibility: melatonin may modulate the α_2-adrenoceptor response (4). This hypothesis implies that a hormone and a neurotransmitter can somehow integrate their own signal messengers into a common signal. The question is where does the interaction occur? Does it take place on the receptor surface and involve an allosteric site, or does it occur somewhere along the transduction pathway, or at some point during the intracellular process that controls pigment aggregation?

Melatonin alone has almost no effect on pigment aggregation in melanophores on isolated scales, but it did enhance pigment aggregation that had been elicited via the α_2-adrenoceptor. Thus, the effect of melatonin must be dependent on the presence of an α_2-adrenoceptor (Figure 3). However, melatonin had the opposite effect when used in combination with B-HT 920 which is another α_2-adrenoceptor agonist (Figure 4). Instead of increasing aggregation as with the other α_2-adrenoceptor agonists, melatonin counteracts the aggregation induced by B-HT 920. A possible hypothesis for the counteracting effect seen with melatonin against B-HT 920 could be that those two substances bind to the same site, a site different from the catecholamine binding site, but with different affinity and efficacy. That is, if melatonin binds with a higher affinity

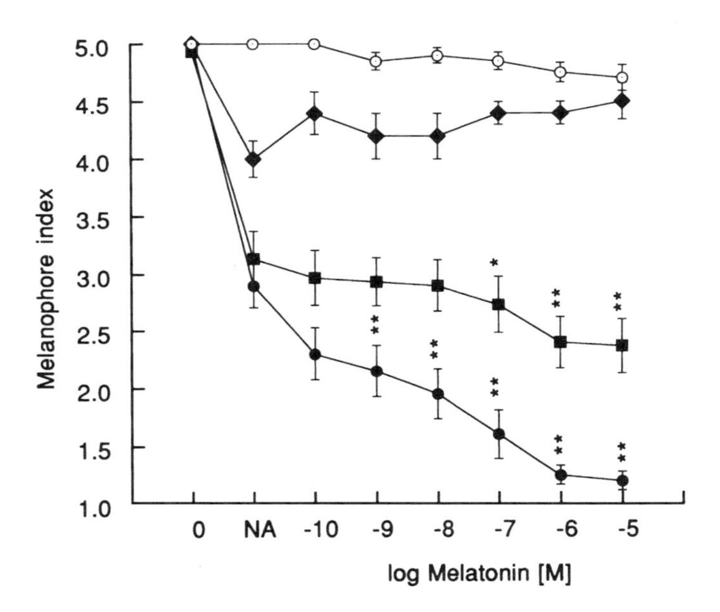

Figure 3. The combined effect of melatonin and noradrenaline (NA) on pigment aggregation in fish melanophores. The pigment granules were partially aggregated by exposure to various concentrations of noradrenaline and the melanophores were subsequently exposed to different concentrations of melatonin (as indicated on the x axis). Symbols: control, i.e. melatonin only (•), n = 10; melatonin and 0.1 mM NA (♦), n = 5; melatonin and 1 mM NA (■), n = 15; melatonin and 10 mM NA (•), n = 10. Each point in the graph represents the mean ± SEM. The statistics were elucidated by Wilcoxon signed-rank test between the NA points, as indicated on the x-axis, and increasing doses of melatonin. The significance levels were set at 95% (*), 99% (**) and 99.9% (***). (With kind permission from Elsevier, ref 8).

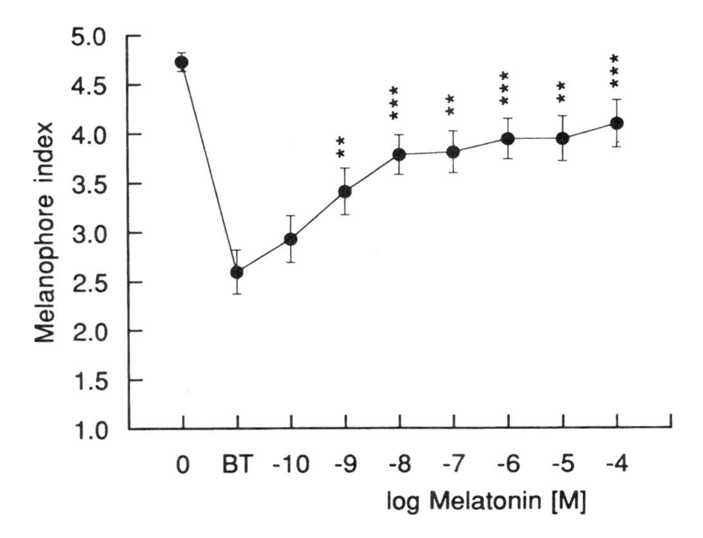

Figure 4. The combined effect of melatonin and B-HT 920 on pigment aggregation in fish melanophores. The pigment granules were partially aggregated by exposure to 10 mM B-HT 920 (BT), n = 15, and the melanophores were subsequently exposed to different concentrations of melatonin, as indicated on the x-axis. (With kind permission from Elsevier, ref 8).

but less efficacy than B-HT 920 does, and when the two drugs are combined, if the one with higher affinity, i.e. melatonin, occupies the binding site more often or more strongly, the overall result will be an inhibition of aggregation. On the other hand, if melatonin is combined with noradrenaline, two sites will be occupied resulting in a synergistic effect, possibly through potentiation of the signal transduction pathway. However, the hypothesis presented above is at odds with the common opinion on how melatonin interacts with receptors, but in our opinion, this is one possibility to interpret the pharmacological results.

3. MELATONIN IN DENERVATED MELANOPHORES

Denervation affects the pigment aggregation response mediated by noradrenaline to become more sensitive and displace the concentration-response curve to the left. Therefore, denervation might then also affect the response mediated by melatonin. Actually, melatonin lost its modulatory effect and became a potent as well as a full agonist (8). Since no functional nerve endings are present in this preparation, that means that the studied receptors have a postsynaptic location.

For decades, it has been known that yohimbine is a selective α_2-adrenoceptor antagonist, and the compound has been used to characterize adrenoceptors. Yohimbine is an effective antagonist against both noradrenaline and B-HT 920, but with slightly different pharmacological profiles. We originally thought that yohimbine would be the negative control to exclude α_2-adrenoceptor involvement in melatonin-induced pigment aggregation. The substance to be tested as antagonist against melatonin, was the α_1-adrenoceptor antagonist prazosin, because it has been reported to have affinity for a second type of melatonin receptors, MT_3. We found that prazosin in no

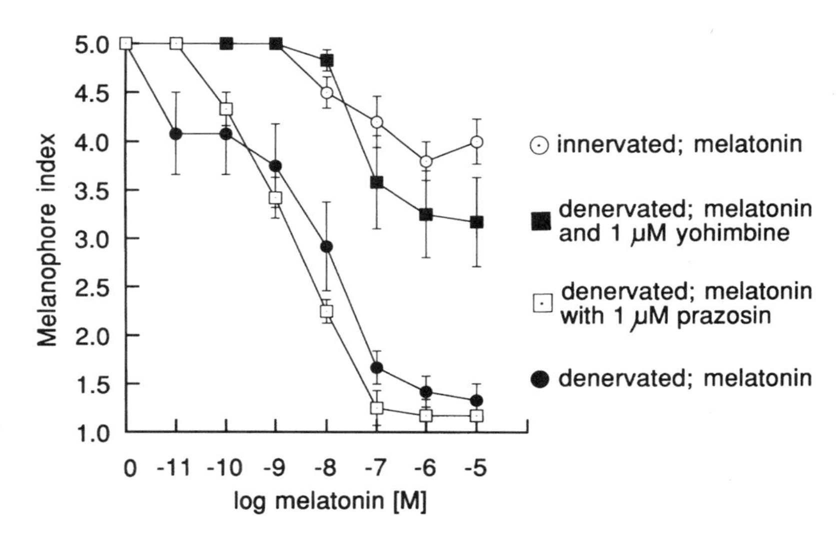

Figure 5. The effects of melatonin on pigment aggregation in innervated (•) and chemically denervated (guanethidin treated) (•) on isolated fish melanophores. Melatonin had only a slight effect on innervated melanophores but caused complete aggregation after chemical denervation. The aggregation could be counteracted by 1 mM yohimbine (■) but not by 1 mM prazosin (□), suggesting that α_2-adrenoceptor feature has been retained. (With kind permission from Elsevier, ref 9.)

way interfered with melatonin, but yohimbine was a most effective antagonist against melatonin (Figure 5). More precisely, yohimbine almost totally inhibited the melatonin-induced pigment aggregation. This further suggests a melatonin-α_2-adrenoceptor interaction.

4. DISCUSSION

The crucial issues are whether or not melatonin and B-HT 920 share the same site within the cuckoo wrasse melanophore assay, and if this site is located on the α_2-adrenoceptor or not.

Possible alternatives are an α_2-adrenoceptor and an mt_1-receptor and the two different types of receptors integrate the melatonin binding somewhere downstream of the receptors, i.e. the second messenger systems, the complex interactions between phosphatases and kinases that lead to an pigment aggregation, or along those proteins that participate in pigment migration. However, this model does not seem to be satisfactory considering that melatonin antagonizes B-HT 920 in innervated melanophores and that yohimbine is a potent antagonist against melatonin-induced pigment aggregation in denervated melanophores.

A highly speculative hypothesis is that there is a new type of a melatonin receptor, MT_4. The criteria for this receptor would be i) affinity for yohimbine (which actually has an indol moiety), ii) affinity for the α_2-adrenoceptor agonist B-HT 920, and iii) an increase in intracellular Ca^{2+} as a possible signal transduction mechanism. The idea is not supported by experimental data from other laboratories or in binding studies.

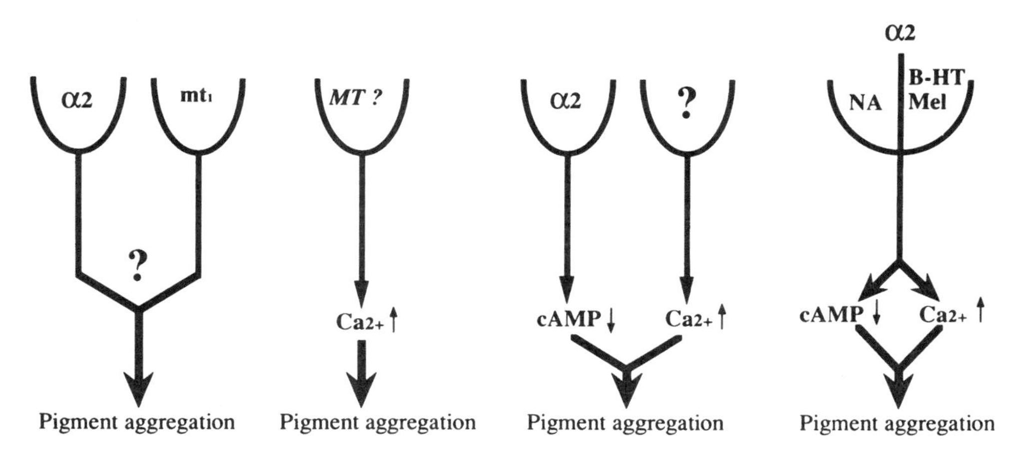

Figure 6. Four different hypotheses that could explain the diversity of the responses mediated by melatonin. The first suggests an α_2-adrenoceptor that cooperates in an unknown way with a mt_1-receptor. The second theory is a new unknown subtype of the melatonin receptor. A third possibility is that an unknown receptor protein mediates an increase in intracellular Ca^{2+}. The fourth hypothesis indicates an α_2-adrenoceptor with two functional sites.

A third possibility is that pigment aggregation is mediated by different receptor proteins that have very similar active sites and recognize B-HT 920, but differentiate between noradrenaline and melatonin. There is, however, no report of such a receptor, perhaps due to the assumption that the α_2-adrenoceptor agonists are highly selective for the presumed receptor type. Nonetheless, it should also be kept in mind that there are no investigations that have ruled out B-HT 920 as a melatonin ligand.

One hypothesis that remains, is the idea of an α_2-adrenoceptor with two active sites, the two-site model. Depending on affinity and efficacy of the ligand that binds to one of the two sites, the response could vary in many ways. Moreover, if a neurotransmitter and a hormone can be influence the cellular response by binding to the same receptor, and if different signaling mechanisms can be induced, it will result in a very dynamic model for receptor activation, cellular signaling, and response.

When scrutinizing several hypotheses (Figure 6), it is impossible to include our findings within an existing pharmacological paradigm. The pharmacology does not suggest a mt_1-receptor, nor a MT_2, or a MT_3 receptor and there is no evidence enough to suggest a new *MT* receptor subtype. What is at hand is an α_2-adrenoceptor-melatonin interaction that implies a possibility for a two-site model.

ACKNOWLEDGMENT

We gratefully acknowledge Kristineberg Marine Research Station for supplying us with experimental fishes and laboratory facilities.

This work was supported by grants from: The Swedish Medical Research Council (04X-4498), Lions foundation, Östergötlands läns landstings forskningsfond, The Royal Swedish Academy for Sciences and The Swedish Society for Medical Research.

REFERENCES

1. Andersson, R.G.G., Karlsson, J.O.G., and Grundström, N. (1984). Adrenergic nerves and alpha2-adrenoceptor system regulating melanosome aggregation within fish melanophores. Acta Physiol. Scand. 121, 173–179.
2. Hogben, L.T. and Slome, D. (1931). The pigmentary effector system VI. The dual character of endocrine coordination in amphibian colour change. Proc. R. Soc. B. 108, 10–53.
3. Karlsson, J.O.G., Elwing, H., Grundström, N., and Andersson, R.G.G. (1988). Pronounced supersensitivity of postjunctional α_2-adrenoceptor after denervation of fish melanophores. J. Pharmacol. Exp. Ther. 246, 345–351.
4. Mårtensson, L.G.E. and Andersson, R.G.G. (1996). A melatonin binding site modulates the α_2-adrenoceptor. Life Sci. 58, 525–533.
5. Svensson, S.P.S., Mårtensson, L.G.E., Grundström, N., Andersson, R.G.G., Cragoe Jr., E.J., and Karlsson, J.O.G. (1991). Antagonistic effect of amiloride on alpha2-adrenoceptor-mediated pigment aggregation: Pharmacological heterogeneity between B-HT 920 and noradrenaline. J. Pharmacol. Exp. Ther., 258, 447–451.
6. Svensson, S.P.S. (1993). Melanophore α_2-adrenoceptors. Linköping University Medical Dissertations No 396.
7. Lundström, K.I., Karlsson, J.O.G., Svensson, S.P.S., Mårtensson, L.G.E., Elwing, H., Ödman, S., and Andersson, R.G.G. (1993). Local and non-local receptor signalling. J. Theor. Biol. 164, 135–148.
8. Mårtensson, L.G.E. and Andersson, R.G.G. (1997). Denervation of pigment cells that leads to a receptor that is ultrasensitive to melatonin and noradrenaline. Life Sci. 60, 1575–1582.

25

SCN CELLS EXPRESSING mt_1 RECEPTOR mRNA COEXPRESS AVP mRNA IN SYRIAN AND SIBERIAN HAMSTERS

C. K. Song,[1,3] T. J. Bartness,[1,2,3,4]* S. L. Petersen,[5] and E. L. Bittman[5]

[1] Department of Biology
[2] Department of Psychology
[3] Department of Neurobiology
[4] Neuropsychology and Behavioral Neuroscience Programs
Georgia State University
Atlanta, Georgia 30303
[5] Department of Biology
Neurosciences and Behavior Program
University of Massachusetts
Amherst, Massachusetts 01003

Many animals, especially mammals living in temperate zones, confine life history events to specific times of the year. For example, species of hamsters and voles bred in the long days of spring/summer, whereas species of sheep and deer breed in the short days (SDs) of fall/winter (1). The most noise-free seasonal cue in their environment is the photoperiod (10). The photoperiod triggers, seasonal responses, including changes in reproductive status mentioned above, as well as cycles of body mass and fat and pelage (1). This photic information is transmitted to the pineal gland via a multi-synaptic circuit beginning with the retina and concluding with the pinealocytes (5). Among the central nervous system structures included in this circuit is the suprachiasmatic nucleus (SCN) of the hypothalamus, the predominant biological clock. The pineal gland, through its primary hormone melatonin (MEL), triggers these and many other seasonal responses (1). MEL is synthesized and released only during the night, therefore, it faithfully codes the duration of the dark portion of the photoperiod. Thus, the durational characteristic of nightly MEL secretion induces these seasonally appropriate responses (3). The SCN not only participates in the transmission of photic information to the pineal gland, but it also appears to be important for the reception of the

---

* Address all correspondence to Dr. Timothy J. Bartness.

Melatonin after Four Decades, edited by James Olcese.
Kluwer Academic / Plenum Publishers, New York, 2000.

durational MEL signals. That is, in pinealectomized Siberian hamsters (*Phodopus sungorus*), lesions of the SCN (SCNx) block the reception of SD MEL signals generated exogenously via s.c. infusions of the hormone (2,9). Similar experiments in SCNx Syrian hamsters (*Mesocricetus auratus*) have shown that an intact SCN is not necessary for the induction of SD-type reproductive responses to exogenously generated SD-like MEL signals (7), but it does seem that the SCN is important for the interpretation of *series* of SD MEL signals (4).

Binding of the melatonin analog 2-[^{125}I] MEL (IMEL) to the SCN of Siberian hamsters supports the apparent critical nature of this structure in the reception of MEL signals in this species (6,12). Moreover, distribution of the mRNA for the MEL receptor that is functionally responsible for the photoperiodic responses (*i.e.*, the mt_1 receptor [*a.k.a.* MEL_{1A} receptor]) shows its expression in the same areas that also exhibit IMEL binding, including the SCN, as well as the paraventricular and reunions nuclei of the thalamus in Siberian hamsters (8,11). Microscopic resolution of cells expressing the mt_1 receptor has not been reported, however. In addition, the neurochemical phenotype of neurons expressing the mt_1 receptor also is unknown. Therefore, we attempted to identify the neurons within the SCN that express the mt_1 receptor in both Siberian and Syrian hamsters, and to begin to typify their neurochemical identity in the former species.

We first used single label *in situ* hybridization for the mt_1 receptor mRNA (probe courtesy of Drs. Steven Reppert and David Weaver). Densely labeled cells in the SCN, as well as other hypothalamic, thalamic and brainstem structures were found using emulsion autoradiography. Specifically for the SCN, cells expressing the mt_1 receptor were distributed throughout the nucleus in Siberian hamsters, with less extensive

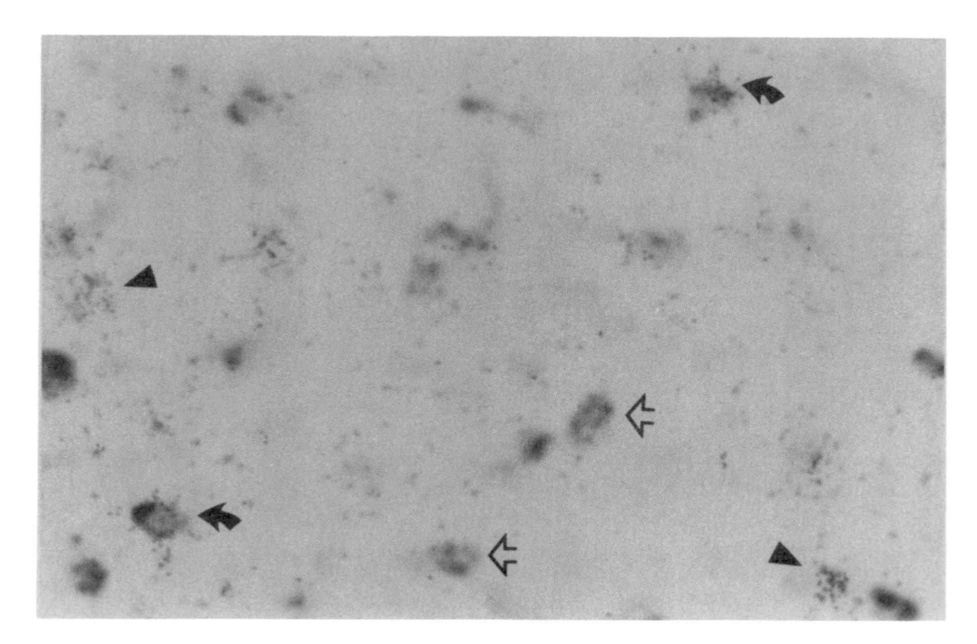

Figure 1. Photomicrograph of Siberian hamster SCN. mt_1 mRNA expressing cells of the SCN also contained AVP mRNA (curved arrow). Single-labeled mt_1 mRNA (triangle) and AVP mRNA (hollow arrows) also are shown.

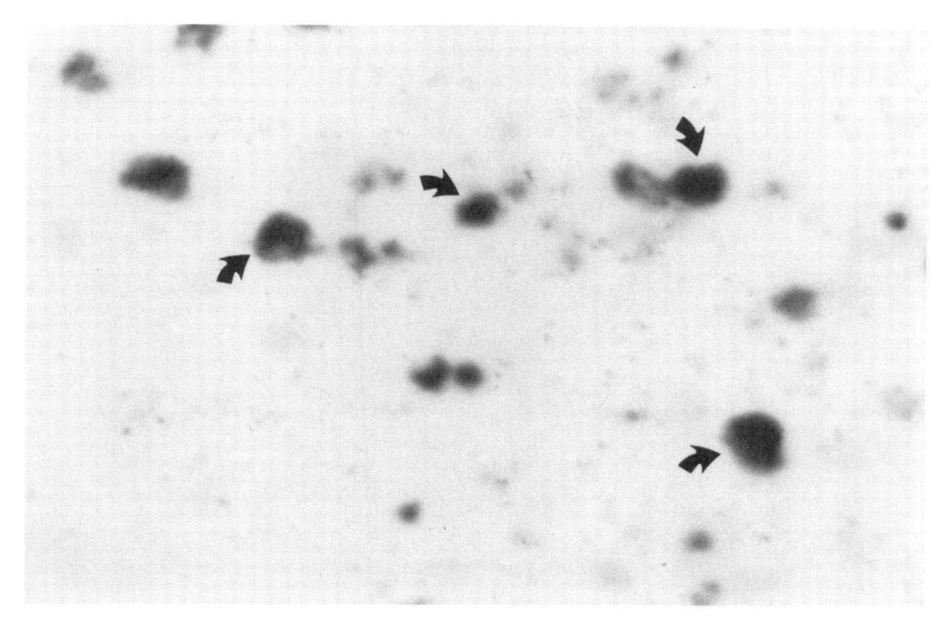

Figure 2. Photomicrograph of Syrian hamster SCN. Co-expression of mt_1 and AVP mRNAs are indicated by curved arrows.

labeling in the Syrian hamster SCN. Then we tested for co-expression of the mt_1 receptor with messages for either arginine vasopressin (AVP), somatostatin (SS) or retinoid Z receptor β (RZRβ; a putative nuclear mt_1 receptor) in Siberian hamster brain only. Cells in the SCN expressing the mt_1 receptor also contained mRNA for AVP (Figures 1 and 2) and RZRβ, but not for SS.

In conclusion, this preliminary report is the first to demonstrate cellular resolution of the mt_1 receptor mRNA in Siberian and Syrian hamster brains. In addition, mt_1 receptor mRNA is co-expressed with AVP and RZRβ mRNA in the SCN of Siberian hamsters. These data suggest that seasonal photoperiodic MEL signals may be received directly by vasopressinergic cells within the SCN.

ACKNOWLEDGMENT

Supported, in part, by NIH RO1 35254, NIMH RO1 48462 and NIMH RSDA KO2 00841 to TJB and NSF IBN-9319653, NIMH RO1 44132 and NIMH RSDA KO2 00914 to ELB.

REFERENCES

1. Bartness, T.J. and B.D. Goldman. Mammalian pineal melatonin: A clock for all seasons. *Experientia* 45:939–945, 1989.
2. Bartness, T.J., B.D. Goldman, and E.L. Bittman. SCN lesions block responses to sytemic melatonin infusions in Siberian hamsters. *Am. J. Physiol.* 260:R102–R112, 1991.

3. Bartness, T.J., J.B. Powers, M.H. Hastings, E.L. Bittman, and B.D. Goldman. The timed infusion paradigm for melatonin delivery: What has it taught us about the melatonin signal, its reception, and the photoperiodic control of seasonal responses? *J. Pineal Res.* 15:161–190, 1993.
4. Grosse, J. and M.H. Hastings. A role for the circadian clock of the suprachiasmatic nuclei in the interpretation of serial melatonin signals in the Syrian hamster. *J. Biol. Rhythms* 11:317–324, 1996.
5. Larsen, P.J., L.W. Enquist, and J.P. Card. Characterization of the multisynaptic neuronal control of the rat pineal gland using viral transneuronal tracing. *Eur. J. Neurosci.* 10:128–145, 1998.
6. Maywood, E.S., E.L. Bittman, F.J.P. Ebling, P. Barrett, P. Morgan, and M.H. Hastings. Regional distribution of iodomelatonin binding sites within the suprachiasmatic nucleus of the Syrian hamster and the Siberian hamster. *J. Neuroendocrinology* 7:215–223, 1995.
7. Maywood, E.S., R.C. Buttery, G.H.S. Vance, J. Herbert, and M.H. Hastings. Gonadal responses of the male Syrian hamster to programmed infusions of melatonin are sensitive to signal duration and frequency but not to signal phase nor to lesions of the suprachiasmatic nuclei. *Biol. Reprod.* 43:174–182, 1990.
8. Reppert, S.M., D.R. Weaver, and T. Ebisawa. Cloning and characterization of a mammalian melatonin receptor that mediates reproductive and circadian responses. *Neuron* 13:1177–1185, 1994.
9. Song, C.K. and T.J. Bartness. The effects of anterior hypothalamic lesions on short-day responses in Siberian hamsters given timed melatonin infusions. *J. Biol. Rhythms* 11:14–26, 1996.
10. Turek, F.W. and C.S. Campbell. Photoperiodic regulation of neuroendocrine-gonadal activity. *Biol. Reprod.* 20:32–50, 1979.
11. Weaver, D.R., C. Liu, and S.M. Reppert. Nature's knockout: The Mel_{1b} receptor is not necessary for reproductive and circadian responses to melatonin in Siberian hamsters. *Mol. Endocrinol.* 10:1478–1487, 1996.
12. Weaver, D.R., S.A. Rivkees, and S.M. Reppert. Localization and characterization of melatonin receptors in rodent brain by in vitro autoradiography. *J. Neurosci.* 9:2581–2590, 1989.

26

SUPRACHIASMATIC NUCLEI, INTERGENICULATE LEAFLET, AND PHOTOPERIOD

P. Pevet, N. Jacob, and P. Vuillez

Neurobiologie des Fonctions Rythmiques et Saisonnières
UMR-CNRS 7518
Université L. Pasteur
Strasbourg, France

The suprachiasmatic nuclei (SCN) of the hypothalamus play an essential role in the generation and maintenance of circadian rhythms in mammals (62). The intrinsic pacemaker activity of the SCN is autonomous in nature and is entrained to the daily 24 h period by environmental factors of which the light-dark cycle is the most effective. This photic information is relayed directly to the SCN by the retinohypothalamic tract (RHT) and indirectly throughout other pathways (7,15,44,46). The best known of such indirect photic pathways is the geniculohypothalamic tract (GHT), a pathway projecting from the intergeniculate leaflet (IGL) to the SCN (45). The IGL is a thin retinorecipient zone, intercalated between the dorsal and ventral nuclei of the lateral geniculate complex (21). The IGL contain a population of neuropeptide Y (NPY)-containing neurons (9,17,18) which project to the SCN. Althought the exact role of the GHT and NPY is at present not fully elucidated, several experiments have shown that NPY fibers are involved in the transmission of both photic and non photic stimuli to the SCN (10,32,45,51,52).

Studies on the role of the SCN have focussed mainly on circadian functions. However, due to its neuronal connections with the pineal gland, the present identified key structure in the transduction of photoperiodic changes (36,52,56,76), the SCN is also known to be involved in the control of seasonal functions. Within the SCN, at least in the rat, NPY levels increase rapidly following a light pulse in a phase dependent

Address correspondence to: P. Pevet, Neurobiologie des Fonctions Rhythmiques et Saisonnières, UMR-CNRS 7518, Université Louis Pasteur, 12 rue de l'Université, 67000 Strasbourg, France. Tél.: 03.88.35.85.09, Fax: 03.88.24.04.61, e.mail: pevet@neurochem.u-strasbg.fr

Melatonin after Four Decades, edited by James Olcese.
Kluwer Academic / Plenum Publishers, New York, 2000.

manner suggesting that IGL neurons convey photic information to the SCN (66). Several studies have also shown that the density of NPY-immunoreactive fibers in the SCN exhibits a daily (8,33,65) as well as a seasonal (39) rhythm. Moreover, it is also known that the IGL is directly involved in the control of the nocturnal metabolic capacity of the rat pineal gland (5,12). These findings suggest that like the SCN, the IGL is also involved in the control of seasonal functions. In the present review data in favor of such a concept will be presented.

I. SCN AND PHOTOPERIODIC INFORMATION

Endogenous rhythms generated by the SCN are known to be entrained by retinal illumination (62). This photic entrainment is accomplished by a daily light-induced phase shift of the clock. Light pulses given during the night induce the expression of immediate early genes, including c-fos, in several cell populations of the SCN (11,37,63,74). In constant environmental conditions such as permanent darkness, circadian rhythms "free run" with a period close to 24 h showing a subjective night (period of activity in nocturnal species such as the hamster, rat or mouse) and a subjective day (inactive period). Fos immunoreactivity (Fos-ir) increases within the SCN only when light pulses are given during the (subjective) night (63). It is also during the (subjective) night that light stimuli are able to induce phase advance or delays of the free-running circadian locomotor activity rhythm (13). The changes in the light sensitivity of different SCN cell populations parallel the phase shifting effects of light on circadian locomotor activity. The pattern of circadian locomotor activity is also clearly dependent upon the photoperiod (24,61,75,80). Recently using light-induced c-fos mRNA expression or Fos protein induction as an *in vivo* marker of intrinsic neuronal activity, it has been demonstrated in the European and Syrian hamster (78) as well as in the rat (71) that the duration of the sensitive phase of the SCN to light depends on the photoperiod. For example, in Syrian hamsters kept in long photoperiod (LP; 14:10, 14 h of light and 10 h of dark) Fos expression in the SCN was induced by light pulses given during the 10 h of darkness. After transfer from long to short photoperiod (SP; 10:14) a 4 h lengthening of the photosensitive phase occurred. The adaptation of the light responsiveness of the SCN to this new photoperiod took 3–4 weeks (78).

All these data have clearly established that the SCN, whose rhythmic activity is known to be regulated by the daily light-dark cycle, is also affected by the photoperiod. Since it has been shown that neither the retina nor other brain structures are required for the phase dependent Fos reaction (63), it must be the functional state of the circadian clock itself which is affected by photoperiod. This raises the question of the site at which the photoperiodic message is elaborated, as well as the question of the mechanism and of neural or neuroendocrine pathways by which the photoperiod influences the SCN.

II. DOES THE CIRCADIAN CLOCK GENERATE THE PHOTOPERIODIC MESSAGE?

Circadian involvement in the photoperiod response hypothesized by Bunning (6), was later strongly supported by numerous experimental data. Based on these,

Pittendrigh and Daan (58) proposed a model for photoperiodic time measurement consisting of a complex circadian pacemaker composed of two mutually coupled components entrained by dusk or dawn, respectively. Interpretation of photoperiod in this system is achieved by recognition of the phase relationship between the two hypothetical components of the pacemaker. The circadian clock could then generate itself the photoperiodic message.

This concept has been further developed for photoperiodic perception in mammals. In the rat, in which the rhythmic production of melatonin (Mel) is known to be controlled by the light entrainable circadian pacemaker, the shift in the evening rise of Mel, or in the morning decline to Mel, observed after photoperiod manipulation are different and opposite in direction (27,28). Seasonal changes in daylength (photoperiod) alter the magnitude of these photic phase shifts as well as the waveform of the Mel rhythm. With longer daylength, light at an earlier dawn advances the morning Mel decline and light at a later dusk delays the Mel rise. The resulting alteration of the nocturnal Mel signal—compressed during long summer days and decompressed during short winter days (changes in duration)—might be the consequence of an adjustment in the timing of the two components of the pacemaker, one entrained by dusk, which controls the time of transition from the diurnal to nocturnal state (e.g. onset of nocturnal Mel secretion) and a second component entrained by light which controls the time of transition from the nocturnal to diurnal state (e.g. offset of nocturnal Mel secretion). This can serve as an endogenous photoperiodic message as suggested by Illnerovà (25).

Importantly photoperiod alters not only the Mel rhythm but also other circadian rhythms such as locomotor activity (24,61,75,80) and, as described above, the circadian rhythm of light responsiveness of the SCN. After transfer from long (LP) to short (SP) photoperiod the time interval enabling Fos-ir photoinduction in the SCN extends gradually, the full extension being achieved after 2 weeks in the rat (71) and 3–4 weeks in the Syrian hamster (78). These periods of time also correspond to the time necessary to get maximum duration of the nocturnal Mel peak in the rat and hamster following transfer from LP to SP. In the same protocol, 3–4 weeks is also the time necessary to get maximal duration of the nocturnal locomotor activity in the Syrian hamster (20). This parallelism in the time course of adaptation of these 3 dependent clock driven phenomena suggests that the SCN is involved in processing photoperiodic information. This also supports the proposed model of a circadian clock consisting of an evening and a morning component (57). The photoperiodic message would then be interpreted by the SCN itself. A consequence of this would be, as far as the pineal is concerned, that the circadian and photoperiodic information is integrated into one signal, the nocturnal Mel secretion. The pineal would then differ little from any other effector driven by the circadian system and its role (still an important one) would only be to distribute, through the Mel rhythm, the photoperiodic message and perhaps also a circadian signal.

Although this hypothesis is attractive and although our results in the hamster (78), like those of Sumova et al. (71) in the rat, can be interpreted using this concept, other possible interpretations cannot be excluded. Data in the literature concerning Mel also suggest other possibilities.

The slow decompression of the duration of the Mel signal in response to short day exposure, as described extensively in the rat, has also been observed in some "true" photoperiodic rodents (for example, the Syrian and Djungarian hamster: 20,22,26). However, this does not occur in all photoperiodic rodents. In the European hamster, for example, the opposite has been observed (77). At the light to dark transition, Mel production commenced 2 h later under long photoperiod (LP; 14:10) than under short photoperiod (SP; 12:12). This observation is not restricted to the European hamster. A similar finding has been made in another rodent species, the *Arvicanthis niloticus* (unpublished data, present lab), as well as in one primate, the lesser mouse lemur (*Microcebus murinus*) (4).

With respect to the morning decline of Mel synthesis in relation to the photoperiod, again data obtained in the rat cannot be generalized to other species. In the rat, Mel production declines spontaneously before light onset even under a very long photoperiod (30). In contrast, in three "true" photoperiodic species, the European, Syrian and Djungarian hamsters, the morning decline in Mel synthesis depends on the photoperiod. Under a long photoperiod the decline is suppressed by the onset of light, while under a short photoperiod, the morning decline, like in the rat, is endogenous and declines before light onset (40,56,77). Moreover, in the European hamster, when comparing the sequence of Mel onset, duration and decline under various natural or experimental photoperiods, although Mel onset is not delayed with shorter photoperiod, the duration of nocturnal Mel production increases linearly. This finding suggests that there is a clear phase relationship between Mel onset and duration and that, at least under short photoperiods, the Mel decline is directly dependent upon the onset of Mel synthesis, rather than on the time of dawn or lights on (77).

Based on all these findings an alternative model can be suggested. The major role of the SCN in photoperiodism is probably related to the essential part it plays in the generation and entrainment of circadian rhythms. Damage to the SCN will disturb the photoperiodic response only because melatonin production is dependent upon the circadian system. The pineal, at least in the hamster species, may receive photoperiodic information through separate mechanisms. Light induces offset of Mel production in LP but, in SP, the pineal could induce itself such offset (e.g. induction of an endogenous inhibitor such as the ICER protein, 69). These two mechanisms would also result in a change in the duration of the nocturnal Mel peak. There is a lot of neurochemical and neuroanatomical evidence in favour of separate subunits within the SCN (47,60,74). Thus there may be regions within the SCN that are specialized and allow dissociation of the effect of light on the circadian response (e.g. phase shifting effect through a direct or indirect effect of light on SCN neurons involved in the circadian mechanism) from the photoperiodic response (e.g. inhibition of nocturnal Mel synthesis through an action of light on neurons which project to a specific neuronal SCN-pineal pathway, such neurons probably also projecting to the circadian zone). Such a model would not need the presence of evening and morning oscillators to explain the photoperiodic changes in the duration of Mel secretion. The fact that in some experiments light has been shown to affect Mel synthesis without affecting the circadian system supports such an idea. For example, Takahashi et al. (73) measured the amount of light necessary for phase shifting the Syrian hamster locomotor activity rhythm and compared it with that needed to suppress Mel synthesis. The curves that describe these two phenomena are quite different, the amount of light required differs 2.5-fold. Suppression of Mel synthesis was much more sensitive to light than phase shifting. Low intensity light known to suppress Mel production (by 50%) is also able to inhibit gonadal regression in the Syrian hamster kept under DD without affecting the circadian system. In other words, the circadian system free-runs despite periodic light pulses that suppress Mel production and affect the reproductive system.

From these studies it is possible to conclude that the observed photoperiodic-induced changes in the duration of the SCN light sensitivity period indicate that the photoperiodic message is read within the circadian system itself. This, in our opinion, might be a overinterpretation as data obtained in the hamster and rat (71,78) simply indicate that the photoperiod acts at the level of the clock. From the present experimental evidence it is not possible to conclude whether this is the consequence of a direct or indirect action. If this is the consequence of an indirect action, how does the SCN perceive the photoperiodic information. Is it through melatonin?

III. THE ROLE OF MELATONIN

For all seasonal rhythms (reproduction, hibernation, daily torpor, moulting, changes in pelage quality and/or colour, body weight regulation, etc . . .) and for all mammalian species studied to date, photoperiodic information is integrated through variation in the duration of the nocturnal peak of Mel, which parallels the length of the night. High affinity melatonin binding sites/receptors have been identified within several brain regions (for review see 41,49,68) including, in most species, the SCN (79). Daily exogenous melatonin administration is known, probably through an action on the SCN, to entrain free-running locomotor activity rhythms of the rat (2,3), Djungarian and Syrian hamster (35) to a 24 h period and to accelerate reentrainment of the locomotor activity rhythm after a shift of the L/D cycle. This action of the SCN has been confirmed by some *in vitro* studies. Neurophysiological recordings from SCN slices have identified neuronal firing rate rhythms which are responsive to Mel (43,70) or to pinealectomy (64). Furthermore, the time course of the lengthening in the photosensitive phase of the SCN after a change from LP to SP is similar to that described to obtain maximal extension of the nocturnal Mel peak. These different observations suggest that, as for other photoperiodic-dependent responses, the photoperiod-induced change of the photosensitive phase of the SCN could be mediated through melatonin in the hamster and rat. To test this hypothesis, the effect of light stimulation on Fos induction in the SCN of animals with no endogenous Mel (pinealectomized) has been studied in the hamster (31) and rat (72) after transfer from LP to SP. The lengthening of the photosensitive phase was investigated using a light stimulation of 15 minutes, 13 hours after the beginning of the night (no light sensitivity at this time in hamsters kept under LP, but light sensitivity when kept under SP for 8 weeks) (78). After the change of photoperiod, the number of light-induced Fos-ir cells in the SCN of pinealectomized hamsters increased progressively until the 25th night. At that time, the total number of Fos-ir cells was similar to that observed after a light stimulus at the same time of the night in animals maintained in SP for 8 weeks. A similar increase in Fos-ir cells was observed for the pinealectomized and non pinealectomized animals. In both groups the number of labeled cells as well as their localization was comparable, irrespective of the night of the light stimulation following the photoperiod change. This result indicates that, like in intact animals, the duration of the photosensitive phase also extends gradually in the SCN of pinealectomized animals. The full extension of the photosensitive phase was reached after 3–4 weeks. A similar observation was reported in the rat (72).

Thus, this photoperiod-dependent extension is independent of the endogenous Mel signal. These results clearly show that in the rat (considered to be a non photoperiodic species) and in the hamster (which has a melatonin-dependent control of some photoperiodic functions), the photoperiod does not act on the SCN via variations in Mel secretion. The SCN appears to be a structure able to integrate photoperiodic information without requiring a hormonal (Mel) signal from the pineal. This observation could mean two things. Firstly, the photoperiod primarily affects the SCN which then interprets the photoperiodic message itself (see details of this concept in the section above). The results obtained strongly support such an interpretation, as originally suggested by Sumova and Illnerovà (72). Secondly, the SCN receives this photoperiodic information through other photoperiod-dependent cues. This alternative model is presented in the following section.

IV. THE ROLE OF OTHER PHOTOPERIODIC CUES

The annual pattern in circadian locomotor activity shows a temporal relationship with photoperiod-induced reproductive changes, at least in the European hamster (80). Moreover, in the Syrian as well as in the European hamster, SP which increases the duration of the light sensitive period of the SCN is also known to inhibit gonadal activity. Although no steroid receptors have been identified within the SCN, such photoperiodic changes in light sensitivity might be induced by the annual change in circulating levels of sex steroids (a well defined seasonal cue in other systems, e.g. hibernation, daily torpor, review in 53).

Changes in neurotransmitter synthesis in neurons which project directly or indirectly to the SCN, induced by changes in circulating levels of sex steroids could explain the data. Steroid-dependent changes (14,67) and seasonal variation (16,38) in the immunoreactivity of some neuropeptides (namely vasopressin, substance P, CCK, VIP) have been described in the SCN. These findings, however, even if they clearly suggest a role for sex steroids in the functioning of the SCN, cannot, in our opinion, explain the SP-induced increase in the duration of the SCN light sensitive period. In the rat this increase was observed when animals were transferred from LP to SP (71). In these conditions, however, no change in reproductive activity was noted. As indicated above, when Syrian hamsters were transferred from LP to SP the maximum extension of the light sensitive period was obtained within 3–4 weeks. After this time in SP, however, the sexual axis in the Syrian hamster is far from being fully regressed (59).

If a possible role for photoperiod-induced changes in sex steroids in the extension of the light sensitive phase is neglected, other photoperiod dependent hormones known to cross the blood brain barrier (e.g. thyroid hormones) might be implicated in photoperiodic/seasonal cues. At present there is no data in the literature to support or refute such an idea.

As discussed in the introduction, the SCN is known to receive photic information from the retina not only through the RHT but also indirectly via other neural pathways. The precise role of these indirect pathways has not yet been well defined. Their possible implication in photoperiodic integration by the SCN has to be considered. Presently three indirect pathways carrying photic information to the SCN have been identified. For two of them, namely the retina-raphe-SCN and retina-pretectum-SCN pathways, there is no data in the literature to suggest their possible role in the photoperiodic response. For the third pathway, the retina—IGL—SCN pathway, more information is available.

V. INTERGENICULATE LEAFLET AND PHOTOPERIODIC INFORMATION

The IGL receives retinal projections and sends NPY-containing fibers to the SCN. Although the role of the IGL in photic entrainment has been the subject of extensive investigation, its exact function is still unknown. The IGL does not appear to be essential for photic-entrainment since, in a number of studies, entrainment was not affected by lesion of the IGL (48,55). However, other data indicate that the IGL may play a role, micro-injections of NPY into the SCN or activation of the IGL by electrical or chemical stimulation inducing either phase advances or phase delays during the subjective night (1,20,34,54).

The IGL is considered to be a component of the circadian system. Whether it also represents a component of the photoperiodic system enabling, for example, the SCN to integrate photoperiodic information is not known.

In the jerboa (39) NPY immunolabeling in the SCN was higher in animals killed in autumn (SP, period of sexual quiescence) than in those killed in the spring to middle of summer (period of sexual activity) (39). These photoperiod-dependent changes are independent of the photoperiod-induced changes in circulating levels of gonadal steroids (39). In the rat, the density of NPY fibers within the SCN exhibits a daily rhythm characterized by an increase at the light/dark transition (8,33) and this rhythm is known to be drastically modulated by photic stimulation and to disappear in constant conditions (66). Moreover, like SCN neurons, specific neurons of the IGL that project to the SCN are responsive to tonic changes in light stimulation (45,81). All these data suggest that the increase in NPY immunoreactivity observed in autumn in the jerboa SCN or under SP in the Siberian hamster might be a consequence of the increased duration of the dark phase. NPY might thus play an important role in the integration of photoperiodic information by the SCN. Are these NPY fibers involved in the photoperiodic-induced changes in the duration of the nocturnal light sensitive period of the SCN?

An increase in the number of NPY mRNA containing neurons in the IGL of the Syrian hamster has been observed after 8 weeks of exposure to SP (31). A densitometric analysis of the concentration of NPY mRNA in each neuron did not show a parallel increase. This findings suggests that SP activates a population of quiescent cells which do not express NPY mRNA in LP. A similar phenomenon has already been described in the human SCN, in which an additional population of neurons express vasopressin in the summer time (23). This SP-induced increase in the number of NPY-neurons indicates that photoperiod controls or modulates the synthesis of NPY within some neurons of the IGL. What could the physiological consequence of this phenomenon be?

Clearly a photoperiodic message is integrated and expressed at the level of the IGL. It may be that it is through this photoperiod dependent change in NPY content in IGL afferent fibers that the circadian clock integrates photoperiodic information. The geniculo-hypothalamic tract (GHT) would then be directly involved in the photoperiodic-induced changes in the duration of the circadian light sensitive period of the SCN. Preliminary data from the present laboratory support such an interpretation. In

IGL-lesioned hamsters, 3 weeks after transfer to SP (a condition in which in non lesioned animals a full extension of the light sensitive period is observed), the number of cells expressing Fos-ir after light exposure was reduced compared to intact animals. This findings indicates that the GHT is involved in the photoperiod-induced SCN changes. However, as the amount of Fos cells in lesioned animals was higher than in the control animals kept under LP, it is also evident that even in the absence of the IGL, the photoperiodic message has been, at least partly, integrated.

A clear photoperiodic message is thus expressed at the level of the IGL but its origin remains to be determined. It could originate directly from the retina, variation in the duration of the night or day being directly transmitted through retinal projections to the IGL. The observation that the increase in number of NPY-neurons is observed 4 days only after the transfer from LP to SP (Figure 1) deeply supports this interpretation. This observation, although melatonin receptors have been described in the IGL of some mammalian species (41), permits to exclude a role of melatonin (3 weeks are needed to obtain the maximal extension of the nocturnal melatonin peak). Since six to eight weeks are needed for the SP to induce gonadal atrophy, it permits also to exclude a role of sex steroids in such phenomenon.

Neurons in the IGL are activated by both photic (55) and non-photic stimuli (32,42,50). Both stimuli involve activation of an immediate early gene, c-fos. Non-photic stimulation (associated with locomotor activity) induces Fos-ir primarily in the NPY neurons. After transfer from LP to SP the locomotor activity period in the hamster is extended. The increase in the number of NPY-mRNA containing neurons in the IGL after exposure to SP may also be the consequence of the photoperiod-dependent increase in the duration of the activity period. Experiments are in progress to test these different hypotheses.

Although it is not a dense tract, SCN neurons also project to the IGL (47). Considering the model proposed by Pittendrigh and Daan (58) and the interpretation of the data by Illnerova's group (see above) and that after an IGL lesion a photoperiodic message can be, at least partly, integrated by the SCN, it might also be that

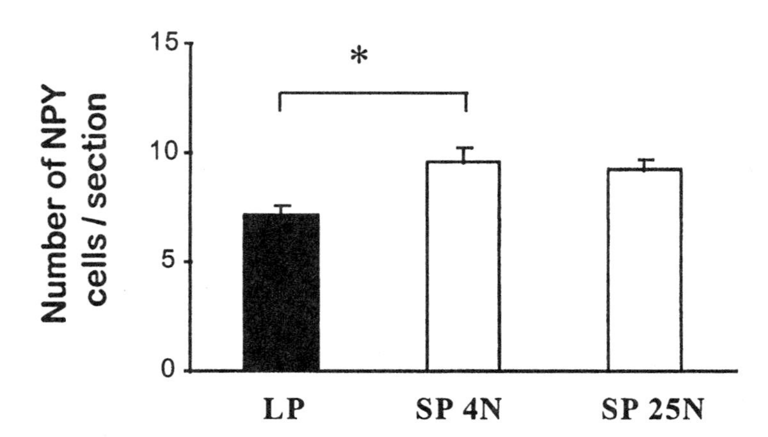

Figure 1. Graph showing the number of NPY mRNA containing cells per section of mid-part IGL of hamsters kept under LP or SP for 25 days. Note that 4 days after transfert under SP the increase in number is already maximal. Means ± SEM. *Significant difference between the two groups ($P < 0.001$, student t-test).

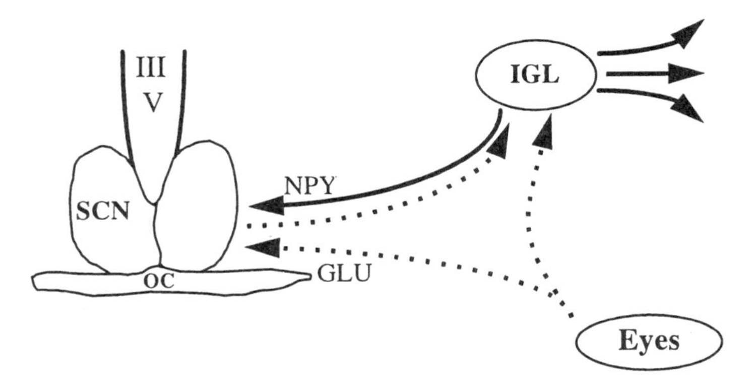

Figure 2. Through its efferent fibers IGL is able to distribute photoperiodic information to different brain structures including the SCN. Whether the photoperiodic signal is directly integrated by the IGL through retinal fibers or originates from the SCN is not known.

the photoperiodic information originates in the SCN itself. If this is true the IGL may be used by the SCN to transmit the photoperiodic message to IGL-connected structures.

CONCLUSION

The duration of the light sensitive period of the SCN is dependent on the photoperiod. This clearly indicates that the photoperiod can affect the functioning of the circadian clock. Contrary to all other photoperiodic responses studied to date, this effect of the photoperiod on the SCN is not mediated through melatonin. Whether the circadian pacemaker system itself interprets the photoperiodic message or integrates the photoperiodic information through indirect neural or neuroendocrine pathways is still an open question.

Through the changes in the number of NPY neurons, the IGL expresses a clear photoperiodic signal. Whether the IGL directly integrates it through direct or indirect neural or neuroendocrine pathways or, like for the pineal, is used by the SCN to distribute its photoperiodic message is also not known. Independent of this important question, the IGL through its NPY projections (47) is able to provide processed photoperiodic information to different brain structures (Figure 2). As the SCN is one of the major efferent areas of the IGL, the effect (feed-back effect?) of this peculiar photoperiodic message on the pacemaker system or on the efferent SCN system is probably important physiologically.

ACKNOWLEDGMENT

The authors are grateful to Dr. D.J. Skene for revising the English and to Miss F. Murro for typing the manuscript.

REFERENCES

1. Albers, H.E. and Ferris, C.F. Neuropeptide Y: role in light-dark cycle entrainment of hamster circadian rhythms. Neurosci. Lett. 50:163–168, 1984.
2. Armstrong, S.M. Melatonin: the internal zeitgeber of mammals. Pineal Res. Rev. 7:157–202, 1989.
3. Armstrong, S.M. and Redman, J.R. Melatonin and circadian rhythmicity. In Melatonin: Biosynthesis, Physiological Effects and Clinical Applications. H.S. Yu, R.J. Reiter, eds, CRC Press—Boca Raton, USA, 187–224, 1993.
4. Aujard, F., Boissy, I., and Claustrat, B. Melatonin secretion in a nocturnal prosimian primate: effect of photoperiod and aging. In Biological clocks. Mechanisms and Applications, Y. Touitou, ed., Elsevier Science, pp. 337–340, 1998.
5. Bartol, I., Skorupa, A.L., Scsalfa, J.H., and Cipolla-Neto, J. Pineal metabolic reaction to retinal photostimulation in ganglionectomized rat. Brain Res. 744:77–82, 1997.
6. Bünning, E. Die endogene Tagesrhythmik als Grundlage der photoperiodischen Reaktion. Berl. Dtsch. Bot. Ges. 54:590–607, 1936.
7. Cagampang, F.R.A., Yamazaki, S., Otori, Y., and Inouye, S.I.T. Serotonin in the raphe nuclei: regulation by light and on endogenous pacemaker. Neuroreport. 4:49–52, 1993.
8. Calza, L., Giardino, L., Zanni, M., Parchi, P., and Marrama, P. Daily changes of neuropeptide Y-like immunoreactivity in the suprachiasmatic nucleus of the rat. Regul. Pept. 127–137, 1990.
9. Card, J.P. and Moore, R.Y. Ventral lateral geniculate nucleus efferents to the rat suprachiasmatic nucleus exhibit avain pancreatic polypeptide-like immunoreactivity. J. Comp. Neurol. 206:390–396, 1982.
10. Challet, E., Jacob, N., Vuillez, P., Pévet, P., and Malan, A. Fos-like immunoreactivity in the circadian timing system of calorie-restricted rats fed at dawn: daily rhythms and light pulse-induced changes. Brain Res. 770:228–236, 1997.
11. Chambille, I., Doyle, S., and Servière, J. Photic induction and circadian expression of Fos-like protein immunohistochemical study in the retina and suprachiasmatic nuclei of hamster. Brain Res. 612:138–150, 1993.
12. Cipolla-Neto, J., Bartol, I., Seraphim, P.M., Afeche, S.C., Scialfa, J.H., and Peracoli, A.M. The effects of lesions of the thalamic intergeniculate leaflet on the pineal metabolism. Brain Res. 691:133–141, 1995.
13. Daan, S. and Pittendrigh, C.S. A functional analysis of circadian pacemaker in nocturnal rodents II. The variability of phase response curves. J. Comp. Physiol. 106:253–266, 1976.
14. De Vries, G.J. Sex differences in neurotransmitter system. J. Neuroendocrinol. 2:1–13, 1990.
15. Foote, W.E., Taber-Pierce, E., and Edwards, C. Evidence for a retinal projection to the midbrain raphe of the cat. Brain Res. 156:135–140, 1978.
16. Fuminier, F., Sicard, B., Boissin-Agasse, L., and Boissin, J. Seasonal changes in the hypothalamic vasopressinergic system of a wild sahalian rodent, *Taterilla petteri*. Cell Tissues Res. 271:309–316, 1993.
17. Harrington, M., Nance, D., and Rusak, B. Neuropeptide Y immunoreactivity in the hamster geniculo-suprachiasmatic tract. Brain Res. Bull. 15:465–472, 1985.
18. Harrington, M., Nance, D., and Rusak, B. Double-labeling of neuropeptide Y-immunoreactive neurons which project from the geniculate to the suprachiasmatic nuclei. Brain Res. 410:275–282, 1987.
19. Harrington, M.E. and Rusak, B. Lesions of the thalamic intergeniculate leaflet alter hamster circadian rhythms. J. Biol. Rhythms 1:309–325, 1986.
20. Hastings, M.H., Walker, A.P., and Herbert, J. Effect of asymmetrical reductions of photoperiod on pineal melatonin locomotor activity and gonadal condition of male Syrian hamsters. J. Endocrinol. 114:221–229, 1986.
21. Hickey, T.L. and Spear, P.D. Retinogeniculate projections in hooded and albino rats: an autoradiographic study. Exp. Brain Res. 24:523–529, 1976.
22. Hoffmann, K., Illnerovà, H., and Vanecek, J. Comparison of melatonin rhythms in young adult and old Djungarian hamsters (*Phodopus sungorus*) under long and short photoperiod. Neurosci. Lett. 56:39–43, 1985.
23. Hofman, M.A. and Swaab, D.F. Seasonal changes in the suprachiasmatic nucleus of man. Neurosci. Lett. 139:257–260, 1992.
24. Hoogenboom, I., Daan, S., Dallinga, J.H., and Schoenmakers, M. Seasonal change in the daily timing of behaviour of the common vole, *Microtus arvalis*. Oecologia 61:18–31, 1984.
25. Illnerovà, H. Circadian rhythms in the mammalian pineal gland. Akademia Ved, Praha, 1986.
26. Illnerovà, H., Hoffmann, K., and Vanecek, J. Adjustment of pineal melatonin and N-Acetyltransferase

rhythms to change from long to short photoperiod in the Djungarian hamster *Phodopus sungorus*. Neuroendocrinology 38:226–231, 1984.
27. Illnerovà, H., Hoffman, K., and Vanecek, J. Adjustment of the rat pineal N-Acetyltansferase rhythm to change from long to short photoperiod depends on the direction of extension of the dark period. Brain Res. 362:403–408, 1986.
28. Illnerovà, H. and Humlova, I. Entrainment of the pacemaker controlling the rhythmic melatonin production depends on photoperiod. In Advances in Pineal Research. J. Arendt, P. Pévet, eds, John Libbey and Co—London, vol. 5:267–272, 1991.
29. Illnerovà, H. and Vanecek, J. The evening rise in the rat pineal N-acetyltransferase activity under various photoperiods. Neurosci. Lett. 36:279–284, 1983.
30. Illnerovà, H. and Vanecek, J. Pineal N-acetyltransferase: A model to study properties of biological clocks. In Fundamentals and Clinics in Pineal Research, G.P. Trentini, C. De Gaetani, P. Pévet, eds, Serono symposia—Raven Press, vol. 54:165–178, 1987.
31. Jacob, N., Vuillez, P., and Pévet P. Photoperiod does not act on the suprachiasmatic nucleus photosensitive phase through the endogenous melatonin, in the Syrian hamster. Neuroscience Lett. 229:117–120, 1997.
32. Janik, D., Mikkelsen, N., and Mrosovsky, N. Cellular colocalization of Fos and neuropeptide Y in the intergeniculate leaflet after nonphotic phase-shifting events. Brain Res. 698:137–145, 1995.
33. Jhanwar-Uniyal, M., Beck, B., Burlet, C., and Leibowitz, S.F. Diurnal rhythm of neuropeptide Y-like immunoreactivity in the suprachiasmatic arcuate and paraventricular nuclei and other hypothalamic sites. Brain Res. 536:331–334, 1990.
34. Johnson, R.F., Moore, R.Y., and Morin, L.P. Lateral geniculate lesions alter activity rhythms in the hamster. Brain Research Bulletin 22:411–422, 1989.
35. Kirsch, R., Belgnaoui, S., Gourmelen, S., and Pévet, P. Daily melatonin infusion entrains free-running activity in Syrian and Siberian hamster. In Light and Biological Rhythms in Man, L. Wetterberg, ed., Wenner-gren International Series, Pergamon Press—Oxford, vol. 63:161–185, 1993.
36. Klein, D.C. and Moore, R.Y. Pineal N-acetyltransferase and hydroxyindole-O-methyltransferase: control by the retinohypothalamic tract and suprachiasmatic nucleus. Brain Res. 174:245–262, 1979.
37. Kornhauser, J.M., Nelson, D.E., Mayo, K.E., and Takahashi, J.S. Photic and circadian regulation of c-fos gene expression in the hamster suprachiasmatic nucleus. Neuron. 5:127–134, 1990.
38. Lakhdar-Ghazal, N., Kalsbeek, A., and Pévet, P. Sexual difference and seasonal variations in vasoactive intestinal polypeptide immunoreactivity in the suprachiasmatic nucleus of jerboa (*Jaculus orientalis*). Neurosci. Lett. 144:29–33, 1992.
39. Lakhdar-Ghazal, N., Oukouchoud, R., and Pévet, P. Seasonal variation in NPY immunoreactivity in the suprachiasmatic nucleus of the jerboa (*Jaculus orientalis*), a desert hibernator. Neurosci. Lett. 193:49–52, 1995.
40. Lerchl, A. and Schlatt, S. Influence of photoperiod on pineal melatonin synthesis, fur colour, body weight and reproductive function in the female Djungarian hamster, *Phodopus sungorus*. Neuroendocrinology 57:359–369, 1993.
41. Masson-Pévet, M., George, D., Kalsbeek, A., Saboureau, M., Lakhdar-Ghazal, N., and Pévet, P. An attempt to correlate brain areas containing melatonin-binding sites with rhythmic functions: a study in five hibernator species. Cell Tiss. Res. 278:97–106, 1994.
42. Maywood, E.S., Smith, E., Hall, S.J., and Hastings, M.H. A thalamic contribution to arousal-induced, non-photic entrainment of the circadian clock of the Syrian hamster. Eur. J. Neurosci. 1739–1747, 1997.
43. McArthur, A.J., Gilette, M.U., and Prossner, R.A. Melatonin directly resets the rat suprachiasmatic circadian clock in vitro. Brain Res. 565:158–161, 1991.
44. Mikkelsen, J.D. and Vrang, N. A direct pretectosuprachiasmatic projection in the rat. Neurosci. 62:497–505, 1994.
45. Moore, R.Y. The enigma of the geniculohypothalamic tract: why two visual entraining pathways. J. Interdiscipl. cycle Res. 23:144–152, 1992.
46. Moore, R.Y. Organization and function of a central nervous system circadian oscillator the suprachiasmatic hypothalamic nucleus. Fed. Proc. 42:2783–2789, 1993.
47. Moore, R.Y. Entrainment pathways and the functional organization of the circadian system. In Hypothalamic integration of circadian rhythms, R.M. Buijs, A. Kalsbeek, H.J. Romijn, C.M.A. Pennarte, M. Mirmiran, eds, Progress in Brain Research, vol. 111—Elsevier Amsterdam, 103–119, 1996.
48. Moore, R.Y. and Card, J.P. Intergeniculate leaflet: an anatomically and functionnally distinct subdivision of the lateral geniculate complex. J. Comp. Neuro. 344:403–430, 1994.
49. Morgan, P.J., Barrett, P., Howell, H.E., and Helliwell, R. Melatonin receptors: localization, molecular pharmacology and physiological significance. Neurochem. Int. 24:101–146, 1994.

50. Mrosovsky, N. A non-photic gateway to the circadian clock of hamsters. Ciba Found. Symp. 183:154–174, 1995.
51. Mrosovsky, N. Locomotor activity and non-photic influences on circadian clocks. Biol. Rev. 71:343–372, 1996.
52. Pévet, P. The role of the pineal gland in the photoperiodic control of reproduction in different hamster species. Reprod. Nutr. Dev. 28:443–458, 1988.
53. Pévet, P. Importance of sex steroids and neuropeptides in the pineal control of seasonal rhythms. In Advances in Pineal Research, J. Arendt, P. Pévet, eds, Libbey—London, vol. 5:219–224, 1991.
54. Pickard, G.E. Entrainment of the circadian rhythm of wheel-running activity is phase shifted by ablation of the intergeniculate leaflet. Brain Res. 494:151–154, 1989.
55. Pickard, G.E., Ralph, M.R., and Menaker, M. The intergeniculate leaflet partially mediates effects of light on circadian rhythms. J. Biol. Rhythms 2:35–56, 1987.
56. Pitrosky, B., Kirsch, R., Vivien-Roels, B., Georg-Bentz, I., Canguilhem, B., and Pévet, P. The photoperiodic response in Syrian hamster depends upon a melatonin-driven circadian rhythm of sensitivity to melatonin. J. Neuroendocrinol. 7:889–895, 1995.
57. Pittendrigh, C.S. Circadian systems: entrainment. In Biological Rhythms. Handbook of Behavioral Neurobiology, J. Aschof, ed., New York, Plenum 4:95–124, 1981.
58. Pittendrigh, C.S. and Daan, S. A functional analysis of circadian pacemakers in nocturnal rodents. J. Comp. Physiol. *A*, 106:333–355, 1976.
59. Reiter, R.J. The pineal and its hormone in the control of reproduction in mammals. Endocrine Rev. 1:109–131, 1980.
60. Romijn, H.J., Slurter, D.A., Pool, C.W., Wortel, J., and Buijs, R.M. Differences in colocalization between Fos and PHI, GRP, VIP and VP in neurons of the rat suprachiasmatic nucleus after a light stimulus during the phase delay versus the phase advance period of the night. J. Comp. Neurology 372:1–8, 1998.
61. Rowsemitt, C.N., Petterborg, L.J., Claypool, L.E., Hoppensteadt, F.C., Negus, N.C., and Berger, P.J. Photoperiodic induction of diurnal locomotor activity in *Microtus montanus*, the montanus vole. Can. J. Zool. 60:2798–2803, 1982.
62. Rusak, B. The mammalian circadian system: models and physiology. J. Biol. Rhythm 4(42):121–134, 1989.
63. Rusak, B., Robertson, H.A., Wisden, W., and Hund, S.P. Light pulses that shift rhythms induce gene expression in the suprachiasmatic nucleus. Science 248:1237–1240, 1990.
64. Rusak, B. and Yu, G.D. Regulation of melatonin-sensitivity and firing-rate rhythms of hamster suprachiasmatic nucleus: pinealectomy effects. Brain Res. 602:200–204, 1993.
65. Shinohara, K. and Inouye, S.I.T. Photic information coded by vasoactive intestinal polypeptide and neuropeptide Y. Neurosci. and Biobehavioral Rev. 19:349–352, 1995.
66. Shinohara, K., Tominaga, K., Isobe, Y., and Inouye, S.I.T. Photic regulation of peptides located in the ventrolateral subdivision of the suprachiasmatic nucleus of the rat: daily variations of vasoactive intestinal polypeptide, gastrin-releasing peptide, and neuropeptide Y. Neurosci. 13:793–800, 1993.
67. Simerly, R.B. and Swanson, L.W. Castration reversibly alters levels of cholexystokinin immunoreactivity within cells of three interconnected sexually dimorphic forebrain nuclei in rat. Proc. Natl. Acad. Sci. USA 84:2087–2091, 1987.
68. Stankov, B., Fraschini, F., and Reiter, R.J. Melatonin binding sites in the central nervous system. Brain Research Rev. 16:245–256, 1991.
69. Stehle, J.H., Foulkes, N.S., Molina, C., Simonneaux, V., Pévet, P., and Sassone-Corsi, P. Adrenergic signals direct rhythmic expression of a transcriptional repressor CREM in the pineal gland. Nature 365:314–321, 1993.
70. Stehle, J., Vanecek, J., and Vollrath, L. Effects of melatonin on spontaneous electrical activity of neurons in rat suprachiasmatic nuclei: an in vitro iontophoretic study. J. Neural Transm 78:173–177, 1989.
71. Sumova, A., Travnickova, Z., Peters, R., Swhartz, W.J., and Illnerovà, H. The rat suprachiasmatic nucleus is a clock for all seasons. Proc. Natl. Acad. Sci. USA 92:7754–7758, 1995.
72. Sumova, A. and Illnerovà, H. Endogenous melatonin signal does not mediate the effect of photoperiod on the rat suprachiasmatic nucleus. Brain Res. 725:281–283, 1996.
73. Takahashi, J.S., De Coursey, P.J., Bauman, L., and Menaker, M. Spectral sensitivity of a novel photoreceptive system mediating entrainment of mammalian circadian rhythms. Nature 308:186–188, 1984.

74. Teclemariam-Mesbah, R., Vuillez, P., Van Rossum, A., and Pévet, P. Time course of neuronal sensitivity to light in the circadian timing system of the golden hamster. Neurosci. Lett. 201:5–8, 1995.
75. Turek, F.W. and Gwinner, E. Role of hormones in the circadian organization of vertebrates. In Vertebrate circadian system, Structure and Physiology, J. Aschoff, S. Daan, G. Groos, eds, Springer—Berlin, 173–182, 1982.
76. Turek, F.W. and Van Cauter E. Rhythms in Reproduction. In the Physiology of Reproduction, E. Knobil, J.D. Neil, eds, Raven Press, Ltd—New York, 487–540, 1994.
77. Vivien-Roels, B., Pitrosky, B., Zitouni, M., Malan, A., Canguilhem, B., Bonn, D., and Pévet, P. Environmental control of the seasonal variations in the daily pattern of melatonin synthesis in the European hamster (*Cricetus cricetus*). General and Comparative Endocrinology 106:85–94, 1997.
78. Vuillez, P., Jacob, N., Teclemariam-Mesbah, R., and Pévet, P. In Syrian and European hamsters, the duration of the sensitive period to light of the suprachiasmatic nuclei depends on the photoperiod. Neurosci. Lett. 208:37–40, 1996.
79. Weaver, D.R., Rivkees, S.A., and Reppert S.N. Localization and characterization of melatonin receptor in rodent brain by in vivo autoradiography. J. Neurosc. 9:2581–2590, 1989.
80. Wollnick, F., Breit, A., and Reinke, D. Seasonal change in the temporal organization of wheel-running activity of the European hamster, *Cricetus cricetus*. Naturwissenschaften 78:419–422, 1991.
81. Zhang, D.X. and Rusak B. Photic sensitivity of geniculate neurons that project to the suprachiasmatic nuclei or contralateral geniculate. Brain Res. 504:161–164, 1989.

27

COMPARISON OF THE PINEAL AND SCN RHYTHMICITY

Effect of Photic and Non-Photic Stimuli, Photoperiod, and Age

Helena Illnerová, Zdeňka Trávníčková, Martin Jáč, and Alena Sumová

Institute of Physiology
Academy of Sciences of the Czech Republic
142 20 Prague 4, Czech Republic

1. INTRODUCTION

In the rat, the rhythmic pineal melatonin production is driven by the circadian rhythm in N-acetyltransferase (NAT; arylalkylamine N-acetyltransferase; EC 2.3.1.87) (18,22). The NAT rhythm, similarly as other overt rhythms, e.g. in locomotor activity, is controlled by a circadian pacemaker located in the suprachiasmatic nucleus (SCN) of the hypothalamus (21). On the bases of both morphological and physiological characteristics, the mammalian SCN can be divided into the ventrolateral (VL) and dorsomedial (DM) part (38).

The VL-SCN receives photic input directly from the retina, via the retinohypothalamic tract, and indirectly via the intergeniculate leaflet of the lateral geniculate body (4,27). Following a light stimulus, in addition to other processes, immediate early genes c-fos and jun-B are transcriptionaly activated, mostly in the VL-SCN (23,30). These genes are believed to function in coupling short-term signals elicited by extracellular events to long-term changes in cellular phenotype by mediating subsequent changes in gene expression (5). Importantly, light induces c-fos and jun-B mRNA expression and elevates c-Fos and Jun-B proteins in the mammalian SCN only during the subjective night, when it also phase shifts circadian rhythmicity (23,29,30,33,36). The rhythm in the light-induced c-fos and jun-B mRNA and c-Fos and Jun-B protein, thus represents an endogenous rhythm in SCN sensitivity to light.

The DM-SCN receives fewer retinal afferents and contains many arginine vasopressin-immunoreactive cells (19). Recently, with a highly sensitive antibody, we described a circadian rhythm in the SCN c-Fos immunoreactivity in darkness, with the

Melatonin after Four Decades, edited by James Olcese.
Kluwer Academic / Plenum Publishers, New York, 2000.

maximum in the morning and trough during the subjective night (35). The spontaneous rhythmic c-Fos induction occurred mostly in the DM-SCN and might indicate an elevated dorsomedial neuronal activity in the early subjective day.

Rhythmic processes in the pineal, though they are controlled by the SCN, might be also partly independent of the SCN. It has been suggested that the pineal itself may have residual clock properties and affect via the cyclic AMP-responsive element modulator (CREM) its rhythmic melatonin production (3). In addition, independent of the SCN, the pineal melatonin might be directly affected by light (10) and via highly sensitive SCN melatonin receptors (40) might itself modulate the SCN rhythmicity.

The present study was undertaken to find out whether and how the rhythmic melatonin production reflects the intrinsic SCN rhythmicity or whether it is partly independent of the rhythmicity. To elucidate this question, the pineal NAT rhythm under various conditions and following various stimuli was compared with the rhythm in the light-induced c-fos mRNA and c-Fos protein which was present mostly in the VL-SCN, and with the spontaneous rhythmic c-Fos induction in darkness which occurred predominantly in the DM-SCN.

2. COMPARISON OF CHANGES OF THE PINEAL NAT RHYTHM WITH CHANGES OF THE SCN RHYTHM IN c-FOS PHOTOINDUCTION

2.1. Rhythm in c-Fos Photoinduction

Induction of c-Fos protein by a 30-min light pulse was followed by the immunocytochemical method, with the primary antiserum generated against the amino acids 2–17 of the N-terminal peptide sequence of c-Fos; the antiserum was kindly provided by D. Hancock, Imperial Cancer Research Fund, London, and generously supplied by M. Hastings (University of Cambridge, U.K.) (33,34,36). Eventually, c-fos mRNA was measured by the *in situ* hybridization method (36).

Rats were maintained under a light-dark regime with 12 h of light (06 h to 18 h) and 12 h of darkness (18 h to 06 h) per day (LD 12:12), unless indicated otherwise, then they were released into darkness and the next night c-Fos photoinduction was followed. After a light pulse at night, c-Fos immunopositive cells were present mostly in the VL-SCN (Figure 1) (35,36). The SCN rhythm in the light-induced c-Fos immunoreactivity had two well defined phase markers, namely the time of the evening rise and the time of the morning decline (Figure 2A) (33). Similarly, the pineal NAT rhythm had also two markers, namely the time of the evening NAT rise and the time of the morning decline (Figure 2B) (9,12,16). Whereas the pineal NAT rise laged by about 2 h the SCN rise in c-Fos photoinduction, the morning decline in both the SCN and pineal variables occurred at about the same time. The gradual evening rise in the number of the light-induced c-Fos immunopositive cells indicates that in the course of the early night more and more SCN neurons begin to be in such a phase that they respond to light.

2.2. Photic Resetting of the SCN Rhythm in c-Fos Photoinduction and of the Pineal NAT Rhythm

When rats were exposed to a light stimulus in the early night and then released into darkness, the next day both the SCN rhythm in c-Fos photoinduction and the pineal

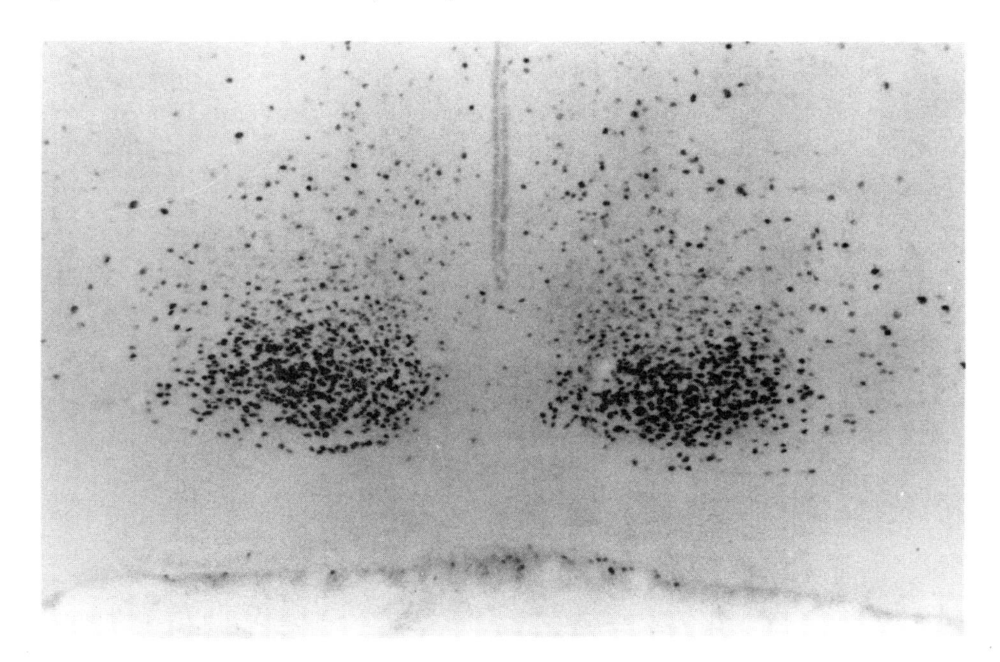

Figure 1. Example of the light-evoked induction of the immunoreactive c-Fos protein in the suprachiasmatic nucleus. A rat was exposed to a 30-min light pulse at 21 h in the night, then returned to darkness and killed 30 min later. Note that the photic c-Fos induction occurred predominantly in the VL-SCN. Adapted from Sumová et al., 1998 (35).

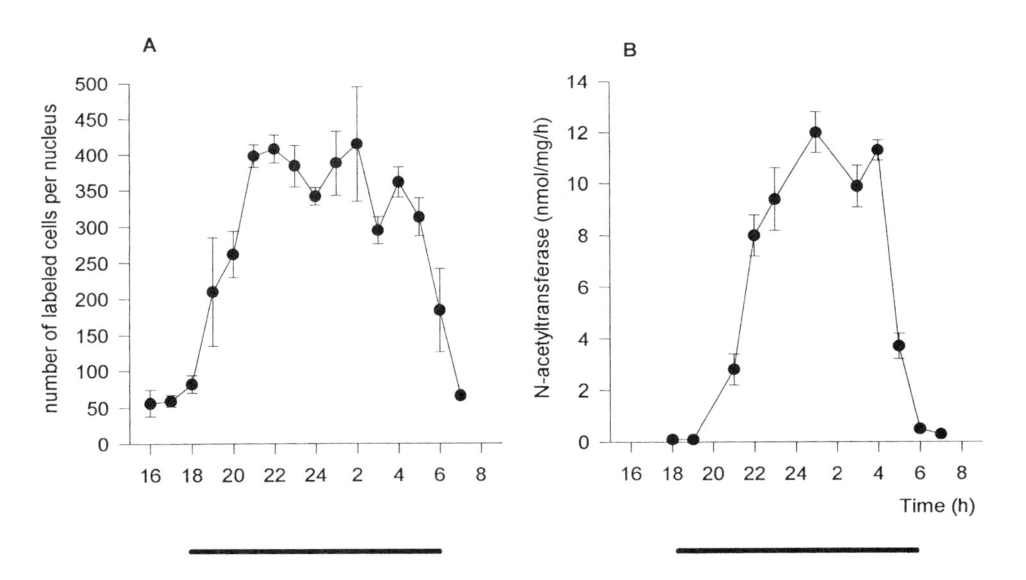

Figure 2. The SCN rhythm in the light-induced c-Fos immunoreactivity (A) and the pineal rhythm in N-acetyltransferase (B). Rats maintained in LD 12:12 were released into darkness and the next day the SCN and the pineal rhythms were followed. Full bars indicate original dark periods. Data from Sumová and Illnerová, 1998 (33) and from Illnerová and Vaněček, 1987 (16), respectively.

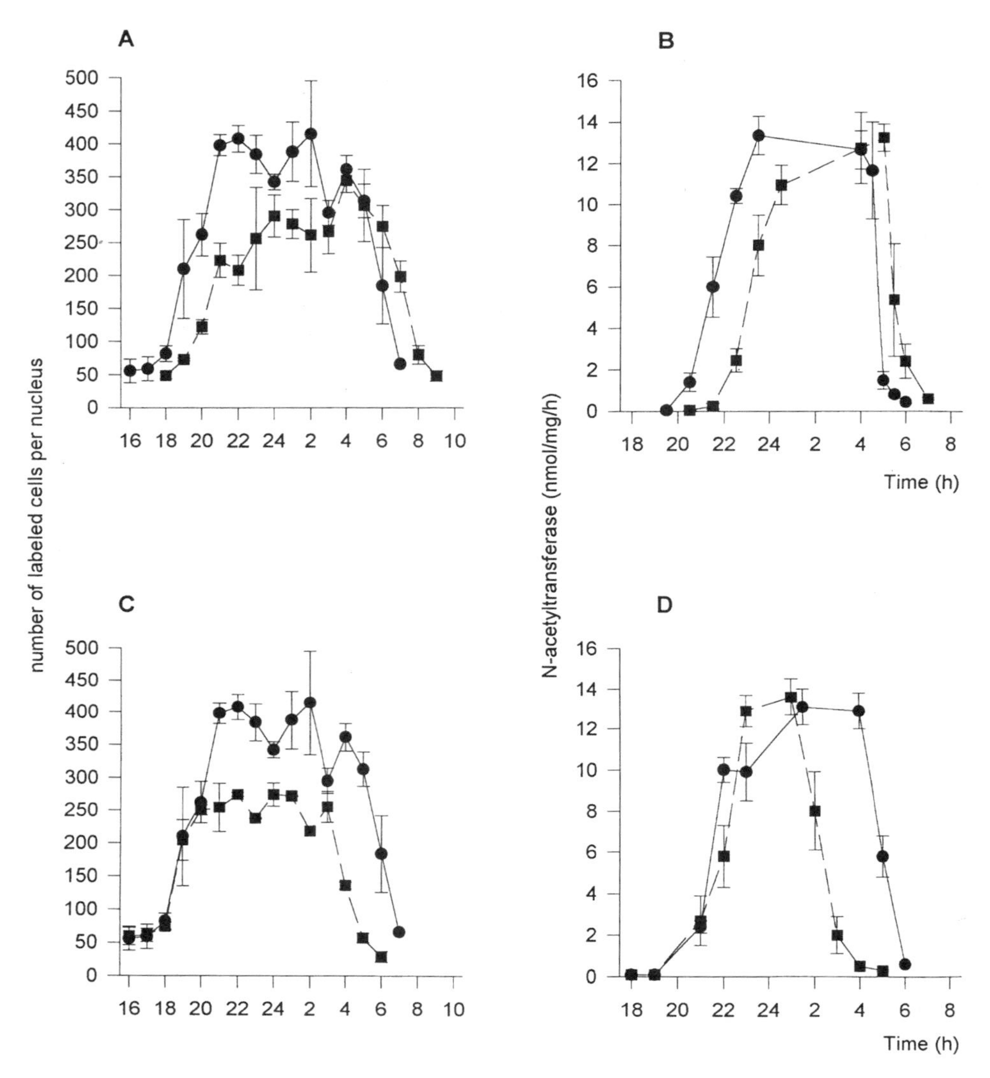

Figure 3. Phase delays of the SCN rhythm in the light-induced c-Fos immunoreactivity (A) and of the pineal N-acetyltransferase rhythm (B) following a photic stimulus in the early night and phase-advances of the SCN (C) and of the pineal rhythm (D) following a photic stimulus in the late night. Rats maintained in LD 12:12 were exposed to a light-stimulus (squares) or left untreated (circles), then they were released into darkness and the next day the SCN and pineal rhythms were followed. Rats were exposed to a 1-h light pulse at 23 h (A), to a 1-min pulse at 22 h (B), to a 1-h light pulse at 02 h (C) and to a 1-min pulse at 03 h (D), respectively. Data from Sumová and Illnerová, 1998 (33) and from Illnerová and Vaněček, 1987 (16), respectively.

NAT rhythms were phase-delayed as compared with the rhythm profiles in control, unexposed animals; in both rhythms, the evening marker was delayed to a larger extent than the morning one (Figure 3A,B) (9,16,17,33). However, when such delays of both rhythms were studied under an extremely long photoperiod, LD 18:6, the morning markers were phase delayed more than the evening ones, due apparently to the state of the underlying pacemaking system (7,33,37). When rats were exposed to a light stimulus in the late night and then released to darkness, the next day, during the first transient cycle, only the morning decline in the SCN c-Fos photoinduction and in the pineal NAT, respectively, was phase advanced, as compared with the decline in control rats, but not the evening rise (Figure 3C,D) (16,17,19,33). Apparently, the evening and the morning SCN and pineal markers do not necessarily phase shift in parallel. The finding suggests a complex nature of the underlying SCN pacemaker where groups of neurons might be first reset together and via coupling might entrain other groups (25,42). The evening NAT rise started to be phase advanced only within four days following a late night light stimulus, and even then to a still lesser extent than the morning decline (16).

2.3. Non-Photic Resetting of the SCN Rhythm in c-Fos Photoinduction and of the Pineal NAT Rhythm by Melatonin

A single evening melatonin administration phase-advanced instantaneously the evening rise in the light-induced SCN c-Fos immunoreactivity (31). Similarly, a single melatonin administration before the evening dark onset phase-advanced instantaneously the evening pineal NAT rise (6). The magnitude of phase shifts of the intrinsic SCN rhythmicity induced by melatonin administration *in vivo* was similar to the magnitude of phase-shifts of the pineal NAT rhythm and was less than half of the magnitude attained *in vitro* experiments following melatonin application to the rat SCN slices during late subjective day (6,26,31).

Importantly, melatonin administration in the late day phase-advanced just the evening NAT rise, but not the morning decline (6). Recently, it has been reported that daily melatonin administration at the time of the former dark onset keeps the rhythm in the pineal melatonin production entrained to the 24-h day just for a few weeks after a release of rats from a LD cycle to constant darkness; thereafter, the whole rhythm begins to free-run (1). But at first, shortly after the release from the LD cycle, it is the morning melatonin decline which starts to free-run, with the ensuing extension of the melatonin signal duration, and only then the entrainment of the whole rhythm breaks. Altogether, the data suggest that melatonin administered in late day entrains primarily an evening component of circadian rhythmicity.

2.4. Effect of Photoperiod

The finding that an early night light stimulus phase delays primarily the evening marker of the SCN and pineal rhythms and a late night light stimulus phase advances primarily the morning marker of both rhythms suggests that on long days light perturbing into the late evening and early morning hours may compress the waveform of the SCN and pineal rhythms; on short days, the waveform may decompress. This actually happens. The interval between the evening rise in c-fos photoinduction and the morning decline under unmasked conditions in darkness as well as that between the evening NAT rise and the morning decline were by about 5 h longer under a short, LD 8:16 photoperiod, than under a long, LD 16:8 photoperiod (Figure 4A,B) (8,9,12,36). In both rhythms, the interval under the short photoperiod was extended assymetricaly, into the morning hours. This indicates a more important role of the morning than of the evening light in entrainment of the rat circadian pacemaking system (9,15). Similarly, in Syrian and European hamsters, the interval between the evening rise and the morning decline in the SCN c-fos photoinduction is also longer on short than on long days (41). Importantly, the photoperiod affected the waveform of the SCN rhythm in

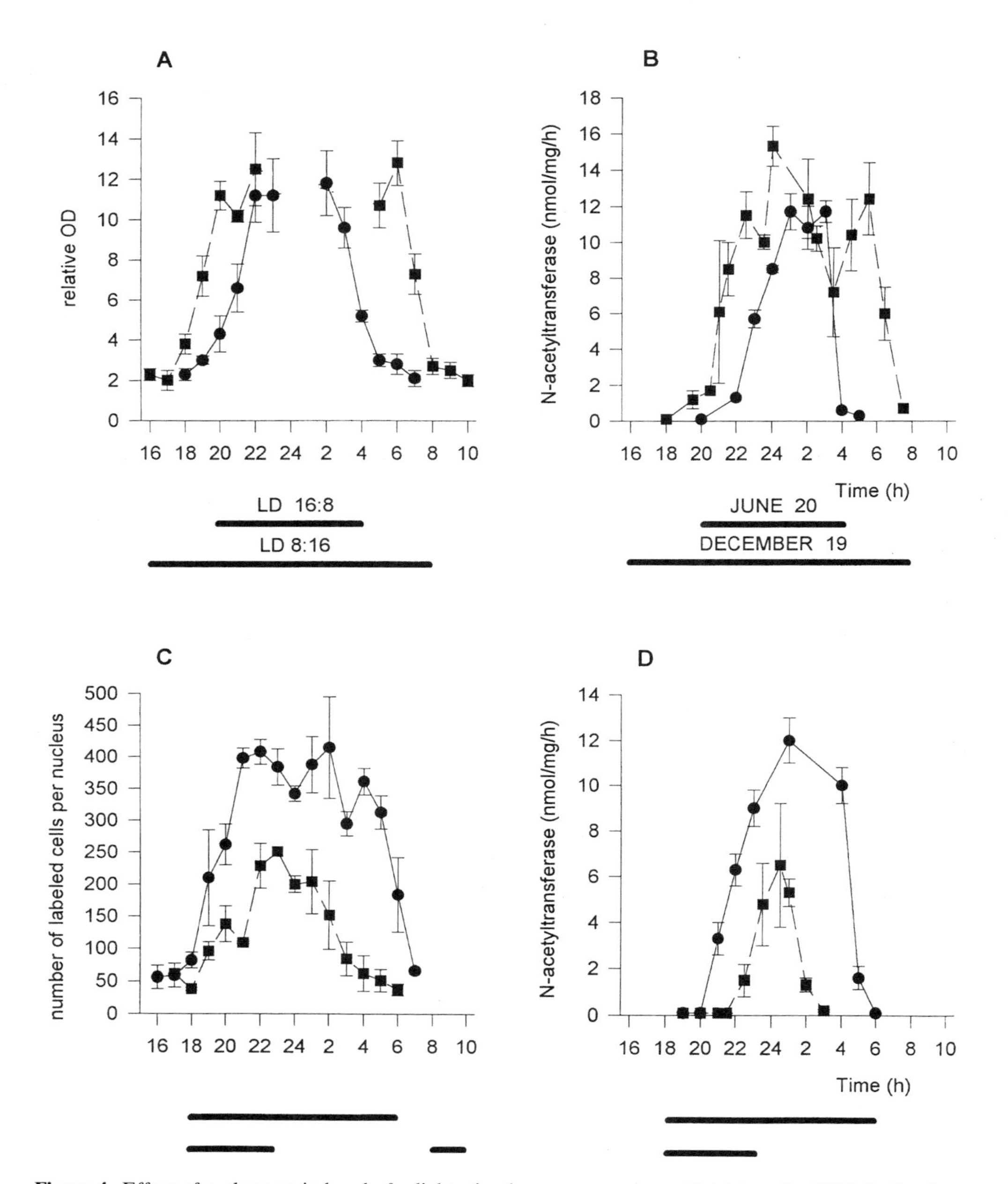

Figure 4. Effect of a photoperiod and of a light stimulus encompassing midnight on the SCN rhythm in c-Fos photoinduction and on the pineal N-acetyltransferase rhythm, respectively. A: Rats were maintained in LD 16:8 (circles) or LD 8:16 (squares) and the SCN rhythm in photic induction of c-fos mRNA (in situ hybridization) was followed in darkness. Full bars indicate original dark periods. B: rats were maintained in natural daylight and the pineal N-acetyltransferase rhythm was followed on June 20 (circles) or on December 19 (squares). Full bars indicate dark periods. C: Rats maintained in LD 12:12 were exposed to a light stimulus from 23 h till 08 h (squares) or left untreated (circles), then released into darkness and the next day the rhythm in the light-induced c-Fos immunoreactivity was followed. Full bar indicates original dark period and the dark period during the night when rats were exposed to a long light stimulus, respectively. D: Rats maintained in LD 12:12 were exposed to bringing forward the light onset to 23 h (squares) or left untreated (circles); the next day, they experienced darkness already since 14 h and the pineal N-acetyltransferase rhythm was followed. Full bar indicates original dark period and the dark period during the night when the light onset was brought forward to before midnight, respectively. Data from Sumová et al., 1995 (36) (A), Illnerová and Vaněček, 1980 (12) (B), Sumová and Illnerová, 1998 (33) (C) and Illnerová and Vaněček, 1987 (17) (D).

the light-induced c-Fos immunoreactivity directly, and not via the pineal melatonin (32).

A long light stimulus encompassing the middle of the night compressed the waveform of the SCN rhythm in c-Fos photoinduction (Figure 4C) and of the pineal NAT rhythm (Figure 4D) in a manner similar to the effect of a long photoperiod, i.e., by phase delaying the evening marker of both rhythms and phase advancing the morning one (9,17,33). When rats were maintained under an extremely long, LD 18:6 photoperiod, even a 5-min or a shorter pulse had such an effect, i.e., it phase-delayed the evening marker, phase-advanced the morning one and further compressed the SCN and the pineal rhythm waveform (15,33).

When rats were transferred from a long, LD 16:8, photoperiod to a short, LD 8:16, photoperiod, the waveform of the SCN rhythm in c-Fos photoinduction (Figure 5A), as well as that of the pineal NAT rhythm (Figure 5B) extended just gradually and it took two weeks before the full extension was achieved (11,34). However, when rats were transferred from a short to a long photoperiod, compression of the interval enabling high SCN c-Fos photoinduction occurred within three days (Figure 5C) (34). It appears that the memory on long but not on short days is stored in the SCN itself.

2.5. Effect of Age

In newborn rats, a light stimulus at night induces already a response in c-Fos immunoreactivity (24). In 3-day-old rats, the interval enabling high c-Fos photoinduction as well as the interval of elevated NAT activity were considerably extended as compared with the SCN and pineal rhythm profiles in adult animals (Trávníčková, Sumová and Illnerová, in preparation). This extension might indicate a loose coupling among groups of the SCN neurons in newborn rats. In 10-day-old animals maintained under a long photoperiod, the SCN rhythm in c-Fos photoinduction was not yet fully entrained to the photoperiod; consequently, the interval enabling high c-Fos photoinduction under the long photoperiod did not differ significantly from that under a short photoperiod (Trávníčková, Sumová and Illnerová, in preparation). Similarly, the pineal NAT and melatonin rhythms are not fully synchronized with a long photoperiod until the age of 12 to 15 days (28,29).

3. MODULATION OF THE SCN RHYTHM IN THE SPONTANEOUS c-FOS INDUCTION IN DARKNESS

3.1. Endogenous Rhythm of c-Fos Immunoreactivity

c-Fos immunoreactivity was determined with a highly sensitive antiserum generated against the amino acids 2–17 of the N-terminal peptide of c-Fos and characterized elsewhere (43); the antiserum was kindly provided by J.D. Mikkelsen, H. Lundbeck, Copenhagen. The spontaneously c-Fos immunopositive cells were present mostly in the DM-SCN (Figure 6) (35).

In rats maintained in LD 12:12 and released to darkness, c-Fos immunoreactivity in the DM-SCN peaked in the early subjective day and then slowly declined; the decrease to low nighttime levels occurred after the expected dark onset (Figure 7). Before the expected light onset, c-Fos started spontaneously to increase.

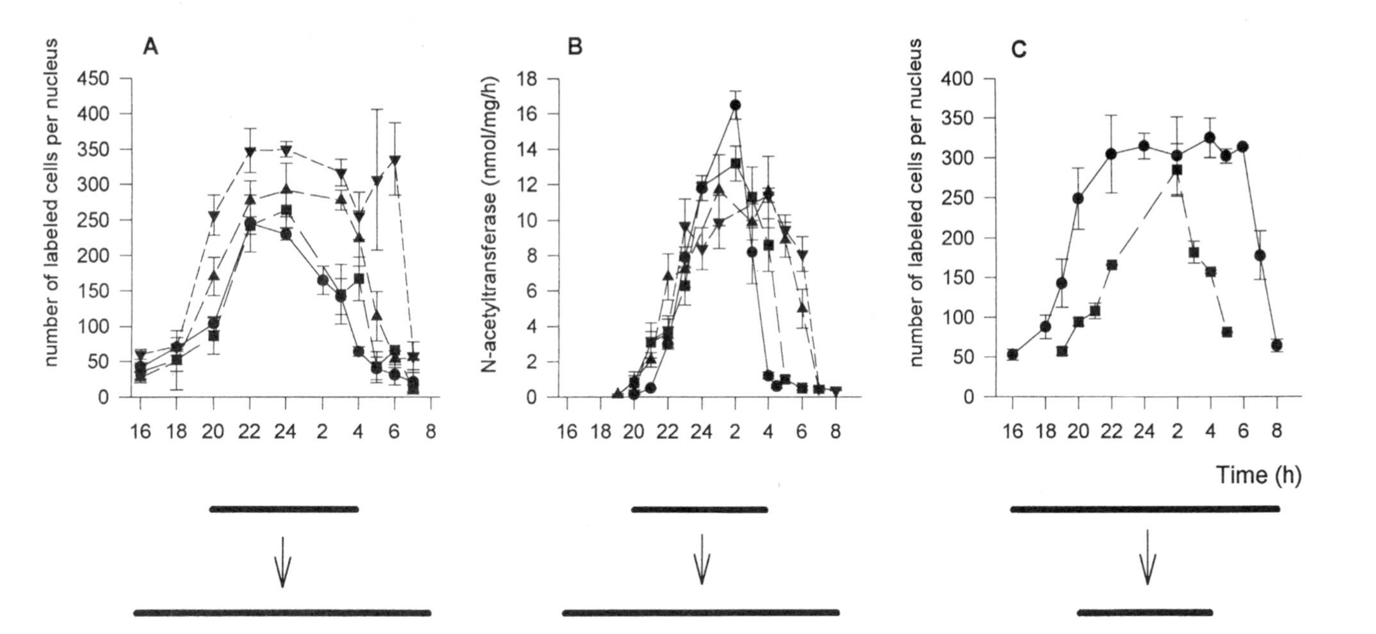

Figure 5. Adjustment of the SCN rhythm in the light-induced c-Fos immunoreactivity and of the pineal N-acetyltransferase rhythm to a change from a long to a short photoperiod and vice versa. A and B: Rats maintained in LD 16:8 (circles) were transferred to LD 8:16 and the SCN (A) and the pineal (B) rhythm were followed after 3 (squares), 6 (up-triangles) and 13 (down-triangles) days, respectively. C: Rats maintained in LD 8:16 (circles) were transferred to LD 16:8 and c-Fos photoinduction under unmashed conditions, i.e. under a 4-h advance of the dark onset and a 4-h delay of the light onset, respectively, was measured after 3 days (squares). Full bars indicate dark periods. Data from Sumová et al., 1996 (34) (A,C) and Illnerová et al., 1986 (11) (B).

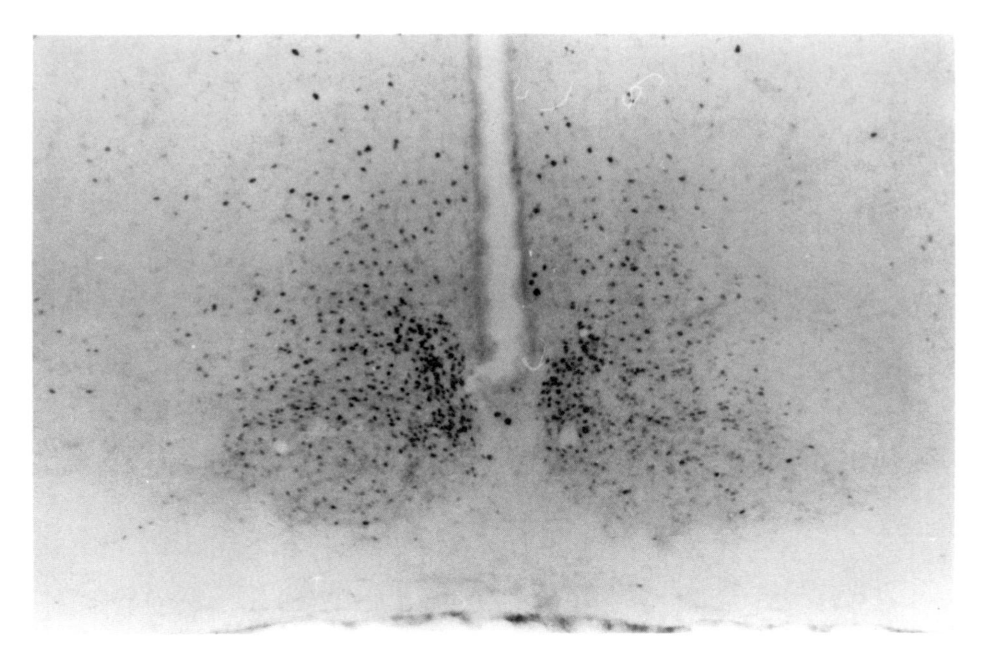

Figure 6. Example of the spontaneous induction of the immunoreactive c-Fos protein in the suprachiasmatic nucleus. A rat was manitained in LD 12:12. In the morning, light was not turned on at 06h as usual and the rat was sacrificed in darkness at 10h. Note, that the spontaneous c-Fos induction occurred mostly in the DM-SCN. Adapted from Sumová et al., 1998 (35).

3.2. Effect of Photoperiod

The interval of the spontaneous c-Fos induction in the SCN, namely in the DM-SCN, was longer in rats released into darkness from a long photoperiod than in those released into darkness from a short photoperiod (Sumová, Trávníčková and Illnerová, in preparation). In other words, low nighttime c-Fos immunoreactivity lasted for a shorter time under the long than under the short photoperiod. Hence even the spontaneous rhythm of c-Fos in the SCN was photoperiod dependent. Under the long photoperiod as well as under the short one, the morning rise in c-Fos immunoreactivity occurred before the expected light onset and was locked to the morning light whereas the evening decline occurred at about the same time under both photoperiods. Hence it appears that not just the rhythm in the light-induced c-Fos immunoreactivity, but also the rhythm in the spontaneous c-Fos immunoreactivity, are entrained mostly by the morning light. When the rhythm of c-Fos immunoreactivity in the DM-SCN was followed directly under the long and under the short photoperiod, respectively, without avoiding the masking effect of the photoperiods, the rhythm was again markedly photoperiod-dependent (Jáč, Sumová and Illnerová, in preparation).

3.3. Effect of Age

In 3-day-old rats maintained under a long and a short photoperiod, respectively, and then released to darkness, there was a marked rhythm in the spontaneous c-Fos immunoreactivity in the SCN, however c-Fos was elevated for a considerably shorter

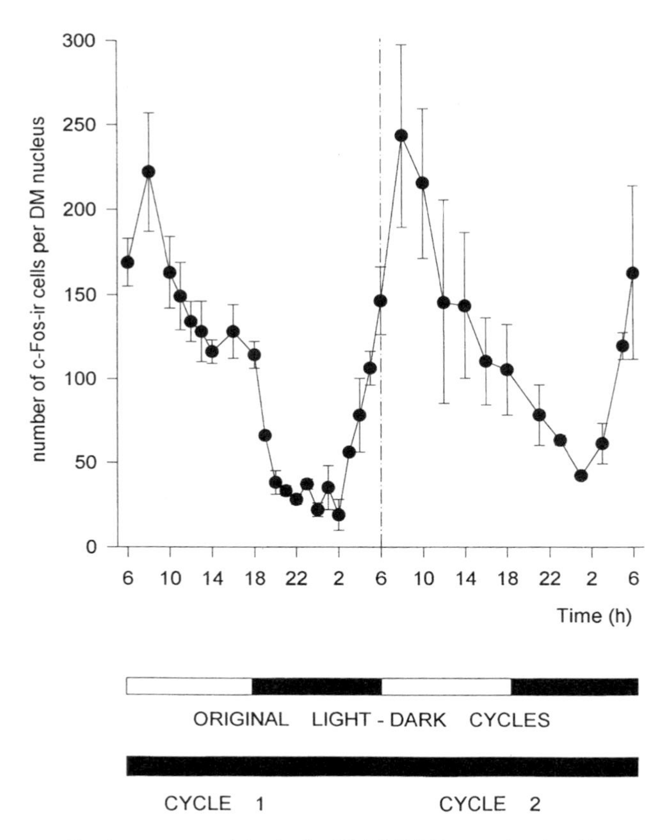

Figure 7. Spontaneous c-Fos immunoreactivity in the DM-SCN. Rats maintained in LD 12:12 were released into constant darkness at the time of the usual light onset at 06 h and Fos-immunoreactivity was followed in darkness for the next two cycles. Full bars denote dark periods. Adapted from Sumová et al., 1998 (35).

time as compared with the elevation in adult animals. This finding indicates a longer duration of subjective night in newborn animals (Trávníčková, Sumová and Illnerová, in preparation). In 10-day-old rats, the spontaneous rhythm in Fos immunoreactivity was present in rats released to darkness from the long as well from the short photoperiod, however there was still no difference in duration of the elevated c-Fos immunoreactivity between these two photoperiods.

4. GENERAL DISCUSSION

A striking similarity exists between resetting of the SCN rhythm in c-Fos photoinduction and resetting of the pineal NAT rhythm. Following photic stimuli in the early, middle and late night, respectively, both the SCN and the pineal rhythms phase shift in a similar way during the first transient cycle. Importantly, the evening markers of both rhythms do not necessarily phase shift in parallel with the morning ones, suggesting a complex nature of the underlying pacemaking system (9,13). Following melatonin administration in late day, the evening rise in the SCN c-Fos photoinduction as well as the pineal NAT rise are phase advanced instantaneously, by about the same amount.

A striking similarity exists also in the response of the SCN rhythm in c-Fos photoinduction and in the response of the pineal NAT rhythm to the photoperiod. Both rhythms are photoperiod-dependent: under a long photoperiod, the interval enabling high c-Fos photoinduction as well as the interval of elevated NAT activity are short and under a short photoperiod they are long. Under LD 16:8, LD 12:12 and LD 8:16, respectively, the morning decline in the light-induced c-Fos occurs at about the same time as that in the NAT activity, whereas the evening NAT rise occurs by 1 to 2 h later than the rise in c-Fos photoinduction. The 1 to 2 h delay in the NAT rise may be explained by the time interval necessary for the NAT mRNA and protein formation (20): following administration of isoproterenol, a beta adrenergic agonist, it takes 1 to 2 h before the NAT activity markedly increases (Figure 8) (14). The gradual evening rise in the number of the light-induced c-Fos immunopositive cells suggests that more and more SCN neurons start to be in a light-responsive phase, and towards the middle of the night all cells capable of c-Fos photoinduction may respond. Similarly, strength of the signal coming from the SCN into the pineal may gradually increase; under an extremely long, LD 20:4 photoperiod, NAT activity rises as rapidly as after isoproterenol administration (Figure 8) (14).

After transition of rats from a long to a short photoperiod, decompression of the waveform of the SCN rhythm in c-Fos photoinduction as well as of the pineal NAT rhythm waveform proceeds just gradually and is roughly completed within two weeks whereas compression of the SCN rhythm after a change from a short to a long photoperiod is roughly achieved already within three days. Memory on long days stored in

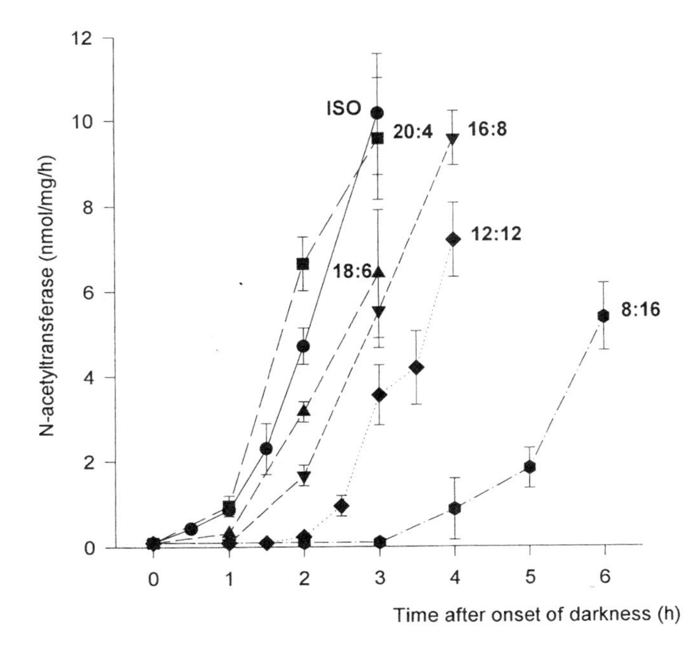

Figure 8. The evening rise in the pineal N-acetyltransferase activity under various photoperiods and following isoprotenerol administration. Rats were maintained in LD 20:4 (squares), LD 18:6 (up-triangles), LD 16:8 (down-triangles), LD 12:12 (rhombs) and LD 8:16 (hexagons), respectively, and killed at various times after the onset of darkness. Isoproterenol (0.5 mg/kg) was administered to rats maintained in LD 12:12 at the time of the usual dark onset; the rats were left in light and killed at various times after isoproterenol administration (circles). Data from Illnerová and Vaněček, 1983 (14).

the SCN pacemaking system may explain "the carry-over phenomenon" when infrequent long day treatment induces long day responses (2).

And finally, a striking similarity exists also between the SCN rhythm in c-Fos photoinduction and the pineal NAT rhythm during ontogenesis. In newborn rats, the interval enabling high c-Fos photoinduction in the SCN as well as the interval of elevated NAT activity in the pineal are markedly extended as compared with those in adult animals. In 10-day-old rats, though the intervals are already reduced, the SCN as well as the pineal rhythm are not yet fully photoperiod-dependent.

The aforementioned data indicate that changes in the pineal NAT rhythm reflect mostly changes of the intrinsic rhythmicity in the VL-SCN. Though we do not have yet enough data which would allow us to compare in detail the rhythm in the spotaneous c-Fos immunoreactivity, which appears mostly in the DM-SCN, with the pineal NAT rhythm, preliminary results suggest that even that rhythm shows similarity with the pineal rhythmicity. First, the rhythm in the spontaneous c-Fos induction in the DM-SCN is photoperiod-dependent similarly as the NAT rhythm. The interval of low c-Fos immunoreactivity which indicates low neuronal activity and at the same time the subjective night, is shorter on long than on short days, similarly as the interval of elevated NAT activity. And second, in newborn rats, the period of a low spontaneous c-Fos immunoreactivity is considerably longer than in adult animals, similarly as the interval of elevated NAT activity is longer in newborn than in adult rats.

5. CONCLUSIONS

The aforementioned data indicate that changes of the rat pineal NAT and hence also of the melatonin rhythm reflect mostly changes of the intrinsic SCN rhythmicity. This holds true for the VL-SCN rhythmicity and it might hold true also for the DM-SCN rhythmicity. If there is an intrinsic time-keeping mechanism in the rat pineal, it is only a marginal one.

ACKNOWLEDGMENTS

This work was supported by the Grant Agency of the Czech Republic, Grant 309-970-512 and by the Grant Agency of the Academy of Sciences of the Czech Republic Grant A-7011508.

REFERENCES

1. Drijhout W.J. 1996 Melatonin-on-line. FEBO Drug BV, Enschede, The Netherlands, pp. 1–226.
2. Ellis D.H. and Follet B.K. 1983 Gonadotropin secretion and testicular function in golden hamsters exposed to skeleton photoperiods with ultrashort light pulses. Biol. Reprod. 29:805–818.
3. Foulkes N.S., Borjigin J., Snyder S., and Sassone-Corsi P. 1996 Transcriptional control of circadian hormone synthesis via the CREM feedback loop. Proc. Natl. Acad. Sci. USA 93:14140–14145.
4. Harrington M.E. and Rusak B. 1986 Lesions of the thalamic intergeniculate leaflet after hamster circadian rhythms. J. Biol. Rhythms 1:309–325.
5. Hughes P. and Dragunow M. 1995 Induction of immediate-early genes and the control of neurotransmitter-regulated gene expression within the nervous system. Pharmacol. Rev. 47:133–178.

6. Humlová M. and Illnerová H. 1990 Melatonin entrains the circadian rhythm in the rat pineal N-acetyltransferase activity. Neuroendocrinology 52:196–199.
7. Humlová M. and Illnerová H. 1992 Entrainment of the circadian clock controlling the pineal N-acetyltransferase rhythm depends on photoperiod. Brain Res. 584:226–236.
8. Illnerová H. 1988 Entrainment of mammalian circadian rhythms in melatonin production by light. Pineal Res. Rev. 6:173–217.
9. Illnerová H. 1991 The suprachiasmatic nucleus and rhythmic pineal melatonin production. In Suprachiasmatic Nucleus. The Mind's Clock, edited by DC Klein, RY Moore and SM Reppert, Oxford University Press, New York, pp. 197–216.
10. Illnerová H., Bäckstrom M., Sääf J., Wetterberg L., and Vangbo B. 1978 Melatonin in the rat pineal gland and serum: Rapid parallel decline after light exposure at night. Neurosci. Lett. 3:189–193.
11. Illnerová H., Hoffmann K., and Vaněček J. 1986 Adjustment of the rat pineal N-acetyltransferase rhythm to change from long to short photoperiod depends on the direction of extension of the dark period. Brain Res. 362:403–408.
12. Illnerová H. and Vaněček J. 1980 Pineal rhythm in N-acetyltrasferase activity in rats under different artificial photoperiods and in natural daylight in the course of a year. Neuroendocrinology 31:321–326.
13. Illnerová H. and Vaněček J. 1982 Two-oscillator structure of the pacemaker controlling the circadian rhythm of N-acetyltransferase in the rat pineal gland. J. Comp. Physiol. 145:539–548.
14. Illnerová H. and Vaněček J. 1983 The evening rise in the rat pineal N-acetyltransferase activity under various photoperiods. Neurosci. Lett. 36:279–284.
15. Illnerová H. and Vaněček J. 1985 Entrainment of the circadian rhythm in rat pineal N-acetyltransferase activity under extremely long and short photoperiods. J. Pineal Res. 2:67–78.
16. Illnerová H. and Vaněček J. 1987 Dynamics of discrete entrainment of the circadian rhythm in the rat pineal N-acetyltransferase activity during transient cycles. J. Biol. Rhythms 2:95–108.
17. Illnerová H. and Vaněček J. 1987 Entrainment of the circadian rhythm in the rat pineal N-acetyltransferase activity by prolonged periods of light. J. Comp. Physiol. A161:495–510.
18. Illnerová H., Vaněček J., and Hoffmann K. 1983 Regulation of the pineal melatonin concentration in the rat (*Ratus norvegicus*) and in the Djungarian hamster (*Phodopus sungorus*). Comp. Biochem. Physiol. 73:155–159.
19. Inouye S.T. and Shibata S. 1994 Neurochemical organization of circadian rhythm in the suprachiasmatic nucleus. Neurosci. Res. 20:109–130.
20. Klein D.C., Coon S.L., Roseboom P.H., Weller J.L., Bernard M., Gastel J.A., Zatz M., Iuvone P.M., Rodriguez I.R., Bégay V., Falcon J., Cahill G.M., Cassone V.M., and Baler R. 1997 The melatonin rhythm-generating enzyme: molecular regulation of serotonin N-acetyltransferase in the pineal gland. Recent Progr. in Hormone Res. 52:307–358.
21. Klein D.C. and Moore R.J. 1979 Pineal N-acetyltransferase and hydroxyindole-O-methyltransferase; Control by the retinal hypothalamic tract and the suprachiasmatic nucleus. Brain Res. 174:245–262.
22. Klein D.C. and Weller J.L. 1970 Indole metabolism in the pineal gland. A circadian rhythm in N-acetyltransferase activity. Science 169:1093–1095.
23. Kornhauser J.M., Mayo K.E., and Takahashi J.L. 1993 Immediate-early gene expression in a mammalian circadian pacemaker: the suprachiasmatic nucleus. In Molecular Genetics of Biological Rhythms, edited by MW Young, Dekker, New York, pp. 271–307.
24. Leard L.E., Macdonald E.S., Heller H.C., and Kilduff T.S. 1994 Ontogeny of photic-induced c-fos mRNA expression in rat suprachiasmatic nuclei. NeuroReport 5:2683–2687.
25. Liu C., Weaver D.C., Strogatz S.H., and Reppert S.M. 1997 Cellular construction of a circadian clock: period determination in the suprachiasmatic nuclei. Cell 91:855–860.
26. McArthur A.J., Gillete M.U., and Prosser R.A. 1991 Melatonin directly resets the rat suprachiasmatic circadian clock in vitro. Brain Res. 565:158–161.
27. Moore R.Y. 1982 The suprachiasmatic nucleus and the organization of a circadian system. Trends Neurosci. 5:404–407.
28. Pelíšek V., Kosař E., and Vaněček J. 1994 Effect of photoperiod on the pineal melatonin rhythm in neonatal rats. Neurosci Lett. 180:87–90.
29. Rusak B., Robertson H.A., Wisden W., and Hunt S.P. 1990 Light pulses that shift rhythms induce gene expression in the suprachiasmatic nucleus. Science 258:1237–1240.
30. Schwartz W.J., Aranin N., Takeuchi J., Bennet M.R., and Peters R.W. 1995 Towards a molecular biology of the suprachiasmatic nucleus: photic and temporal regulation of c-fos gene expression. Semin. Neurosci. 7:53–60.

31. Sumová A. and Illnerová H. 1996 Melatonin instantaneously resets intrinsic circadian rhythmicity in the rat suprachiasmatic nucleus. Neurosci. Lett. 218:181–184.
32. Sumová A. and Illnerová H. 1996 Endogenous melatonin signal does not mediate the effect of photoperiod on the rat suprachiasmatic nucleus. Brain Res. 725:281–283.
33. Sumová A. and Illnerová H. 1998 Photic resetting of intrinsic rhythmicity of the rat suprachiasmatic nucleus under various photoperiods. Am. J. Physiol. 274 (Regulatory Integrative Comp. Physiol. 43):R857–R863.
34. Sumová A., Trávníčková Z., and Illnerová H. 1995 Memory on long but not on short days is stored in the rat suprachiasmatic nucleus. Neurosci. Lett. 200:191–194.
35. Sumová A., Trávníčková Z., Mikkelsen J.D., and Illnerová H. 1998 Spontaneous rhythm in c-Fos immunoreactivity in the dorsomedial part of the rat suprachiasmatic nucleus. Brain Res. 801:254–258.
36. Sumová A., Trávníčková Z., Peters R., Schwartz W.J., and Illnerová H. 1995 The rat suprachiasmatic nucleus is a clock for all seasons. Proc. Natl. Acad. Sci. USA 92:7754–7758.
37. Trávníčková Z., Sumová A., Peters R., Schwartz W.J., and Illnerová H. 1996 Photoperiod-dependent correlation between light-induced SCN c-fos expression and resetting of circadian phase. Am. J. Physiol. 271 (Regulatory Integrative Comp. Physiol. 40):R825–R831.
38. Van den Pol A.N. 1991 The suprachiasmatic nucleus: morphological and cytochemical substrates for cellular interaction. In Suprachiasmatic Nucleus. The Mind's Clock, edited by D.C. Klein, R.Y. Moore, and S.M. Reppert, Oxford Univ. Press. New York, pp. 17–50.
39. Vaněček J. and Illnerová H. 1985 Effect of short and long photoperiods on pineal N-acetyltransferase rhythm and on growth of testes and brown adipose tissue in developing rats. Neuroendocrinology 41:186–191.
40. Vaněček J., Pavlík A., and Illnerová H. 1987 Hypothalamic melatonin receptor sites revealed by autoradiography. Brain Res. 435:359–362.
41. Vuillez P., Jacob N., Teclemariam-Mesbah R., and Pevet P. 1996 In Syrian and European hamsters, the duration of the sensitive period to light of the suprachiasmatic nuclei depends on the photoperiod. Neurosci. Lett. 208:37–40.
42. Welsh D.K., Logothetis D.M., Meister M., and Reppert S.M. 1995 Individual neurons dissociated from rat suprachiasmatic nucleus express independently phased circadian firing rhythms. Neuron 14:697–706.
43. Woldbye D.P.D., Griesen M.H., Bolwig T.G., Larsen P.J., and Mikkelsen J.D. 1996 Prolonged induction of c-fos in neuropeptide Y- and somatostatin-immunoreactive neurons of the rat dentate gyrus after electroconvulsive stimulation. Brain Res. 720:111–119.

28

MELATONIN NORMALIZES THE RE-ENTRAINMENT OF SENESCENCE ACCELERATED MICE (SAM) TO A NEW LIGHT-DARK CYCLE

Shigenobu Shibata,[1] Makoto Asai,[1] Itsuki Oshima,[3] Masayuki Ikeda,[2] and Toru Yoshioka[2]

[1]Department of Pharmacology and Brain Science
[2]Department of Cellular and Molecular Biology
School of Human Sciences
Waseda University
Tokorozawa, Saitama 359-1192, Japan
[3]Center for Exp. Animals Develop
Shionogi Co. Ltd., Japan

1. INTRODUCTION

Aging alters various components of the circadian locomotor rhythm of animals, such as decreased amplitude in the number of locomotion, the phase response curve to light and free-running periods under constant conditions. In addition, age-related changes occur within the suprachiasmatic nucleus (SCN), which is the circadian pacemaker (14). Senescence-accelerated mice (SAMP8) are a murine model of accelerated aging and memory dysfunction. SAMP8 exhibit age-related emotional and learning memory changes as compared with SAMR1 controls (10). In addition, SAMP8 show an age-related abnormality of activity rhythms under light-dark (LD) cycle; i.e., the motor activity and water intake in SAMP8 during daytime are higher than that of SAMR1. Thus, it has been suggested that the SAMP8 strain is a useful animal model for studying not only learning and memory impairments, but also circadian rhythm disorders in patients with dementia (10). However, it is not yet known whether aging affects free-running periods or the time course for synchronization after a phase advance or phase delay of the LD cycle in SAMP8 mice. Interestingly, it was recently reported that 4-month-old SAMP8 animals re-entrained sooner than did SAMR1 after a 6 hr phase advance of the LD cycle (11). Daily melatonin administration can entrain

Melatonin after Four Decades, edited by James Olcese.
Kluwer Academic / Plenum Publishers, New York, 2000.

activity rhythms in free-running rats and hamsters. This effect is temporally restricted; melatonin administered to rats at circadian time (CT) 10–12 (CT 12: activity onset time of rodents under free-running condition) significantly advances behavioral rhythms (2). Armstrong (1) also reported a single incident of phase advance at CT 22 during late subjective night in rats. McArthur et al. (9) reported that melatonin caused advances in the firing rate rhythm in the SCN at near subjective dusk (CT 10–14) and dawn (CT 23–0) in rat SCN slices. A melatonin injection induced Fos expression in the SCN only when injected at CT22 (7). These findings suggested that melatonin can entrain circadian rhythms in rats at CT10 as well as at CT22, and that it acts at the level of the SCN. It was reported that continuous melatonin administration through implanted silastic tubing accelerated the re-entrainment of rat locomotor activity rhythms to new LD cycles (8). Interestingly, melatonin is involved in the normal aging process. The results of a study by Sandyk (12) suggested that the rate of ageing and the time of onset of age-related disease in rodents could be retarded by the administration of melatonin. Therefore, we investigated whether an administration of melatonin in the drinking water can change re-entrainment to a new LD cycle in SAMP8, Ca2+ influx into their SCN cells and the mRNA expression of the mouse mPer1 mRNA in the SCN.

2. RE-ENTRAINMENT TO 8-HR ADVANCED LD CYCLE

Male SAMP8 and SAMR1 (4, 12 and 24 months of age) were used in the re-entrainment experiment. All mice were transferred to individual mouse cages (32 × 20 × 13 cm) for measuring locomotor activity rhythm. These cages were kept in a box equipped with an infrared sensor (F5B, Omron, Osaka, Japan) and a fluorescent lamp. The movements of each mouse in its cage were counted by the infrared sensor, and total activity counts were collected every 6 min by a computer. For the examination of the effect of melatonin on the speed of re-entrainment to the 8-h advanced LD, the drinking water of some of the SAMP8 and SAMR1 was supplemented with melatonin (13.3 μg/ml with 0.2% ethanol as the vehicle). All of the mice had free access to the standard mouse chow and drinking water with or without melatonin. One week after the start of their access to melatonin-water or vehicle-water, the LD cycle was advanced for 8 hr for some mice. The phase-advanced time (hr) was defined as the advanced time of activity onset on Day 3 after the LD shift.

In the first experiment, we compared the phase-advanced time three days after the initiation of the LD shift in SAMP8 with that in SAMR1. The phase-advanced times for 12- and 24-month-old mice were lower than that for the 4-month-old mice ($p < 0.05$ and $p < 0.01$, ANOVA, respectively). In the comparison of the shift in the phase-advanced (i.e., re-entrainment) time among 4-, 12- and 24-month-old mice, the re-entrainment in the SAMP8 (2.62 hr) occurred about twice as fast as it did in the SAMR1 (1.16 hr). Thus, aging may affect SAMP8 rather than SAMR1 (Figure 1).

In the next experiment, we examined the effect of the drinking water administration of melatonin on an impairment of re-entrainment to a new LD cycle in aging mice. In the SAMR1, melatonin significantly ameliorated the impairment of re-entrainment in 4-, 12- and 24-month-old mice ($p < 0.05$, Student's t-test), whereas in the SAMP8, it did so only in 12-month-old mice ($p < 0.01$, Student's t-test) (Figure 1).

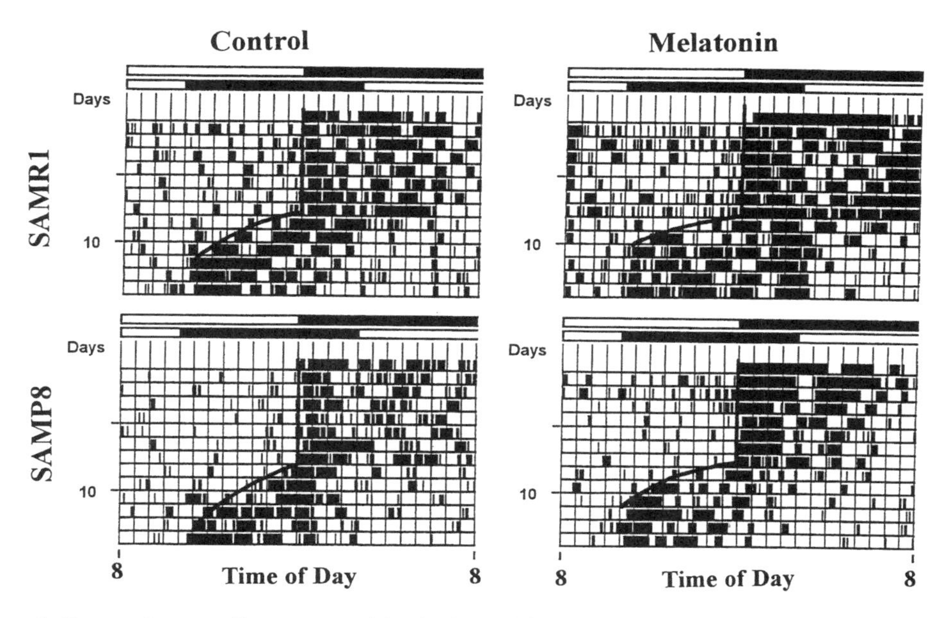

Figure 1. Re-entrainment of locomotor activity rhythm to 8-hr advanced light-dark cycles and effect of melatonin drinking on re-entrainment in SAMP8 and SAMR1 mice. Top vertical bar shows the schedule of the light-dark cycle. The light-dark cycle was advanced on Day8. Re-entrainment occurred more quickly in the SAMR1 than in the SAMP8 (left panel). Melatonin drinking facilitated the re-entrainment to the advanced light-dark cycle (right panel).

3. CALCIUM IMAGING

In this experiment, male SAMP8 (4, 19 and 27 months of age) and SAMR1 (9, 19 and 27 months of age) were used. The drinking water administration of melatonin was performed for about 2 weeks prior to the experiment. The mice were then anesthetized with ether, and their brains were removed immediately. Coronal slices (200 µm) containing the SCN were made by a vibrating microtome (microslicer DTK-1000, Dosaka, Kyoto, Japan). The slices were incubated for 1 hour in artificial cerebrospinal fluid (ACSF) containing (in mM): NaCl (138.6), KCl (3.35), $NaHCO_3$ (21), $NaH_2PO_42H_2O$ (0.6), D-glucose (9.9), $CaCl_2$ (2.5), and $MgCl_2$ (1), bubbled with 95% O2–5% CO_2 gas mixture. Fura2-AM (10 µM) was then loaded on the slices for 1 hour with cremophore EL (0.001%) by dissolving in the ACSF. After a further 30-min incubation in the Mg^{2+}-free ACSF, one slice was submitted for fluorometric recordings of intracellular calcium ($[Ca^{2+}]i$). In the course of the recording, the slice was perfused with the Mg^{2+}-free ACSF at a flow rate of 2 ml/min. The SCN, optic chiasma and anterior hypothalamic area were identified by an optical microscope (Axioplan2, Zeiss, Oberkochen, Germany) with a ×10 immersion objective (LUMPlan, Olympus, Tokyo). To determine whether melatonin affects the SCN and to evaluate functional differences among ionotropic glutamate receptors, N-methyl-D-aspartate (NMDA), α-amino-3-hydroxy-5-methylisoxszole-4-propionic acid (AMPA) and glutamate or melatonin were administered in the perfusate for 1 min. Changes in the intracellular Ca^{2+} concentration were estimated from the ratio of emission intensities excited by consecutive pairs of 340/380 nm light pulses (dichronic mirror: 430 nm; emission: 510 nm, long pass). All fluorescent images were sampled every 10 seconds by a Ca^{2+} imaging system with a cooled charge-coupled device camera (C4880, Hamamatsu Photonics, Hamamatsu, Japan). The on-line observation and analysis of the intracellular Ca^{2+} concentration were done with a personal computer base system (Argus50CA, Hamamatsu Photonics).

We examined whether melatonin altered the NMDA-, AMPA- or glutamate-induced Ca^{2+} influx into SCN cells using the Fura2 imaging technique. Melatonin application alone (0.1–1.0 mM) caused a small but significant Ca^{2+} influx (Figure 2A). The melatonin–induced increase of Ca^{2+} influx was also observed in the SCN of the melatonin-treated mice. The application of NMDA, AMPA or glutamate caused a concentration-dependent increase of the Ca^{2+} influx in the SCN cells (Figure 2B). However, the drinking-water administration of melatonin did not affect the NMDA-, AMPA- or glutamate-induced Ca^{2+} influx. The baseline Ca^{2+} levels in the SCN were lower in the aged (19- or 27-month-old) mice. Interestingly, the baseline Ca^{2+} levels were significantly higher in the melatonin-treated SAM of both strains compared to the control vehicle-treated animals. The total Ca^{2+} influx into the SCN cells induced by the excitatory amino acids was augmented in these SAM mice.

4. RE-ENTRAINMENT OF LOCOMOTOR ACTIVITY AND EXPRESSION OF mPer1 mRNA TO SUCCESSIVE 8-HR ADVANCED LD CYCLE

As described above, we found that the drinking-water administration of melatonin ameliorated the delay of re-entrainment to an 8-hr advanced LD cycle in both SAMP8 and SAMR1. Thus, melatonin may accelerate the re-entrainment to LD shifts in mice, as observed in melatonin-containing tube-implanted rats (8). However, we do not know whether the melatonin-induced acceleration means the acceleration of an overt rhythm such as locomotion or that of clock gene expression. Expression of *mPer1* mRNA in the SCN is under the control of the clock genes and this expression is good marker for the circadian rhythm (13). Therefore, we expected to examine the reduction of the mRNA expression of mPer1 in the SCN at ZT4 (the peak time of the expression of *mPer1* mRNA in normal LD cycle, 13) in successive 8-hr advanced mice. We expected to observe the normalization of the expression of *mPer1* mRNA at ZT4 in the melatonin-treated animals, because melatonin accelerated the re-entrainment.

The experimental schedule of re-entrainment is shown by a horizontal bar in Figure 3A. The LD cycle in the housing room was advanced by 8 hr. At this schedule, the phase advance was conducted 3 times with 3-day intervals from Day 2; therefore, LD cycle returned to the previous initial LD cycle after 10 days. Melatonin (13.3 μg/ml, 0.2% ethanol) or vehicle (0.2% ethanol) was administered in the animals' drinking water. In this experiment, we used young ddY mice (6–8 weeks old). The mice had free access to food and the drinking water. The measurement of locomotion was the same as described earlier. LD-shifted mice exhibited significantly low locomotor activity during the last 3 days (Figure 3A and 3B, $p < 0.05$ vs. non-shifted animals, Dunnett's test). In contrast, the melatonin-treated LD-shifted mice showed a recovery of the reduction of locomotion.

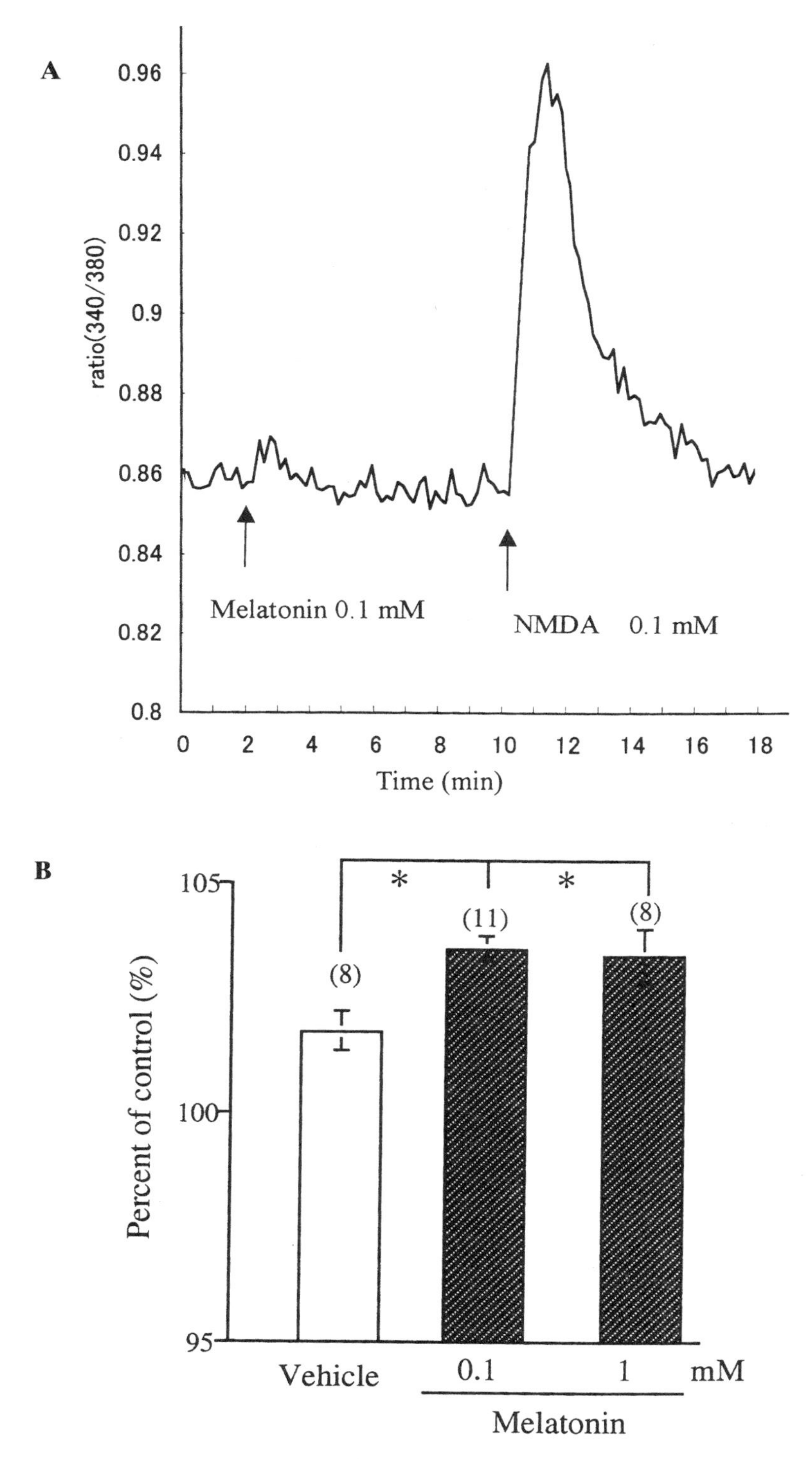

Figure 2. Effect of melatonin application on Ca^{2+} influx into the SCN cells. A shows the time course of changes in the ratio (340 nm/380 nm) and B shows the summarized data. The numbers of SCN slices are shown in parentheses; the columns and bars means and S.E.M., respectively. Melatonin (0.1 and 1 mM) significantly increased the Ca^{2+} influx (Dunnett's test, $*p < 0.05$).

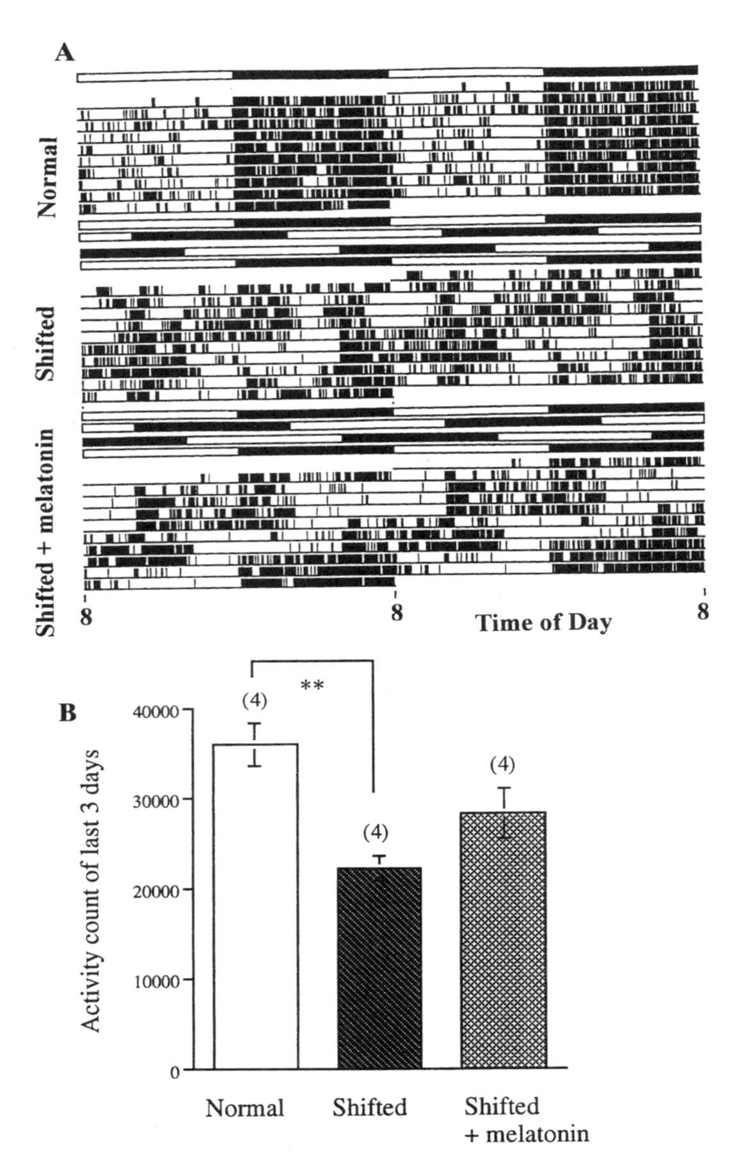

Figure 3. Re-entrainment of locomotor activity rhythm to three-times 8-hr advanced light-dark cycles and the effect of melatonin drinking water on re-entrainment in ddY mice. The second and third vertical bars show the schedule of the light-dark cycle. Melatonin ameliorated the deficit of re-entrainment in shifted animals (third actogram in A and B). In panel B, the numbers of mice are shown in parentheses; the columns and bars means and S.E.M., respectively. $^{**}p < 0.01$ (Student's t-test).

4.1. Competitive RT-PCR Analysis

The effect of melatonin drinking on *mPer1* expression in the SCN at ZT4 was examined by competitive reverse transcription-polymerase chain reaction (RT-PCR). Control mice of the ddY strain were entrained to a normal LD cycle for 10 days. The mice were then transferred to an 8 hr-advanced LD cycle 3 times with 3-day intervals from Day 2. Some mice were administered melatonin in their drinking water. On the 8 day of the treatment, the mice were sacrificed and their brains were removed and placed in ice-cold saline at ZT4. Slices (0.5 mm thick) of mice brain that contained the SCN were frozen on dry ice and the both SCN were punched out. Total RNA from the SCN ($n = 4$) was extracted by Trizol solution (GIBCO BRL, Gaithersburg, MD, USA). An mRNA selective PCR Kit (Takara, Osaka, Japan) was used for the reverse transcription of approximately 100 ng of RNA, and *mPer1* and β-actin cDNA were quantified by competitive PCR. PCR reactions were carried out for 18 and 23 cycles for β-actin and *mPer1*, respectively. The primer pairs used for the amplification of each product were as follows: 5′-AGG ACT CCT ATG TGG GTG ACG A-3′ and 5′-CAG CCT GGA TGG CTA CGT ACA A-3′ (β-actin); 5′-CAA GTG GCA ATG AGT CCA ACG G-3′ and 5′-GAC ACA GGC CAG AGC CGT ACT G-3′ (*mPer1*). The competitor DNA fragments for *mPer1* and β-actin was constructed with a DNA Competitor Construction Kit (Takara). The sizes of the PCR products of *mPer1*, *mPer1* competitor, β-actin and β-actin competitor were 362, 292, 267 and 223 bp, respectively. The PCR products were run on 3% agarose gels, and DNA in the appropriate bands were detected with an EDAS-120 system (Kodak, Rochester, USA).

The reduction of *mPer1* induction in the SCN after shifted-LD cycles was quantified by RT-PCR. Before this quantification, the RT products were tested for possible genomic DNA contamination. A gel analysis showed bands of expected lengths. The *mPer1* mRNA levels were low after shifted-LD cycle, whereas the levels were recovered in the melatonin-treated shifted mice (Figure 4). These results suggest that the phenotypic effects of melatonin treatment on the LD shift-induced decrease in locomotor activity may be mediated by the recovery of *mPer1* expression in the SCN.

5. DISCUSSION

In this study, we observed that the number of days until re-entrainment to a new LD cycle was age-dependent. The phase-advanced times for 12- and 24-month-old mice were lower than that for the 4-month-old mice ($p < 0.05$ and $p < 0.01$, ANOVA, respectively). Little can be said about the mechanisms responsible for the age-related reduction of re-entrainment. One possible explanation is that SCN neurons of aging rodents may exhibit weak responses to light exposure. Indeed, the NMDA-induced Ca^{2+} influx into the SCN and light-induced Fos production in the SCN were attenuated in aging mice and rats (our unpublished observation). These deteriorating changes may affect the light-induced re-entrainment. An alternative explanation is that aging extends the free-running period of locomotor activity (our unpublished observation), and this may delay the re-entrainment to 8-hr advanced LD cycles in aging animals.

Interestingly, the re-entrainment to 8-hr advanced LD cycles was faster in the SAMP8 than in the SAMR1 among the 4-month-old mice. Sanchez-Barcelo et al. (11) reported that re-entrainment to a 6-hr phase-advance was faster in SAMP8 than in

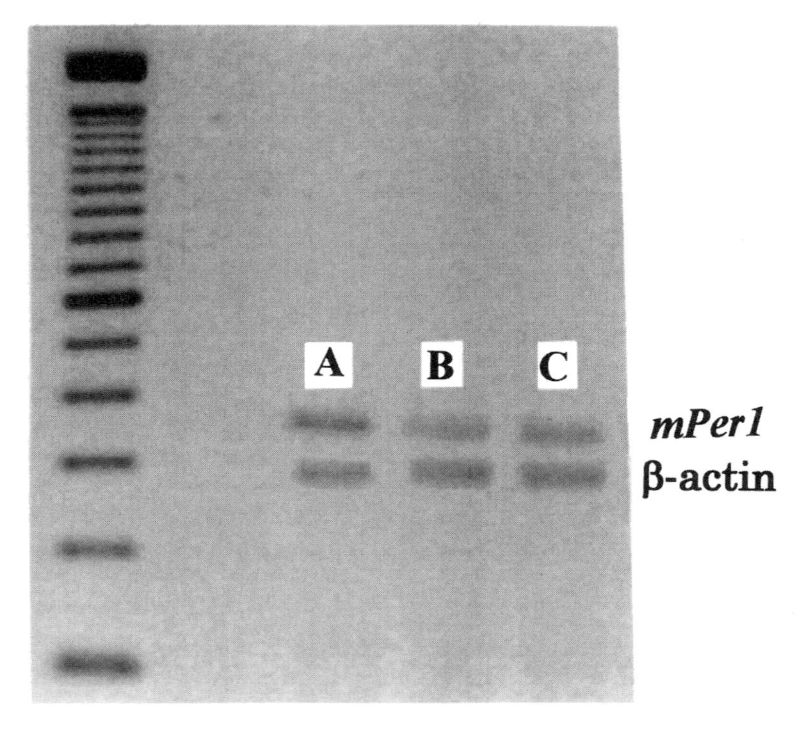

Figure 4. Effects of melatonin in the drinking water on the decrement of SCN *mPer1* mRNA expression in ddY mice with consecutive 8-hr advanced light-dark cycle. Lane A is RT-PCR products from control non-shifted animals; lane B is from shifted animals and lane C is from shifted melatonin-treated animals. Total RNA was isolated and *mPer1* and β-actin cDNA were quantified by a competitive PCR method.

SAMR1. Thus, it seems that the present result may have been caused by characteristics of SAMP8 and not by aging.

In the experiment measuring locomotor activity rhythm during drinking-water administration of melatonin, melatonin accelerated the re-entrainment to 8-hr advanced LD cycles in SAMP8 and SAMR1 mice, except in the SAMP8 at 4 and 24 months of age. The phase-advanced time in 4-month-old SAMP8 was 7.42 ± 0.45 hours, which is close to the maximum values (8 hours). Therefore, melatonin did not accelerate re-entrainment in the SAMP8 at 4 months of age. Regarding the SAMP8 at 24 months of age, we do not know the reason for the failure of the melatonin effect.

The accelerating effect of melatonin on re-entrainment after a phase-advance of 8 hours seems to be due to a direct effect of melatonin on the SCN neurons. Melatonin administered in vitro significantly enhanced intracellular Ca^{2+} in the SCN in this study. Melatonin synthesis occurs in the mouse strain AKR, which is ascent strain of SAMP8 and SAMR1 (3). It is well known that melatonin injection at CT10 causes the phase advance of the activity rhythm (2,4) and SCN firing rhythm (9). These lines of evidence have suggested that melatonin can advance circadian rhythms at CT10 as well as at CT22 (1,7,9). Interestingly, an intraperitoneal injection of melatonin to SAMP8 at ZT0 but not at ZT12 caused a potentiation of re-entrainment to 8-hr advanced LD cycles (our unpublished observation). Chronic exposure to melatonin receptor agonists does not alter the effect of an inhibition on SCN neurons (15). In the present study, the baseline Ca^{2+} levels were significantly higher in the melatonin-treated SAMP8 and SAMR1 than in the vehicle group. The total Ca^{2+} influx into SCN cells induced by the excitatory amino acids, which are NMDA, AMPA and glutamate, was augmented. Therefore, the change of Ca^{2+} homeostasis produced by melatonin drinking may be related to the acceleration of re-entrainment. These results suggest that the drinking of water containing melatonin increases the melatonin concentration in the brain at around dawn and results in the induction of an advance of activity rhythm. This effect may occur even in aging mice such as aging SAMP8 and SAMR1, because the density of high-affinity sites of melatonin binding in the rat SCN does not change with age (5).

It has been reported that both NMDA and AMPA receptors are involved in the light-induced phase shifts (6). In the present experiment, however, melatonin drinking did not affect the Ca^{2+} influx into the SCN induced by NMDA, AMPA or glutamate application. Thus, melatonin failed to influence the light-induced advances.

Another possible explanation is that melatonin antagonizes the extended free-running period in aging SAMP8 and SAMR1 mice, and results in the facilitation of re-entrainment. However, melatonin drinking did not cause any changes of free-running periods (our unpublished observation).

We observed the reduction of mPer mRNA expression in shifted mice and the recovery of mRNA expression in melatonin-treated shifted animals. Thus, the melatonin treatment recovered not only the overt rhythm but also the mPer1 gene expression level.

REFERENCES

1. Armstrong, S.M. Melatonin and circadian control in mammals. Experientia. 45:932–938, 1989.
2. Cassone, V.M., Chesworth, M.J., and Armstrong, S.M. Entrainment of rat circadian rhythms by daily injection of melatonin depends upon the hypothalamic suprachiasmatic nuclei. Physiol. Behav. 36:1111–1121, 1986.
3. Conti, A. and Maestroni, G.J.M. HPLC validation of a circadian melatonin rhythm in the pineal gland of inbred mice. J. Pineal. Res. 20:138–144, 1996.
4. Dubocovich, M.L., Benloucif, S., and Masana, M.I. Melatonin receptors in the mammalian suprachiasmatic nucleus. Behav. Brain Res. 73:141–147, 1996.
5. Duncan, M.J. and Davis, F.C. Developmental appearance and age related changes in specific 2-[125I]iodomelatonin binding sites in the suprachiasmatic nuclei of female Syrian hamsters. Dev. Brain Res. 73:205–212, 1993
6. Inouye, S.T. and Shibata, S. Neurochemical organization of circadian rhythm in the suprachiasmatic nucleus. Neurosci. Res. 20:109–130, 1994.
7. Kilduff, T.S., Landel, H.B., Nagy, G.S., Sutin, E.L., Dement, W.C., and Heller, H.C. Melatonin influences Fos expression in the rat suprachiasmatic. Mol. Brain Res. 16:47–56, 1992.
8. Marumoto, N., Murakami, N., Kuroda, H., and Murakami, T. Melatonin accelerates reentrainment of circadian locomotor activity rhythms to new light-dark cycles in the rat. Japanese J. Physiol. 46:347–351, 1996.
9. McArthur, A.J., Hunt, A.E., and Gillette, M.U. Melatonin action and signal transduction in the rat suprachiasmatic circadian clock: Activation of protein kinase C at dusk and dawn. Endocrinology. 138:627–634, 1997.
10. Miyamoto, M., Kiyota, M., Yamazaki, N., Nagaoka, A., Nagawa, Y., and Takeda, Y. Age-related changes in learning and memory in the senescence-accelerated mouse (SAM). Physiol. Behav. 38:399–406, 1986.

11. Sanchez-Barcelo, E.J., Megias, M., Verduga, R., and Crespo, D. Differences between the circadian system of two strains of senescence-accelerated mice (SAM). Physiol. Behav. 62:1225–1229, 1997.
12. Sandyk, R. The forum for scientific controversies. Int. J Neurosci. 55:143–144, 1990.
13. Shigeyoshi, Y., Taguchi, K., Yamamoto, S., Takekida, S., Yan, L., Tei, H., Moriya, T., Shibata, S., Loros, J.J., Dunlap, J.C., and Okamura, H. Light-induced resetting of a mammalian circadian clock is associated with rapid induction of the *mPer1* transcript. Cell 91:1043–1053, 1997.
14. Watanabe, A., Shibata, S., and Watanabe, S. Circadian rhythm of spontaneous neuronal activity in the suprachiasmatic nucleus of old hamster in vitro. Brain Res. 695:237–239, 1995.
15. Ying, S.W., Rusak, B., and Mocaer, E. Chronic exposure to melatonin receptor agonists does not alter their effects on suprachiasmatic nucleus neurons. Eur. J. Pharmacol. 342:29–37, 1998.

29

DEVELOPMENTAL EXPRESSION OF BOTH MELATONIN RECEPTOR mt_1 mRNA AND MELATONIN BINDING SITES IN SYRIAN HAMSTER SUPRACHIASMATIC NUCLEI

François Gauer,* Carole Schuster, Vincent-Joseph Poirel, Paul Pevet, and Mireille Masson-Pevet

"Neurobiologie des Fonctions Rythmiques et Saisonnières"
CNRS-UMR 7518
Université Louis Pasteur
12 Rue de l'Université, 67000 Strasbourg
France

Melatonin has been demonstrated to act as a neuroendocrine message involved in the regulation of both mammalian circadian and seasonal biological rhythms (16,17). Over the past ten years extensive studies have been conducted to locate the central structures responding to the melatonin message (for review see 10,21). The most consistently labeled structures in the mammalian species studied so far are the suprachiasmatic nuclei (SCN) and the pars tuberalis of the anterior pituitary (PT). While the PT seems to be implicated in mediating some of the melatonin photoperiodic effects on the neuroendocrine system (14), the SCN are well known to contain the mammalian circadian biological clock (15) which controls a wide variety of circadian biological rhythms (13) and especially the day/night melatonin synthesis rhythmicity. Since 1994, different cDNA subtypes encoding melatonin receptors have been cloned in lower and higher vertebrates (19). Particularly, a melatonin receptor cDNA designated as the Mel_{1a} receptor subtype (now called mt1; 4) has been isolated in mammals (18). When expressed in mammalian cells, this subtype exhibits the pharmacological and the functional characteristics of the 2-iodo-melatonin binding sites described so far. Furthermore, combination of both in situ hybridization and receptor autoradiography experiments revealed that 2-iodo-melatonin binding sites and mt_1

* Corresponding author, E-mail: gauer@neurochem.u-strasbg.fr

Melatonin after Four Decades, edited by James Olcese.
Kluwer Academic / Plenum Publishers, New York, 2000.

melatonin receptor mRNA are present in an overlapping pattern including an expression in the PT and SCN (18). At the present time, the mt_1 mRNA is, therefore in both the PT and SCN, the only mRNA that can be correlated to 2-iodo-melatonin binding sites.

The presence of melatonin receptors within the SCN strongly supports the idea that melatonin is involved in the regulation of the SCN circadian activity. Indeed, in mice, rats and Siberian hamsters, melatonin has been demonstrated to either phase-shift the circadian activity rhythms, to directly affect the SCN metabolism or to reset the electrical activity of SCN slices cultured in vitro (see 1,2,3,11,22). All these experimental data are strongly in favor of a feed-back control of the circadian rhythmicity by a melatonin direct action on the SCN clock.

On the other hand, in adult Syrian hamsters, manually injected melatonin or vehicle both resulted in a phase advance of the locomotor activity rhythm, while when delivered remotely through chronic cannulae, neither melatonin nor vehicle had any effect upon the free-running activity rhythm (9). These results then failed to support the idea that, in adult Syrian hamster, melatonin injections could either phase shift or resynchronize SCN activity. Interestingly, melatonin receptor density in adult Syrian hamster SCN has consistently been reported to be very low in contrast with high melatonin receptor density in mice, rat or Siberian hamster SCN (5,12,23). Therefore, it has been proposed that the SCN non-response to melatonin injection in adult Syrian hamsters, could be correlated to age-related changes in the SCN melatonin binding capacities (8). The goal of the present study was to obtain a complete pattern of the post natal expression of the melatonin binding in the SCN of the Syrian hamster and to investigate whether a postnatal regulation of the expression of the mt_1 mRNA could generate a decline in SCN melatonin receptor density. Therefore, both SCN mt_1 mRNA expression and 2-iodo-melatonin binding were followed from birth to adulthood through the ages of 8, 21 and 30. The same experiments were also performed in the PT of the same animals as a control. The first part of the study was however dedicated to generate Syrian hamster mt_1 clone by cross-species homology based PCR amplification.

1. MATERIALS AND METHODS

1.1. Cloning of a Partial cDNA Encoding mt_1 SCN Melatonin Receptors

Female Syrian hamsters (*Mesocricetus auratus*) from our own colony were sacrificed during daytime. Blocks of basal hypothalamus weighing approximately 5 mg and containing either the SCN or the PT were immediately dissected and frozen on dry ice. Total RNA was extracted by using TRIzol (Gibco, BRL). One µg of total RNA was reverse transcribed by 200 units of M-MLV reverse transcriptase. Afterwards, cDNAs were PCR amplified with various universal primers designed from Mice, Siberian hamster and rat Mt_1 sequence information. (18,20,24). The whole RT-PCR procedure is described in details in Gauer et al., (7). The amplified bands of the predicted size were purified, blunt-ended, ligated into the pCR-Script Amp SK(+) cloning vector according to the manufacturer's protocol and then introduced in XL1-Blu MRF' supercompetent cells. After purification of the plasmid both strands of all the obtained cDNA were dideoxysequenced using both vectors primers and internal primers derived from sequence information.

1.2. Developmental Regulation of both the Melatonin Receptor Density and the mt_1 mRNA in the PT and SCN of the Syrian Hamster

Animals: New born Syrian hamsters (*Mesocricetus auratus*) from our own colony were kept from birth in long photoperiod (LP, 14L/10D). Animals were sacrificed at various times after birth from birthday to postnatal day 60 (PN 60). All the animals were sacrificed at 5 pm (1 hour before lights off). The entire head (PN 0 and PN 8) or brains (PN 21, PN 30 and PN 60) of the animals were rapidly dissected out and frozen in −30°C isopentane, and maintained at −80°C until cryosectioning (20 μm thick).

In situ hybridization: After pre-hybridization treatments (6), SCN and PT tissue sections were hybridized in a humid chamber with 2×10^7 cpm/ml of either the sense or the antisense mt_1 cRNA $\alpha[^{35}S]$-UTP-labeled riboprobe in a solution containing 50% deionized formamide, 10% dextran sulfate, 50 mM DTT, 1× Denhardt's solution, 2 × SSC, 1 mg/ml salmon sperm DNA, 1 mg/ml yeast RNA, at 54°C for 16 hours. Afterwards, sections were rinsed for 5 min at RT in 2 × SSC before being treated with ribonuclease A (2 μg/ml) in a 10 mM Tris pH 7.4, 0.5 M NaCl, 10 mM EDTA buffer. Slides were then rinsed 3 times, 10 min at RT in 2 × SSC and then finally washed 15 min, at 54°C, in 0.5 × SSC before being dehydrated in a graded ethanol series (70%, 90%, 95% and 100%, 1 min each) and air-dried. The sections were subsequently exposed to X-ray films for 5–10 days concomitantly with microscale standards. Quantitative analysis of the autoradiograms was performed as previously described (6).

Autoradiography: The 2-iodo-melatonin ligand was synthesized as previously described. Autoradiographic binding procedure, as well as quantitative analysis of the autoradiograms were performed as previously described (6).

2. RESULTS

Specific oligonucleotides derived from rat, mice, and Siberian hamster sequences allowed amplification of DNA products with the predicted size, in both the PT and the SCN. Sequence analysis revealed that the 977 bp Syrian hamster mt_1 cDNA sequence (after excluding both 3′ and 5′ ends corresponding to the universal primer sequences) presents 86.5%, 88.8% and 94.7% of sequence identity with the rat, mouse and Siberian hamster mt_1 sequences, respectively (7). Furthermore, the Syrian hamster deduced amino acid sequence presents a 96.9 and 92.3% similarity with the Siberian hamster and mouse deduced amino acid sequences, respectively. According to Siberian hamster and mouse amino acid sequences, this Syrian hamster amino acid sequence would include the seven presumed transmembrane domains characterizing G protein-coupled melatonin receptors, and a portion of both the first extracellular and the last intracellular domains (Figure 1).

Quantitative autoradiography revealed that in the PT, the melatonin receptor density declined from PN 0 to PN 8. Indeed, at PN 8, binding was only 40% of the value measured at birth (Figure 2 and Table 1). Afterwards, PT melatonin receptor density slightly went up until it got stabilized around 25 fmol/mg of protein (Figure 3A and Table 1). Similarly, the mt_1 mRNA expression decreased from PN 0 to PN 8 (Figure 2 and Table 1), and slightly went up from PN 8 to PN 60 (Figure 3A and Table 1). Regression analysis revealed that mt_1 mRNA expression and melatonin binding values were highly correlated ($r = 0.98$, $P < 0.003$, between PN 0 and PN 60).

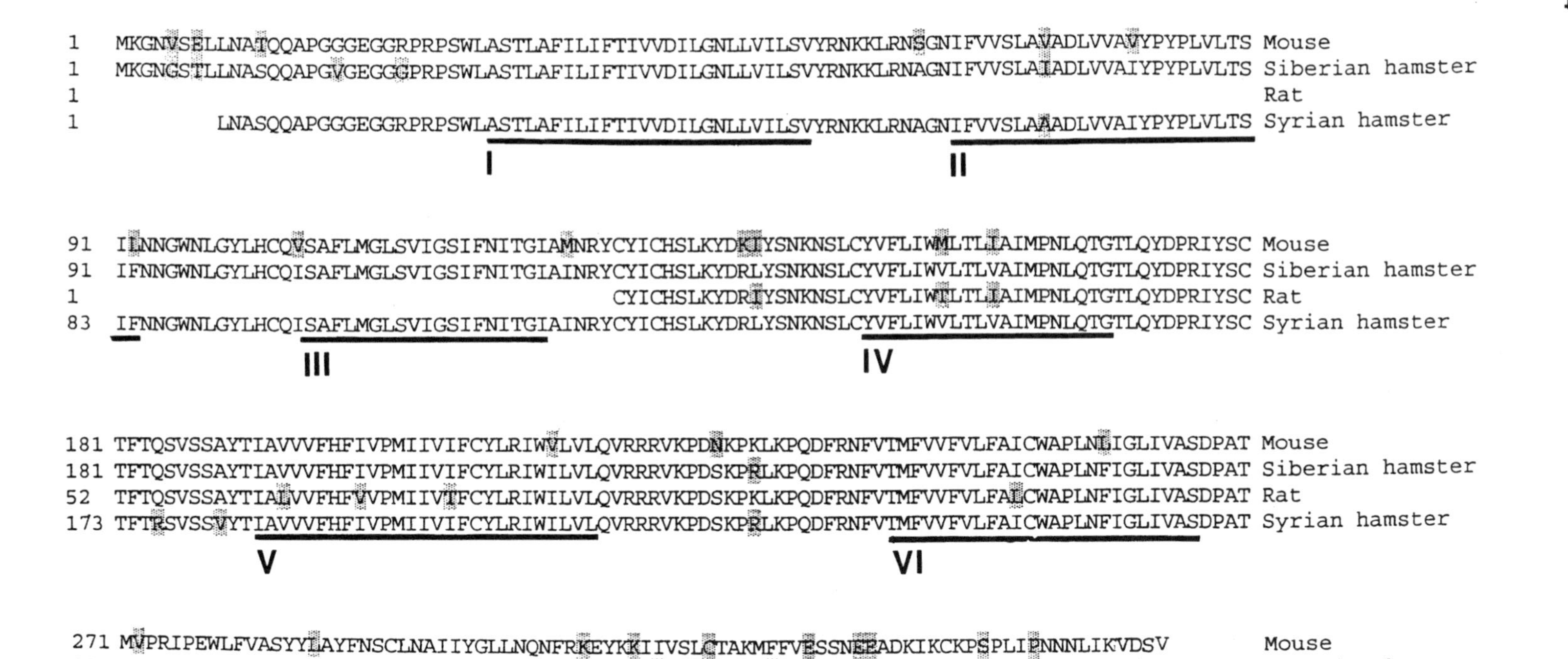

Figure 1. Alignment of the deduced amino acid sequences of the mt_1 melatonin receptor subtype in Syrian hamster, Siberian hamster, mouse and rat. The shaded residues are those who differ from the consensus. Analysis were performed using Lasergene Softwares (Dnastar USA). The seven presumed transmembrane domains (I–VII) are highlighted on the deduced amino acid sequence by solid bars. The sequence has been deposited in GenBank under the AF061158 accession number.

Table 1. Post natal development of both the specific 2-iodo-melatonin binding (in fmol/mg protein) and the specific hybridization of an antisense mt_1 cRNA riboprobe (in DPM) in both the PT (A) and the SCN (B) between PN 0 and PN 60 (adulthood). Each value is the mean ± SEM of 5 animals

Age	PN 0	PN 8	PN 21	PN 30	PN 60
PT 2-iodo-melatonin specific binding (fmol/mg prot)	50.1 ± 1.1	18.5 ± 0.7^{++}	19.3 ± 0.8$^{**,+}$	24.4 ± 1.4**	26.1 ± 1.3**
PT mt_1 mRNA expression (DPM)	130 ± 6	52 ± 3^{++}	63 ± 4$^{**,+}$	59 ± 2$^{**,+}$	79 ± 4**
SCN 2-iodo-melatonin specific binding (fmol/mg prot)	12.1 ± 0.3	4.5 ± 0.1^{++}	1.7 ± 0.3**	1.1 ± 0.1**	0.9 ± 0.1**,*
SCN mt_1 mRNA expression (DPM)	52 ± 3	30 ± 2**	30 ± 2**	30 ± 2**	22 ± 1.2**

**: $P < 0.001$ when compared with values obtained at PN 0; ++: $P < 0.001$ when compared with values obtained at PN 0 and PN 60; +: $P < 0.01$ when compared with values obtained at PN 60; *: $P < 0.05$ when compared with values obtained at PN 21.

In the SCN of the same animals, the melatonin receptor density was 12.1 ± 0.3 fmol/mg protein at PN 0, and then dramatically decreased until a basal value was reached between PN 30 and PN 60 (Figs. 2 and 3B, Table 1). The mt_1 mRNA expression decreased rapidly from PN 0 to PN 8 (Fig. 2 and Table 1), plateaued between PN 8 and PN 30, and afterwards declined slightly until adult age (Fig. 3B and Table 1). In contrast with PT data, after PN 8, the dramatic decrease in SCN binding capacities did not correlate to the mt_1 mRNA expression variations ($r = 0.45$, $p < 0.55$, between PN 8 and PN 60).

3. DISCUSSION

This study shows that in adult Syrian hamster SCN 2-iodo-melatonin binding undergoes a dramatic loss within the first 2 months of life, as at PN 60 the melatonin

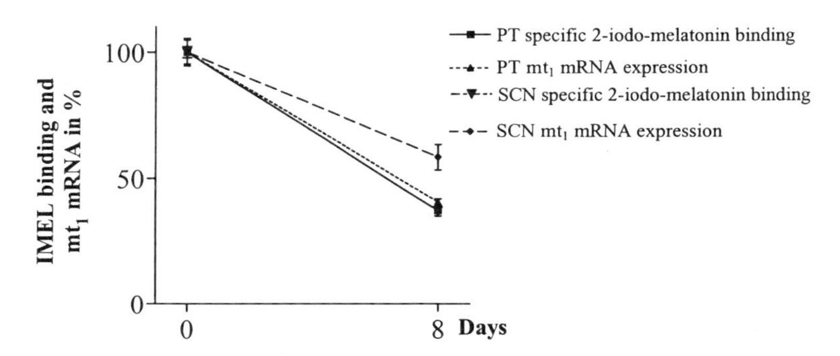

Figure 2. Post natal development of the specific 2-iodo-melatonin binding and the specific hybridization of an antisense mt_1 cRNA riboprobe in the PT and the SCN of Syrian hamster between PN 0 and PN 8. All values are expressed in percentage by comparison to the PN 0 Value (100%). Each time point is the mean ± SEM of 5 animals.

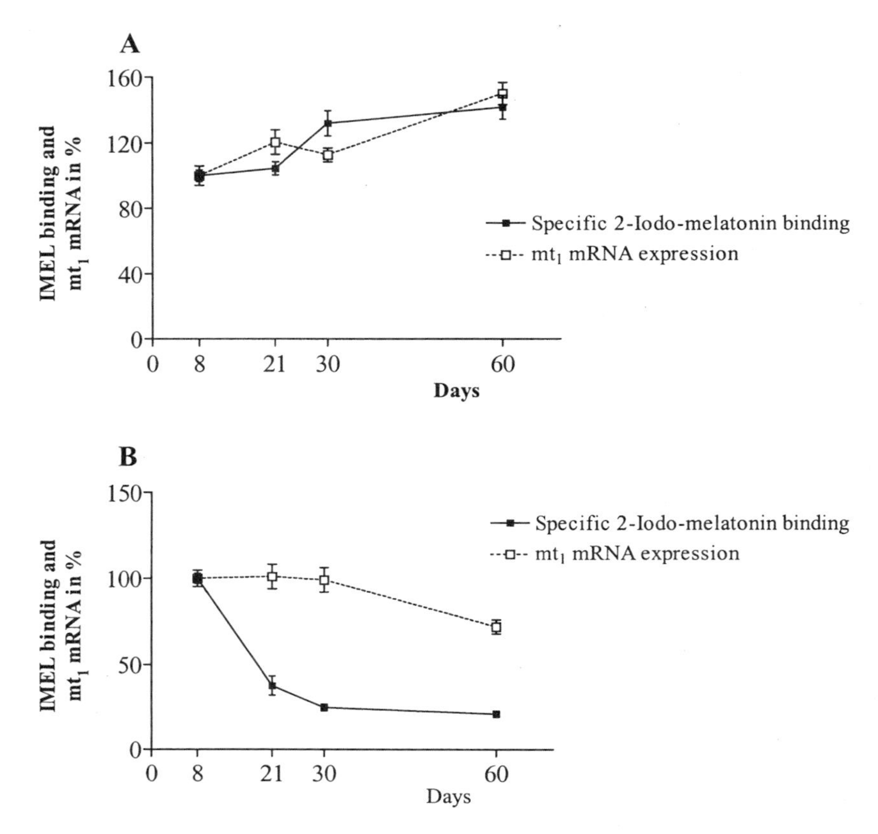

Figure 3. Post natal development of both the specific 2-iodo-melatonin binding (and the specific hybridization of an antisense mt_1 cRNA riboprobe in both the PT (A) and the SCN (B) between PN 8 and PN 60 (adulthood). Values are expressed in percentage by comparison to PN 8 value (100%). Each time point is the mean ± SEM of 5 animals.

receptor density is only 7.7% of the PN 0 value. We used a newly subcloned mt_1 Syrian hamster cDNA to investigate the postnatal developmental expression of the mt_1 mRNA within the SCN and the PT of the Syrian hamster to test whether a postnatal regulation of the expression of the mt_1 mRNA could generate the observed decline in SCN melatonin receptor density. At a first glance what one could have expected is a dramatic slow-down of mt_1 mRNA expression preceding and then leading to low binding capacities in adulthood. Actually, the mechanisms underlying this binding drop appear to be more complicated. Indeed, when considering the PT as a control, postnatal variations of both specific 2-iodo-melatonin binding and mt_1 mRNA expression are highly correlated ($r = 0.98$, $P < 0.003$) from PN 8 to PN 60. This strongly suggests that the observed postnatal variations of PT melatonin receptor density are, in the PT, a direct consequence of the postnatal mt_1 mRNA expression variations.

This conclusion can also be proposed to explain the initial drop of the SCN melatonin binding capacity between PN 0 and PN 8, as during this first step the binding capacity decline correlates with mRNA expression slow-down. However after PN 8, in contrast to what was observed in the PT, the melatonin receptor density dramatically followed its drop in the SCN while the mRNA expression level got stabilized at 40%

of the PN 0 value. Therefore, the dramatic post natal loss of melatonin receptor in the SCN can not be attributed to an inhibition of the mRNA expression, but certainly to a post transcriptional blockade of the mt_1 receptor expression.

This study provides the first evidence of a regulation of the melatonin receptor involving both transcriptional (before PN 8) and post-transcriptional (after PN 8) mechanisms that may directly influence a key role of melatonin: its feed-back effect on SCN circadian functions. In free running rodents, like the rat or the mouse, which present a robust melatonin receptor density, the locomotor activity rhythm is either phase shifted or entrained by melatonin injections. Furthermore, experiments showing a direct melatonin resetting activity of rat SCN electrical activity in vitro clearly support the hypothesis that this exogenous melatonin directly acts on the rat SCN through melatonin receptors (see 1,2,3,11,22). According to this model, the dramatic decline in SCN melatonin receptor density observed in the Syrian hamster should obviously affect the ability of the SCN to read the melatonin humoral message and therefore would explain the absence of melatonin specific effect of melatonin injections on the locomotor activity of free-running hamsters (8,9). Our results thus strongly suggest that a post transcriptional blockade of the expression of the mt_1 melatonin receptor in the SCN would take place after PN 8 and consequently would induce a dramatic loss of SCN melatonin receptors that might be responsible for the SCN non response to acute melatonin injections.

These results however do not necessarily imply that Syrian hamster circadian functions should be totally unresponsive to both exogenous and endogenous melatonin on a long term basis. Indeed, exogenous melatonin administered by infusion during long periods of time has been reported to entrain pinealectomized Syrian hamsters locomotor activity [Schuler et al., this issue]. Therefore, the understanding of the neuronal circuitry involved in the regulation of biological circadian rhythms like the locomotor activity requires more than ever fundamental investigations into the physiological role of the melatonin secretion in terms of daily and seasonal functions.

REFERENCES

1. Armstrong, S.M., Cassone, V.M., Chesworth M.G., and Redman, J.R. Synchronization of mammalian circadian rhythms by melatonin. J. Neural Trans. 21:371–375, 1986.
2. Benloucif, S. and Dubocovich, M.L. Melatonin and light induce phase shifts of circadian activity rhythms in the C3H/HeN mouse. J. Biol. Rhythms. 11:113–125, 1996.
3. Cassone, V.M. Effects of melatonin on vertebrate circadian system TINS. 13:457–464, 1990.
4. Dubocovich, M.L., Cardinali, D.P., Guardiola-Lemaitre, B., Hagan, R.M., Krause, D.N., Sugden, D., Vanhoutte, P.M., and Yocca, F.D. Melatonin receptors. The IUPHAR-Melatonin Receptor Nomenclature and classification Subcommittee (MRNC) (June 11, 1998). Compendium of receptor Characterization and Classification. IUPHAR Media, London, 1998 pp. 188–193.
5. Duncan, M.J. and Davis, F.C. Developmental appearance and age related changes in specific 2-[^{125}I]iodomelatonin binding sites in the suprachiasmatic nuclei of female Syrian hamsters. Dev. Brain Res. 73:205–212, 1993.
6. Gauer, F., Schuster, C., Pévet, P., and Masson-Pévet, M. Effect of a light pulse on melatonin receptors density and mRNA expression in the siberian hamster suprachiasmatic nuclei. Neurosci. Lett. 233:49–52, 1997.
7. Gauer, F., Schuster, C., Poirel, V.J., Pévet, P., and Masson-Pévet, M. Cloning experiments and developmental expression of both melatonin receptor Mel_{1a} mRNA and melatonin binding sites in the Syrian hamster suprachiasmatic nuclei. Mol. Brain Res. 60:193–202, 1998.
8. Grosse, J., Velickovic, A., and Davis, F.C. Entrainment of Syrian hamster circadian activity rhythms by neonatal melatonin injections. Am. J. Physiol. 270:R533–R540, 1996.

9. Hastings, M.H., Mead S.M., Vindlacheruvu, R.R., Ebling, F.J.P., Maywood, E.S., and Grosse, J. Non-photic phase shifting of the circadian activity rhythm of Syrian hamsters: the relative potency of arousal and melatonin. Brain Res. 591:20–26, 1992.
10. Masson-Pévet, M., George, D., Kalsbeek, A., Saboureau, M., Lakhdar-Ghazal, N., and Pévet, P. An attempt to correlate brain areas containing melatonin-binding sites with rhythmic functions: a study in five hibernator species. Cell Tissue Res. 278:97–106, 1994.
11. MacArthur, A.J., Hunt, A.E., and Gilette, M.U. Melatonin action and signal transduction in the rat suprachiasmatic circadian clock: activation of protein kinase C at dusk and dawn. Endocrinology. 138:627–634, 1997.
12. Maywood, E.S., Bittman, E.L., Ebling, F.J.P., Barrett, P., Morgan, P., and Hastings, M.H. Regional distribution of iodomelatonin binding sites within the suprachiasmatic nucleus of the Syrian hamster and the Siberian hamster. J. Neuroendocrinol. 7:215–223, 1995.
13. Meijer, J.H. and Rietveld, W.J. Neurophysiology of the suprachiasmatic circadian pacemaker in rodents. Physiol. Rev. 69:671–707, 1989.
14. Morgan, P.J., Webster, C.A., Mercer J.G., Ross, A.W., Hazlerigg, D.G., MacLean, A., and Barrett, P. The ovine pars tuberalis secretes a factor(s) that regulate(s) gene expression in both lactotropic and non lactotropic pituitary cells. Endocrinology. 137:4018–4026, 1996.
15. Ralph, M.R., Foster, R.G., Davis, F.C., and Menaker, M. Transplanted suprachiasmatic nucleus determines circadian period. Science. 247:975–978, 1990.
16. Reiter, R.J. Pineal melatonin: Cell biology of its synthesis and of its biological interactions. Endocrine Rev. 12:151–180, 1991.
17. Reiter, R.J. The melatonin rhythm: both a clock and a calendar. Experientia. 49:654–664, 1993.
18. Reppert, S.M., Weaver, D.R., and Ebisawa, T. Cloning and characterization of a mammalian melatonin receptor that mediates reproductive and circadian responses. Neuron. 13:1177–1185, 1994.
19. Reppert, S.M. and Weaver, D.R. Melatonin madness. Cell. 83:1059–1062, 1995.
20. Roca, A.L., Godson, C., Weaver, D.R., and Reppert, S.M. Structure, Characterization and expression of the gene encoding the mouse Mel_{1a} melatonin receptor. Endocrinology. 137:3469–3477, 1996.
21. Stankov, B., Fraschini, F., and Reiter, R.J. Melatonin binding sites in the central nervous system. Brain Res. Rev. 16:245–256, 1991.
22. Sumova, A. and Illnerova, H. Melatonin instantaneously resets intrinsic circadian rhythmicity in the rat suprachiasmatic nucleus. Neurosci. Lett. 218:181–184, 1996.
23. Weaver, D.R., Rivkees, S.A., and Reppert, S.M. Localization and characterization of melatonin receptors in rodent brain by in vitro autoradiography. J. Neurosci. 9:2581–2590, 1989.
24. Weaver, D.R., Liu, C., and Reppert, S.M. Nature's knockout: the Mel_1b receptor is not necessary for reproductive and circadian responses to melatonin in Siberian hamsters. Mol. Endocrinol. 10:1478–1487, 1996.

30

ENTRAINMENT OF RAT CIRCADIAN RHYTHMS BY DAILY ADMINISTRATION OF MELATONIN

Influence of the Mode of Administration

H. Slotten,[1,2] B. Pitrosky,[1] and P. Pévet[1]

[1] Neurobiologie des fonctions rythmiques et saisonnières
CNRS-UMR 7518
Université Louis Pasteur
F-67000 Strasbourg France
[2] Department of Psychology
NTNU, N-7055 Trondheim Norway

INTRODUCTION

In the rat, the suprachiasmatic nuclei (SCN) are known to be involved in the expression of circadian rhythms such as locomotor activity and body temperature (1). When the animals are held in constant darkness, the rhythmic functions will free-run with a period different from 24 hour. Under natural conditions, photic information from the light/dark-cycle synchronizes these functions to a 24-hour period, but daily administration of melatonin (MLT) also causes such entrainment (2). The aim of this study was to explore different potential ways of administering MLT.

MATERIAL AND METHODS

Male Long Evans rats were entrained to an LD 12:12 cycle, and thereafter received different treatments.

One group was fitted with chronically placed subcutaneous catheters that allowed timed infusion (see (3) for technical details). For this group the daily infusion lasted 1 hour.

Melatonin after Four Decades, edited by James Olcese.
Kluwer Academic / Plenum Publishers, New York, 2000.

In a second group, the daily consumption of drinking water was measured. For 61 consecutive days the water bottles were presented to the animals only two hours daily (ZT 18–20, defined in the beginning of the experiment). This restricted drinking water regime (DWR) allowed us to control the dose of MLT as well as the duration of substance administration.

All animals were then transferred to constant dim (<3 lux) red light. When the rats showed free-running rhythms, the first group received a MLT infusion (0.5 mg MLT/kg in a 1% alcohol/Ringer-vehicle), while for the other group MLT was added to the drinking water (1 mg MLT/kg in 0.1% alcohol/tap water).

Body temperature, wheel-running activity, and general activity were recorded online using Mini-Mitter telemetry devices and the Dataquest III-system.

RESULTS

We found that MLT entrained all free-running rhythms measured in both of the experimental groups (see figure). Despite group differences in substance administration, the onset of running-wheel activity occured approximately at the same

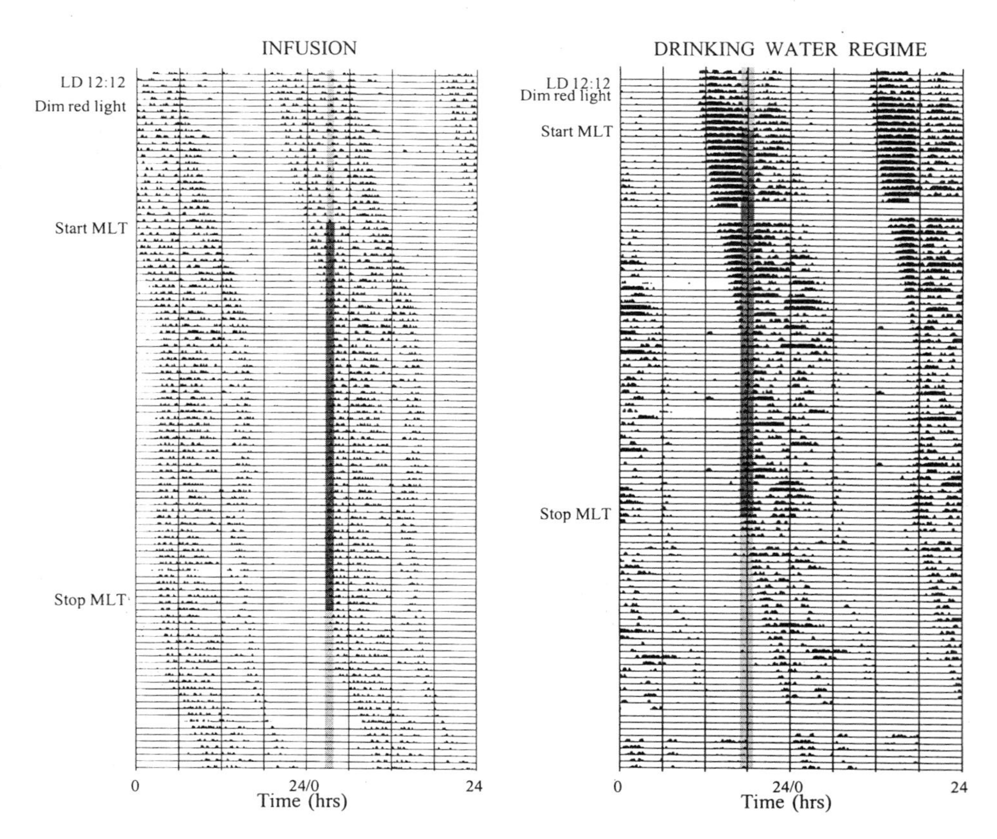

Figure 1. Representative double plotted running-wheel record of an infused rat (left panel), and of an animal submitted to a DWR (right panel). Successive days are plotted from top to bottom. Vertical bars indicate hour of infusion, or presence of water bottle for the animal kept under a DWR, respectively. Period of MLT administration is marked in black.

time in both groups, i.e. a few minutes after the beginning of MLT administration. The resulting phase angle between melatonin administration and onset of activity was very short for both groups. A free-running pattern in the measured rhythms was re-established when MLT was no longer present.

CONCLUSION

The results show that MLT, and not the manipulation alone, was able to entrain the rhythms. Because entrainment globally occured at the same zone for the two differently treated groups, a common mechanism seems to underlie the synchronizing effect of MLT. We conclude that MLT administration via drinking water and via infusion (1 hour) represent efficient ways to synchronize free-running rhythms in rats.

REFERENCES

1. Meijer, J.H. and Rietveld, W.J. Neurophysiology of the suprachiasmatic circadian pacemaker in rodents. Physiological Reviews 69:671–707, 1989.
2. Armstrong, S.M. Melatonin and circadian control in mammals. Experientia 45:932–938, 1989.
3. Kirsch, R., Belgnaoui, S., Gourmelen, S., and Pévet, P. Daily melatonin infusion entrains free-running rhythm activity in Syrian hamsters and Siberian hamsters. In Wetterberg, L. (Ed.). Light and biological rhythms in man. Oxford: Pergamon Press, pp. 107–120, 1993.

MELATONIN AS A CHRONOBIOTIC FOR CIRCADIAN INSOMNIA

Clinical Observations and Animal Models

Stuart Maxwell Armstrong

Brain Sciences Institute
Swinburne University of Technology
PO Box 218, Hawthorn, Victoria 3122
Australia

1. CIRCADIAN INSOMNIAS

Over the last two decades, circadian rhythm sleep disturbances have risen rapidly in prominence as evidenced by their full inclusion and description in the modern International Classification of Sleep Disorders (63). These disorders are classified into six subgroups (NODS is not pertinent)(Figure 1), but can be regrouped in different ways. Previously, we have grouped the disorders according to the type of chronobiotic action required to readjust or synchronize the disturbed rhythm (9). Here, the disorders are first grouped according to "natural" versus "artificial occurrence" and second according to the natural occurrence with age. Time Zone Change and Shift Work Sleep Disorder are clearly artificially induced, and will not be dealt with here. Delayed and Advanced Sleep Phase Syndromes (DSPS and ASPS) and Irregular Sleep-Wake Pattern (ISWP) occur naturally while Non-24-Hour Sleep-Wake Disorder can occur naturally or by accident as in many cases of the totally visually impaired. These represent the circadian insomnias and it is their recognition that has risen in importance in the context of the sleep disorders clinic. In the circadian insomnias the basic sleep mechanism is essentially functionally intact, but there is a temporal misalignment between the timing of the sleep-wake cycle and the norms of society for work and rest. This contrasts with the situation of the classic insomnias which are found under the headings of two forms of dyssomnia: Intrinsic and Extrinsic Sleep Disorders depending upon whether the cause is due to endogenous or exogenous causes. However, what both intrinsic and extrinsic groupings have in common is that the disturbed sleep is due to changes to the sleep mechanisms *per se*.

Melatonin after Four Decades, edited by James Olcese.
Kluwer Academic / Plenum Publishers, New York, 2000.

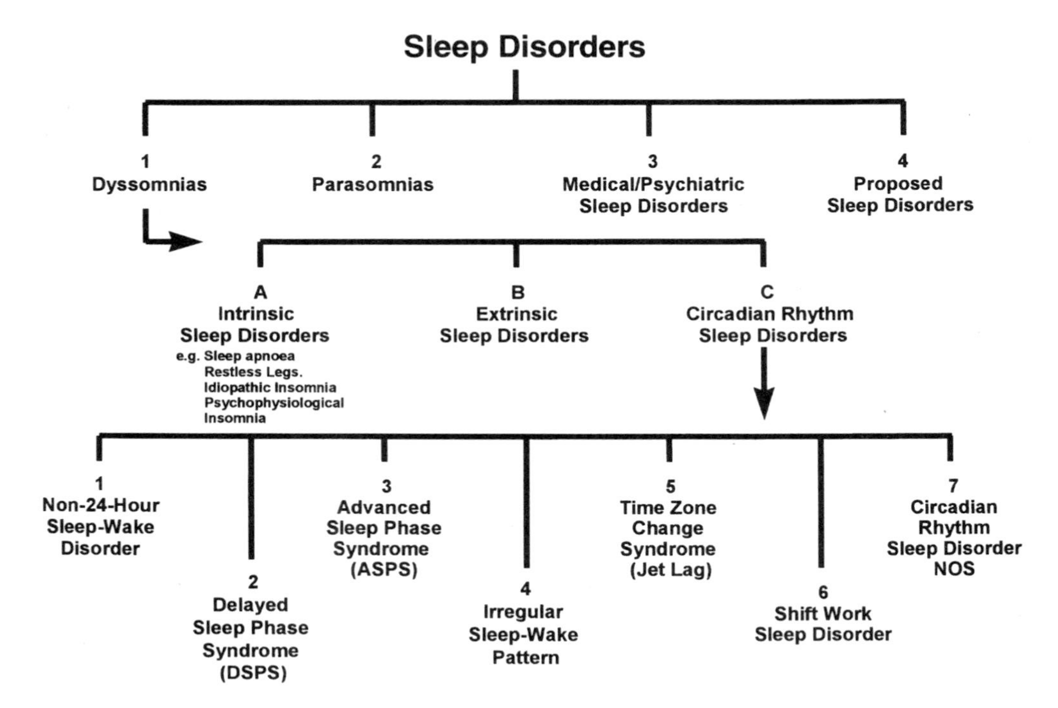

Figure 1. Diagrammatic summary of the 1990 International Classification of Sleep Disorders.

1.1. Description and Misdiagnosis

Traditionally, psychophysiological and idiopathic insomnias are divided into three groups depending on the time of wake occurrence: sleep onset, sleep maintenance and terminal insomnia. It is now clear that these three types of classic insomnia can be readily confused with the circadian insomnias particularly at the level of primary care physician. Clinical observation suggests strongly that circadian insomnias are far more extensive than hitherto believed.

1.1.1. Delayed Sleep Phase Syndrome (DSPS). In DSPS, sleep onset and wake times occur far later than normal (e.g. sleep 0300 to 1100 hours), there is no difficulty in maintaining sleep once initiated and sleep times are regular from day to day. Enforcing conventional sleep-wake times fails to advance sleep phase. DSPS is easily misdiagnosed as sleep onset insomnia, hence the extent of the disorder is likely to be more widespread than acknowledged at present. Furthermore, evening-types (*owls*) who prefer late night retirement may represent a milder form of DSPS and has been termed *sub-DSPS* (9). Since patients may have to rise in the morning at conventional times in order to go to work, a sleep debt accumulates as the week progresses.

Anecdotal evidence from the sleep disorders clinic indicates that many cases of Idiopathic Insomnia (childhood insomnia) may be due to delayed circadian phase and this may explain why Idiopathic Insomnia has been so hard to treat. The origin of childhood DSPS may not be due to circadian clock abnormality per se. A prolonged period of late nocturnal sleep onset due to family or environmental reasons (e.g. fear of the dark, noise from television, arguments between parents, etc.) may be enough to delay sleep as a learned phenomenon which subsequently becomes "phase locked" to a later than socially acceptable time.

1.1.2. Non-24-Hour Sleep-Wake Disorder. Occasionally, patients attending the sleep disorders clinic who report DSPS symptoms are found subsequently not to be synchronized at all but free-run; they have Non-24-Hour Sleep-Wake Disorder. Thus, as their sleep-onset and wake times occur later (delay) on successive days the individuals sleep-wake schedule becomes progressively asynchronised to real clock time by an average amount daily.

As with previous reports (30,41), these "normal" individuals may free-run while living in society under "normal" exposure to the day-night cycle. When such an individual's sleep span is out-of-phase with societal sleep-wake conventions, the major complaint is of difficulty in nocturnal sleep and daytime sleepiness. When the individual is in-phase with societal sleep-wake conventions there is usually no complaint; therefore, reports of sleeping problems are periodically intermittent. In the symptomatic periods, psychological and physical performance is impaired and mood changes may be noted.

This problem of lack of entrainment is more commonly found in blind individuals, in addition to astronauts and sub-mariners living under artificial LD cycles for extended periods of time.

1.1.3. Advanced Sleep Phase Syndrome (ASPS). In ASPS, which is far less common than DSPS, sleep onset and wake times are intractably earlier than convenient for the normal work-leisure schedule. Early morning awakening takes place between 0300 and 0500 hours and sleep onset between 2000 and 2100 hours. ASPS may be misdiagnosed as terminal insomnia. It is found frequently in the elderly, in major depression as well as occasionally in young normal adults. While DSPS patients tend to be long sleepers, ASPS tend to be short sleepers.

1.1.4. Irregular Sleep-Wake Pattern Disorder (ISWP). ISWP refers to the loss of the overt circadian rhythm of sleep waking and the loss of response to the zeitgeber, resulting in a disorganized and variable sleep pattern, usually broken up onto several short blocks. Daytime napping is frequent but irregular and bed rest is excessive, while nocturnal insomnia is typical. However, the total amount of sleep per 24 hours may be normal for the age of the individual. Other circadian functions may flatten and lose their rhythmicity e.g. core temperature and endocrine rhythms, the pattern of normal daily schedules and activities, such as endocrine rhythms, the pattern of normal daily schedules and activities, such as meal times. The loss in amplitude of internal circadian rhythms and the fragmentation of the sleep pattern are highly suggestive of an uncoupling of the oscillatory systems that comprise the circadian pacemaker.

ISWP may be misdiagnosed as sleep maintenance insomnia, particularly in the elderly. Conversely, voluntary or involuntary long durations in bed will fragment sleep and this fragmented sleep pattern misdiagnosed as ISWP. Many elderly use bed as an escape from life; 10 hours in bed for someone needing 6 hours sleep is a formula for insomnia.

1.2. Life-Span and the Circadian Insomnias

One obvious feature of the circadian insomnias apparently not commented on previously is their relationship to age. In Figure 2 the insomnias are arranged not in the order presented in the diagnostics manual (63) but according to time of life. It becomes clear that there is a progressive change with age that parallels changes to the circadian pacemaker evident from animal studies though not necessarily yet demonstrated in humans. A review of sleep disturbances, changes to circadian rhythmicity and the chronobiotic potential of MLT in retarding aging has been given elsewhere (12,42). Briefly, a large body of evidence, mainly animal but some human, indicates that with age: circadian period is altered, amplitude is reduced, alpha-rho loses its definition, stability to environmental challenge is reduced (e.g. phase shifts of the zeitgeber), internal desynchrony (humans) and splitting (rodents) increases. In addition, a number of changes to the circadian organization of the sleep-wake cycle (as distinct from sleep mechanisms *per se*) occur. Taken together, the single underlying variable tying all these appears to be amplitude of the circadian clock. It is also interesting to note the interrelationship between the occurrence of circadian insomnias, changes with age of circadian variables (Figure 2) and the occurrence of depressive illness (Figure 3).

The classic cases of DSPS occur in adolescence and teenage years but could occur earlier (the possible confounding of childhood DSPS as Idiopathic Insomnia has been raised above). In contrast, ASPS is found commonly in seniors. Clinical observations suggest exaggerated evening tiredness and dozing (particularly in front of TV) starts in the mid-forties and fifties, and by the sixties early morning awakening is apparent. ASPS is rare in younger individuals (unless they are suffering from diagnosed/undiagnosed major depression where terminal insomnia is common). Thus, there is a trend from a delayed clock in youth to an advanced clock in old age; a phase-lag of "normal" bedtime becomes a phase-lead.

Whereas ASPS occurs in seniors, ISWP is more likely found in the nursing home and geriatric population. The fragmented sleep of the demented, nocturnal wanderings and Sundowner's syndrome are all suggestive of a fragmented sleep-wake cycle due to a low amplitude clock. However, non-demented nursing home residents are also obvious targets for ISWP. Indeed, in the nursing home, lack of bright light input to the circadian clock could exacerbate reduced clock amplitude as would lack of strong secondary zeitgebers. In addition to lack of light, an important health parameter would be change in visual sensitivity to light with age.

In assessing clock amplitude in the elderly, it is advantageous to measure more than one variable, e.g. core temperature nadir and circulating MLT levels. Impaired thermoregulation on the one hand, and potential pathology to the pineal gland and the circuitous pathway from the suprachiasmatic nuclei (SCN) (including autonomic innervation) on the other, makes hazardous single variable amplitude estimation in the aged.

It is well established that after puberty MLT levels reduce with age and therefore the changes summarized in Figure 2 covary with a reduction in circulating MLT levels.[1] The pertinent question is: What do reduced MLT levels do, if anything, to the integrity of the SCN pacemaker? One can speculate on possible changes by observing the effects of Px on rest-activity rhythms. This is best done, not by looking primarily at mammals but by reverting to our knowledge of birds and lizards.

[1] If the approach outlined here is substantially correct, it only accounts for sleep phase changes post-puberty: it implies that the relationship between the SCN and circulating MLT feedback changes dramatically after puberty. The highest levels of melatonin are found pre-puberty, yet, as far as is known, these high levels have no major influence on sleep. Perhaps the dynamics of Process S and C (19) are different in pre-puberty with the homeostatic aspect dominating.

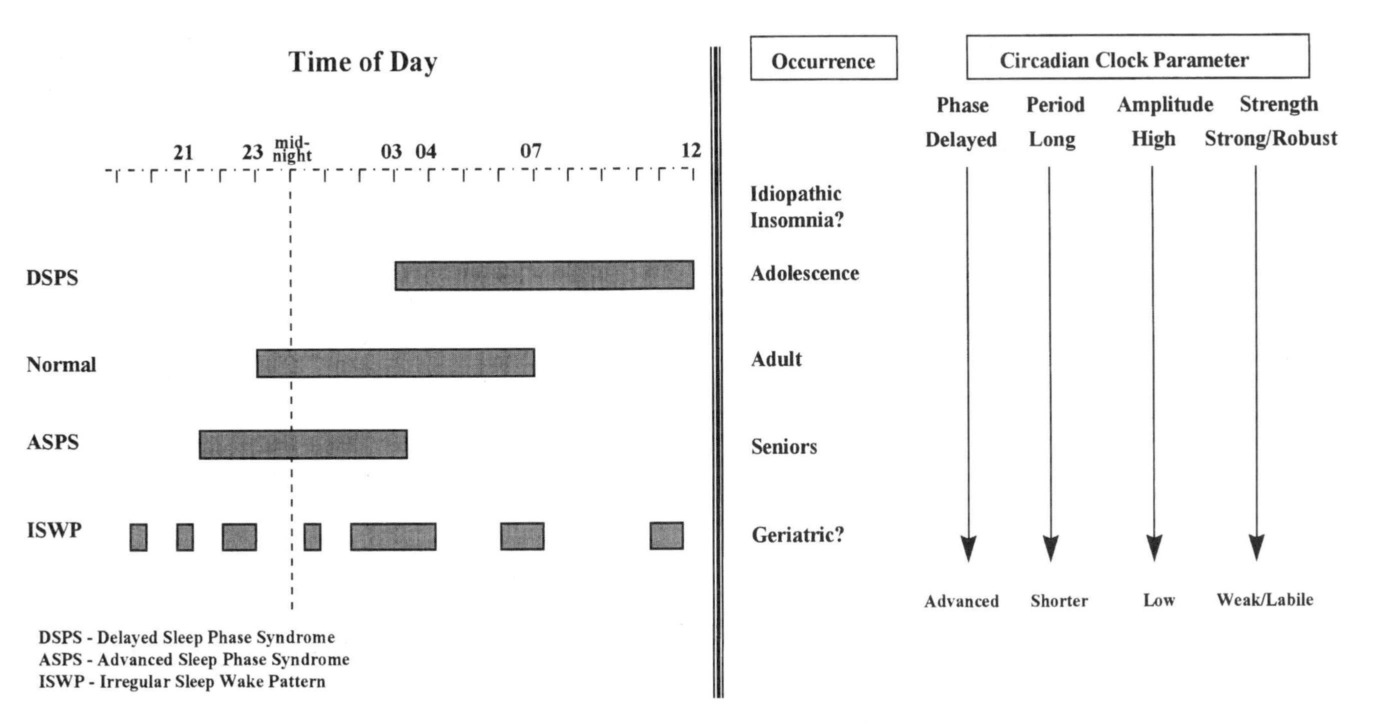

Figure 2. Relationship between the circadian insomnias and changes to the circadian pacemaker with age in the normal population.

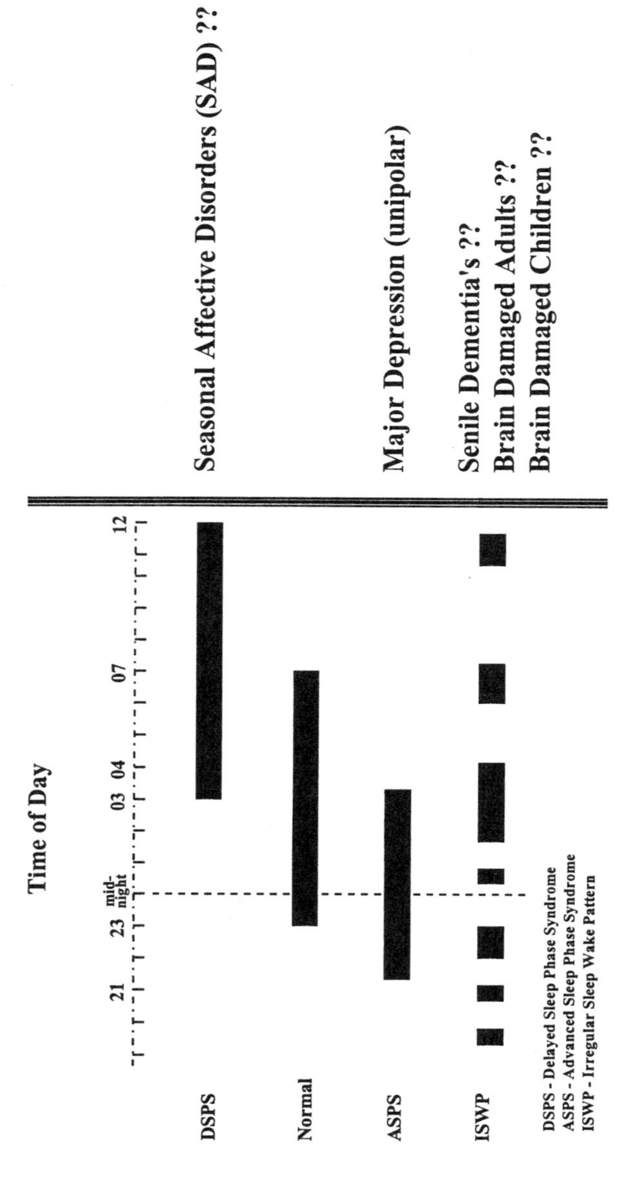

Figure 3. Circadian Sleep Disorders Occurrence in Psychiatric Populations.

1.3. Aging, Sleep, and Endogenous MLT

One way of conceptualizing the changes in sleep phase over the life span comes from an understanding of the importance of the pineal in circadian locomotor activity in non-mammalian vertebrates (7,13). Briefly, results from pinealectomy (Px) studies in both birds (24) and lizards (65) point to an important role for MLT in regulating the phase and amplitude of the circadian pacemaker.

In birds, three categories of effect have been recorded: in passerines such as the house sparrow (*Passer domesticus*) (23), as well as in three other species of sparrow, Px induces arrhythmicity in constant conditions. This contrasts markedly with the lack of effect of Px in gallinaceous birds such as the Japanese quail (61) and domestic chicken. Lying between these extremes is the effect of Px on the European starling, *Sturnus vulgaris* (24,51) (See Figure 4 A–C). In individual starlings Px may be: [a] ineffective, [b] shorten the free-running period or [c] produce arrhythmicity. In order to account for these three outcomes within one species it was proposed that differences in circadian organization between species must be quantitative and not qualitative (24,25); both intra- and interspecies differences to Px can be accounted for by differences in coupling strength between self-sustained oscillators comprising the circadian pacemaker system (Figure 4D). Where coupling is weak, such as in sparrows, Px leads to loss of mutual entrainment between oscillators so that behavioral activity arrhythmia results. At the other extreme in the quail, coupling would be strong and Px would have little effect on rest-activity. The circadian rhythm of MLT secretion was suggested to function as an internal synchronizing agent on one or more or the self-sustained oscillators (25). A caveat must be introduced here. Although there is a lack of effect of Px on quail rest-activity cycles, Px plus enucleation results in arrhythmia just as Px does in the sparrow; over half of the circulating MLT in the quail comes from the retinae (67). This finding does not invalidate the present thesis, since individual differences after Px still exist in one species, the European starling. In *S. vulgaris*, individual differences may reflect changes to visual sensitivity (26). A parallel between this and human aging is obvious (see Section 2.5).

On the basis of rodent studies, it is generally thought that Px does not affect circadian rhythmicity (reviewed elsewhere (7,13)). However, when appropriate experimental paradigms are employed, for example environmental insults such as phase-shifts (11) or exposure to bright constant light (16), the importance of the pineal on circadian parameters can be demonstrated. Thus, the circadian clock of rodents is influenced by endogenous MLT but strong coupling between oscillators make ingenious experimentation mandatory to demonstrate this.

To go further, in order to account for changes in direction of free-running period after Px in *S. vulgaris*, one would envisage a change in the balance/ratio strength between two main clusters of oscillators: dawn and dusk. The evidence for E and M oscillators (45), the changes to split rhythms by unilateral SCN lesions (44), the changes to rhythms of c-fos photoinduction (28) all in rodents provide a foundation for this, as does the splitting of the sleep span in humans confined to 14 hours of darkness (70). In humans an indication of the ratio strength between E and M may eventually be determined when large enough numbers of older (ASPS) and younger (DSPS) subjects are compared for sleep phase and MLT profile. Presumably, from the covariation with age, high MLT levels induce phase delays while low MLT levels induce phase advances. This suggests a preferential influence of MLT on E and M oscillators in humans. Receptor density distribution is an unknown variable here.

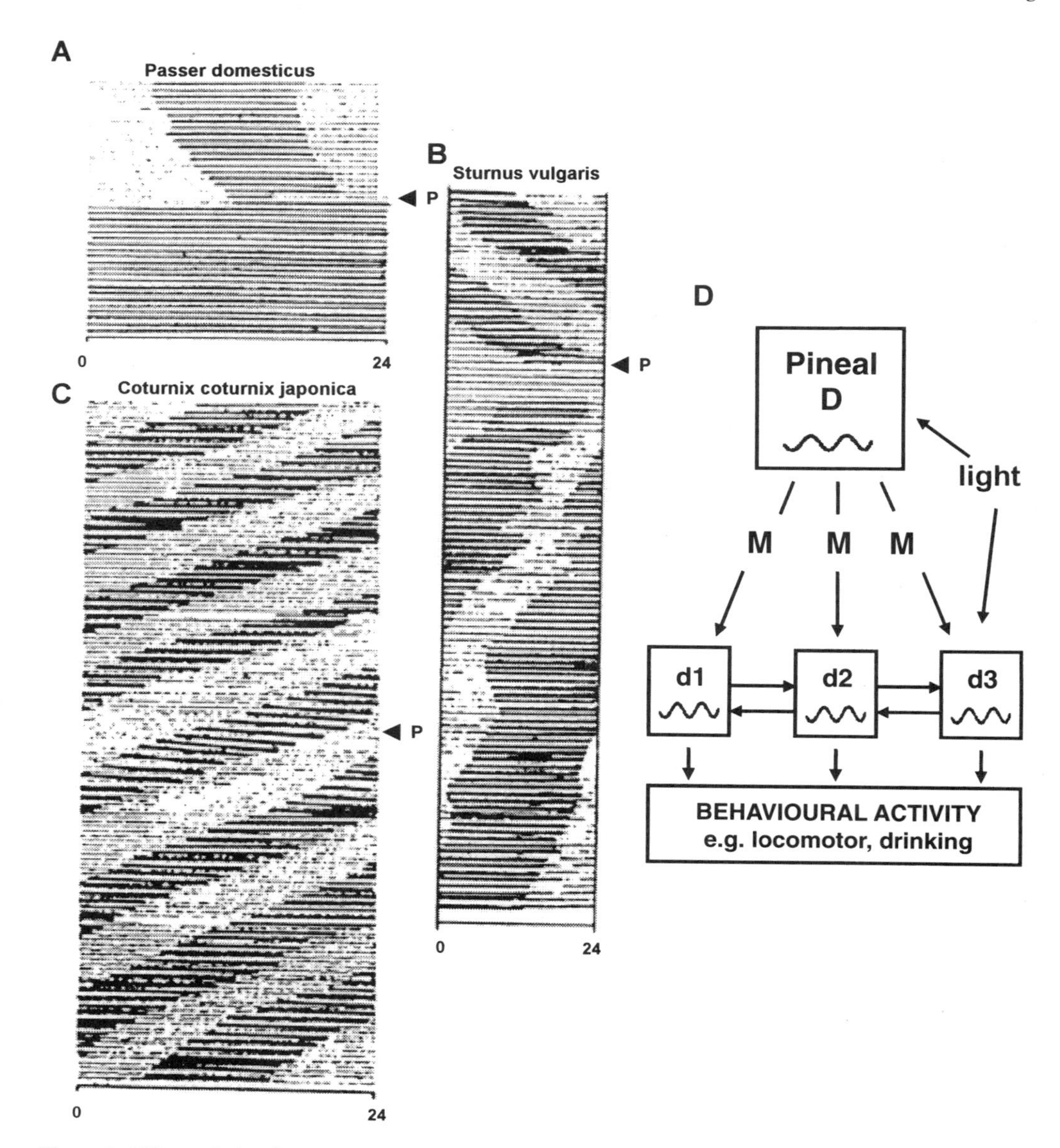

Figure 4. Effects of pinealectomy on free-running locomotor rhythms of three species of birds. Pinealectomy (indicated by P) of the house sparrow [A] results in arrhythmicity, European starling [B] change in tau and Japanese quail [C] no effect. [D] Gwinner's (24,25) model of the avian circadian pacemaker system. (Redrawn and modified (7).)

It is interesting that the bird and lizard models, when generalized to humans, allow us to interpret changes to sleep in the elderly in a meaningful way as depicted in Table 1. Sleep in the elderly can be categorized into three types depending upon coupling strength of the oscillators in the pacemaker: quail-like, starling-like and sparrow-like. The timing of MLT therapy as a "sleeping pill" in the elderly will be very different depending on the underlying typology.

Table 1. Seep-wake patterns of elderly humans: Avian analogies

[1] Px Quail-like pattern[#]
1. Low amplitude rhythm in pineal melatonin leads to low circulating nocturnal melatonin levels.
2. Normal coupling strength between SCN self-sustained oscillators.
3. No changes to tau
4. Sleep disorder: NONE (NORMAL).

[2] Px Starling-like pattern
1. Low amplitude rhythm in pineal melatonin leads to low circulating nocturnal melatonin levels.
2. Reasonable but unbalanced coupling strength between SCN self sustained oscillators leads to changes in tau, manifesting as a shortening in humans.
3. Since tau is less than 24h this produces to a phase-lead of the rest –activity cycle under entrained light-dark conditions.
4. Sleep disorder: ADVANCED SLEEP PHASE SYNDROME

[3] Px Sparrow-like pattern
1. Low amplitude rhythm in pineal melatonin leads to low circulating melatonin levels
2. Weak coupling strength between SCN self-sustained oscillators leads to arrhythmicity of the rest –activity cycle.
3. Sleep disorder: IRREGULAR SLEEP-WAKE PATTERN

Note: Px plus enucleation in the quail results in complete arrhythmia.

In summary, one may conceptualize MLT acting as the "cement" of the clock, reflected in the strength of the coupling of the oscillators and sub-oscillators. As coupling strength diminishes, amplitude is reduced, phase changes with respect to the LD cycle, a strong, robust rhythm becomes labile and eventually at the lowest amplitude the whole rest-activity cycle moves from its circadian to an ultradian domain.

Against this background, the ability of exogenous MLT administration to reverse circadian phase and amplitude changes and therapeutically improve circadian insomnias can be assessed. The prognosis is good, but the vexing question of pharmacology versus physiology remains.

2. CHRONOBIOTICS AND INSOMNIA THERAPY

2.1. Definition of Chronobiotic

A chronobiotic was first defined loosely as a drug that specifically affects some aspect(s) of biological time structure (60) and then more precisely as: a chemical substance capable of therapeutically re-entraining short-term dissociated or long-term desynchronized circadian rhythms, or prophylactically preventing their disruption following environmental insult (59) (see reviews elsewhere (20,55)).

MLT may be an internal zeitgeber (7,8) and as a rule of thumb, its effects on circadian timing are the opposite to that of light i.e. when light phase advances a rhythm, MLT administration phase delays and vice-versa. Since the ability of MLT to entrain rat circadian rhythms depends upon the integrity of the SCN (17) and since both the rodent and human SCN contain high affinity MLT receptors (50,69), it would be reasonable to expect that exogenous MLT administration should be therapeutic for treatment of circadian insomnias.

With respect to MLT and treatment of circadian sleep disorders, the chronobiotic properties of the substance should be distinguished from its soporific/hypnotic effects

(56,73). In many studies claiming hypnotic effects, these have been confounded with chronobiotic effects. The extents to which the hypnotic properties are reliant on the hypothermic effects are still debatable (31,38).

2.2. Delayed Sleep Phase Syndrome (DSPS)

An animal model of DSPS has been developed in laboratory rats (10). By maintaining rats in DD for several months and then returning them to a conventional 12:12LD cycle, the onset of activity (α) often lags behind the onset of darkness by amounts of three to four hours. Since rodents are nocturnally active species, the shifting of α by MLT administration is equivalent to shifting of ρ (the inactive phase) in diurnally active species such as humans. Injecting MLT or the agonist, S20098 (Servier, France) just prior to dark onset can eliminate this negative phase angle difference. Over approximately one week, the onset of α advances until it reaches dark onset and can be held there by continuing daily injections. When activity ceases, α drifts back to lag dark onset again (10). Daily MLT injections also phase advance the locomotor activity rhythms of individual Siberian hamsters; however, in this seasonal breeding species, the photoperiodic conditions are crucial to attaining a phase adjustment to MLT injections (46).

Oral, daily administration of 2 mg MLT at 1700 hours for one month increased early evening fatigue and sleep onset, and in five out of 12 human subjects advanced the endogenous MLT rhythm by 1–3 hours (4,5,74). The phase advance of the endogenous MLT rhythm in normal sighted subjects was confirmed and elaborated when MLT (8 mg) was administered at 2200 hours for only four days (39).

In the course of administering 5 mg MLT orally to free running blind subjects phase advances of the endogenous MLT rhythm as well as the circadian rhythm of cortisol were found. One totally blind subject entrained at an abnormal delayed phase was able to be advanced (52,54). In two independent studies, two blind subjects with DSPS were also successfully treated with 5 mg MLT (58,64). In the latter study the timing of treatment was crucial to a successful phase advance of sleep onset, and is predictable from the human PRC for MLT (34,36). Inconsistent results of some previous studies on the blind may be best explained by the need for crucial timing in MLT administration as summarized elsewhere (53).

Against the preliminary background in blind subjects, the demonstration of a successful resetting of sleep in sighted DSPS sufferers by 5 mg MLT administrated over four weeks in a double-blind placebo cross-over trail was exciting. MLT was taken at 2200 hours, five hours before sleep onset (21). Subsequently, five hours before Dim Light MLT Onset was confirmed as being a better marker for successful DSPS treatment (32,62). In patients with winter depression, thought to be phase delayed, low doses of afternoon administered MLT significantly decreased depressive ratings, presumably by phase advancing the endogenous clock (35).

2.3. Non-24-Hour Sleep-Wake Disorder

MLT injected at the same clock time each day to rats free-running in DD entrains the circadian locomotor rhythm (49). Free running rhythms of lizards (68) and European starlings (25) entrain to daily injections but at a different phase to rats. In the nocturnal rat, entrainment takes place when the beginning of α coincides with injection time while in the diurnal starling and lizard it is the beginning of ρ that synchronizes to the injection. While daily infusions entrain both Siberian (*Phodopus sungaris*) and Syrian (*Mesocricetus auratus*) hamsters the phase angle can be peculiar (29) compared to rats, presumably demonstrating the differential effect of infusion versus injection and not a species difference.

Since exogenous MLT entrains representatives from three classes of terrestrial vertebrates (lizards, birds and rodents) one would predict that the likelihood of it entraining humans and other primates would be high. However, a negative finding on primates (*Sagerinus fuscicollis* the saddleback tamarin) comes from an experiment conducted under an exotic LD regime (33). The LD cycle was steadily phase advanced by one hour per day and MLT administered at the same clock time each day failed to compete with the LD cycle.

In a temporal isolation study on humans in which the fractional desynchronisation procedure was used, MLT failed to strengthen the lengthening LD zeitgeber and prevent internal desynchronisation between the rhythm in the temperature and the sleep-wake cycle (71). Nevertheless, a synchronizing effect on the fatigue alertness rhythm was reported (71). However, from PRC to MLT in humans (34,36) it is evident that MLT was administered at entirely the wrong time to maintain entrainment.

There are several reports of successful entrainment of circadian rhythms of individual blind human subjects by MLT treatment. These include in single subjects, synchronization of a disturbed sleep-wake cycle (5 mg) (3), advancing sleep onset, stabilizing sleep and reducing day time naps (5 mg) (58) and entrainment (7.5 mg) (57). In the first subject (3), the free-running rhythms in both temperature and urinary cortisol were on further analysis found not to be entrained (22) whereas in the third subject (57) entrainment of the endogenous MLT rhythm was demonstrated, indicating true entrainment of the underlying circadian pacemaker. In a blind, retarded boy with Non-24-Hour Sleep Wake Disorder, MLT corrected sleep phase (43).

Thus the evidence to date in humans is ambiguous; synchronization of human sleep is reported, but in many subjects it is less likely that the clock is entrained, as plasma MLT rhythm may continue to free-run (6). The sleep-wake pattern of humans evidently is not equivalent to the rest-activity locomotor rhythm of rodents. However, since "feelings of well being" seem to depend on synchronizing sleep phase to nighttime (30), this may be all that is needed in clinical practice for an efficacious result.

2.4. Advance Sleep Phase Syndrome (ASPS)

In our early work, rats in our laboratory only phase advanced to exogenous MLT. Therefore, without phase delays we had no animal model of ASPS. In the free-running rat, single injections given approximately every two to three weeks at different circadian times produced a PRC to MLT (7,14) with a "gate of circadian frequency" or narrow "window of sensitivity" between CT 9 to CT 11 when phase advances occur. This situation contrasted with that of the lizard in which both the PRC to MLT had both large delay and advance portions (66) and was similar to a PRC to dark pulses in nocturnal rodents. More recently, other rodents show findings which contrast to our rats. In C3H/HeN mice (15), CBA mice (48) and the diurnal Indian palm squirrel, *Funambulus pennati*, there appear to be two times of day when the circadian clock is sensitive to MLT (47). As yet, there is no clear explanation for these species differences in PRC's.

It is important to note that *in vitro* stimulation of rat hypothalamic slices with intact SCN shows changes to neuronal firing at subjective dawn as well as subjective dusk (40). Interestingly, when the mouse SCN is stimulated directly, the size of the phase shifts are larger (several hours) than those observed in the wheel running behavior of the whole animal (37).

There appear to be no clinical trials on the therapeutic affects of MLT administration in human ASPS, but clearly early morning administration would be optimal for phase delaying the sleep-wake cycle. The published PRC to MLT in humans (34,36) clearly shows phase delays to morning MLT administration that would be sufficient to combat ASPS.

2.5. Irregular Sleep-Wake Pattern [ISWP]

Since rats take up to five years to age in our laboratory, a model of a severely disrupted circadian rest-activity rhythm was produced by utilizing an exotic lighting regime in young adult animals (18). By increasing photoperiod and decreasing scotoperiod by 30 minutes per week, instability of the rat circadian system starts at about LD 19:5. The rhythm either breaks away and free-runs, or internally desynchronizes, resulting in severe disruption of the rest-activity rhythm. Daily injections of MLT resynchronize the disrupted components and this reorganization suggests that MLT acts on the coupling of sub-oscillators that constitute the circadian pacemaker.

From the approach taken above (Section 1.1, 1.2), ISWP would be found in geriatric humans with highly fragmented sleep, a group not evaluated for MLT therapy. MLT has been tested in the elderly with sleep maintenance insomnia with disappointing results (27). However, these subjects were healthy seniors not geriatrics with diseases, and not reported were the number and duration of individual nocturnal waking and their correlation with endogenous melatonin levels.

3. CONCLUSIONS

Previously it has been concluded that amplitude is the key variable to be focused upon in chronopathology in general and aging in particular (12). According to Aschoff (1) the stability of the circadian system is positively correlated with its amplitude: circadian systems with labile phases can be expected to have small amplitudes. If phase lability is symptomatic of a certain illness, then measures taken to improve amplitude are likely to be beneficial (2). Along the same lines, Wever (72) concluded that amplitude of circadian rhythms is their most relevant parameter because it determines stability: the larger the amplitude the smaller the cycle to cycle variability, and the more rigid the rhythm against external or irregular environmental perturbations.

The present thesis is that these statements are particularly relevant to the development and treatment of circadian insomnias.

REFERENCES

1. Aschoff, J. A survey of biological rhythms. In Handbook of Behavioral Neurobiology, 4, Biological Rhythms, eds. J. Aschoff, Plenum Press, New York, pp. 3–10, 1981.
2. Aschoff, J. Disorders of the circadian system as discussed in psychiatric research. In Circadian Rhythms in Psychiatry, eds. T.A. Wehr and F.K. Goodwin, Boxwood Press: California, 1983.

3. Arendt, J., Aldhous, M., and Wright, J. Synchronisation of a disturbed sleep-wake cycle in a blind man by melatonin treatment. Lancet 2:772–773, 1988.
4. Arendt, J., Bojowski, C., Folkard, S., Franey, C., Marks, V., Minors, D., Waterhouse, J., Wever, R.A., Wildgruber, C., and Wright, J. Some effects of melatonin and the control of its secretion in humans. In Photoperiodism, melatonin and the pineal. Ciba Foundation Symposium 117, eds D. Evered and S. Clark, Pitman Publishing, UK, pp. 266–279, 1985.
5. Arendt, J., Borbély, A.A., Franey, C., and Wright, J. The effects of chronic, small doses of melatonin given in the late afternoon on fatigue in man: a preliminary study. Neurosci. Lett. 45:317–321, 1984.
6. Arendt, J., Skene, D.J., Middleton, B., Lockley, S.W., and Deacon, S. Efficacy of melatonin treatment in jet lag, shift work and blindness. J. Biol. Rhythms, 12:604–617, 1997.
7. Armstrong, S.M. Melatonin: the internal zeitgeber of mammals? In Pineal Research Reviews. 7, ed. R.J. Reiter, New York: Alan R. Liss, pp. 157–202, 1989.
8. Armstrong, S.M. Melatonin and circadian control in mammals. Experientia 45:932–938, 1989b.
9. Armstrong, S.M. Treatment of sleep disorders by melatonin administration. In Advances in Pineal Research: 6, ed. A. Foldes and R.J. Reiter, J. Libbey and Co., pp. 263–274, 1991.
10. Armstrong, S.M., McNulty, O.M., Guardiola-Lemaitre, B., and Redman, J.R. Successful use of S20098 and melatonin in an animal model of delayed sleep-phase syndrome (DSPS). Pharmac. Biochem. Behav. 46:45–49, 1993.
11. Armstrong, S.M. and Redman, J. Melatonin administration: effects on rodent circadian rhythms. In Photoperiodism, Melatonin and the Pineal, ed. D. Evered and S. Clark, London: Pitman, pp. 188–207, 1985.
12. Armstrong, S.M. and Redman, J.R. Melatonin: A chronobiotic with anti-aging properties. Med. Hypotheses 34:300–309, 1991.
13. Armstrong, S.M. and Redman, J.R. Melatonin and Circadian Rhythmicity. In Melatonin: Biosynthesis, Physiological Effects, and Clinical Applications, eds. H-s Yu and R.J. Reiter, Boca Raton: CRC Press, pp. 187–224, 1992.
14. Armstrong, S.M., Thomas, E.M.V., and Chesworth, M.J. Melatonin induced phase-shifts of rat circadian rhythms. In advances in Pineal Research, ed. R.J. Reiter and S.F. Pang, J. Libbey and Co., London, pp. 265–290, 1989.
15. Benloucif, S. and Dubocovich, M.L. Melatonin and light induce phase shifts of circadian activity rhythms in the C3H/HeN mouse. J. Biol. Rhythms 11:113–125, 1996.
16. Cassone, V.M. The pineal gland influences rat circadian activity rhythms in constant light. J. Biol. Rhythms 7:27–40, 1996.
17. Cassone, V.M., Chesworth, M.J., and Armstrong, S.M. Entrainment of rat circadian rhythms by daily injections of melatonin depends upon the hypothalamic suprachiasmatic nuclei. Physiol. Behav. 36:1111–1121, 1986.
18. Chesworth, M.J., Cassone, V.M., and Armstrong, S.M. Effects of daily melatonin injections on activity rhythms of rats in constant light. Am. J. Physiol. 253:R101–R107, 1987.
19. Daan, S., Beersma, D.G.M., and Borbely, A.A. The timing of human sleep: recovery process gated by a circadian pacemaker. Am. J. Physiol. 246:R161–R178, 1984.
20. Dawson, D. and Armstrong, S.M. Chronobiotics- drugs that shift rhythms. Pharmacol. Ther. 69:15–36, 1996.
21. Dahlitz, M.J., Alvarez, B., Vignau, J., English, J., Arendt, J., and Parkes, J.D. Delayed sleep-phase syndrome: response to melatonin. Lancet 337:1121–1124, 1991.
22. Folkard, S., Arendt, J., Aldhous, M., and Kennett, H. Melatonin stabilises sleep onset times in a blind man without entrainment of cortisol or temperature rhythms. Neurosci. Lett. 113:193–198, 1990.
23. Gaston, S. and Menaker, M. Pineal function: the biological clock in the sparrow? Science 160:1225–1227, 1968.
24. Gwinner, E. Effects of pinealectomy on circadian locomotor activity rhythms in European starlings, Sturnus vulgaris. J. comp. Physiol. 126:123–129, 1978.
25. Gwinner, E. and Benzinger, I. Synchronisation of a circadian rhythm in pinealectomised European starlings by daily injections of melatonin. J. comp. Physiol. 127:209–213, 1978.
26. Hau, M. and Gwinner, E. Continuous melatonin administration accelerates resynchronization following phase shifts of a light-dark cycle. Physiol. Behav. 58:89–95, 1995.
27. Hughes, R.J., Sack, R.L., and Lewy, A.J. The role of melatonin and circadian phase in age-related sleep-maintenance insomnia: assessment in a clinical trial of melatonin replacement. Sleep 21:52–68, 1998.
28. Illnerova, H. and Samova, A. Photic entrainment of the mammalian rhythm in melatonin production. J. Biol. Rhythms. 12:547–555, 1997.

29. Kirsch, R., Belgnaoui, S., Gourmelen, S., and Pevet, P. Daily melatonin infusion entrains free-running activity in Syrian and Siberian hamsters. In Light and biological rhythms in man, ed. L. Wetterberg, Pergamon Press, pp. 107–120, 1993.
30. Kokkoris, C.P., Weitzman, E.D., Pollak, C.P., Spielman, A.J., Czeisler, C.A., and Bradlow, H. Long-term ambulatory temperature monitoring in a subject with a hypernychthermeral sleep-wake cycle disturbance. Sleep 1:177–190, 1978.
31. Krauchi, K., Cajochen, C., and Wirz-Justice, A. A relationship between heat loss and sleepiness: effects of postural change and melatonin administration. J. Appl. Physiol. 83:134–139, 1997.
32. Laurant, M.W., Nagtegaal, J.E., Krekhof, G.A., Coenen, A.M.L., and Smits, M.G. Quality of life before and after treatment with melatonin in patients with delayed sleep phase syndrome. In Sleep-wake research in the Netherlands, vol 8, eds. D.G.M. Beersma, J.B.A.M. Arends, A. Lex van Bemmel, W.F. Hofman and G.S.F. Ruigt, Dutch Soc. For Sleep-wake Research, Leiden, The Netherlands, pp. 83–86,1997.
33. Lerchl, A. and Küderling, I. Gonadal steroid excretion and daily locomotor activity patterns in saddle back tamarins (Sagerinus fuscicollis; Callitrichidae; Primates) influence of melatonin and light-dark cycles. In Advances in Pineal Research: 4, eds. S.F. Pang and R.J. Reiter, J. Libbey: London, pp. 271–276, 1989.
34. Lewy, A.J., Bauer, Ahmed, Thomas, Cutler, Singer, Moffit, and Sack, R.L. The human phase response curve (PRC) to melatonin is about 12 hours out of phase with the PRC to light. Chronobiol. Int. 15:71–83, 1998.
35. Lewy, A.J., Bauer, V., Cutler, N.L., and Sack, R.L. Melatonin treatment of winter depression: a pilot study. Psychiat. Res. 77:57–61, 1998.
36. Lewy, A.J., Saeeduddin, A., Latham-Jackson, J.M., and Sack, R.L. Melatonin shifts human circadian rhythms according to a phase response curve. Chronobiol. Int. 9:380–392, 1992.
37. Liu, C., Weaver, D.R., Jin, X., Shearman, L.P., Pieschl, R.L., Gribkoff, V.K., and Reppert, S.M. Molecular dissection of two distinct actions of melatonin on the suprachiasmatic circadian clock. Neuron 19:91–102, 1997.
38. Lushington, L., Pollard, K., Lack, L., Kennaway, D.J., and Dawson, D. Daytime melatonin administration in elderly good and poor sleepers: effects on core temperature and sleep latency. Sleep 20:1135–1144, 1997.
39. Mallo, C., Zaidan, R., Faure, A., Brun, J., Chazot, G., and Claustrat, B. Effects of a four-day nocturnal melatonin treatment on the 24 h plasma melatonin, cortisol and prolactin profiles in humans. Acta Endocrinol. 119:474–480, 1988.
40. McArthur, A.J., Hunt, A.E., and Gilette, M.U. Melatonin action and signal transduction in the rat suprachiasmatic circadian clodk: activation of protein kinase C at dusk and dawn. Endocrinology 138:627–634, 1997.
41. McArthur, A.J., Lewy, A.J., and Sack, R.L. Non-24-hour sleep-wake syndrome in a sighted man: circadian rhythm studies and efficacy of melatonin treatment. Sleep 19:544–553, 1996.
42. Myers, B.L. and Badia, P. Changes in circadian rhythms and sleep quality with aging: mechanisms and interventions. Neurosci. Biobehav. Rev. 19:553–571, 1995.
43. Palm, L., Blenow, G., and Wetterburg, L. Correction of non-24-hour sleep-wake cycle by melatonin in a blind retarded boy. Ann. Neurol. 29:336–339, 1991.
44. Pickard, G.E. and Turek, F.W. Splitting of the circadian rhythm of activity is abolished by unilateral lesions of the suprachiasmatic nuclei. Science 215:1119–1121, 1982.
45. Pittendrigh, C.S. and Daan, S. A functional analysis of circadian pacemakers in nocturnal rodents. V. Pacemaker structure: a clock for all seasons. J. comp. Physiol. 106:333–355, 1976.
46. Puchalski, W. and Lynch, G.R. Daily melatonin injections affect the expression or circadian rhythmicity in Djungarian hamsters kept under a long-day photoperiod. Neuroendocrinology, 48:280–286, 1988.
47. Rajaratnam, S.M.W. and Redman, J.R. Effects of daily melatonin administration on circadian activity rhythms in diurnal Indian palm squirrel (Funambulus pennanti). J. Biol. Rhythms. 12:339–347, 1997.
48. Redman, J.R. Circadian entrainment and phase shifting in mammals with melatonin. J. Biol. Rhythms 12:581–587, 1997.
49. Redman, J.R., Armstrong, S.M., and Ng, K.T. Free-running activity rhythms in the rat: Entrainment by melatonin. Science 219:1089–1091, 1983.
50. Reppert, S.M., Weaver, D.R., Rivkees, S.A., and Stopa, E.G. Putative melatonin receptors in a human biological clock. Science 242:78–81, 1988.

51. Rutledge, J.T. and Angle, M.J. Persistence of circadian activity rhythms in pinealectomized European starlings (Sturnus vulgaris). J. Exp. Zool. 202:333–337, 1977.
52. Sack, R.L. and Lewy, A.J. Melatonin and major affective disorders. In Melatonin, Clinical Perspectives, eds A. Miles, D. Philbrick and C. Thompson, New York: Oxford Medical Publications, pp. 205–227, 1988.
53. Sack, R.L. and Lewy, A.J. Melatonin as a chronobiotic: treatment of circadian desynchrony in night workers and the blind. J. Biol. Rhythms. 12:595–603, 1997.
54. Sack, R.L., Lewy, A.J., and Hoban, T.M. Free-running melatonin rhythm in blind people: Phase shifts with melatonin and Triazolam administration. In Temporal Oscillatory Systems, eds. L. Rensing, U. an der Heiden, and M.C. Mackey, Heidelberg: Springer Verlag, pp. 219–224, 1987.
55. Sack, R.L., Lewy, A.J., Hughes, R.J., McArthur, A.J., and Blood, M.L. Melatonin as a chronobiotic drug. Drug News Perspective 9:325–332, 1996.
56. Sack, R.L., Hughes, R.J., Edgar, D.M., and Lewy, A.J. Sleep-promoting effects of melatonin: at what dose, in whom, under what conditions, and by what mechanisms. Sleep 20:908–915, 1997.
57. Sack, R.L., Stevenson, J., and Lewy, A.J. Entrainment of a previously free-running blind human with melatonin administration. Sleep Res. 19:404, 1990.
58. Sarrafzadeh A., Wirz-Justice, A., and Arendt, J. Melatonin Stabilises sleep onset in a blind man. In Sleep '90, ed. J. Horne, pp 51–54. Bochum: Patengel Press, 1990.
59. Short, R.V. and Armstrong, S.M. Method for minimizing disturbances in circadian rhythms and bodily performance and function. Australian Patent Application PG 4737, April 27, 1984,
60. Simpson, H.W. Chronobiotics: selected agents of potential value in jet-lag and others dyschronisms. In Chronobiology: Principles and Applications to Shifts in Schedules, eds. L.E. Scheving and F. Halberg, Netherlands: Sijthoff and Noordhoff, pp. 433–446, 1980.
61. Simpson, S.M. and Follett, B.K. Pineal and hypothalamic pacemaker: their role in regulating circadian rhythmicity in Japanese quail. J. comp. Physiol. 144:381–389, 1981.
62. Smits, M.G., Nagtegaal, J.E., and Kerkhof, G.A. Melatonin in delayed sleep phase syndrome. Chronobiol. Int. 14:159, 1997.
63. Thorpy, M.J. (Chairman). The International Classification of Sleep Disorders. Diagnostic and Coding Manual, Rochester, USA: The American Sleep Disorders Association, 1990.
64. Tzischinsky, O., Pal, I., English, J., Epstein, R., and Lavie, P. Effects of melatonin in a blind subject: The importance of the timing treatment. 20th International Conference on Chronobiology, Tel Aviv (Israel), June 16–21, pp. 17–8, 1991.
65. Underwood, H. The pineal and circadian rhythms. In The Pineal Gland, ed. J.R. Reiter, Raven Press: New York, pp. 221–251, 1984.
66. Underwood, H. Circadian rhythms in lizards: phase response curve for melatonin. J. Pineal Res. 3:1987–196, 1986.
67. Underwood, H., Binkley, S., Scopes, T., and Mosher, K. Melatonin rhythms in the eyes, pineal bodies, and blood of Japanese quail (Coturnix coturnix japonica). Gen. comp. Enocrinol. 56:70–81, 1984
68. Underwood, H. and Harless, M. Entrainment of the circadian activity rhythms of a lizard to melatonin injections. Physiol. Behav. 35:267–270, 1985.
69. Vanecek, J., Pavlik, A., and Illnerova, H. Hypothalamic melatonin receptor sites revealed by autoradiography. Brain Res. 435:359–362, 1987.
70. Wehr, T.A. Melatonin and seasonal rhythms. J. Biol. Rhythms. 12:518–527, 1997.
71. Wever, R.A. Characteristics of circadian rhythms in human functions. In Melatonin in Humans, eds. R.J. Wurtman and F. Waldhauser, Wien: Springer-Verlag, pp. 323–373, 1986.
72. Wever, R.A. Order and disorder in human rhythmicity: possible relations to mental disorders. In Biological Rhythms and Mental Disorders, eds. D.J. Kupfer, T.H. Monk and J.D. Barachas, Guilford Press, New York, pp 253–346,1988
73. Wirz-Justice, A. and Armstrong, S.M. Melatonin: Nature's soporific? J. Sleep Res. 5:137–141, 1996.
74. Wright, J., Aldhous, M., Franey, C., English, J., and Arendt, J. The effects of exogenous melatonin on endocrine function in man. Clin. Endoc. 24:375–382, 1986.

32

MELATONIN AND CARDIOVASCULAR FUNCTION

Diana N. Krause, Greg G. Geary, Suzanne Doolen, and Sue P. Duckles

Department of Pharmacology
College of Medicine
University of California
Irvine, California 92697

1. INTRODUCTION

During the past 40 years, relatively few studies have addressed possible cardiovascular effects of melatonin, although daily rhythms in cardiovascular function are well known. At night, when plasma melatonin levels are high, humans exhibit a decrease blood pressure, heart rate and cardiac output and an increase in total peripheral vascular resistance (51). Melatonin levels fall to undetectable levels by morning, which is the time of highest risk for acute vascular disorders such as stroke and myocardial infarction (2,23,32). Low levels of nocturnal melatonin were found in patients with coronary heart disease (5), ischaemic stroke (22) and migraine (6). While these correlations are intriguing, the role of melatonin in cardiovascular function and disease has yet to be established. Early studies suggested melatonin could affect blood pressure. More recently, melatonin binding sites, receptor mRNA, and/or functional responses have been demonstrated in certain arteries and other tissues involved in cardiovascular regulation. These findings, summarized briefly below, suggest that melatonin acts both systemically and regionally, e.g., on cerebral perfusion and peripheral thermoregulatory arteries, to alter cardiovascular function.

2. SYSTEMIC EFFECTS

2.1. Blood Pressure

Early work with gland extracts provided the first indication that the pineal influences cardiovascular function. In both animals and humans, pineal extracts were found to decrease blood pressure (31). Conversely, pinealectomy resulted in an increase in

Melatonin after Four Decades, edited by James Olcese.
Kluwer Academic / Plenum Publishers, New York, 2000.

blood pressure of rats (12,29,57). Administration of melatonin lowers blood pressure in normal, pinealectomized, or spontaneously hypertensive rats (11,26,28,30) as well as in hypertensive patients (3) and healthy young women (7). When infused i.v. into rats, however, melatonin has no immediate effect on either systolic or diastolic blood pressure (43).

The hypotensive effect of melatonin may be mediated, in part, through an action on brain areas that regulate blood pressure and sympathetic outflow (11,28,52,54). The known action of melatonin on the suprachiasmatic nucleus would impact both circadian and sympathetic regulation of blood pressure (33). In addition, melatonin may influence the release of vasopressin from this nucleus (41). The area postrema of rat brain also exhibits melatonin receptor binding; and, interestingly, spontaneously hypertensive rats show a higher density of 2-[^{125}I]-iodomelatonin binding in this region as compared to normotensive animals (54).

2.2. Other Hemodynamic Effects

Other hemodynamic effects of melatonin have not been well studied. Melatonin appears to have little or no effect on heart rate (7,43,49), although some changes in cardiac function have been reported (4). In vitro, melatonin (0.3 nM) modulated the contractility of rat papillary muscle by counteracting the effects of isoproterenol, and this effect was blocked by N-acetyltryptamine (1). Melatonin may also affect the levels of cholesterol and other plasma lipids (8,20), but again this area needs further investigation.

3. VASCULAR EFFECTS OF MELATONIN

There is now compelling evidence that melatonin acts directly on blood vessel receptors in certain vascular beds, such as the brain, heart and skin. Initial autoradiographic studies using 2-[^{125}I]iodomelatonin indicated a restricted localization of vascular melatonin receptors. Specific binding was detected in cerebral and tail (caudal) arteries, but not in other arteries of the rat, including the aorta (53). Both vasoconstrictor and vasodilator effects have been reported for melatonin in isolated arteries, and recent studies have begun to elucidate the underlying receptors and mechanisms.

The function of vascular melatonin receptors has been studied mainly in the rat tail artery, a thermoregulatory vessel. In isolated tail arteries, nanomolar concentrations of melatonin produce constrictor effects; and the pharmacological characteristics of the response are consistent with known melatonin receptors (21,25,35,50,53, see below). Vasoconstrictor responses to melatonin also have been characterized in rat cerebral arteries (24,39, see below) and in pig and human coronary arteries (39). Recent evidence from the tail artery suggests melatonin also relaxes blood vessels via a second receptor subtype (14).

In contrast to cerebral and tail arteries, we have found neither constrictor nor dilator responses to melatonin in the aorta or femoral and mesenteric arteries isolated from rat (19). The latter arteries also did not exhibit specific 2-[^{125}I]iodomelatonin binding (19,53). Our findings suggest that melatonin receptors may not mediate the dilation produced by higher concentrations of melatonin (10^{-5}–10^{-3} M) in rat and rabbit aorta (44,45,56) and in iliac, renal and basilar arteries (44,47). This issue, however, should be re-examined given the current surge in our understanding of melatonin receptor subtypes (17). As summarized below, receptor mechanisms underlying vascular effects of melatonin are currently being characterized in both cerebral and skin arteries.

4. CEREBRAL CIRCULATION

4.1. Presence of Melatonin Receptors

High affinity 2-[^{125}I]-iodomelatonin binding has been demonstrated in cerebral arteries of rat (46,53,54), non-human primates (48), and humans (48). Initially, binding was found in arteries associated with the circle of Willis, e.g. anterior and middle cerebral arteries (53). Subsequently, 2-[^{125}I]-iodomelatonin binding was observed in inferior cerebellar, vertebral, spinal, and internal carotid arteries as well (48), suggesting a widespread distribution of melatonin receptors in the cerebral circulation. Interestingly, the density of 2-[^{125}I]-iodomelatonin binding sites in rat cerebral arteries is modified during the female estrous cycle (46), postnatal development (38) and in genetic hypertension (54).

Saturation analysis of binding to rat anterior cerebral arteries indicated two affinity states for 2-[^{125}I]-iodomelatonin (Kd of 13 pM and 823 pM) and a sensitivity to guanine nucleotides (9). In slices of rat cerebral arteries, melatonin inhibited forskolin-stimulated cyclic AMP production via a pertussis toxin sensitive mechanism (9). These findings are consistent with the characteristics of known Gi/o-protein coupled melatonin receptors, but do not indicate the identity of the receptor subtype(s) involved (17,34).

The presence of melatonin receptor subtypes is supported by recent demonstrations of appropriate receptor mRNA in cerebrovascular tissue. Using RT-PCR, cDNA for the mt_1 receptor subtype could be amplified from human middle cerebral and cortical arteries (37). Expression of both mt_1 and MT_2 receptor mRNA has been detected in rat and human cerebral arteries using RT-PCR and in situ hybridization (40).

4.2. Contractile Effects *in Vitro*

Melatonin acts on cerebrovascular receptors to modulate smooth muscle contractility, as shown in studies of rat cerebral arteries, pressurized in vitro (24,39,55). Melatonin decreased lumen diameter of middle cerebral artery segments in a concentration-dependent manner (EC_{50} = 2.7 nM, Figure 1). This effect was inhibited by luzindole, a competitive melatonin receptor antagonist, and by pertussis toxin pretreatment (24).

The magnitude of the constrictor response to melatonin appears to depend on the level of arterial pressure; in vitro responses were greater at higher levels of transmural pressure (24). When compared with other vasoconstrictors at the same pressure, the effects of melatonin were modest relative to those of serotonin, but similar in nature to those of tetraethylammonium (TEA) and charybdotoxin, two blockers of the large conductance, calcium-activated potassium (BK_{Ca}) channel (24). Smooth muscle BK_{Ca} channels play an important role in determining arterial tone and are activated at the membrane potentials found in pressurized cerebral arteries (42). In this preparation, melatonin-induced contractions were attenuated in the presence of either TEA or charybdotoxin, but unaffected by apamin, an inhibitor of small conductance K_{Ca}

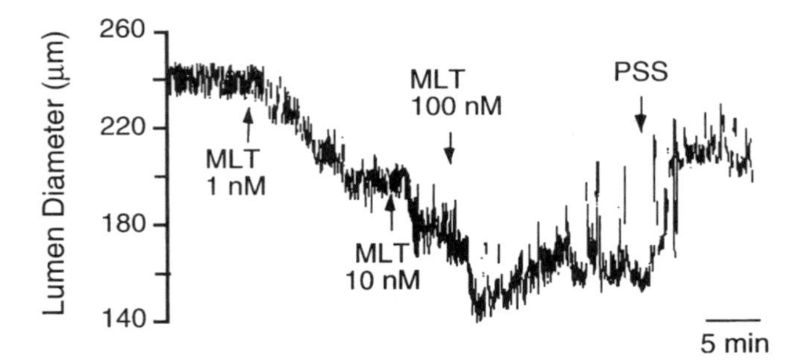

Figure 1. Representative tracing of the vasoconstrictor effect of melatonin (MLT). Increasing concentrations of melatonin decrease the lumen diameter of a segment of rat middle cerebral artery, pressurized to 60 mm Hg in vitro. (PSS, physiological salt solution) From Ref. 24, with permission.

channels. Melatonin also reversed dilations by NS-1619, an activator of BK_{Ca} channels. Based on these findings, it was proposed that melatonin-induced constriction involves closure of BK_{Ca} channels, perhaps through inhibition of the cyclic AMP-protein kinase A pathway (24).

We have obtained preliminary evidence that melatonin also has a direct vasodilatory effect in cerebral arteries (13; Figure 2). In a small pressurized branch of rat middle cerebral artery (<180 μm diameter), melatonin (1–100 nM) induced an increase in lumen diameter that was reversibly blocked by luzindole, which antagonizes both mt_1 and MT_2 melatonin receptors. At this point, the factors determining whether dilation

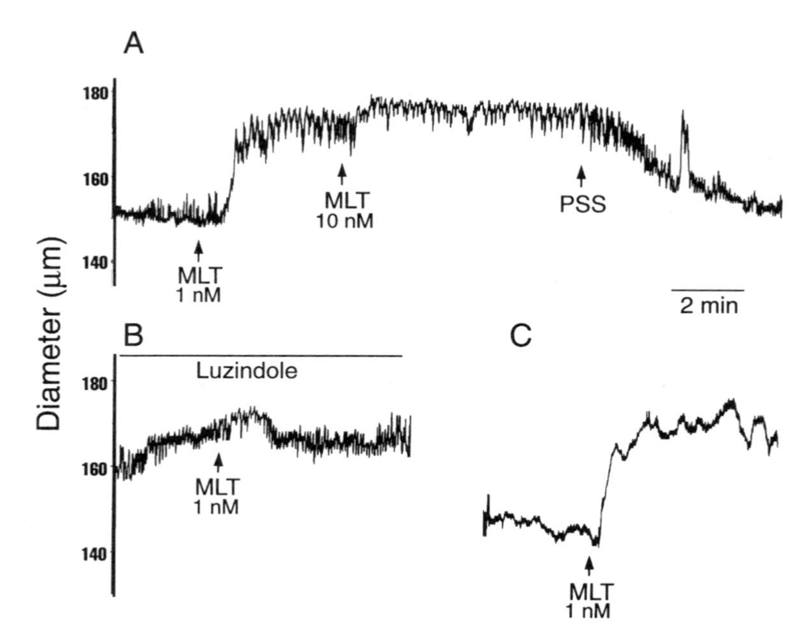

Figure 2. The vasodilator effect of melatonin (MLT). In this segment of rat middle cerebral artery, melatonin increased lumen diameter (A). Subsequent exposure to luzindole inhibited the effect of melatonin (B), which returned following washout of luzindole (C).

or contraction is elicited by melatonin are not known, but possibilities include the presence of different receptor subtypes and/or different receptor distributions in various cerebral arteries.

4.3. Functional Effects *in Vivo*

Cerebral blood flow was decreased in rats following acute administration of melatonin (10). A subsequent study showed that melatonin-induced cerebral vasoconstriction in rats was accompanied by an improvement in the vasodilatory response to hypercapnia (43). In addition, melatonin shifted the lower limit of cerebral blood flow autoregulation to a lower level of arterial blood pressure, thus increasing the security margin for maintaining cerebral blood flow during hypotensive hemorrhage. The authors suggest that melatonin may act on the cerebrovasculature to diminish the risk of hypoperfusion-induced cerebral ischemia.

5. RAT TAIL ARTERY

The rat tail artery has provided a useful model for characterizing vascular melatonin receptors. In vitro, melatonin generally enhances the contractile response induced by either adrenergic nerve stimulation or direct vasoconstrictors such as norepinephrine (21,25,35,50,53). These responses are consistent with the presence of specific 2-[^{125}I]-iodomelatonin binding sites (19,53) and melatonin receptor mRNA in the smooth muscle layers of this artery (40).

Recent advances in the understanding of melatonin receptors, including elucidation of two molecular subtypes (mt_1 and MT_2), have provided a basis for defining vascular melatonin receptors. The recent development of selective antagonists for the MT_2 receptor (17,18) permits investigation of possible receptor heterogeneity in vascular tissue. As detailed below, we now have evidence for at least two functional receptor subtypes in the tail artery, one mediating constriction and one causing dilation. Although little is known regarding regulation of melatonin receptors, we have found that estrogen status is one factor that alters vascular reactivity to melatonin, apparently by altering the balance of melatonin-induced constriction and dilation.

5.1. Constrictor Effects *in Vitro*

In most isolated preparations of rat tail artery, melatonin does not produce contraction by itself but potentiates the action of other vasoconstrictors (25,35,50,53, Figure 3). However, in pressurized tail arteries from juvenile rats, melatonin could directly constrict the vessel segments (21), similar to what is seen in pressurized cerebral arteries (24). In perfused, non-pressurized tail arteries, melatonin also caused contraction following the addition of NS-1619 to open BK_{Ca} channels (25). We have hypothesized that the varying nature of in vitro responses to melatonin may reflect different resting states of the isolated arterial preparations (25). Evidence suggests the effect of melatonin may depend on factors such as membrane potential, open state of potassium channels, and/or levels of cyclic AMP production (9,24,25). Consequently, significant changes in these factors, in vitro or in vivo, would influence the ability of melatonin to elicit a response.

Our initial study to pharmacologically characterize functional melatonin responses indicated that mt_1/MT_2-like melatonin receptors mediate potentiation of tail

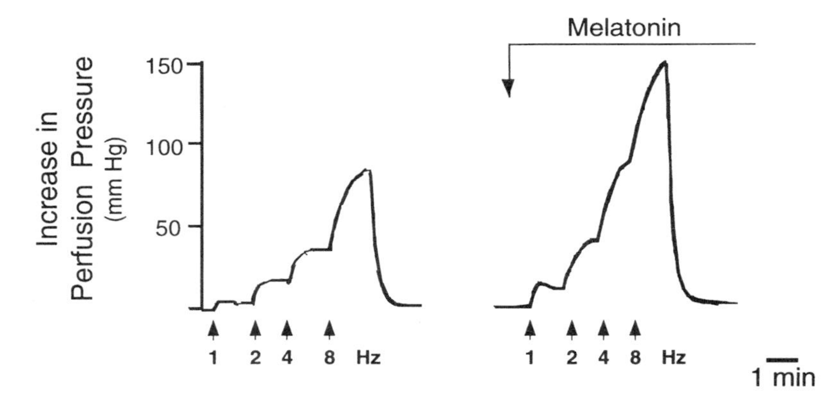

Figure 3. Melatonin (100 nM) potentiates the contractile effect of adrenergic nerve stimulation in an isolated, perfused segment of rat tail artery. Representative tracing shows increase in arterial pressure in response to increasing frequencies of adrenergic nerve stimulation (Hz). From Ref. 25, with permission.

artery constriction (35). This conclusion was based on the relative potencies of 2-iodomelatonin, melatonin and N-acetylserotonin and competitive inhibition by luzindole, a known melatonin receptor antagonist (34). The constrictor response in tail artery has been further characterized using a naphthalenic series of mt_1/MT_2 agonists, partial agonists and antagonists and a pair of indole-based stereoisomers, (+) and (−)AMMTC (50). These findings also suggest that melatonin is acting on mt_1/MT_2-like melatonin receptors to increase vascular tone.

Most of the drugs tested to date, however, do not discriminate between the mt_1 and MT_2 melatonin receptor subtypes. The affinity of luzindole for recombinant human mt_1 receptors is only about 15-fold lower than its affinity for the recombinant mt_2 subtype (18). However, the K_B estimated for luzindole in the rat tail artery (3×10^{-7} M; 35,36) correlates well with the affinity of this antagonist for the recombinant mt_1 subtype (17,18). We also find the relative agonist potency of 6-Cl-melatonin to be consistent with the mt_1 subtype, i.e, it is ten-fold less potent than melatonin (36). Constrictor effects of melatonin are not inhibited by 4-P-ADOT (4-phenyl-2-acetamidotetralin) at concentrations selective for the MT_2 receptor (10^{-8} M) nor are they mimicked by the selective MT_3 agonist, MCA-NAT (5-methoxy-carbonyl-amino-N-acetyltryptamine) (36). These data, together with the finding of mt_1 mRNA in tail artery smooth muscle (40), suggest that mt_1 receptors mediate vasoconstriction. The development of a selective mt_1 antagonist, however, is needed to confirm this hypothesis.

5.2. Role of the Endothelium

Melatonin appears to act on receptors located in the vascular smooth muscle to produce contractile effects (14,16,24,35). Our initial study of tail artery found that melatonin potentiated constriction in ring segments lacking an intact endothelial layer (35). However, an additional influence of the endothelium became apparent when we studied cannulated artery segments that were perfused through the lumen (25). We found, to our surprise, that melatonin had no vasomotor effects in this preparation when either the endothelium was removed or endothelial nitric oxide (NO) synthesis was inhibited with L-NAME (NG-nitro-L-arginine methyl ester) (25).

Thus, constrictor effects of melatonin in the isolated perfused artery depended on the presence of nitric oxide released by the endothelium. However, melatonin did not affect relaxation responses to either acetylcholine in endothelium-intact arteries or to sodium nitroprusside in endothelium-denuded arteries (25). These findings indicate that melatonin does not directly affect endothelial NO synthase or the smooth muscle action of NO. Fluid perfusion and shear stress on the lumen are known to induce NO formation through a specific mechanism that is inhibited by potassium channel blockers (27). Thus, we hypothesized that melatonin may act on the endothelium in a similar way to specifically modulate flow-induction of NO. Consistent with this hypothesis, L-NAME had no signficant effect on melatonin-induced constriction in a non-perfused preparation of cerebral arteries (24). Under in vivo conditions, however, both smooth muscle and endothelial actions would contribute to the vasoactive response to melatonin.

5.3. MT_2 Receptors

We recently re-examined responses to melatonin in precontracted tail artery segments, denuded of endothelium, and found that luzindole not only shifted the melatonin concentration-response curve to the right, as expected, but also enhanced the maximal constrictor effect of melatonin (36, Figure 4). Because this antagonist acts on all three melatonin receptor subtypes, we tested 5-MCA-NAT to see if MT_3 receptors were involved (17). No significant effects (potentiation, constriction nor dilation) of this MT_3 selective agonist were observed in the tail artery (36). Selective MT_2 antagonists, 4-P-ADOT and 4-P-PDOT (4-phenyl-2-propionamidotetralin), however, enhanced the vasoconstriction elicited by melatonin (14,16,36). Using concentrations designed to maximally block MT_2 receptors without affecting mt_1 receptors, these antagonists not only increased the maximal effect of melatonin, but shifted the melatonin concentration-response curve to the left (14,36, Figure 4). These data are consistent

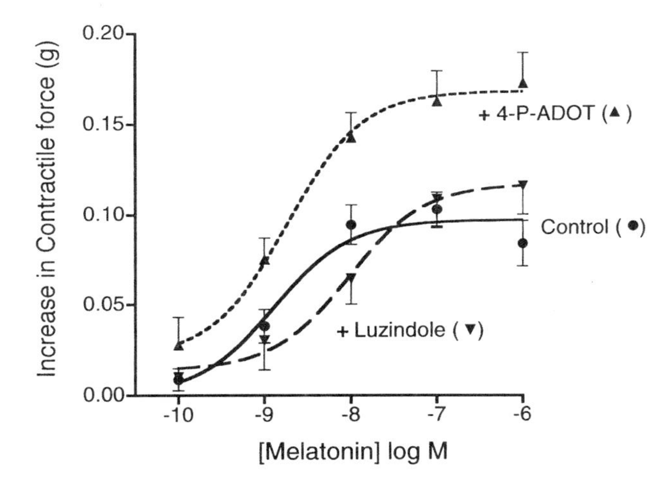

Figure 4. The effect of melatonin antagonists in endothelial-denuded segments of rat tail artery, preconstricted with phenylphrine. Concentration-response curves are shown for melatonin alone (control) or in the presence of either the mt_1/MT_2 antagonist, luzindole (3×10^{-6} M), or the selective MT_2 antagonist, 4-P-ADOT (10^{-8} M).

with the activation of two melatonin receptor subtypes in the rat tail artery. We hypothesize that melatonin acts on both mt_1-like receptors to elicit constrictor responses and MT_2 receptors to cause dilation (14). When the latter effect is blocked by 4-PADOT or 4-P-PDOT, vascular tone is increased. The presence of both mt_1 and MT_2 receptor mRNA in tail artery smooth muscle (40) further supports this hypothesis.

5.4. Effect of Estrogen

One of our most intriguing recent observations is the change that occurs in vascular effects of melatonin during the female rat estrous cycle and following ovariectomy (15,16). In tail arteries isolated from female rats, melatonin potentiated contractions produced by either transmural nerve stimulation or phenylephrine, consistent with our results from male rats. However, in arteries from females in proestrus (when estrogen is high), constrictor responses to melatonin were smaller than those measured at other stages of the estrous cycle or following ovariectomy (15,16). Administration of 17β-estradiol to ovariectomized females also resulted in decreased constriction of isolated arteries to melatonin (16).

The selective MT_2 receptor antagonist, 4-phenyl-2-propionamidotetralin (4-P-PDOT), enhanced constrictor responses to melatonin in arterial segments from intact females, consistent with inhibition of MT_2 receptor-mediated relaxation. In contrast, 4-P-PDOT had no significant effect in arteries from ovariectomized females (16,36). We propose that estrogen exposure may up-regulate the function of MT_2 receptors in vascular smooth muscle (Figure 5).

5.5. *In Vivo* Relevance

In the rat, the tail artery plays an important role in thermoregulation by modulating heat loss and conservation. The opposing responses of melatonin receptor subtypes may help optimize tail blood flow to control body temperature as melatonin levels

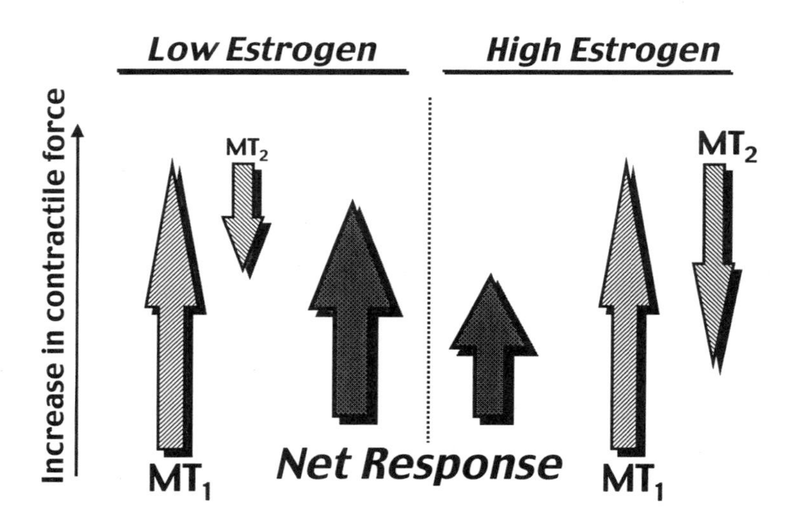

Figure 5. Schematic representation of the proposed relative contributions of mt_1-like and MT_2 melatonin receptors in the rat tail artery under different levels of circulating estrogen.

rise and fall under different physiological and environmental conditions. Circulating estradiol enhances the MT_2-mediated relaxation component of the net vascular response to melatonin, which may be important for appropriate temperature regulation during the rat estrous cycle. Melatonin action on the tail artery no doubt reinforces adaptive circadian, reproductive and seasonal changes in thermoregulation in the rat (16,53). The recent finding that mt_1 receptor mRNA is expressed in human subcutaneous arteries (37) suggests that melatonin may regulate peripheral thermoregulation in humans as well.

6. CONCLUSIONS AND POTENTIAL CLINICAL RELEVANCE

With the limited data at hand, it is still difficult to discern the overall role of melatonin in vascular regulation. From the few studies which address tissue distribution, it appears that melatonin may target only certain vascular beds. This strategy may allow for local adjustment in the face of global circadian changes in the cardiovascular system. The functional effects of melatonin observed so far in vitro differ among the various vascular preparations examined, however. These findings may reflect different melatonin receptor subtypes, receptor heterogeneity and/or multiple cellular targets, different receptor coupling and/or signal transduction mechanisms, variations in the resting tone of the vessel and/or interactions with endothelial factors or other vasoactive agents. The various possibilities are areas for future investigation.

Using rat cerebral and tail arteries, our laboratory has obtained evidence for two receptor-mediated contractile responses to melatonin. Constriction is produced via mt_1-like melatonin receptors located in arterial smooth muscle that appear to signal through Gi/o proteins to inhibit cyclic AMP and modulate calcium-dependent potassium channels. Constrictor effects of melatonin appear to be sensitive to membrane potential, arterial pressure and perfusion shear stress. Melatonin also can produce relaxation. The latter response appears to be mediated by the MT_2 melatonin receptor subtype because it is blocked by the selective antagonists, 4-P-ADOT and 4-P-PDOT. Consistent with these data, rat and human arteries express mRNA for both the mt_1 and MT_2 receptor subtypes. Interestingly, MT_2 receptor-mediated dilation in vascular smooth muscle is enhanced by estrogen, suggesting that the relative proportion of dilator MT_2 and constrictor mt_1-like receptors may be modulated under different physiological conditions to optimize vascular responses to melatonin.

Given the current use of melatonin as a popular over-the-counter jet lag and sleeping aid, it is important to understand the potential for cardiovascular effects in people self-administering melatonin. Possible interactions with estrogen are additional considerations for women taking melatonin during puberty, childbearing years and hormone replacement therapy after menopause. In addition, defining the role of melatonin in vascular regulation may lead to novel approaches for treating migraine, stroke and cardiovascular disease.

ACKNOWLEDGMENTS

This work was supported by National Heart, Lung, and Blood Institute Grant RO1-HL-50775.

REFERENCES

1. Abete, P., Bianco, S., Calabrese, C., Napoli, C., Cacciatore, F., Ferrara, N., and Rengo, F. Effects of melatonin in isolated rat papillary muscle. FEBS Lett. 412:79–85, 1997.
2. Behar, S., Halabi, M., Reicher-Reiss, H., Zion, M., Kaplinsky, E., Mandelzweig, L., and Goldbourt, U. Circadian variation and possible external triggers of onset of myocardial infarction. SPRINT Study Group. Am. J. Med. 94:395–400, 1993.
3. Birau, N., Petterssen, U., Meyer, C., and Gottschalk, J. Hypotensive effect of melatonin in essential hypertension. IRCS Med. Sci. 9:906.
4. Bosman, H., Dormehl, I.C., Hugo, N., Redelinghuys, I.F., and Theron, J.J. The effect of intravenous administration of melatonin on cardiovascular parameters of the *baboon* (*Papio ursinus*). J. Pineal Res. 11:179–181, 1991.
5. Brugger, P., Marktl, W., and Herold, M. Impaired nocturnal secretion of melatonin in coronary heart disease. The Lancet 345:1408, 1995.
6. Brun, J., Claustrat, B., Saddier, P., and Chazot, G. Nocturnal melatonin excretion is decreased in patients with migraine without aura attacks associated with menses. Cephalalgia 15:136–139, 1995.
7. Cagnacci, A., Arangino, S., Angiolucci, M., Maschio, E., and Melis, G.B. Influences of melatonin administration on the circulation of women. Am. J. Physiol. 43:R335–R338, 1998.
8. Chan, T.Y. and Tang, P.L. Effect of melatonin on the maintenance of cholesterol homeostasis in the rat. Endocrine Res. 21:681–696, 1995.
9. Capsoni, S., Viswanathan, M., de Oliveira, A.M., and Saavedra, J.M. Characterization of melatonin receptors and signal transduction system in rat arteries forming the circle of Willis. Endocrinol. 135:373–378, 1994.
10. Capsoni, S., Stankov, B.M., and Fraschini, F. Reduction of regional cerebral blood flow by melatonin in young rats. NeuroReport 6:1346–1348, 1995.
11. Chuang, J.I., Chen, S.S., and Lin, M.T. Melatonin decreases brain serotonin release, arterial pressure and heart rate in rats. Pharmacol. 47:91–97, 1993.
12. Cunnane, S.C., Manku, M.S., Oka, M., and Horrobin, D.F. Enhanced vascular reactivity to various vasoconstrictor agents following pinealectomy in the rat: role of melatonin. Can. J. Physiol. Pharmacol. 58:287–293, 1980.
13. Doolen, S., Duckles, S.P., Geary, G.G., and Krause, D.N. Melatonin mediates vascular dilation via a distinct melatonin receptor sub-type. Soc. Neurosci. Abstr. 23:1517, 1997.
14. Doolen, S., Krause, D.N., Dubocovich, M.L., and Duckles, S.P. Melatonin mediates two distinct responses in vascular smooth muscle. Eur. J. Pharmacol. 345:67–69, 1998a.
15. Doolen, S., Krause, D.N., and Duckles, S.P. Gender differences in response to melatonin. Soc. Neurosci. Abstr. 21:1406, 1995.
16. Doolen, S., Krause, D.N., and Duckles, S.P. Estradiol modulates vascular response to melatonin in rat caudal artery. Am. J. Physiol. 276:H1281–1288, 1999.
17. Dubocovich, M.L., Cardinali, D.P., Guardiola-Lemaitre, B., Hagan, R.M., Krause, D.N., Sugden, D., Yocca, F.D., and Vanhoutte, P.M. Melatonin Receptors. *The IUPHAR Compendium of Receptor Characterization and Classification*. IUPHAR Media, London, 1998, pp. 187–193.
18. Dubocovich, M.L., Masana, M.I., Iacob, S., and Sauri, D.M. Melatonin receptor antagonists that differentiate between the human Mel_{1a} and Mel_{1b} recombinant subtypes are used to assess the pharmacological profile of the rabbit retina ML_1 presynaptic heteroreceptor. Naunyn-Schmiedeberg's Arch. Pharmacol. 355:365–375, 1997.
19. Duckles, S.P., Barrios, V.E., Doolen, S., and Krause, D.N. Melatonin receptors potentiate contractile responses to adrenergic nerve stimulation in rat caudal artery. Proc. West. Pharmacol. Soc. 38:101–102, 1995.
20. Esquifino, A., Agrasal, C., Velazquez, E., Villanua, M.A., and Cardinali, D.P. Effect of melatonin on serum cholesterol and phospholipid levels, and on prolactin, thyroid-stimulating hormone and thyroid hormone levels, in hyperprolactinemic rats. Life Sciences 61:1051–1058, 1997.
21. Evans, B.K., Mason, R., and Wilson, V.G. Evidence for direct vasoconstriction of melatonin in "pressurized" segments of isolated caudal artery from juvenile rats. Naunyn-Schmiedebergs Arch. Pharmacol. 346:362–365, 1992.
22. Fiorina, P., Lattuada, G., Ponari, O., Silvestrini, C., and Dall'Aglio, P. Impaired nocturnal melatonin excretion and changes of immunological status in ischaemic stroke patients. The Lancet 347:692–693, 1996.
23. Gallerani, M., Manfredini, R., Ricci, L., Cocurullo, A., Goldoni, C., Bigoni, M., and Fersini, C. Chronobiological aspects of acute cerebrovascular diseases. Acta Neurol. Scand. 87:482–487, 1993.

24. Geary, G.G., Krause, D.N., and Duckles, S.P. Melatonin directly constricts rat cerebral arteries through modulation of potassium channels. Am. J. Physiol. 273:H1530–H1536, 1997.
25. Geary, G.G., Duckles, S.P., and Krause, D.N. Effect of melatonin in the rat tail artery: Role of K^+ channels and endothelial factors. Br. J. Pharmacol. 123:1533–1540, 1998.
26. Holmes, S.W. and Sugden, D. The effect of melatonin on pinealectomy-induced hypertension in the rat. Br. J. Pharmacol. 56:360P, 1976.
27. Hutcheson, I.R. and Griffith, T.M. Heterogeneous population of K^+ channels mediate EDRF release to flow but not agonists in rabbit aorta. Am. J. Physiol. 266:H590–H596, 1994.
28. K.-Laflamme, A., Wu, L.Y., Foucart, S., and de Champlain, J. Impaired basal sympathetic tone and alpha 1-adrenergic responsiveness in association with the hypotensive effect of melatonin in spontaneously hypertensive rats. Am. J. Hypertension 11:219–229, 1998.
29. Karppanen, H., Airaksinen, M.M., and Sarkimaki, I. Effects in rats of pinealectomy and oxypertine on spontaneous locomotor activity and blood pressure during various light schedules. Ann. Med. Exp. Biol. Fenn. 51:93, 1973.
30. Kawashima, K., Miwa, Y., Fujimoto, K., Ochata, H., Nishino, H., and Koike, H. Antihypertensive action of melatonin in the spontaneously hypertensive rat. Clin. Exp. Hypertens. Ther. Pract. A9:1121–1131, 1987.
31. Kitay, J.I. and Altschule, M.D. *The Pineal Gland, A Review of the Physiologic Literature*, Harvard University Press, Cambridge, MA, pp. 69–70, 1954.
32. Kelly-Hayes, M., Wolf, P.A., Kase, C.S., Brand, F.N., McGuirk, J.M., and D'Agostino, R.B. Temporal patterns of stroke onset. The Framingham Study. Stroke 26:1343–1347, 1995.
33. Krause, D.N. and Dubocovich, M.L. Regulatory sites in the melatonin system of mammals. Trends Neurosci. 13:464–470, 1990.
34. Krause, D.N. and Dubocovich, M.L. Melatonin receptors. Ann. Rev. Pharmacol. Toxicol. 31:549–568, 1991.
35. Krause, D.N., Barrios, V.E., and Duckles, S.P. Melatonin receptors mediate potentiation of contractile responses to adrenergic nerve stimulation in rat caudal artery. Eur. J. Pharmacol. 276:207–213, 1995.
36. Krause, D.N., Doolen, S., and Duckles, S.P. Vascular melatonin receptor subtypes: function, pharmacology and effect of estrogen. Naunyn-Schmiedeberg's Arch. Pharmacol. 358 (Suppl. 1):R550, 1998.
37. Krause, D.N., Dubocovich, M.L., Edvinsson, L., Geary, G.G., Masana, M.I., and Duckles, S.P. ML_{1A} melatonin receptors in human and rat cerebral and peripheral arteries: mRNA expression and functional pharmacology. Soc. Neurosci. Abstr. 22:1400, 1996.
38. Laitinen, J.T., Viswanathan, M., Vakkuri, O., and Saavedra, J.M. Differential regulation of the rat melatonin receptors: selective age-associated decline and lack of melatonin-induced changes. Endocrinology 130:2139–2144, 1992.
39. Mahle, C.D., Goggins, G.D., Agarwal, P., Ryan, E., and Watson, A.J. Melatonin modulates vascular smooth muscle tone. J. Biol. Rhythms 12:690–696, 1997.
40. Masana, M.I., Al-Ghoul, W., and Dubocovich, M.L. MT_1 and MT_2 melatonin receptor expression in mammalian vasculature and superior cervical ganglion. Naunyn-Schmiedeberg's Arch. Pharmacol. 358 (Suppl. 1):R72, 1998.
41. Maywood, E.S., Bittman, E.L., Ebling, F.J., Barrett, P., Morgan, P., and Hastings, M.H. Regional distribution of iodomelatonin binding sites within the suparchiasmatic nucleus of the Syrian hamster and the Siberian hamster. J. Neuroendocrinol. 7:215–223, 1995.
42. Nelson, M.T. and Quayle, J.M. Physiological roles and properties of potassium channels in arterial smooth muscle. Am. J. Physiol. 268:C799–C822.
43. Regrigny, O., Delagrange, P., Scalbert, E., Atkinson, J., and Lartaud-Idjouadiene, I. Melatonin improves cerebral circulation security margin in rats. Am. J. Physiol. 275:H139–H144, 1998.
44. Satake, N., Shibata, S., and Takagi, T. The inhibitory action of melatonin on the contractile response to 5-hydroxytryptamine in various isolated vascular smooth muscles. Gen. Pharmacol. 17:553–558, 1986.
45. Satake, N., Oe, H., Sawada, T., and Shibata, S. The mode of vasorelaxing action of melatonin in rabbit aorta. Gen. Pharmacol. 22:219–222, 1991.
46. Seltzer, A., Viswanathan, M., and Saavedra, J.M. Melatonin-binding sites in brain and caudal arteries of the female rat during the estrous cycle and after estrogen administration. Endocrinology 130:1896–1902, 1992.
47. Shibata, S., Satake, N., Takagi, T., and Usui, H. Vasorelaxing action of melatonin in rabbit basilar artery. Gen. Pharmacol. 20:677–680, 1989.
48. Stankov, B., Capsoni, S., Lucini, V., Pauteck, J., Gatti, S., Gridelli, B., Biella, G., Cozzi, B., and Fraschini,

F. Autoradiographic localization of the putative melatonin receptor in the brains of two old world primates: Cercopithecus wethiops and Papio ursinus. Neuroscience 52:459–468, 1993.
49. Terzolo, M., Piovesan, A., Puligheddu, B., Torta, M., Osella, G., Paccoti, P., and Angeli, A. Effects of long-term, low-dose, time-specified melatonin administration on endocrine and cardiovascular variables in adult men. J. Pineal Res. 9:113–124, 1990.
50. Ting, K.N., Dunn, W.R., Davies, D.J., Sugden, D., Delegrange, P., Guardiola-Lemaitre, B., Scalbert, E., and Wilson, V.G. Studies on the vasoconstrictor action of melatonin and putative melatonin receptor ligands in the tail artery of juvenile Wistar rats. Br. J. Pharmacol. 122:1299–1306, 1997.
51. Veerman, D.P., Imholz, B.P., Wieling, W., Wesseling, K.H., and van Montfrans, G.A. Circadian profile of systemic hemodynamics. Hypertension 26:55–59, 1995.
52. Viswanathan, M., Hissa, R., and George, J.C. Suppression of sympathetic nervous system by short photoperiod and melatonin in the Syrian hamster. Life Sci. 38:73–79, 1986.
53. Viswanathan, M., Laitinen, J.T., and Saavedra, J.M. Expression of melatonin receptors in arteries involved in thermoregulation. Proc. Natl. Acad. Sci. USA 87:6200–6203, 1990.
54. Viswanathan, M., Laitinen, J.T., and Saavedra, J.M. Differential expression of melatonin receptors in spontaneously hypertensive rats. Neuroendocrinology 56:864–870, 1992.
55. Viswanathan, M., Scalbert, E., Delagrange, P., Guardiola-Lemaitre, B., and Saavedra, J.M. Melatonin receptors mediate contraction of a rat cerebral artery. Neuroreport 8:3847–3849, 1997.
56. Weekley, L.B. Pharmacological studies on the mechanism of melatonin-induced vasorelaxation in rat aorta. J. Pineal Res. 19:133–138, 1995.
57. Zanaboni, A. and Zanaboni-Muciaccia, W. Experimental hypertension in pinealectomized rats. Life Sci. 6:2327–2331, 1967.

THE EFFECT OF MELATONIN ON VASOPRESSIN RELEASE UNDER STRESS CONDITIONS IN PINEALECTOMIZED MALE RATS

Marlena Juszczak, Ewa Bojanowska, Jan W. Guzek, Bozena Stempniak, and Ryszard Dabrowski

Department of Pathophysiology
Medical University of Lodz
ul. Narutowicza 60, 90-136 Lodz
Poland

ABSTRACT

The findings here reported showed that the response of vasopressinergic neurons to immobilization stress is augmented by melatonin. The effectiveness of melatonin in functional modification of these neurons' activity under conditions of stress changes after pineal removal.

INTRODUCTION

Pinealectomy as well as melatonin treatment are known to affect the vasopressin (AVP) secretion under some pathological conditions; they are also supposed to be of some importance during the endocrine response to chronic stress (1–2, also for references). Stress response involves the activation of the hypothalamo-pituitary-adrenal (HPA) axis, i.e., the increased secretion of corticotropin-releasing hormone, ACTH and corticosterone as well as the activation of the magnocellular supraoptic and paraventricular neurons (3). Activation of the HPA axis during stress response is usually accompanied with the enhanced release of the neurohypophysial hormones into general circulation (4). Therefore, this study was designed to investigate the possible effects of pineal removal and melatonin treatment on AVP secretion under immobilization stress.

Melatonin after Four Decades, edited by James Olcese.
Kluwer Academic / Plenum Publishers, New York, 2000.

MATERIAL AND METHODS

Male rats of the Wistar strain were kept in regulated light-dark cycle (lights on from 0600 to 1800 h). Animals were subjected to sham-operation or pinealectomy according to the procedure described by Kuszak and Rodin (5). Six weeks after the surgery they were divided as follows: A—rats injected intraperitoneally over two weeks, once daily, with vehicle (1% ethanol in 0.9% NaCl), B—rats similarly injected with melatonin solution (MLT; N-Acetyl-5-methoxytryptamine; Sigma Chemical Co., St. Louis) in a dose of 50 µg/100 g bw. The injections were administered in the late afternoon (an hour before lights off). In each group three further subgroups were set up: 1—rats with free access to food and water; 2—rats with no access to food and water for 24 hr before decapitation; 3—rats as (2) and additionally immobilized (i.e., singly transferred into cages so narrow that they could not move freely) for 24 hr. On the next day after last injection of vehicle or MLT solution animals were killed by decapitation between 0900–0930 h. The pituitary neurointermediate lobes (NIL) were removed and prepared for AVP radioimmunoassay (RIA) as previously described (2). For the determination of blood plasma AVP level, hormone was extracted from plasma using C18 "Sep-pak" columnes (Water Associates Ltd, Northwick, UK).

The NIL as well as plasma AVP levels were determined by double-antibody specific RIA as previously described (6) with an AVP-antiserum provided by Calbiochem (rabbit antiserum to [Arg^8]-Vasopressin, 969115, Lot 466892). Arginine vasopressin ([Arg^8]-Vasopressin; Peninsula Laboratories Europe Ltd.) was used for standard curve preparation as well as for iodination with ^{125}I using the chloramine-T method. The sensitivity of anti-AVP serum was 1.25 pg/ml. Intra-assay coefficient of variation for the AVP assay was 2.5%. All results were reported as mean ± standard error of the mean (S.E.M.). Significance of the differences between means was assessed by non-parametric analysis of variance (ANOVA) followed by Mann-Whithney "U" test. $P < 0.05$ was used as the minimal level of significance.

RESULTS AND DISCUSSION

In sham-operated rats (Figure 1A), immobilization stress decreased the content of AVP in the NIL both in vehicle- and MLT-treated animals. Plasma AVP concentration was higher in MLT-treated stressed rats (Figure 1B). In sham-operated rats, MLT decreased the NIL AVP content in animals of all experimental subgroups (Figure 1A), while in pinealectomized animals MLT treatment had an opposite effect (Figure 2A). After the pineal removal, plasma AVP concentrations were increased in MLT-treated rats of subgroups 2 and 3 (Figure 2B). The efficiency of the experimental immobilization procedure as an stressor was confirmed by an increased ACTH concentration in blood plasma (the ACTH plasma levels in these animals are here not shown [see: 2]).

In the present study we demonstrated that the pineal and MLT are involved in the functional mechanisms related to the neurohypophysial response to immobilization stress. Moreover, we have confirmed that the NIL AVP content decreased under immobilization stress as it was earlier reported (7). Also, the finding that MLT injection decreased the NIL AVP content was consistent with results described previously (8, also for references). However, the mechanisms of the MLT-induced enhancement of AVP release (as noted during this study) remain to be investigated. In this regard, the interactions of MLT with its brain receptors (9) and/or neurotransmitters that

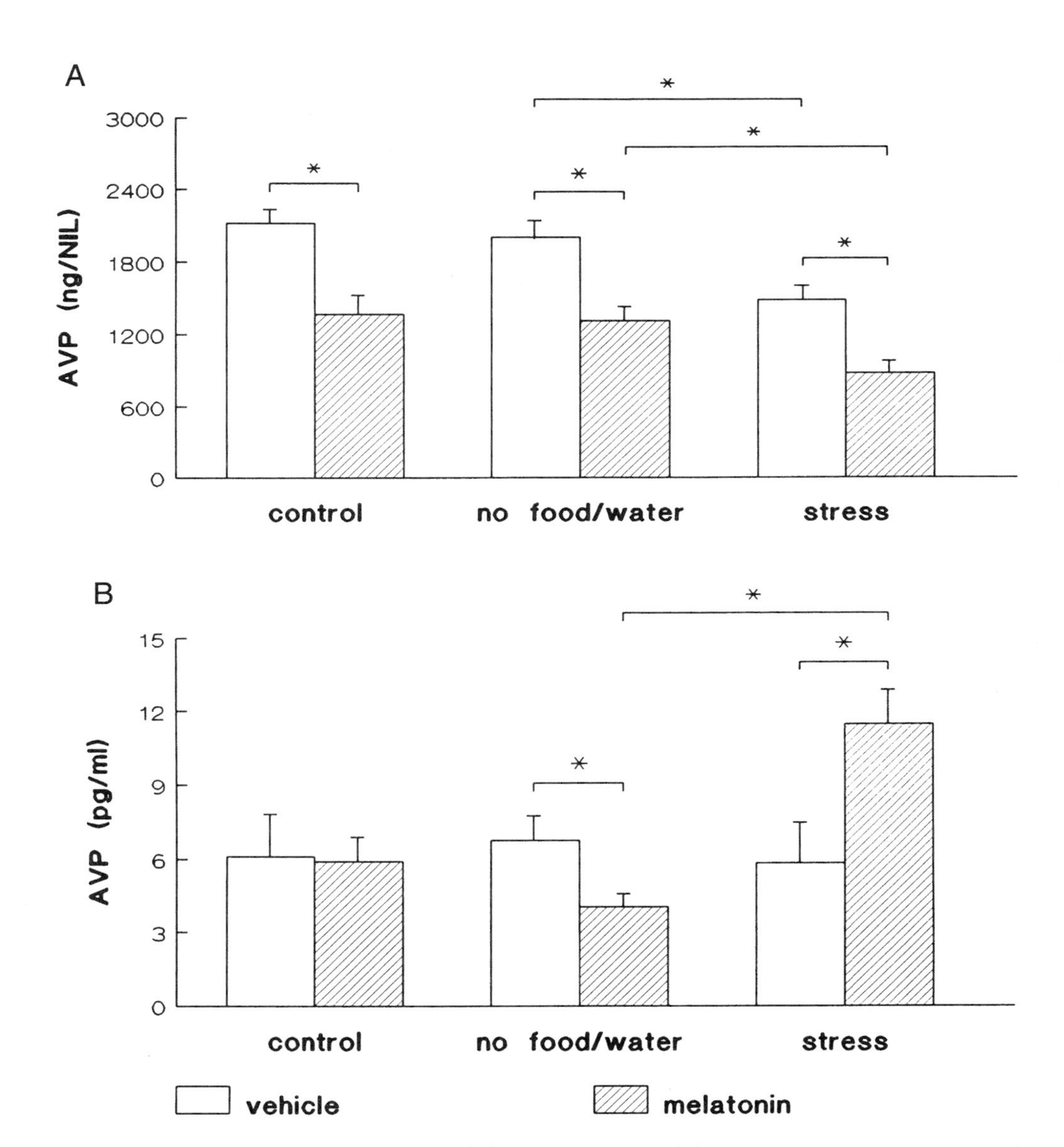

Figure 1. The effect of melatonin on the NIL (A) and blood plasma (B) AVP levels in sham-operated male rats under immobilization stress (mean ± S.E.M.; n = 8–10).

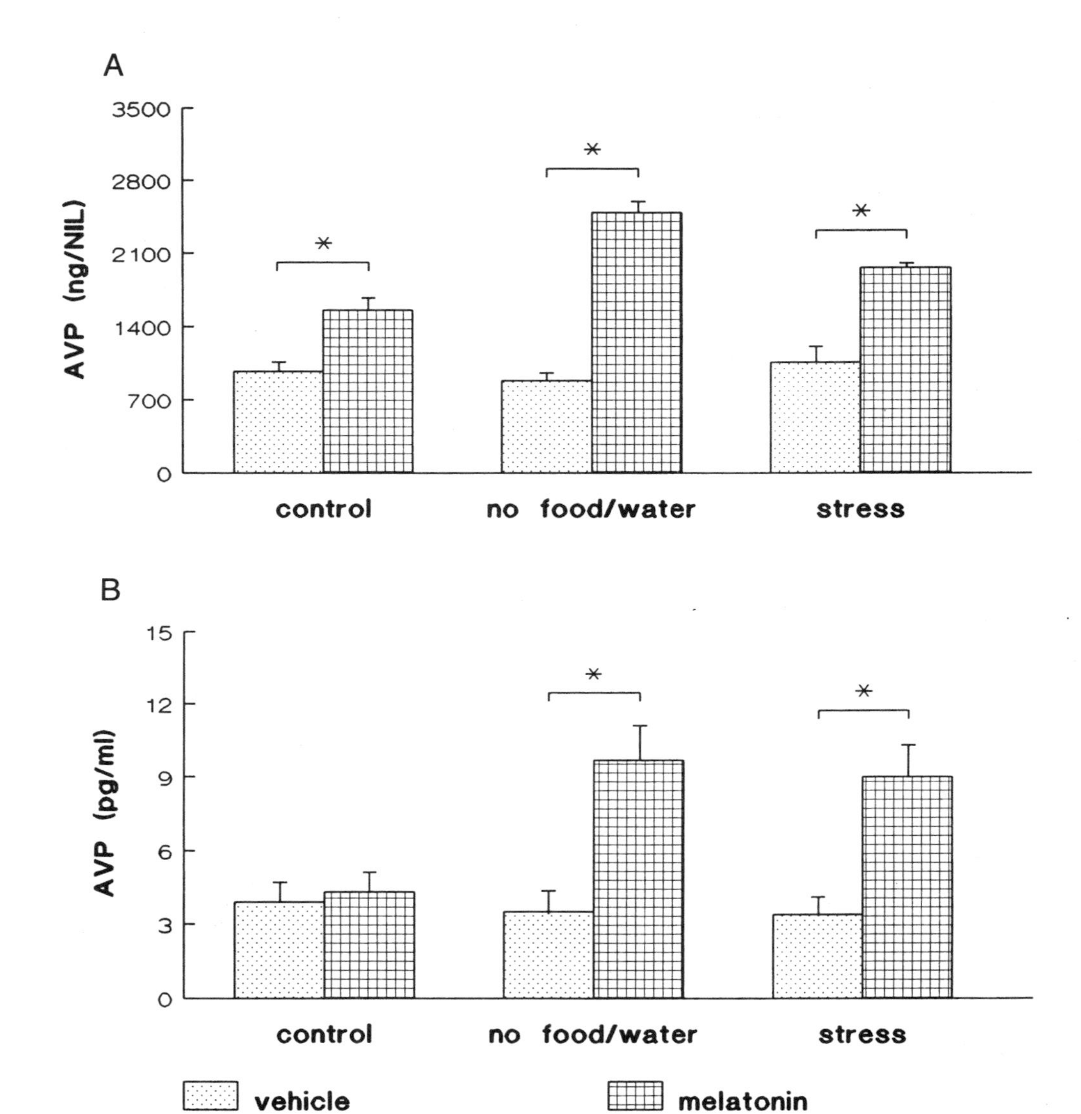

Figure 2. The effect of melatonin on the NIL (A) and blood plasma (B) AVP levels in pinealectomized male rats under immobilization stress (mean ± S.E.M.; n = 7–8).

regulate AVP release (10) are conceivable as to both sham-operated and pinealectomized rats. It is concluded that under stress conditions MLT modifies both AVP (this study) as well as OT, PRL and ACTH (2) release. The fact that MLT attenuated the stress-induced ACTH secretion implied possible anti-stress activity of the pineal.

ACKNOWLEDGMENTS

Supported by the State Committee for Scientific Research (Poland), grant No. 4.PO5A.104.12 (contract No. 546/PO5/97/12).

REFERENCES

1. Milin, J., Demajo, M., and Todorovic, V. Rat pinealocyte reactive response to a long-term stress inducement. Neuroscience 3:845–854, 1996.
2. Juszczak, M. Melatonin affects the oxytocin and prolactin responses to stress in male rats. J Physiol Pharmacol 49:151–163, 1998.
3. Stratakis, C.A. and Chrousos, G.P. Neuroendocrinology and pathophysiology of the stress system. Ann NY Acad Sci 771:1–18, 1995.
4. Jezova, D., Skultetyova, I., Tokarev, D.I., Bakos, P., and Vigas, M. Vasopressin and oxytocin in stress. Ann NY Acad Sci 771:192–203, 1995.
5. Kuszak, J. and Rodin, M. A new technique of pinealectomy for adult rats. Experientia (Basel) 33:283–284, 1977.
6. Juszczak, M., Kaczorowska-Skora, W., and Guzek, J.W. Vasopressin and oxytocin release as affected by constant light in pinealectomized male rats. Endocrine Regul 29:163–170, 1995.
7. Olczak, S., Guzek, J.W., and Stempniak, B. The hypothalamic and neurohypophysial vasopressor and oxytocic activities in stressed rats treated with haloperidol. Exp Clin Endocrinol 89:145–152, 1987.
8. Juszczak, M., Debeljuk, L., Stempniak, B., Steger, R.W., Fadden, C., and Bartke, A. Neurohypophyseal vasopressin in the Syrian Hamster: response to short photoperiod, pinealectomy, melatonin treatment or osmotic stimulation. Brain Res Bull 42:221–225, 1997.
9. Williams, L.M., Hannah, L.T., Hasting, M.H., and Maywood, E.S. Melatonin receptors in the rat brain and pituitary. J Pineal Res 19:173–177, 1995.
10. Bojanowska, E. and Forsling, M.L. The effects of melatonin on vasopressin secretion in vivo: interactions with acetylcholine and prostaglandins. Brain Res Bull 42:457–461, 1997.

MOTILITY AND PASSIVE AVOIDANCE MODULATION BY SEPTAL VASOPRESSIN IS DEPENDENT ON THE PINEAL GLAND

Helmut Schwarzberg* and Edgar Appenrodt

Institute of Neurophysiology
Otto von Guericke University
D-39120 Magdeburg, Germany

INTRODUCTION

The neuropeptide, arginine vasopressin (AVP) is well known to modulate several behavioural paradigms (5). Septal AVP has been demonstrated to be involved in spatial learning, social recognition, and pole jumping (3) in varied modes.

The pineal gland and its hormone, melatonin have been found to exert an influence on various physiological systems including behavioural processes. Initial first findings have been obtained about pineal-dependent effects of exogenous AVP and/or melatonin on passive avoidance behaviour (4), showing that (i) intraperitoneally applied AVP as well as melatonin facilitated passive avoidance performance in rats and (ii) pinealectomy blocked this effect.

In order to ascertain whether or not endogenous septal AVP is involved in passive avoidance behaviour as well as locomotor activity and whether this involvement may be influenced by the pineal gland experiments were performed to measure the passive avoidance response and motility after application of exogenous AVP or blockade of the endogenous AVP by administration of a V1/V2 receptor antagonist (AAVP) into the medio-lateral septum of pinealectomized, sham-operated, and non-operated rats.

---

* Address: Institut für Neurophysiologie, der Otto-von-Guericke-Universität Magdeburg, Medizinische Fakultät, Leipziger Str. 44, D-39120 Magdeburg, Phone: (+49) 391 67 15891, Fax: (+49) 391 67 15886.

Melatonin after Four Decades, edited by James Olcese.
Kluwer Academic / Plenum Publishers, New York, 2000.

METHODS

Male Wistar rats were pinealectomized (PE, n = 29) or sham-operated (SO, n = 28) under pentobarbital anesthesia. The third group of 43 rats were non-operated controls (C). After a recovery period of about 3 weeks all animals were implanted an intracerebral guide cannula (CMA/12, CMA/microdialysis AB, Stockholm) into the medio-lateral septum and fixed to the skull with dental cement and acrylate. Animals were allowed to recover for at least 1 week prior to commencement of behavioural experiments. A CMA 12 microdialysis probe (membrane length 2 mm) was used for application of the substances. The dialysis probe before being lowered into the septum via the guide cannula, was connected to a perfusion pump and the perfusion solution pumped through the probe at a rate of 2 µl/min, perfusion time 30 min. Over the 30 min dialysis period a total amount of approximately 200 pg AVP or approximately 40 ng of the V1/V2 receptor antagonist AAVP [d(CH_2)$_5$Thyr(Et)VAVP (Sigma)] were delivered into the brain tissue via the dialysis membrane.

The motility was recorded at a 30-min time interval during brain microdialysis as cumulated activity counts from a motility meter (TSE Mot V1.2, TSE Inc., Bad Homburg, Germany).

The passive avoidance behaviour was studied by employing a one-trial learning paradigm in a step-through type situation. The first retention of the passive avoidance response was tested 24 h after the single learning trial and 30 min after starting application of the substances by microdialysis and measuring of motility. The second retention was tested 48 h after the learning trial.

RESULTS

The results are summarized in Table 1. The blockade of vasopressinergic neurotansmission or neuromodulation in the septal area by AAVP decreased the motility in both pineal-intact groups, whereas AVP was without effect. In PE rats during AVP administration an increased motility was found, but AAVP was without effect.

In SO animals application of AVP and AAVP was without effect on passive avoidance response. In PE rats administration of both AVP and AAVP facilitated the passive avoidance latency in the first and second retention trial.

Table 1. The effect of AVP, AAVP and artificial CSF on motility and passive avoidance latency in non-operated (control), sham-operated (SO), and pinealectomized (PE) rats

		Control	SO	PE
Motility	artif. CSF	80 ± 12	46 ± 10	44 ± 10
(activity counts)	AVP	74 ± 23	45 ± 8	101 ± 23*
	AAVP	41 ± 8*	22 ± 4*	47 ± 9
Avoidance	artif. CSF	81 ± 20	53 ± 11	26 ± 8
latency (s),	AVP	68 ± 15	54 ± 8	117 ± 29*
first retention	AAVP	57 ± 22	55 ± 15	125 ± 23**

Mean ± SE; * $p < 0.05$; ** $p < 0.01$.

DISCUSSION

The present finding that AAVP reduced the motility in both pineal-intact groups indicates that endogenous AVP under physiological conditions seems to be involved in the regulation of the motility. The observation that, unlike SO, in PE rats AAVP failed to reduce but AVP enhanced the motility cannot be explained in detail; however, it is known that pinealectomy influenced the AVP content in the hypothalamus and neurohypophysis of rats (1). Moreover, since it is known that the pineal hormone, melatonin influenced several brain neurotransmitter systems (2) which can also be involved in locomotor activity, a more complex situation should be assumed.

The effect of AVP applied into various brain areas in passive avoidance behaviour has been studied by a number of workers (5). In the present study, intraseptal AVP application facilitates the avoidance response only in PE rats. The following mechanisms of the lengthy prolongation of avoidance latency in PE rats after AVP and AAVP treatment might be a reasonable explanation. On the one hand it has been reported that antagonists may also have agonistic properties, change receptor characteristics or cross-react with related receptors (6). On the other hand, both enhanced (AVP) and reduced (AAVP) influence of septal AVP in PE rats may have effects on passive avoidance behaviour. In pinealectomized animals, AVP in the medio-lateral septum may be involved in passive avoidance behaviour through lengthy, more indirect mechanisms, but, these mechanisms might be masked in pineal-intact rats.

ACKNOWLEDGMENT

This research was supported by grants of the Land of Saxony-Anhalt.

REFERENCES

1. Appenrodt, E., Bojanowska, E., Janus, J., Stempniak, B., Guzek, J.W., and Schwarzberg, H. Effects of methylphenidate on oxytocin and vasopressin levels in pinealectomized rats during light-dark cycle. Pharmacol. Biochem. Behav. 58(2):415–419; 1997.
2. Alexiuk, N.A.M. and Vriend, J.P. Melatonin reduces dopamine content in the neurointermediate lobe of male Syrian hamsters. Brain Res. Bull. 32:433–436; 1993.
3. Engelmann, M., Wotjak, C.T., Neumann, I., Ludwig, M., and Landgraf, R. Behavioral consequences of intracerebral vasopressin and oxytocin: Focus on learning and memory. Neurosci. Biobehav. Rev. 20(3):341–358; 1996.
4. Juszczak, M., Drobnik, J., Guzek, J.W., and Schwarzberg, H. Effect of pinealectomy and melatonin on vasopressin-potentiated passive avoidance in rats. J. Physiol. Pharmacol. 47:621–627; 1996.
5. Kovács, G.L., Bohus, B., Versteeg, D.H.G., De Kloet, E.R., and De Wied. D. Effect of oxytocin and vasopressin on memory consilidation: Sites of action and catecholaminergic correlates after local microinjection into limbic-midbrain structures. Brain Res. 175:303–314; 1979.
6. Landgraf, R. Mortyn Jones Memorial Lecture. Intracerebrally released vasopressin and oxytocin: Measurement, mechanisms and behavioural consequences. J. Neuroendocrinol. 7:243–253; 1995.

35

EFFECTS OF MELATONIN AND ITS RELATION TO THE HYPOTHALAMIC-HYPOPHYSEAL-GONADAL AXIS

Olga Ianăş,[1] Dana Manda,[1] D. Câmpean,[2] Mariana Ionescu,[1] and Gh. Soare[1]

[1] "C.I.Parhon" Institute of Endocrinology
34–36, Bd. Aviatorilor, R-79660, Bucharest
and Romania.
[2] ΦPharmex, Bucharest, Romania

1. INTRODUCTION

The principal secretory product of the pineal gland, melatonin, has been found to reverse many of the effects of pinealectomy (6,9,16,18). Melatonin administration to hamsters, rats, mouse can also mimic the regressive effects of short days on the reproductive system (hypothalamic-pituitary-gonadal axis) (14,15,18).

In addition to its described inhibitory influence on reproduction and sexual development, melatonin can also have a stimulatory influence under certain circumstances (1,15).

This "progonadal" or counter-antigonadotropic influence has been described most frequently for the hamster which is a long day breeder (16).

Some data indicated that the mother's pineal gland and melatonin treatment can act on foetal development and influence the postnatal ontogeny of the hormones involved in the neuroendocrine-reproductive axis in developing rats (2,4,7).

In the laboratory rat, melatonin was found to have an inhibitory influence on neuroendocrine-reproductive axis in young animals (5,13) but no effect was seen in the intact adult rat (5,9,16,13).

In humans, the role of the pineal gland in the reproductive physiology is still unclear, however there is an increasing evidence that melatonin plays a role in human reproduction (10,17).

The pathologic picture of male reproduction and sexuality is dominated by sterility and sexual insufficiency. The pathogenic mechanism of male sterility is still difficult

Melatonin after Four Decades, edited by James Olcese.
Kluwer Academic / Plenum Publishers, New York, 2000.

to clear up in view of the numerous negative influences with which the adult organism is faced (8).

In order to clarify if melatonin acts to influence the reproductive axis, we re-examined the role of melatonin upon hypothalamic-pituitary-gonadal axis in adult rats and studied the correlations between melatonin (urinary 6-sulphatoxymelatonin) and the status of the reproductive system in humans.

2. MATERIALS AND METHODS

2.1. Experimental Studies

Adult male Wistar rats were used in three different sets of experiments.

In the first experiment 36 male rats were divided into two groups: control (N = 18) and melatonin treated (N = 18). Melatonin treatment was given for ten days, at the end of the light phase by daily subcutaneous injections of 50 μg of melatonin (MLT). Controls received the same volume (0.5 ml) of vehicle alone. On the day following the last injection, six animals of each group were killed at three time points, 10 a.m., 10 p.m., 2 a.m. and blood was collected. Plasma was stored for hormone assay. Plasma testosterone concentrations were measured by radioimmunoassay with an in house kit, after ethylether extraction (12).

In the second experiment, 32 adult male rats were divided into four groups: intact (control), sham-operated, castrated and castrated + testosterone replacement. The last group received daily subcutaneous testosterone injections (5 μg/animal/day) immediately after operation, for three days. After 17 hrs from the last injection, all animals were sacrificed at 2 a.m., the pineal glands were removed in 0.1 ml phosphate buffer (0.01 M, pH 7.4) and sonicated in 1 ml phosphate buffer. Pineal melatonin was assayed by RIA using a commercially available kit (RK-MEL2 Melatonin RIA, Bühlmann Laboratories AG, Switzerland).

In the third experiment, 32 male rats were assigned to one of the following four groups: (1)-control rats, (2)- GnRH treated rats and (3),(4)- MLT + GnRH treated rats. Rats from groups 3 and 4 received two subcutaneous melatonin injections: 25 μg/injection (group 3), and 50 μg/injection (group 4). Groups 1 and 2 received the same volume (0.5 ml) of vehicle alone. Melatonin was injected in the evening (5 p.m.) and night (1 a.m.). In order to stimulate the gonadotropin secretion the rats were injected i.p. with a single dose of 5 μg/kg body weight of GnRH (supplied by SERONO) or vehicle (group 1) (1:30 a.m.). After half an hour, animals were sacrificed and blood was collected. The concentrations of FSH and LH were measured by a double antibody RIA method using reagents donated by the National Institute of Health (NIAMDD, Bethesda, USA). The assays were validated in our laboratory.

2.2. Clinical Studies

Fourteen males, aged 21–44 years with infertility and sexual insufficiency participated in our study. The following investigations were made: seminal fluid parameters, gonadotropic and gonadal hormones and 6-sulphatoxy-melatonin excretion.

Urine LH, FSH, were assayed by IRMA kits from Serono, serum LH, FSH, PRL, testosterone and estradiol were assayed by FIA using Delfia kits (Wallac Oy, Finland).

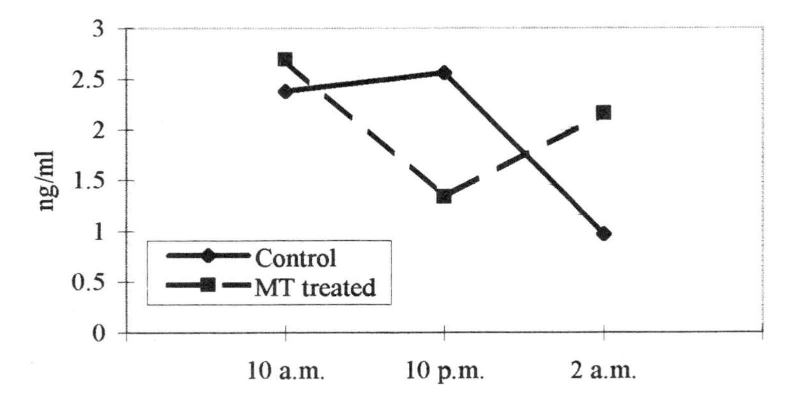

Figure 1. Effect of melatonin on plasma testosterone in male rats.

Nocturnal 6-sulphatoxymelatonin excretion was measured by a RIA kit from CIDTech, Canada and corrected for creatinine.

2.3. Statistical Analysis

Results were expressed as mean ± SEM. Data comparison among groups were determined by Student's *t* test.

3. RESULTS

3.1. Experimental Studies

Experiment 1. Figure 1 illustrates the effect of melatonin treatment on plasma testosterone. Testosterone concentration (ng/ml) shows a significant circadian change after melatonin treatment. After melatonin injections, at the beginning of night (10 p.m.), testosterone plasma concentrations were significantly decreased ($p = 0.02$), while at 2 a.m. (when melatonin production is maximum) the levels of testosterone increased ($p = 0.01$). Melatonin administration did not change the morning testosterone concentrations (10 a.m.).

Experiment 2. Effect of orchidectomy and testosterone treatment on melatonin production are shown in Figure 2. Pineal melatonin content (pg/pineal) in castrated rats ($p = 0.007$) was more lowered than in sham-orchidectomized rats. Reduced melatonin production was also observed in sham castrated rats ($p = 0.003$) compared to intact group but its circadian rhythm was preserved. After testosterone administration in castrated rats, melatonin levels returned to the levels of sham castrated rats.

Experiment 3. Serum LH and FSH concentrations significantly increased after 30 min from GnRH administration (group 2) as compared to control group (1) (Figure 3). A significant decrease was observed in both 3 and 4 groups after melatonin administration. The effect of melatonin was dose-dependent, mean serum FSH and

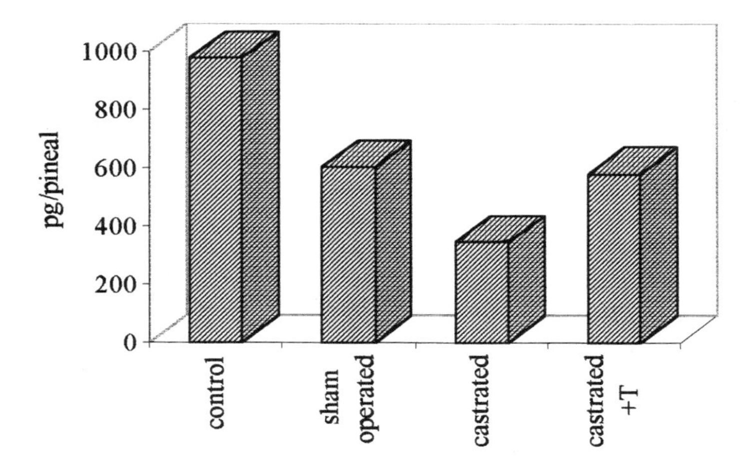

Figure 2. Effect of castration and testosterone replacement on pineal melatonin content in rats.

LH concentrations were lower in group (4) (50μg MLT) than those in group (3) (25μg MLT).

3.2. Clinical Studies

The sterility-associated troubles of sexual dynamics were usually accompanied by oligospermia, azoospermia, loss of erection and premature ejaculation. 60% of the cases presented a normal gonadotropic status associated with troubles of sexual dynamics and alteration of seminal fluid.

In 11 males, the levels of urinary 6-sulphatoxymelatonin were significantly lower as compared to the control group.

The high levels of 6-sulphatoxymelatonin were associated with low concentrations of FSH, LH and testosterone. Figures 5 and 6 show the urine levels of LH and FSH in patients with infertility and sexual deficiency expressed in mIU/h.

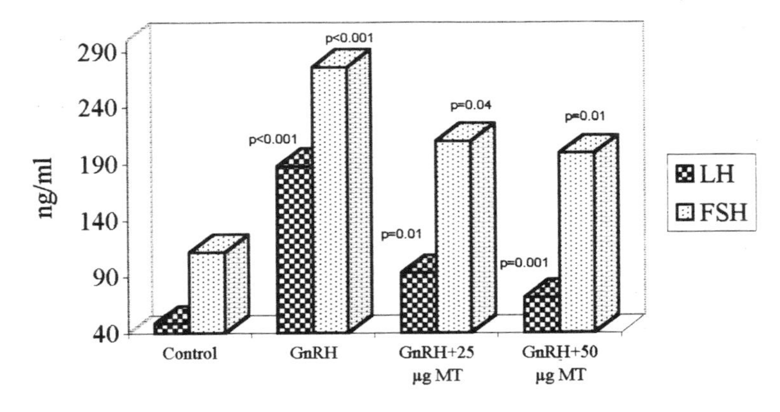

Figure 3. Effect of different melatonin doses on serum LH and FSH levels in adult GnRH-trated rats.

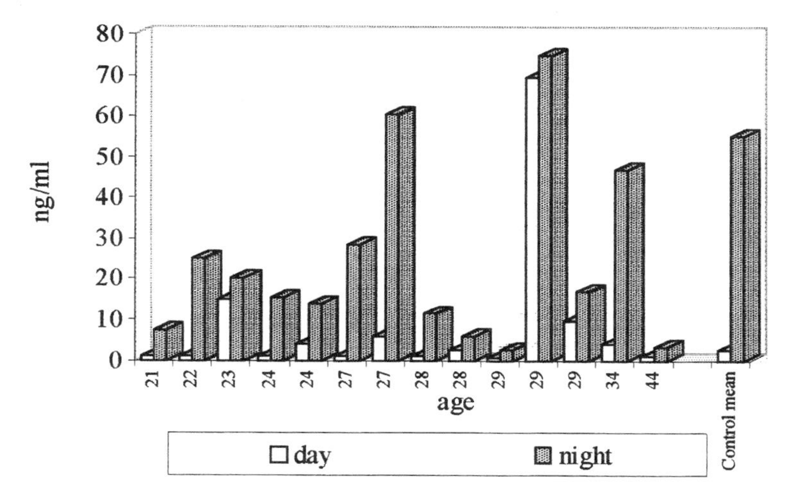

Figure 4. Urinary 6-sulphatoxymelatonin values in sexual dysfunction.

4. DISCUSSIONS

Previous researchers failed to document an effect of melatonin on reproductive axis of adult rats(9,16). Moreover, the studied day/night moments, different from those we used, may have determined melatonin's failure to influence testosterone and gonadotropin secretion. In our study, after melatonin treatment, the plasma testosterone concentrations at 10 a.m. were similar to control group (Figure 1), while at 10 p.m. and 2 a.m., were significantly changed. Melatonin treatment induced a biphasic response of testosterone secretion; at 10 p.m., 2 hours after the onset of darkness, the high testosterone levels were significantly reduced by melatonin treatment, while at 2 a.m. the low testosterone levels were significantly stimulated. Our studies on adult rats,

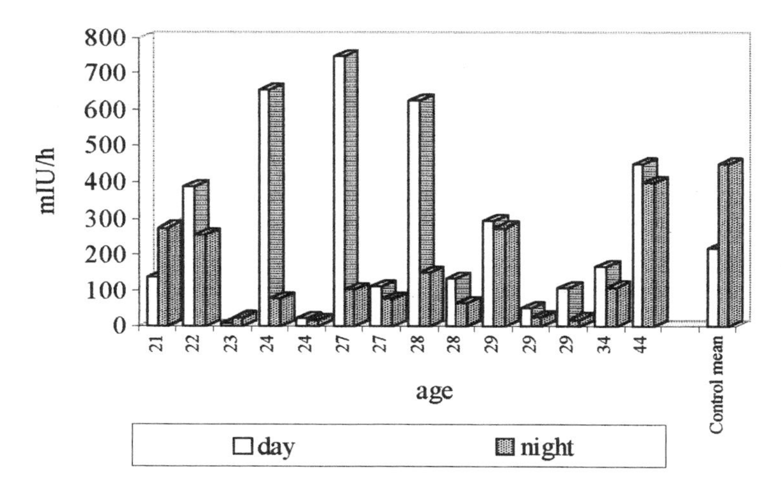

Figure 5. Urinary LH (mIU/h) values in sexual dysfunction.

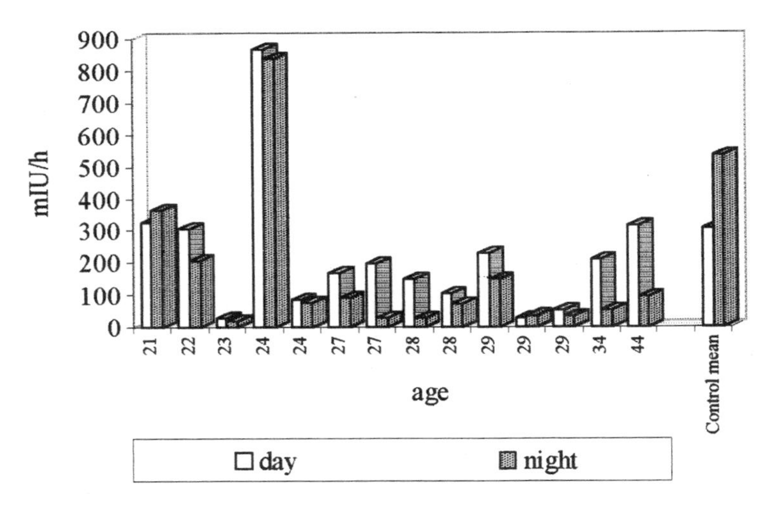

Figure 6. Urinary FSH (mIU/h) values in sexual dysfunction.

have shown that the selected moments of the light/dark cycle are critical for the evidence of melatonin action, suggesting an interaction between the light/dark cycle, melatonin secretion and the responsiveness to melatonin. The biphasic response to exogenous melatonin after the onset of night could be related with endogenous melatonin nocturnal activity.

This progonadal or antigonadal influence of melatonin suggests that melatonin is able to modulate reproductive activity. However, the mechanism by which melatonin modulates testosterone secretion cannot be explained by a simple down-regulation of melatonin receptors.

Our previous studies have shown that melatonin administration in the morning did not alter the testosterone and raised levels of LH and FSH induced by GnRH treatment (3). The present study showed that melatonin administration 2 h after the onset of darkness and at 1 a.m. significantly decreased the high LH and FSH levels induced by GnRH treatment. The antigonadotropic effect of melatonin was dose-dependent.

These results support the hypothesis that melatonin acts at hypothalamic and/or pituitary level (5,7,11,13). Melatonin treatment during the juvenile and pubertal periods decrease the hypothalamic GnRH and suppress the pubertal peak of pituitary GnRH receptor number. It also reduces pituitary LH content and almost completely suppresses the pubertal peak of FSH secretion (5,7,9,13).

However, an additive direct action of melatonin on gonadal steroidogenesis cannot be excluded. Our studies on castrated and testosterone treated rats have shown changes in pineal melatonin levels. At 62 hours after castration, the pineal melatonin concentrations were significantly decreased, nocturnal values were close to those measured during the day. Testosterone administration significantly reduced the inhibitory effect of castration upon melatonin production. The melatonin concentrations returned to the melatonin levels in sham-castrated rats. Interestingly, the pineal melatonin was lower in sham-castrated rats as compared to intact rats but the circadian pattern was preserved. The decrease of melatonin production in sham castrated rats supports the findings that stress induces changes in melatonin production.

Our data confirm the existence of a reciprocal modulation between the hypothalamic-pituitary-gonadal axis and pineal gland. Also, they show that the timing of studied points within the light/dark cycle is critical for the evidence of melatonin action in adult male rats and also suggest an interaction between the light/dark cycle, melatonin secretion and responsiveness to MLT.

As regarding the clinical studies, our results support the evidence of correlations between melatonin and the reproductive system status in humans. Considering our results regarding the sexual maturation and the present results, melatonin production seems to be an important hormonal signal for the reproductive system. In addition, the oral treatment with 2.5 mg of melatonin taken before bedtime (10–11 p.m.) for a month, yielded the improvement of sexual dynamics and spermatogenesis in subjects with sleep disorders.

Taken together these studies again convincingly establish that the reproductive system is pineal dependent. These results support the concept that melatonin is an evolutionary stable timing signal to which each species has adapted the timing of physiological processes.

REFERENCES

1. Bartness, T.J., Powers, J.B., Hastings, M.H., Bittman, E.L., and Goldman, B.D. The timed infusion paradigm for melatonin delivery: what has it taught us about the melatonin signal, its reception, and the photoperiodic control of seasonal responses?. J. Pineal. Res. 15:161–190, 1993.
2. Colmenero, M.D., Diaz, B., Miguel, J.L., Gonzales, M.L., Esquifino, A., and Marin, B. Melatonin administration during pregnancy retards sexual maturation of female offspring in the rat, J. Pineal Res., 11:23–27, 1991.
3. Damian, E., Ianăş O., Badescu, I., and Oprescu, M., Gonadotropin Inhibiting Activity of Melatonin-free Pineal Extract, Advances Biosciences. 29:171–175, 1980.
4. Diaz-Lopez, B., Colmenero, M.D., Diaz-Rodriguez, M.E., Arce, Fraguas A., Esquifino, Parras, A., Marin, and Fernandez B. Effect of pinealectomy and melatonin treatment during pregnancy on the sexual development of the female and male rat offspring. Eur. J. Endocrinol. 132:765–770, 1995.
5. Esquifino, A.I., Arce, A., Villanua, M.A., and Cardinali, D.P. Twenty-four hour rhythms of serum prolactin, growth hormone and luteinizing hormone levels, and of medial basal hypothalamic corticotropin-releasing hormone levels and dopamine and serotonin metabolism in rats neonatally administered melatonin. J. Pineal Res. 22:52–58, 1997.
6. Grosse, J., Maywood, E.S., Ebling, F.J.P., and Hastings, M.H. Testicular regression in pinealectomized Syrian hamsters following infusion of melatonin delivered on non-circadian schedules. Biol. Reprod. 49:666–674, 1993.
7. Hattori, A., Herbert, D.C., Vaughan, M.K., Yaga, K., and Reiter, R.J. Melatonin inhibits luteinizing hormone releasing hormone (LHRH) induction of LH release from fetal rat pituitary cells. Neurosci. Lett. 184:109–112, 1995.
8. Irvine, D.S. Epidemiology and aetiology of male infertility. Hum. Reprod. 13:33–44, 1998
9. Langu, U. Melatonin, and Puberty. Pin. Res. Rev. 4:199–254, 1986.
10. Luboshitzky, R., Wagner, O., Lavi, S., Herer, P., and Lavie P. Abnormal melatonin secretion in hypogonadal men: the effect of testosterone treatment. Clinical Endocrinology. 47:463–469,1997.
11. Malpaux B., Daveau, A., Maurice-Mandon, F., Duarte, G., and Chemineau, P. Evidence that melatonin acts in the premammillary hypothalamic area to control reproduction in the ewe: Presence of binding sites and stimulation of luteinizing hormone secretion by in situ microimplant delivery. Endocrinology. 139:1508–1516, 1998.
12. Neacšu, E., Simionescu, L., Zimel, A., and Caragheorgheopol, A. The development of a radioimmunoassay system for testosterone and dihydrotestosterone. Rev. Roum. Méd- Endocrinol. 28:127–132, 1990.
13. Okatani, Y., Watanabe, K., Morioka, N., Hayashi, K., and Sagara, Y. Nocturnal changes in pineal melatonin synthesis during puberty: relation to estrogen and progesterone levels in female rats. J. Pineal Res. 22:33–41, 1997

14. Petterborg, L.J. Photoperiod, pineal, and reproduction in the white-footed mouse. Advan. Pin. Res. 1:177–183, 1986.
15. Pitrosky, B., Kirsch, R., Vivien-Roels, B., Georg-.Bentz, I., Canguilhem, B., and Pevet, P. The photoperiodic response in Syrian hamster depends upon a melatonin-driven circadian rhythm of sensitivity to melatonin. J. Neuroendocr. 7:889–895, 1995.
16. Reiter, J.R. The pineal, and its hormones in the control of reproduction in mammals. Endocrine Reviews. 1:109–131, 1980.
17. Reiter, R.J. Melatonin, and human reproduction. Ann. Med. 30:103–108,1998.
18. Tamarkin, L., Baird, C.J., and Almeuda, O.F.X. Melatonin: a coordinating signal for mammalian reproduction? Science. 227:714–720, 1985.

36

MELATONIN INFLUENCE UPON OVARY DURING AGEING

A Morphometric Study

B. E. Fernández, *E. Diaz, *C. Fernández, and *B. Díaz

Dptos. Biología Celular
*Biología Funcional, Area Fisiología
Fac. Medicina. Univ. Oviedo, Spain

1. OOCYTE MORPHOLOGY

In the ovary the microscopic anatomy of maturational follicular progression correlates with the physiology of gametogenesis and hormogenesis (4). The primordial follicle is the structure composed of the dictyate oocyte, a single granulosa layer, and a surrounding basement membrane. The distinguishing feature of the follicle is its development of a fluid-filled cavity or antrum. The first sign of follicular recruitment is cuboidal differentiation. The oocyte enlarges and secretes a glycoprotein-containing mucoid substance called the zona pellucida that surrounds the oocyte (1). During each ovarian menstrual cycle, primordial follicles depart the resting pool and begin a well-characterized pattern of growth and development. In a given cycle a few (or at last one) members of the follicular cohort continue to develope and escape atresia, until one becomes the preovulatory (graafian) follicle (2). Abnormalities of the zona pellucida may be characterized by fractures, especially prevalent among postmature eggs. Occasionally such abnormalities of the gamete are associated with mature, normal-appearing granulosa cells (5).

1.1. Morphometric Study

Cytometric studies of the oocytes were performed on the semithin sections of the ovaries. Oocytes were evaluated using an automatic image analysis system (7). In the present paper the oocyte volume, Circular form factor (Form Circ) and Regular form factor (Form Rug) were measured. Data were calculated according to the Computer

Melatonin after Four Decades, edited by James Olcese.
Kluwer Academic / Plenum Publishers, New York, 2000.

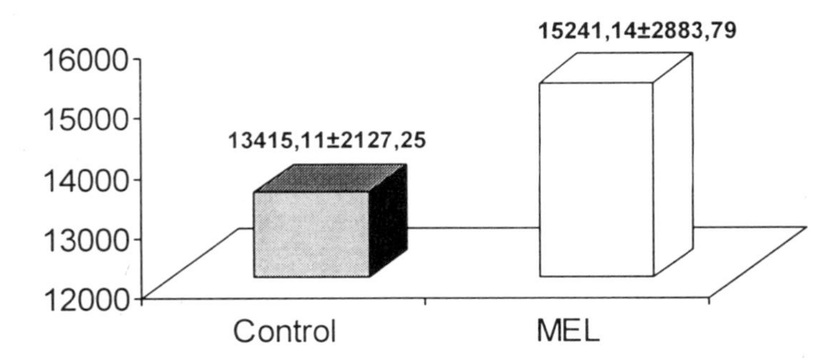

Figure 1. Cellular volume from oocytes of middle-aged rats control and MEL-treated (150 mg/100 g BW), during two months. Values are mean ± SEM.

Assisted Microscopy Program (6,3). Spherical volume (V. Sphere) was determined from measurements of diameters. From the diameter of a measured area, the equivalent volume is obtained with the formula:

$$\text{V. Sphere} = 2/3\ \text{Area} \times \text{Spherical Diameter}$$

Form Circ (FC): This is a dimensionless factor which reflects the irregularity of a structure in relation to a circle. It is calculated from the area and perimeter of the structure by the formula:

$$\text{Form Circ} = (4\ \Pi\ \text{Area})/(\text{Perimeter})^2$$

Form Rug (FR): Is used to define regular or irregular structures. For a regular circle, the value is 1; for irregular structures, values are smaller than 1. This is calculated by the formula:

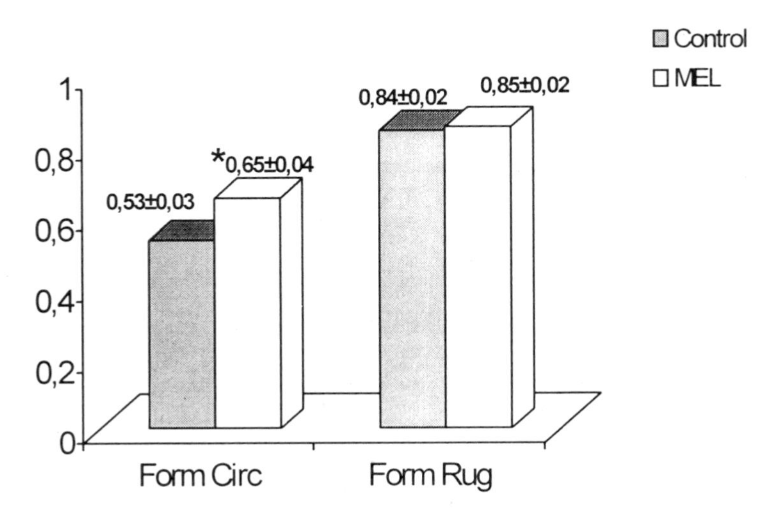

Figure 2. Form Circ and Form Rug from oocytes of middle-aged rats control and MEL-treated (150 mg/100 g BW), during two months. Values are mean ± SEM. * $p < 0.05$ vs. control.

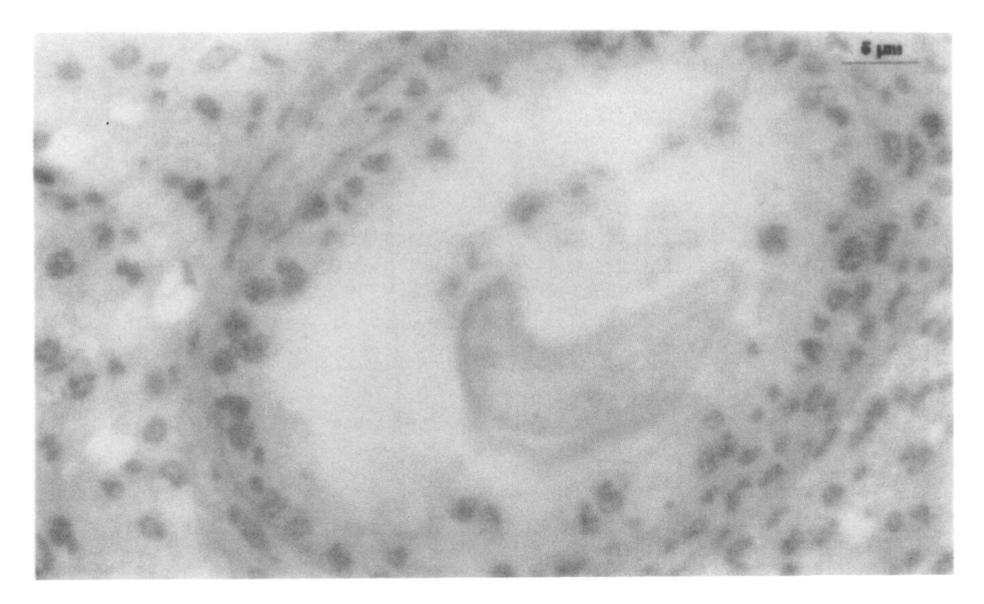

Figure 3. Oocyte from middle-aged control rat.

$$\text{Form Rug} = \text{Convex Perimeter}/\text{Perimeter}$$

1.1.1. Influence of melatonin on oocyte. In the present paper we evaluate the cytometry of oocytes from middle-aged female rats (11 months-old), by using the above mentioned parameters, that allows us to evaluate the functional stage of the oocyte. Although oocyte volume was higher in melatonin-treated rats, not significant

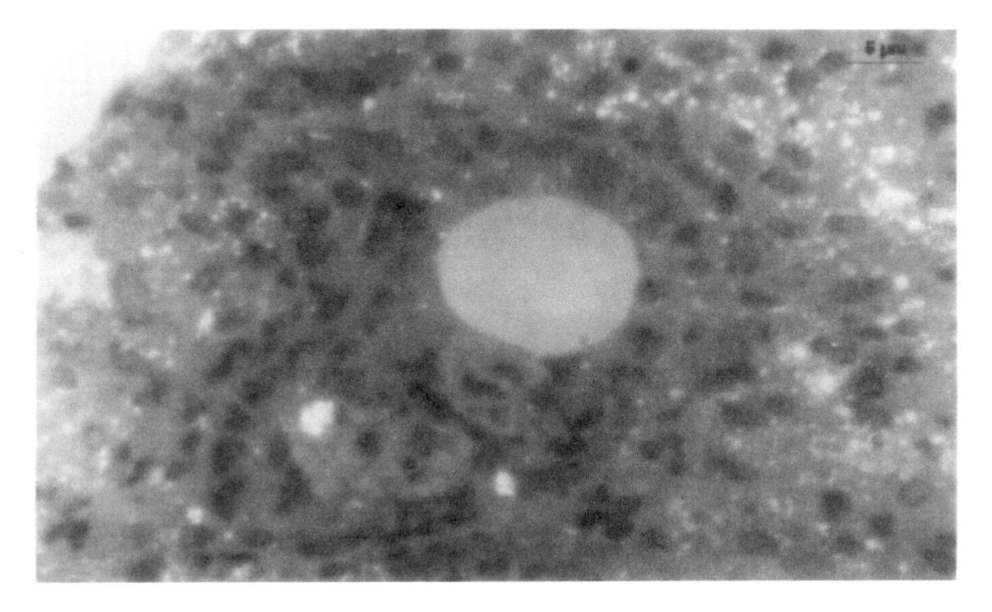

Figure 4. Oocyte from middle-aged melatonin-treated (150µg/100g BW) rat.

differences were found when compared to controls, similar results were obtained for Form Rug. Form Circ values were significantly increased ($p < 0.05$) in oocytes from melatonin-treated rats resulting in an increased tendency to oocyte circularity, pointing out to a more regular functional stage.Supported by Spanish Ministry of Health. Fondo de Investigación Sanitaria, FIS N° 97/0988.

REFERENCES

1. Dombar, B.S. Morphological, biochemical and inmunochemical characterization of the mammalian zone pellucid. Hartmann J. Ed. Mechanism and control of Animal Fertilization. London: Academic: 140–175, 1983.
2. Ericson, G.F. Normal ovarian function. Clin. Obstet. Ginecol. 21:31, 1978.
3. Fernández, B., Díaz, E., Colmenero, M.D., and Díaz, B. Maternal pineal gland participates in prepubertal rats' ovarian oocyte development. The Anatomical Record 243:461–465, 1995.
4. Goodman, AL. and Hodgen, G.D. The ovarian triad of the primate menstrual cycle. Recent Prog. Horm. Res. 39:1, 1983.
5. Kenigsberg, D., Rosenwaks, Z., and Hodgen, G.D. The ovary development and control of follicular maturation and ovulation. Leslie J. De Groot Ed. Endocrinology. W.B. Saunders Company. London: 1915–1928, 1989.
6. Russ, J.C. Computer-Assisted Microscopy. The Measurement and Analysis of Images. Plenum Press, New York, 1991.
7. Vega, J.A., Bengocheaga, M.E., and Pérez-Casas, A. The quantification of the forms: an introduction to the methodology and measurements systems-semiautomatic analysis. Scripta Med. 61:495–502, 1988

37

MELATONIN EFFECT DURING AGING ON REPRODUCTIVE HORMONES OF FEMALE RATS THROUGH THE ESTROUS CYCLE

B. Díaz,[1] E. Díaz,[1] C. Fernández,[1] P. O. Castrillón,[2] A. I. Esquifino,[2] and B. Marín[1]

[1] Dpto. Biología Funcional
Area Fisiología, Fac. Medicina
Universidad de Oviedo
Oviedo, Spain
[2] Dpto. Bioquímica y Biología Molecular III Fac. Medicina, Universidad Complutense
Madrid, Spain

1. MELATONIN DURING AGING

The pineal gland, through its hormone melatonin is involved in animal and human reproduction. It has been suggested that melatonin may be involved in the aging process (8). Melatonin is produced and secreted into the blood in a circadian rhythm with maximal production always during the dark phase of the day. The 24h rhythm of melatonin production is very robust in young animals, but this circadian rhythm deteriorates with age. Thus, in old animals the amounts of melatonin secreted are lower than that of young individuals and the supplemental administration of melatonin may be beneficial in delaying age-related degenerative conditions (7). Long-term melatonin treatment in rats can postpone the age related decrease in survival rates, circulating sex steroids and ^{125}I-melatonin binding sites in the brain (5).

1.1. Reproductive Hormones during Aging

It has been previously described that 12 month-old Wistar rats showed disturbances of estrous cycles (3), which are indicative of reduced activity of the hypothalamic-pituitary ovary axis during aging (2, 4). In this paper we described

Melatonin after Four Decades, edited by James Olcese.
Kluwer Academic / Plenum Publishers, New York, 2000.

Table 1. Concentrations of plasma LH, FSH, PRL and estradiol at stages of the estrous cycle of 11–13-month-old female rats, control and melatonin treated (150µg/100g BW) for two months

		Diestrus	Proestrus AM	Proestrus PM	Estrus	Metaestrus
LH	Control	583.68 ± 69.62	817.92 ± 70.31	4537.47 ± 1474.47	775.06 ± 70.06	892.1 ± 139.45
(pg/ml)		(21)	(16)	(20)	(21)	(13)
	MEL	506.05 ± 77.61	677.85 ± 71.79	25193.93 ± 6159.77*	753.97 ± 139.98	921.92 ± 182.99
		(14)	(14)	(6)	(13)	(11)
FSH	Control	3610.44 ± 319.39	2083.99 ± 215.1	3219.28 ± 278.62	4397.05 ± 473.06	4332.52 ± 713.54
(pg/ml)		(22)	(17)	(21)	(18)	(15)
	MEL	3623.1 ± 552.13	3072.3 ± 368.45**	21115.82 ± 5064.76*	6256.73 ± 601.83**	5686.57 ± 989.99
		(15)	(14)	(13)	(11)	(12)
PRL	Control	19974.65 ± 2639.21	27195.71 ± 4183.57	22395.24 ± 2164.64	22846.29 ± 4889.25	35177.4 ± 2639.21
(pg/ml)		(20)	(17)	(23)	(19)	(13)
	MEL	20574.7 ± 5041.49	35169.88 ± 4865.62	39275.53 ± 4762.12*	38652.63 ± 9099.24	33383.99 ± 6340.17
		(12)	(14)	(11)	(10)	(13)
ESTRADIOL	Control	108.49 ± 6.93	109.26 ± 10.81	159.13 ± 25.61	146.49 ± 10.21	100.24 ± 10.78
(pg/ml)		(9)	(10)	(10)	(9)	(13)
	MEL	143.18 ± 43.39	147.04 ± 13.78**	153.76 ± 20.84	136.42 ± 23.81	102.32 ± 9.27
		(10)	(10)	(10)	(10)	(12)

Values are Mean ± SEM. (Number in parenthesis are number of cases). * $p < 0.01$; ** $p < 0.05$ vs. control group.

the effect of long-term (2 months) melatonin treatment (150 μg/100 g BW) to 12-month-old female Wistar rats on plasma reproductive hormones through the sexual cycle phases: diestrus, proestrus AM, proestrus PM, estrus and metaestrus. We evaluated the concentrations of the following hormones: luteinizing hormone (LH), follicle stimulating hormone (FSH), prolactin (PRL) and estradiol before and after melatonin treatment.

1.1.1. Effect of Melatonin on Reproductive Hormones. Previous studies have shown that the timing of melatonin injections is critical to produce an inhibitory effect on LH secretion during estrous in young cycling rats (10,9). However in middle-aged rats we found significantly increased LH levels in the afternoon of proestrus in melatonin treated rats. Similar result was also found in FSH secretion. The present data suggest that melatonin treatment to middle-aged rats, in which melatonin secretion is decreased (6) could improve LH as well as FSH secretion. These findings are also supported by the effect of melatonin on PRL secretion. Melatonin restored PRL peak values in the afternoon of proestrus. It is known that the pattern of PRL secretion during estrous cycle of the rat is characterized by low levels except for a brief period on the afternoon and evening of proestrus (1). Increased estradiol levels in the morning of proestrus were found after melatonin treatment. Our results indicate that treatment with melatonin to middle-aged rats showed a positive influence on the preovulatory mechanism, with increased estradiol values in the morning of proestrus, which preceded the gonadotropin and PRL surges observed in the afternoon of proestrus.

ACKNOWLEDGMENT

This work was supported by Spanish Ministry of Health. Fondo de Investigación Sanitaria, FIS N° 97/0988.

REFERENCES

1. Anselmo-Franci, J.A., Franci, C.R., Krulich, L., Antunes-Rodrigues, J., and McCann, S.M. Locus coeruleus lesions decrease norepinephrine input into the medial preoptic area and medial basal hypothalamus and block the LH, FSH and prolactin preovulatory surge. Brain Res. 767:289–296, 1997.
2. Arias, P., Carbone, S., Szwarcfarb, B., Feleder, C., Rodríguez, M., Scacchi, P., and Moguilevsky, J.A. Effects of aging on N-methyl-D-aspartate (NMDA)-induced GnRH and LH release in female rats. Brain Res. 740:234–238, 1996.
3. Aschheim, P. Aging in the hypothalamic-hypophyseal ovarian axis in the rat. In: Hypothalamus pituitary and aging. Everitt, A.V. and Burgess, J.A. (Eds.) Springfield, Illinois 376–418, 1976.
4. Fujimoto, V.Y., Klein, D.C., Battaglia, D.E., Bremer, W.J., and Soules, M.R. The anterior pituitary response to gonadotropin-releasing hormone challenge test in normal older reproductive-age women. Fert. Steril. 85:539–544, 1996.
5. Oaknin-Bendahan, S., Anis, Y., Nir, I., and Zisapel, N. Effects of long-term administration of melatonin and a putative antagonist on ageing rats. Neuroreport 6:785–788, 1995.
6. Pang, S.F., Tang, F., and Tang, P.L. Negative correlation of age and the levels of pineal melatonin, pineal N-acetylserotonin, and serum melatonin in male rats. J. Exp. Zool. 229:41–47, 1984.
7. Reiter, R.J. Pineal function during aging: attenuation of the melatonin rhythm and its neurobiological consequences. Acta Neurobiol. Exp. 54:31–39, 1994.

8. Reiter, R.J. The aging pineal gland and its physiological consequences. Bioassays 14:167–175, 1992.
9. Rivest, R.W., Lang U., Aubert, M.L., and Sizonenko, P.C. Daily administration of melatonin delays rat vaginal opening and disrupts the first estrous cycle: Evidence that these effects are synchronized by the onset of light. Endocrinology 116:779–787, 1985.
10. Ying, S-Y. and Greep, R.O. Inhibition of ovulation by melatonin in the cyclic rat. Endocrinology 92:333–336, 1973.

NEW INSIGHTS INTO MELATONIN REGULATION OF CANCER GROWTH

David E. Blask, Leonard A. Sauer, Robert T. Dauchy,
Eugene W. Holowachuk, and Mary S. Ruhoff

Bassett Research Institute
Mary Imogene Bassett Hospital
Cooperstown, New York, 13326

INTRODUCTION

Numerous studies have confirmed the ability of melatonin to inhibit cancer growth at both physiological and pharmacological concentrations both *in vivo* and *in vitro* in experimental model systems utilizing either spontaneous, carcinogen-induced or transplantable murine neoplasms or murine and human cancer cell lines. These model systems have provided important information supporting an antineoplastic role for melatonin at all stages of the tumorigenic process including initiation, promotion, progression and metastasis (1). However, very little is known about the mechanisms by which either the endogenous physiological melatonin signal or the administration of pharmacological doses of melatonin inhibit the various stages of carcinogenesis.

Over the past several years, research in our laboratory has focused on the hypothesis that melatonin is an oncostatic neurohormone that inhibits tumor growth promotion. Melatonin inhibits the mitogenic action of a number of growth factors including estradiol (E_2), prolactin (PRL) and epidermal growth factor (EGF), on cancer cell proliferation *in vitro* (1). Additionally our research has addressed the important relationship between dietary fat and cancer, particularly the role of the uptake and metabolism of linoleic acid, an essential fatty acid important in the promotion of tumor growth (2–5) (see below). By combining these two hypotheses, we have formulated a novel hypothesis that melatonin inhibits tumor growth by inhibiting the tumor uptake and metabolism of linoleic acid (6–8). In this article, we will discuss the current evidence from our laboratory that has emerged from the use of the "tissue-isolated" tumor preparation, hepatoma 7288CTC, that supports this unique postulate.

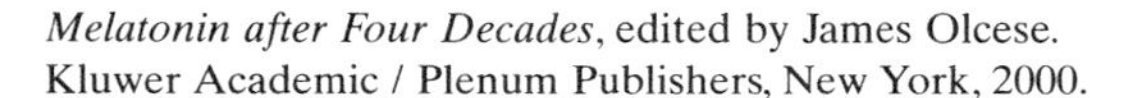
Melatonin after Four Decades, edited by James Olcese.
Kluwer Academic / Plenum Publishers, New York, 2000.

TISSUE-ISOLATED HEPATOMA 7288CTC MODEL SYSTEM

In this model system, a 3-mm cube of Morris rat hepatoma is attached to and grown on the end of a vascular stalk composed of the truncated superficial epigastric artery and vein in the inguinal region of adult male Buffalo rats. The tumor implant and adjacent pedicle are then enclosed in a sterile parafilm envelope containing penicillin. The arterial supply to and venous drainage from the tumor is thus exclusively via the superficial epigastric vessels. Tumor attachment to other host tissues or vasculature is blocked by the parafilm envelope. Following replacement of the tumor implant into the inguinal fossa and closure of the skin, the tumor is allowed to grow in a "tissue-isolated" manner. This arrangement allows the cannulation of the epigastric vessels for the direct perfusion of the tumor itself with physiological and/or pharmacological agents and the measurement of arteriovenous differences across the tumor of various biochemical factors and products important in tumor growth and metabolism. All this is accomplished while the tumor is maintained in a physiological state as reflected by the monitoring of blood gases, blood flow, pH, glucose uptake and lactate release (9). We have taken advantage of such a model system to investigate for the first time, using a totally integrative approach, the mechanisms by which melatonin inhibits tumor growth *in vivo* from the systemic to the biochemical and molecular levels within a chronobiological context.

THE ROLE OF LINOLEIC ACID IN TUMOR METABOLISM AND GROWTH

The Morris rat hepatoma 7288CTC is among a variety of transplantable tumors that are characterized by a unique growth requirement for linoleic acid, an essential polyunsaturated fatty acid (ω6,C18:2). Linoleic acid is the major polyunsaturated fatty acid consumed in the human diet and numerous investigations have shown that high fat diets, particularly those containing linoleic acid as the major fatty acid, increase the growth rates of transplantable tumors in murine species (10). More specifically, Sauer and Dauchy (2–5) have shown that the increased blood concentrations of linoleate following its dietary intake results in an increased arterial supply to and uptake of this fatty acid by tissue-isolated hepatoma 7288CTC. They further demonstrated that the increased tumor uptake of linoleate directly stimulates the growth of this tumor and inferred from these results that either the fatty acid itself or a metabolite initiated or enhanced specific tumor growth processes.

These results implied to these investigators that linoleate is more than merely an energy source for tumor growth and may function as a specific tumor growth signal transduction molecule (11). Support for this hypothesis comes from evidence showing that linoleic acid is oxidized intracellularly to 13-hydroxy-9, 11-octadecadienoic acid (13-HODE) by an n-6 lipoxygenase which is most likely part of the 15-lipoxygenase family of enzymes. The activity of this lipoxygenase is regulated by the tyrosine kinase moiety of the EGF receptor such that binding of EGF to its cognate receptor stimulates lipoxygenase activity and the metabolism of linoleic acid to 13-HODE. Following its formation, 13-HODE apparently enhances the EGF-induced autophosphorylation of tyrosine kinase of the EGF receptor as well as the tyrosine phosphorylation of key downstream signal transduction proteins such as GTPase activating protein and mitogen-activated protein kinase (MAPK). Recent evidence indicates that 13-HODE up-regulates EGF-dependent tyrosine phosphorylation, and thus EGF-induced mitogenesis, by inhibiting the dephosphorylation of the EGF receptor presumably by altering the interaction of tyrosine phosphatases with the EGF receptor. Therefore, linoleic appears to play a critical role in transducing the EGF-induced mitogenic signal from the cell surface to the nucleus of cancer cells (11,12).

In tissue-isolated hepatoma 7288CTC, the rate of 13-HODE released by this tumor is directly dependent upon the rate of tumor linoleic acid uptake. Furthermore, the lipoxygenase inhibitor, nordihydroguairetic acid (NDGA), suppressed 13-HODE production and tumor growth without affecting linoleic acid uptake. Moreover, the addition of 13-HODE to linoleic acid deficient donor blood perfusing tumors of linoleic acid-deficient recipient animals resulted in a dose-dependent increase in [^{3}H]-thymidine incorporation into DNA (13). Taken together, these results indicate that 13-HODE is the mitogenic signal responsible for linoleate-dependent growth in hepatoma 78288CTC *in vivo*.

EFFECTS OF MELATONIN ON THE GROWTH AND LINOLEIC ACID METABOLISM OF HEPATOMA 7288CTC

We initially determined the sensitivity of hepatoma 7288CTC to the oncostatic effects of pharmacological doses of melatonin (50 to 200 µg) injected s.c. into tumor-bearing rats, maintained on a 12L:12D light:dark cycle, every afternoon one to two hours prior to lights off (PM). Injections with melatonin or vehicle began one week prior to tumor implantation and continued until the end of the experiment three to four weeks later. Melatonin treatment was effective in delaying the appearance and suppressing the growth of hepatoma 7288CTC at all doses tested. Furthermore, the tumor uptake, content and release of linoleic acid, total fatty acids and 13-HODE, respectively, was suppressed in animals receiving PM melatonin therapy in a dose-dependent manner. Interestingly, in a study in which melatonin (200 µg to 1 mg) was injected in the morning (AM) two to three hours following lights on, there was no effect on tumor growth or linoleic acid uptake and metabolism (6–8; unpublished results). However, unlike animals maintained in diurnal lighting, tumor growth and linoleic acid uptake and metabolism in rats maintained under constant light conditions were equally inhibited regardless of whether melatonin was injected during either the subjective AM or PM (unpublished results). We concluded from these results that melatonin's inhibitory effect on the growth of and linoleic acid uptake and metabolism by hepatoma 7288CTC was both dose- and circadian-time dependent. It is unclear at this point what mechanism is responsible for the apparent diurnal rhythm of tumor sensitivity to melatonin in rats maintained on diurnal lighting. However, the elimination of this sensitivity rhythm of hepatoma 7288CTC to melatonin in constant light strongly suggests that the normal endogenous melatonin rhythm itself may drive this tumor rhythm under diurnal lighting conditions.

We next turned our attention to the issue of whether the physiological nocturnal melatonin signal itself exerted an oncostatic effect on tumor growth and metabolism. Previous studies *in vivo* had demonstrated that pinealectomy or constant light exposure increased the incidence of carcinogen-induced mammary cancer in rats while *in vitro* studies have shown that physiological concentrations of melatonin inhibit cancer cell proliferation (1). We approached this question *in vivo* by extinguishing the nocturnal melatonin peak by either pinealectomy or exposure of tumor-bearing animals to either constant bright light or a 12L:12D lighting regimen in which the dark phase was "contaminated" by low intensity light (0.2 lux) (14). In each of these scenarios, the onset of palpable tumors was substantially advanced and the tumor growth rate was accelerated by two-fold over that of 12L:12D intact or sham-pinealectomized controls. Additionally, the tumor uptake of linoleic acid and its conversion to 13-HODE were also markedly elevated in these animals providing strong evidence that the physiological nocturnal melatonin signal is critical for restraining the growth of these tumors, presumably by inhibiting linoleic acid uptake and metabolism to 13-HODE.

We reasoned that if the circadian melatonin signal itself is a critical inhibitory regulator of tumor growth and metabolism as well as a driving force for a diurnal rhythm of tumor sensitivity to exogenous melatonin, then hepatoma 7288CTC should evince a circadian rhythm of linoleic acid uptake and metabolism that is the "mirror-image" of the melatonin rhythm. Arteriovenous difference measurements made across tumors at six different circadian time points showed that linoleic acid uptake and oxidation to 13-HODE were highest during the light phase when circulating melatonin levels were low and lowest when melatonin levels were at their peak. This circadian rhythm of tumor metabolism was completely eliminated by pinealectomy (8; unpublished results). These results make a convincing argument for a melatonin-driven circadian rhythm of linoleic acid uptake and metabolism.

DIRECT EFFECTS OF MELATONIN ON TUMOR LINOLEIC ACID UPTAKE AND METABOLISM AND THE SIGNAL TRANSDUCTON MECHANISMS INVOLVED

In the tumor growth experiments cited above, it was not completely clear whether the inhibitory action of melatonin on tumor growth was the result or cause of the inhibition of linoleic acid uptake and its conversion to 13-HODE. To more precisely address the question of whether melatonin directly inhibits tumor fatty acid uptake and metabolism, we perfused tissue-isolated hepatoma 7288CTC *in situ* with melatonin. Following a 60 minute perfusion with donor whole blood from 48-hr fasted rats with elevated circulating fatty acid levels, melatonin was added to the whole blood perfusate at near physiological nocturnal peak levels. Approximately 40 minutes following the addition of melatonin to the perfusate, linoleic acid uptake decreased by nearly 70% from the steady-state control perfusion levels. Over this same time-course, tumor 13-HODE production declined to undetectable levels in response to tumor perfusion with melatonin. The suppressive effects of melatonin on linoleic acid uptake and metabolism to 13-HODE were reversible following the removal of melatonin from the perfusate. Neither N-acetylserotonin nor 6-hydroxymelatonin had any effect on tumor linoleate uptake and metabolism in this perfusion system (6–8; unpublised results). These findings indicate that melatonin's inhibition of linoleic acid uptake and metabolism is a rapid and specific effect exerted directly on the tumor itself and is likely the cause of tumor growth inhibition rather than its result.

To further clarify that melatonin's inhibition of tumor growth was ultimately due to its ability to halt the production of the mitogenic signaling molecule 13-HODE, we examined the incorporation of [^{3}H]-thymidine into DNA in hepatoma 7288CTC in response to perfusion with melatonin either alone or in combination with 13-HODE. The incorporation of [^{3}H]-thymidine by hepatoma 7288CTC was reduced by approximately 43% in response to perfusion with melatonin (1 nM) for two hours; linoleic acid uptake was reduced by 80% and 13-HODE production was completely negated. When melatonin was co-perfused with 13-HODE (12 μg/ml) the tumors took up 50% of the 13-HODE resulting in inhibition of [^{3}H]-thymidine incorporation by melatonin though even linoleic acid uptake was still completely blocked (unpublished results). These data strongly support the hypothesis that melatonin inhibition of the production of the mitogenic metabolite 13-HODE, via the inhibition of linoleic acid uptake, is responsible for the inhibition of tumor growth.

The rapid time-course and the circadian nature of melatonin inhibition of tumor uptake and metabolism of linoleic acid suggested that high affinity melatonin receptors (i.e., mt_1 and/or MT_2) (15) mediated the first step of a signal transduction cascade that ultimately culminated in the inhibition of tumor growth. Moreover, since tumor linoleate uptake and metabolism to 13-HODE could be rapidly and reversibly regulated by melatonin, this suggested that a recently cloned fatty acid transport protein (FATP) (16) may be involved via a functional link to melatonin receptors. Indeed, Northern blot analysis revealed the overexpression of mRNA transcripts for FATP while RT-PCR demonstrated the presence of both mt_1 and MT_2 receptor mRNAs in hepatoma 7288CTC (8; unpublished results). This provided us with compelling evidence that the molecular substrates were present in hepatoma 7288CTC for facilitated linoleate transport and conversion to 13-HODE and their rapid inhibition by melatonin. Interestingly, neither daily afternoon melatonin injections nor constant light exposure altered the expression of either FATP or melatonin receptor mRNAs.

Functional melatonin receptors negatively coupled to adenylate cyclase via a pertussis toxin (PTX)-sensitive G protein such that melatonin suppresses cAMP have been demonstrated in a number of tissues (15). It stood to reason that if high affinity melatonin receptors mediated melatonin's blockade of linoleate uptake and metabolism to 13-HODE via a suppression of cAMP, then such an effect should be reversible with either PTX, forskolin or cAMP itself. High constitutive levels of cAMP have been observed in several types of malignancies including liver and mammary adenocarcinoma (17,18). In our tumor perfusion system, we found that PTX, forskolin and 8-bromo-cAMP completely reversed the melatonin-induced inhibition of linoleate uptake and metabolism to 13-HODE. Additionally, NF-023, an inhibitor of inhibitory G proteins, also reversed the suppressive action of melatonin on linoleic acid uptake and metabolism (8; unpublished results). These results argued strongly for the hypothesis that melatonin inhibits tumor growth by suppressing linloeate uptake and conversion to 13-HODE via a suppression of cAMP through PTX-sensitive melatonin receptors that may be functionally linked to FATP.

CONCLUSIONS

The tissue-isolated rat hepatoma model has provided us with an unprecedented opportunity to address the role of melatonin in the regulation of cancer progression from the organismal to the molecular level in the context of circadian biology. Our results have confirmed that the oncostatic effects of physiological and pharmacologial concentrations of melatonin are indeed circadian time-dependent. Furthermore, circadian time-dependent inhibition of tumor growth is mediated via melatonin's ability to inhibit the tumor uptake of dietary linoleic acid and its conversion to 13-HODE, an amplifying signal for EGF-induced mitogenesis (11,12). Such a mechanism may explain our earlier reported results showing the melatonin inhibits EGF-induced mitogenesis in human breast cancer cells in culture (19). Moreover, a signal transduction cascade involving melatonin receptor-induced suppression of cAMP may mediate melatonin's ability to suppress the function of FATP and prohibit the entry of linoleate into tumor cells thereby obstructing the transduction of the EGF-induced mitogenic signal from the tumor cell surface to the nucleus.

REFERENCES

1. Blask, D.E. Melatonin in oncology. in *Melatonin—Biosynthesis, Physiological Effects, and Clinical Implications* (eds Yu H.S. and Reiter, R.J.) (CRC Press, Boca Raton), 447–475 (1993).
2. Sauer, L.A. and Dauchy, R.T. Identification of linoleic and arachidonic acids as the factors in hyperlipemic blood that increase [^{3}H]thymidine incorporation in hepatoma 7288CTC perfused in situ. Cancer Res. 48:3106–3111 (1988).
3. Sauer, L.A. and Dauchy, R.T. Uptake of plasma lipids by tissue-isolated hepatoma 7288CTC and 7777 in vivo. Br. J. Cancer 66:290–296 (1992).
4. Sauer, L.A. and Dauchy, R.T. The effect of omega-6 and omega-3 fatty acids on [^{3}H]thymidine Incorporation in hepatoma 7288CTC perfused in situ. Cancer Res. 66:297–303 (1992).
5. Sauer, L.A., Dauchy, R.T., and Blask, D.E. Dietary linoleic acid intake controls the arterial blood plasma concentration and the rates of growth and linoleic acid uptake and metabolism in hepatoma 7288CTC in Buffalo rats. J. Nutrition 127:1412–1421 (1997).
6. Blask, D.E., Sauer, L.A., and Dauchy, R.T. Melatonin suppression of tumor growth *in vivo*: a novel mechanism involving inhibition of fatty acid uptake and metabolism. in *Therapeutic Potential of Melatonin, Frontiers of Hormone Research, vol. 23*, (eds Maestroni, G.J.M., Conti, A., and Reiter, R.J.) (Basel, Karger) 107–114 (1997).
7. Blask, D.E., Sauer, L.A., and Dauchy, R.T. Melatonin regulation of tumor growth and the role of fatty acid uptake and metabolism. Neuroendocrinol. Lett. 18:59–62 (1997).
8. Blask, D.E., Sauer, L.A., Dauchy, R.T., Holowachuk, E.W., and Ruhoff, M.S. Circadian rhythm in experimental tumor carcinogenesis and progression and the role of melatonin. in *Biological Clocks—Mechanisms and Applications*, (ed. Touitou, Y.) (Amsterdam, Elsevier) 469–474 (1998).
9. Dauchy, R.T. and Sauer, L.A. Preparation of "tumor-isolated" rat tumors for perfusion: a new surgical technique that perserves continuous blood flow. Lab. Animal Sci. 36:678–681 (1986).
10. Klurfield. D.M. Fat effects on experimental tumorigenesis. J. Nutr. Biochem. 6:201–205 (1995).
11. Eling, T.E. and Glasgow, W. Cellular proliferation and lipid metabolism: importance of lipoxygenases in regulating epidermal growth factor-dependent mitogenesis. Cancer Metast. Rev. 13:397–410 (1994).
12. Glasgow, W.E., Hui, R., Everhart, A.L., Jayawickreme, S.P., Angerman-Stewart, J., Han, B.B., and Eling, T.F. The linoleic acid metabolite (13 S)-hydroxyoctadecadienoic acid, aug-ments the epidermal growth factor receptor signaling pathway by attenuation of receptor dephosphorylation. J. Biol. Chem. 272:19269–19276 (1997).
13. Sauer, L.A., Dauchy, R.T., Blask, D.E., Armstrong, B.J., and Scalici, S. 13-Hydroxyoctadecadienoic acid is the mitogenic signal for linoleic acid -dependent growth in rat hepatoma 7288CTC *in vivo*. Cancer Res. 59:4688–4692 (1999).
14. Dauchy, R.T., Sauer, L.A., Blask, D.E., and Vaughan, G.M. Light contamination during the dark phase in "photoperiodically controlled" animal rooms: effect on tumor growth and metabolism in rats. Lab. Animal Sci. 47:511–518 (1997).
15. Reppert, S.M. and Weaver, D.R. Melatonin madness. Cell 83:1059–1062 (1995).
16. Schaffer, J.E. and Lodish, H.F. Expression cloning and characterization of a novel adipocyte long chain fatty acid transport protein. Cell 79:427–436 (1994).
17. DeRubertis, F.R. and Craven, P.A. Sequential alterations in the hepatic content and metabolism of cyclicAMP and cyclic GMP induced by DL-ethionine: evidence for malignant transformation of liver with a sustained increase in cyclic AMP. Metabolism 25:1611–1625 (1976).

18. Cho-Chung, Y.S. Role of cyclic AMP receptor proteins in growth, differentiation, and suppression of malignancy: new approaches to therapy. Cancer Res. 50:7093–7100 (1990).
19. Cos, S. and Blask, D.E. Melatonin modulates growth factor activity in MCF-7 human breast cancer cells. J. Pineal Res. 17:25–32 (1994).

39

MELATONIN SYNERGIZES WITH RETINOIC ACID IN THE PREVENTION AND REGRESSION OF BREAST CANCER

Steven M. Hill,[1,2] Stephenie Teplitzky,[3] Prahlad T. Ram,[1] Todd Kiefer,[4] David E. Blask,[5] Louaine L. Spriggs,[1,2] and Kristin M. Eck[1]

[1] Department of Anatomy
[2] Tulane Cancer Center
[3] Department of Surgery
[4] Graduate Program in Molecular and Cellular Biology
Tulane University School of Medicine
1430 Tulane Avenue, New Orleans, Louisiana 70112
[5] Mary Imogene-Bassett Research Institute
Cooperstown, New York 13326

1. INTRODUCTION

Breast cancer is the second leading cause of cancer-related deaths in the United States, and the incidence of and mortality from breast cancer is increasing worldwide (1). Breast cancer represents approximately 31% of all cancers in women, and current predictions are that one in eight women will develop breast cancer at some point during her lifetime (2). In general, breast cancer is a uniquely endocrine-responsive neoplasm, and over the last two decades, considerable progress has been made in the clinical management of hormone-responsive, estrogen receptor (ER)-positive breast cancer. However, greater than one third of all women who develop metastatic breast cancer do not respond to standard endocrine therapies, and will succumb to the disease (3,4). Breast cancer cells, as well as breast epithelial cells, respond to a variety of hormones, including androgens, estrogens, insulin, progesterone, prolactin, and retinoic acid, as well as growth factors such as epidermal growth factor (EGF), insulin-like growth factor (IGF-1), and transforming growth factors (TGF) α and β (5,6). The growth of endocrine-responsive breast tumors depends heavily on the hormonal milieu presented

Melatonin after Four Decades, edited by James Olcese.
Kluwer Academic / Plenum Publishers, New York, 2000.

to the tumor cells and the cross-talk between various signaling pathways activated by the different hormones and growth factors.

Within the last several decades, tremendous advances have been made with respect to the elucidation of the functions of the pineal gland and it major hormone, melatonin (MLT), particularly those relating to photoperiodicity and the regulation of seasonal reproductivity. With the recent cloning of the MLT receptors, we are beginning to define the cellular and molecular mechanisms of MLT action. However, advances in our understanding of the pineal gland's impact on human health and disease have been considerably slower. One of the more exciting aspects of current pineal research concerns the role of the pineal gland and, more specifically, MLT's role in neoplastic disease.

The emergence of MLT as an important neuroendocrine regulator of neoplastic growth has stimulated a new phase of research into its mechanisms of action at the systemic, cellular and molecular levels. A number of experimental neoplasms have been studied with respect to the ability of MLT and its analogs to influence cancer growth both *in vivo* and *in vitro*. These neoplasms include tumors of the breast, prostate, uterus, cervix, ovary, pituitary, skin, neural tissue, lung, liver, colon, and connective tissue (7–9). However, the most studied and one of the more responsive neoplasms to the anti-tumor effects of MLT is breast cancer.

2. MELATONIN AND BREAST CANCER

2.1. *In Vivo* Effects of MLT on Mammary Cancer

Melatonin, the major hormonal product of the pineal gland, has repeatedly been shown to exert a negative influence on the development and growth of hormone-responsive breast cancer (10,11). Most of the work to date on the oncostatic effects of MLT *in vivo* has been conducted in the 7,12-dimethylbenzanthracene (DMBA)- and N-nitroso-N-methylurea (NMU)-induced hormone-responsive rat mammary tumor models. The DMBA model has been studied with respect to the effects of pinealectomy, photoperiod, and MLT administration on the growth and regression of mammary tumors. Blask et al. (10) found that DMBA-induced mammary tumorigenesis is inhibited in light-deprived female rats, and that pinealectomy not only prevents this growth inhibition but actually enhances tumor growth. The advantage of the NMU model over the DMBA model is that the growth characteristics of the NMU-induced tumors more closely resemble those of human breast cancer cells. The tumor cells of this model are primarily responsive to the mitogenic effects of estrogen, while the tumors in the DMBA model are primarily prolactin-responsive. Pinealectomy followed by NMU administration results in enhanced mammary tumor formation compared to intact controls, and daily late afternoon injections of MLT significantly decreases the latency to onset of the tumors, tumor size, and tumor number (10). Studies investigating the effects of MLT administered in conjunction with other compounds have also been very promising. Kothari et al. (12) have shown that combined treatment with MLT and the anti-estrogen, tamoxifen, is highly effective in the NMU-induced mammary tumor model in suppressing mammary tumor incidence, and in increasing the latency to onset of tumor appearance.

2.2. *In Vitro* Effects of MLT on Breast Cancer

In addition to animal models, the effects of MLT on breast cancer have been investigated through *in vitro* studies using several different human breast tumor cell lines. Examination of MLT's effects on various breast cancer cell lines shows that MLT suppresses the growth of cells which express the estrogen receptor (ER-positive cell lines), but has no effect on ER-negative cell lines (13). One of the most extensively studied cell lines is the ER-positive MCF-7 human breast cancer cell line derived from the pleural effusion of a woman with metastatic adenocarcinoma of the breast (14). This breast cancer cell line expresses the ER and the estrogen-inducible progesterone receptor (PgR), as well as other steroid receptors, such as the androgen (AR) and glucocorticoid (GR) receptors (15,16). Studies investigating the effects of MLT on MCF-7 cells have shown that 10^{-9} M MLT induces a 60%–80% inhibition of cell proliferation after 7 days of treatment (10). The effective growth-inhibitory concentration of MLT ranges from 10^{-9} M to 10^{-11} M, which corresponds to peak physiological night and daytime values, respectively (17). The antiproliferative effects of MLT appear to be limited to a very narrow range; both sub-physiological and supra-physiological concentrations of MLT are ineffective in suppressing MCF-7 cell proliferation (8).

Since MLT has been shown to exert antiproliferative effects only on ER-positive cell lines, it is possible that MLT's effects are mediated through modulation of the estrogen response pathway. Molis et al. (18) reported that 10^{-9} M MLT, a concentration previously shown to inhibit MCF-7 cell proliferation, is also able to significantly repress ER mRNA expression and ER protein levels as early as 6h after treatment. It has also been shown that MLT does not compete with labeled estrogen for binding to the ER, demonstrating that MLT's inhibitory effects on cell proliferation do not result from interference with estrogen binding (19).

The antiproliferative effects of MLT have also been shown to be serum-dependent, suggesting that interaction between MLT and specific serum components may be necessary in order to produce its growth-inhibitory effects (10). Two hormones, estrogen and prolactin, both of which exert a mitogenic effect on MCF-7 cells, are present in the serum, and MCF-7 cells express receptors for both hormones. It is possible that MLT may exert its antiproliferative effects by interfering with estrogen and/or prolactin-stimulated growth. In support of this hypothesis, Blask and Hill (10) reported that MLT inhibits estrogen-stimulated growth in MCF-7 cells, and that the loss of MLT's growth-inhibitory effects seen in serum-free, defined media can be partially reconstituted by the addition of exogenous prolactin. These findings suggest that the growth-inhibitory effects of MLT may involve important interactions between MLT and other hormones and growth factors present in serum.

Studies using flow cytometry show that MLT inhibits MCF-7 cell proliferation by delaying the G1 to S transition in the cell cycle, indicating that the antiproliferative effects of MLT may be cell cycle-specific (20,21). Cos et al. (22) have further shown that MLT significantly increases the duration of the cell cycle of MCF-7 cells from 20.36 h to 23.48h. These data support the belief that MLT exerts its antiproliferative effects, at least in part, through a cell cycle-specific mechanism that delays the entry of MCF-7 cells into mitosis. Cos et al. (23) have also shown that MLT inhibits DNA synthesis in MCF-7 cells, which suggests that the antiproliferative effects of MLT on breast cancer cells may also depend on its inhibitory actions on DNA synthesis.

2.3. Clinical Studies of MLT and Breast Cancer

In addition to the studies in the rat model, the effects of MLT have also been investigated in human breast cancer. It has been reported that the nocturnal rise in MLT levels is significantly reduced in women with ER-positive breast tumors compared to women with ER-negative breast tumors and healthy age-matched controls (24). Another clinical study assessing pineal function in patients with breast cancer found that post-menopausal women with advanced breast cancer have diminished urinary levels of MLT compared to healthy controls (25). Based on these data, it has been suggested that depressed MLT secretion may be a predisposing factor for the development of breast cancer in humans (26). It has also been reported that a two-fold higher MLT level is associated with breast tumors with a low proliferative index compared to those with a high index, suggesting that hypersecretion of MLT may predict a more favorable prognosis (27). Clinical trials investigating the combined effects of MLT with tamoxifen are also promising. Lissoni et al. (28) initiated a clinical phase II study to assess the efficacy of combined treatment using tamoxifen and MLT in women with metastatic breast cancer who were unresponsive to tamoxifen therapy alone. Twenty-eight percent of the patients exhibited a partial response to treatment with MLT and tamoxifen, indicating that MLT may amplify the effects of tamoxifen in women with metastatic breast cancer, and may induce tumor regression in patients who have previously not responded to tamoxifen therapy. However, despite this positive data on the oncostatic effects of MLT, very few clinical trials have been conducted investigating the role of MLT in human breast cancer.

2.4. Melatonin Receptors

Melatonin is a highly lipophilic molecule that is capable of diffusing through both cell and nuclear membranes to enter intracellular compartments. Recently, two distinct MLT receptors were identified—a cell membrane, G-protein linked receptor, and a nuclear MLT receptor. The membrane-bound MLT receptor was first characterized by Reppert et al. (29), and is termed the Mel_{1a} (mt_1) receptor. The mt_1 receptor binds MLT with a K_d of approximately 0.06 nM, and affects cellular activity by transmitting its signal through second messenger G-proteins (30). A second MLT receptor, termed the Mel_{1b} (MT_2) receptor, has also been described. The MT_2 receptor binds MLT with an affinity similar to that of mt_1. The MT_2 receptor is also a membrane-associated, G-protein linked receptor with 60% homology to the mt_1 receptor at the amino acid level (31).

Becker-André et al. (32) recently reported that MLT is able to specifically bind to the retinoid Z receptor (RZR) α and β sub-types and the retinoid orphan receptor α (RORα). However, these receptors have a 100–1000 fold lower binding affinity for MLT than the mt_1 receptor. The RZR/ROR receptors belong to the steroid/thyroid hormone receptor superfamily, and show sequence homology to the retinoic acid receptors, RAR and RXR (32,33). The RZR/RORs appear to bind DNA as monomers or homodimers, unlike most members of the steroid hormone receptor superfamily which bind target DNA sequences as heterodimers and homodimers (33). Our laboratory (34), as well as other research groups (31), has subsequently been unable to confirm the binding studies performed by Becker-André et al. (32), thus, it is generally believed that the mt_1 receptor is the functional MLT receptor.

3. RETINOIDS AND BREAST CANCER

The retinoids comprise a family of polyisoprenoid lipids that includes vitamin A (retinol) and structurally related compounds. Since its discovery early this century, vitamin A has been shown to play an essential biological role (35–37). It has now been established that the active derivatives of vitamin A, such as *all-trans* retinoic acid (*at*RA) and *9-cis* retinoic acid (*9c*RA) and their synthetic analogs function as hormone-like signaling molecules that affect important biological processes at the level of gene transcription. The ability of retinoids to regulate growth and development, cell proliferation, and differentiation has made them attractive therapeutic candidates for the treatment of cancer.

3.1. Retinoid Receptors

The biological activity of retinoids can be modified by changes in the molecule's oxidation state or by *cis/trans* isomerization. Retinoid activity is also dependent on the level of specific types of retinoid-binding proteins which exist in extracellular, cytosolic, and nuclear compartments. It is now well established that retinoid signals are mediated by specific nuclear receptors (38). The three retinoid acid receptors (RARα, β, γ) and three retinoid X receptors (RXRα, β, γ) are part of a large family of regulatory proteins that also includes steroid and thyroid hormone receptors. The structural domains of the RARs and RXRs have been reviewed elsewhere in detail (39). Briefly, retinoid receptors contain a highly conserved DNA binding domain (DBD) that allows specific DNA recognition and protein-protein interactions. The second general feature of these receptors is a ligand binding domain (LBD) at the carboxy terminal half of the receptor. This LBD also contains a transcriptional activation function and a strong dimerization domain. In addition to the six major retinoid receptor sub-types, isoforms for each of the receptors have been identified that differ in their amino terminal region, which can also contain a transactivation function.

Currently, the RARs and RXRs represent the only receptors that are known to directly interact with or bind retinoids. However, an increasing number of other receptors have been and are being identified that interact with the retinoid receptors either by the formation of heterodimers, or by competing for the same DNA response element—the retinoic acid response element (RARE). These receptors include the thyroid hormone receptor (TR), vitamin D_3 receptor (VDR), peroxisome proliferator activator receptor (PPAR), and a number of "orphan receptors", such as COUP-TF and RORαs (40).

In contrast to the steroid hormone receptors that bind as homodimers, both RARs and RXRs alone bind poorly to DNA (40–42). Since effective DNA binding is required for receptor function, these data suggest that RARs and RXRs function primarily as heterodimers. RARs have been shown to interact with their response elements as heterodimers complexed with RXR. The *in vitro* data have been confirmed by gene "knockout" experiments which demonstrate that the RAR/RXR heterodimer is an *in vivo* transducer of the retinoid signal (43,44).

In the presence of *at*RA or *9c*RA, the RAR/RXR heterodimers function as activatiors of transcription. However, not only do RXRs enhance the binding of RARs to DNA, but they are also required for efficient DNA binding of TR, VDR, PPAR (42,45), and the *v-erbA* oncogene, a mutated form of the TR (46), as well as several orphan receptors (47). These various RXR-containing heterodimers bind to response elements that can have overlapping specificities. Thus, one class of retinoid receptors, the RXRs, play a central role as co-regulators of several other nuclear receptors that are activated by structurally unrelated hormones and signaling molecules.

In addition to their role as co-receptors, RXRs can also function as homodimers. The RXR homodimer has efficient DNA binding activity even in the absence of RARs or other receptors (46). Nine-*cis* retinoic acid is a natural ligand for RXR, inducing the formation of both RAR/RXR heterodimers and RXR/RXR homodimers. The RXR homodimer selectively binds to a distinct subset of RAREs (48), and has been found to mediate a distinct retinoid signaling pathway compared to the RAR/RXR heterodimer. The unique homo- and heterodimerization capacity of RXR not only greatly expands the repertoire of regulatory diversity and specificity of these nuclear hormone receptors, but also allows the RXR and its ligand, 9*c*RA, to serve as important regulators of the interaction between retinoids and other signal transduction pathways. For example, some of the dominant-negative effects of *v-erb*A and the unliganded TR on retinoid action could be achieved at the receptor level through their competion with RARs for heterodimerization with RXR (46,49,50). The formation of RXR homodimers can lead to an additional level of hormonal interaction induced by the RXR ligand. For instance, 9*c*RA can repress thyroid hormone (51) or vitamin D (52) activity through its ability to shift RXR from heterodimer to homodimer formation. The interaction of retinoids with other signaling pathways can also be achieved by the overlap of RARE with other transcription factor-binding sequences. An overlap of RARE with the AP-1 binding site, which mediates tetradecanoyl phorbol acetate (TPA) responsiveness, has been shown to mediate the interaction between nuclear hormone and cellular signal transduction pathways (53). Similary, a single response element can function both as a RARE and vitamin D response element, mediating the cross-talk between RA and vitamin D (52).

3.2. Retinoids and Breast Cancer

Studies using human breast cancer cell lines show that retinoids inhibit the growth of ER-positive but not ER-negative cell lines (54,55). This may be due to the fact that ER-positive breast cancer cells express higher levels of RARα than ER-negative breast cancer cells. For example, ER-positive MCF-7 breast cancer cells are RARα and β-positive and respond to the growth-inhibitory effects of RA; while ER-negative MDA-MB-231 breast cancer cells are RARα and β-negative and are unresponsive to the anti-mitogenic actions of RA. Several growth-regulatory pathways are affected when breast tumor cell lines are exposed to RA (56). For example, in MCF-7 cells, RA increases the activity of insulin-like growth factor binding protein one (IGF-BP-1), thereby reducing the mitogenic efficacy of IGF-1 (57). In addition, treatment of ER-positive MCF-7 and T47D human breast cancer cells with RA induces secretion of growth-related proteins, including TGF-β (55), and suppresses the expression of several key growth-regulatory proteins, including the ER, progesterone receptor (PgR) and TGF-α (58,59). Differentiation and loss of Her-2/*neu* expression may also be induced by RA (60).

Both natural and synthetic retinoids have been used in large doses *in vivo* to prevent DMBA- or NMU-induced rat mammary carcinogenesis (61,62). Retinoid treatment delays the appearance of mammary tumors and decreases the incidence and multiplicity of tumors. In addition, studies have shown that, under certain conditions, these carcinogen-induced tumors can be induced to regress. Castration of rats prior to administration of NMU prevents tumor formation but does not routinely induce the regression of established tumors (63). However, Lacroix et al. (62) found that *at*RA in combination with ovariectomy can effectively inhibit the growth of established NMU-induced mammary tumors. The majority of studies have examined the effects of *at*RA and *13-cis*-RA, both of which have been shown to have moderate efficacy in the prevention of tumors in the NMU-mammary tumor model (64). More recently, *9c*RA has been shown to have greater efficacy than *at*RA in the prevention of both the incidence and multiplicity of rat mammary carcinoma (65). A report by Anzano et al. (66) demonstrated that *9c*RA, in combination with tamoxifen is more effective that *9c*RA alone in preventing mammary tumor formation. Retinoids have also been found to be effective chemotherapeutic agents in human cancer clinical trials (67).

Although highly efficacious in the treatment of breast cancer, the clinical use of retinoids is severely limited by its toxicity (67). To circumvent retinoid-induced toxicity, pharaceutical companies have begun to develop ligand-specific synthetic retinoids. However, this approach has met with only limited success. Recent reports from Gottardis et al. (68) and Bischoff et al. (69) found that an RXR-selective ligand, LGD1069 (Targretin), is very effective in preventing the development and inducing the regression of rat mammary tumors. Other synthetic ligands have not been found to be as useful. Since the growth of endocrine-responsive breast tumors depends heavily on the hormonal milieu presented to the tumor cells and the cross-talk between various signaling pathways activated by the different hormones and growth factors, it seems likely that other hormones or growth factors, when used in combination with RA, might lead to increased efficacy without the risk of toxicity. For example, we have found that a sequential treatment regimen of MLT for 24h followed by *at*RA results in the decreased expression of the ER and a significant induction of TGF-β, and that the combination of MLT and *at*RA induces apoptosis (programmed cell death) of hormone-responsive breast tumor cells. We have also demonstrated that the *in vivo* combination of MLT and *9c*RA is more efficacious in the prevention of NMU-induced rat mammary tumors that either MLT or *9c*RA alone.

4. EFFECTS OF THE COMBINATION OF MLT AND RA ON BREAST CANCER

Recent studies have suggested that MLT may be a ligand for the RORα receptors (13,57), and that the RORα and RARα receptors may cross-talk at the level of the hormone response element (70). We initiated a series of studies to examine the possible additive or synergistic effects of combined treatment with MLT and RA on breast cancer. As shown in Figure 1, ER-positive MCF-7 breast tumor cells showed significant growth-suppression after five days of treatment with either 10^{-9}M MLT (64% of control) or 10^{-9}M *at*RA (62% of control). Surprisingly, the simultaneous treatment of the cells with MLT and *at*RA had no inhibitory effect on cell proliferation. However, a sequential regimen of MLT (10^{-9}M) followed 24h later by *at*RA (10^{-9}M) resulted in a cytocidal effect, decreasing the cell number to below the initial plating density after 5 days of treatment. Similar results were seen with the ER-positive T47D cell line in which both MLT and *at*RA, when used alone, inhibited cell proliferation by 60% and 61% of control, respectively. However, in this cell line, the simultaneous administration of MLT and *at*RA also inhibited cell growth (54% of control). The specificity of the cytocidal effect was demonstrated by the fact that the sequential MLT and *at*RA regimen had no effect on ER-negative MDA-MB-231 breast cancer cells.

The cytotoxic effects of the sequential MLT and RA treatment could be the result of either cellular necrosis or apoptosis. Apoptosis can be differentiated from cellular necrosis by a unique series of ultrastructural changes, including chromosomal and cytoplasmic condensation, nuclear fragmentation, membrane blebbing, an increased number of lysosomal bodies, and the formation of membrane-bound apoptotic bodies. The pattern of DNA oligomerization in MCF-7 tumor cells treated with the sequential regimen of MLT followed by *at*RA demonstrated the development over time of a ladder of nucleosomal oligomers (Figure 2). This laddering is characteristic of many cell types undergoing apoptosis. It should also be noted that there was no evidence of complete DNA degradation, which would be expected if the cells were undergoing

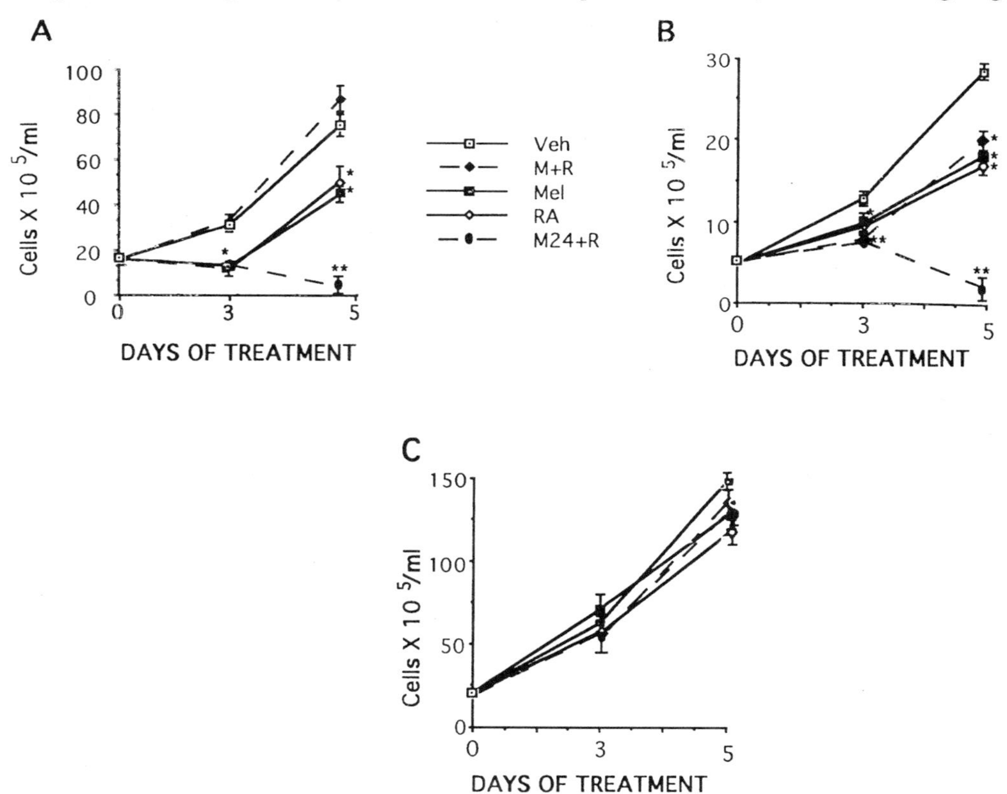

Figure 1. Effects of MLT and *at*RA on the proliferation of (A) MCF-7, (B) T47D, and (C) MDA-MB-231 cells. MCF-7 and MDA-MB-231 cells were seeded at a density of 2×10^6 cells/ml and T47D at 5×10^5 cells/ml in Costar 6-well dishes in IDMEM medium supplemented with 10% CS-FBS. Five hours after seeding (Day 0), MLT or *atRA*, both MLT and *at*RA, or MLT followed 24h later by *at*RA were added as 1000-fold concentrates to the appropriate wells. Ethanol vehicle was added to the control plates such that the final concentration was 0.001%. On Days 1, 3, and 5, cells were harvested by brief trypsinization, and viable cells were counted on a hemacytometer using the trypan blue dye exclusion method. Each point represents the mean cell count ± SEM from six plates containing either the vehicle (Veh), 10^{-9} M MLT (Mel), 10^{-9} M *at*RA (RA), MLT plus *at*RA (M + R), or MLT followed 24h later by *at*RA (M24 + R); * $p < 0.05$, ** $p < 0.01$ vs vehicle-treated controls.

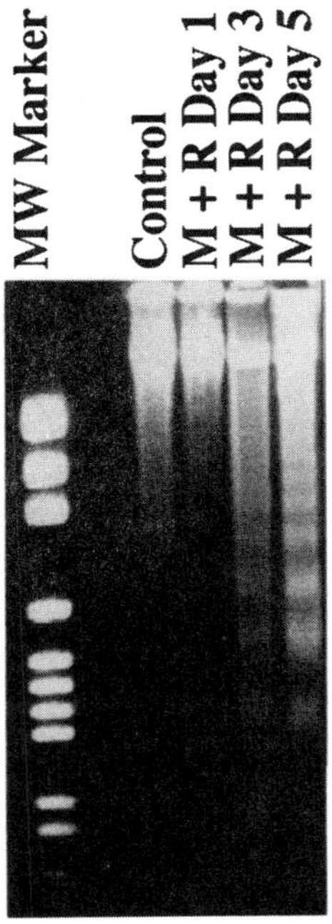

Figure 2. Electrophoretic analysis of DNA isolated from MCF-7 cells grown in IDMEM supplemented with 10% FBS was performed after treatment with the timed regimen of MLT and *at*RA. The molecular weight marker in this figure is a combination of λ DNA digested with Hind III and Φ 174 digested with Hae III. MCF-7 cells were treated for 1, 3, or 5 days with the sequential regimen of melatonin and *at*RA (M + R), after which high molecular weight DNA was isolated as described in Materials and Methods. Twenty micrograms of DNA were run on a 2.0% agarose gel. This is a representative picture of three separate experiments.

cellular necrosis. The onset of apoptosis was further substantiated by morphological criteria such as membrane blebbing, increased lysosomal formation, perinuclear chromatin condensation, and the formation of apoptotic bodies (71).

4.1. Effects of MLT and *at*RA on ER and TGF-β mRNA Levels

Previous work by our laboratory has demonstrated that MLT can down-regulate the expression of the ER in human breast cancer cells (18), while others have shown that RA also suppresses ER gene expression (59). It has also been shown that both MLT (72) and RA (72) can alter the expression of estrogen-regulated genes, such as TGFβ, which are involved in breast tumor cell proliferation. We, therefore, examined the steady state levels of mRNA encoding the ER and TGF-β1 in MCF-7 cells by Northern blot analysis following 48h of treatment with either MLT or *at*RA alone, or with the sequential regimen of MLT and *at*RA. Figure 3 shows that both *at*RA and MLT alone significantly decreased the steady state level of ER mRNA by 62% and 79%, respectively ($p < 0.01$ vs control). However, the sequential regimen of MLT and *at*RA reduced ER mRNA expression to almost undetectable levels ($p < 0.001$ vs MLT or *at*RA alone). In addition, both *at*RA and MLT alone enhanced the steady state level of TGF-β1 mRNA by 40% and 53%, respectively; while the sequential regimen produced a super-induction of TGF-β1 mRNA levels (91% increase over control, and 65% and 52% increase over *at*RA or melatonin treatment, respectively). These results suggest that the sequential treatment of MCF-7 cells with MLT followed by *at*RA results in an additive or synergistic induction of TGF-β1 mRNA expression.

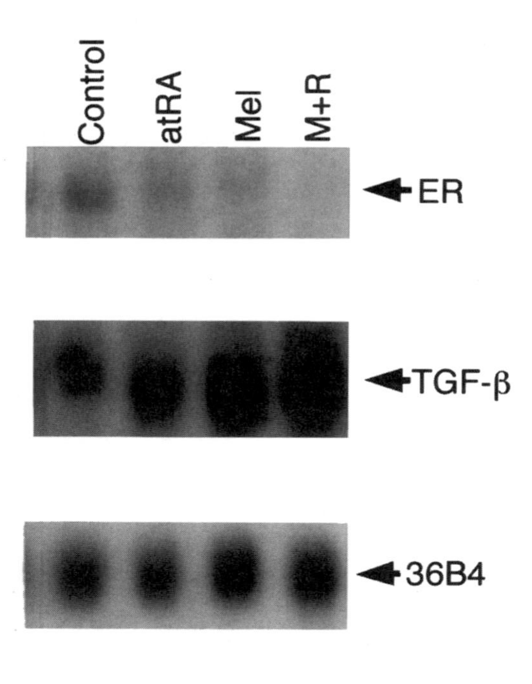

Figure 3. Effects of treatment with MLT or *at*RA alone, versus the sequential regimen of MLT and *at*RA on steady state ER and TGF-β1 mRNA levels in MCF-7 cells cultured in medium supplemented with 5% CS-FBS. MCF-7 cells were incubated with ethanol diluent (control), 10^{-9} M MLT (Mel), 10^{-9} M *at*RA (*at*RA), or a regimen of MLT followed 24 h later by *at*RA (M + R). For each time point, 50 μg of total RNA was fractionated on denaturing 1% agarose gels and blotted. Northern blots were probed with [^{32}P]-labeled human ER and human TGF-β1 cDNAs. The 36B4 cDNA was used to monitor RNA loading. The autoradiograph of this Northern blot analysis is representative of three independent experiments. ER and TGF-β1 mRNA expression was quantified by scanning densitometry and normalized to 36B4 mRNA.

4.2. Temporal Expression of Bcl-2 and Bak in MCF-7 Cells in Response to the Sequential Treatment with MLT and *at*RA

Numerous studies have demonstrated the involvement of the family of Bcl-2 proto-oncogenes in the apoptosis of breast cancer cells and breast epithelium (73). It has also been shown that estrogen treatment of MCF-7 breast cancer cells induces expression of Bcl-2, a death suppressor gene, and that this increased Bcl-2 expression protects cells from undergoing apoptosis in response to chemotherapy (74). Based on the previous observations that both MLT and RA suppress ER expression, we examined the possibility that the combinatorial treatment with MLT and RA is able to repress the levels of Bcl-2 while increasing the expression of the "death inducers" of the Bcl-2 family (i.e., Bax or Bak). Figure 4 shows the composite results of the densitometric analysis of Western blots of Bcl-2 and Bak expression after pre-treatment with MLT followed 24 h later by *at*RA. The levels of Bcl-2 exhibited the greatest divergence from control levels on Days 3 and 4, at which time Bcl-2 expression was reduced by 64% and 66%, respectively, compared to controls. Conversely, Bak expression was increased compared to controls beginning on Day 2 of treatment. To summarize, these results showed that the sequential treatment regimen of MLT followed by *at*RA inhibited the expression of the "death suppressor", Bcl-2, and enhanced the expression of the "death inducer", Bak, on Days 2, 3, 4, and 5.

4.3. Effects of MLT on RAR Transactivation

Melatonin, acting through its G-protein coupled receptor, mt_1, has been shown by our laboratory to modulate the estrogen response pathway, as measured by several parameters (34). For example, we have demonstrated that MLT pre-treatment of MCF-7 cells transiently transfected with an ERE-luciferase reporter gene for 5 min followed by EGF transactivates the ER in a ligand-independent manner. Recent experiments using this system have also demonstrated that MLT pre-treatment is able to blunt the estrogen-induced transactivation of the ER in a time-dependent manner. To determine if MLT's effects are specific to the ER, or if MLT can modulate the activity of other members of the steroid/thyroid hormone receptor superfamily, we examined the possible cross-talk between the MLT and RA signaling pathways. The preliminary data from transient transfection experiments using a RARE-luciferase reporter gene construct indicates that MLT pre-treatment of MCF-7 cells followed by *at*RA results in a time-dependent enhancement of the reporter gene transcription compared to *at*RA alone (Figure 5). These results, although still preliminary, suggest that MLT may be acting as a biological modifier of steroid hormone receptor action, leading to a blunting of estrogen stimulation, while amplifying the inhibitory effects of RA.

4.4. Effects of MLT and RA on NMU-induced Rat Mammary Tumors

Both MLT and RA have been reported to individually repress the development and growth of carcinogen-induced rat mammary tumors (11,62). To determine if the antiproliferative and apoptotic effects of the sequential regimen of MLT and RA which we observed *in vitro* could be translated to *in vivo* conditions, we conducted a series of pre-clinical trials using the NMU-induced rat mammary tumor model to examine

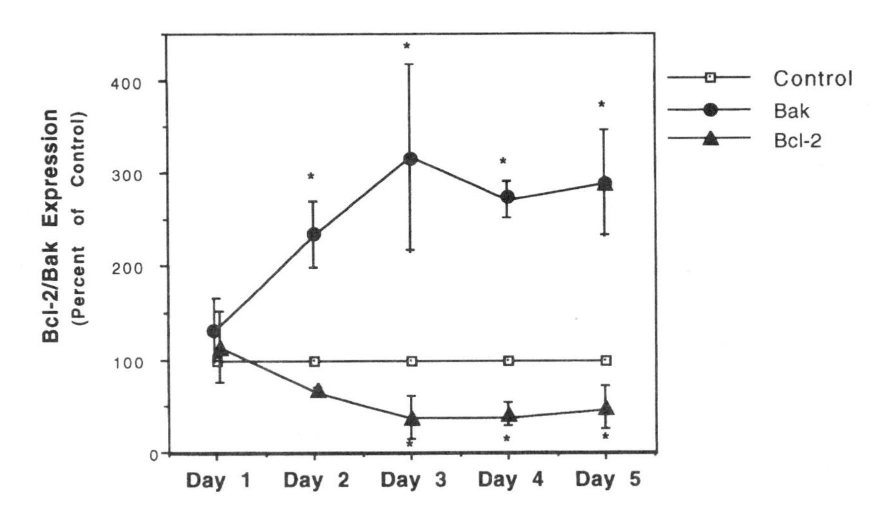

Figure 4. The temporal effects of the sequential regimen of MLT and *at*RA on Bcl-2 and Bak protein expression in MCF-7 cells. MCF-7 cells were incubated with diluent (control) or a regimen of MLT followed 24 h later by *at*RA (M + *at*RA) for 1, 2, 3, 4, or 5 days. For each time point, 25 μg of total cellular protein per lane was fractionated on 12.5% polyacrylamide gels and transferred to nitrocellulose membranes. Western blots were probed with polyclonal antibodies specific for Bcl-2 and Bak, and proteins were visualized after incubation with a horseradish peroxidase-conjugated secondary antibody and chemiluminescent substrate. Actin protein levels were used to monitor protein loading. Fluorographs from Western blot analyses of the time-course of Bak and Bcl-2 proteins in response to the sequential treatment of melatonin and *at*RA were quantified by scanning densitometry and normalized to actin protein levels. Results are presented graphically as percent of control (n = 3 independent experiments). * $p < 0.05$ vs controls.

the effects of MLT and RA on the development and growth of mammary tumors (Figure 6). Five treatment groups (n = 20 rats per group) were studied: control, MLT (500 μg/day s.c.), 9*c*RA (30 mg/kg chow), MLT and 9*c*RA daily (Mel = 9*c*RA), and MLT alternating every other day with 9*c*RA (Mel/RA). All treatments were initiated one day following NMU administration, and the animals were palpated for tumor development each week. Rats receiving 9*c*RA developed fewer mammary tumors (26%) compared to diluent-treated control animals (55%). Although we have previously reported that MLT suppresses NMU-induced mammary tumor formation (11), no difference in tumor incidence was seen between MLT-treated animals (60%) and control animals (55%) in this particular study. In animals receiving 9*c*RA and MLT every other day (Mel/RA), tumor incidence was further suppressed (to 20%) past that seen with 9*c*RA alone. A significant decrease in tumor incidence to only 5% (one animals with one tumor) was seen with the daily administration of MLT (late afternoon injections) and 9*c*RA. In addition, the latency to onset of tumor formation was significantly lengthened from 18 to 21 or 22 weeks in both groups receiving the combination of MLT and 9*c*RA over controls. Thus, it appears that the *in vitro* suppressive effects of MLT and RA on breast tumor cell growth is supported by the data from these *in vivo* pre-clinical studies. Although the most efficacious regimen for

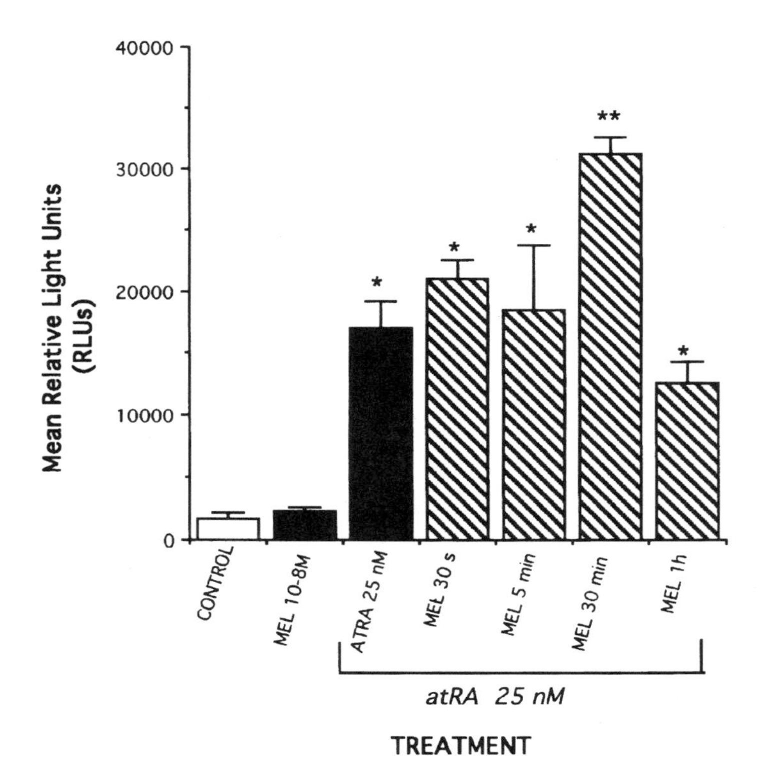

Figure 5. RAR transactivation assay of MCF-7 cells treated with MLT and/or *at*RA. Cells were grown in serum-free medium for 3 days, and treated with either *all-tran*-retionic acid (*at*RA) [25 nM] or MLT (1 nM), or treated simultaneously with MLT and *at*RA, or pre-treated with MLT for the times indicated (5 min, 30 min, or 1 h) followed by *at*RA. Cells were incubated for 18 h following the final treatment to allow for expression of the reporter gene. Control cells were treated with ethanol (0.01%). Statistical analysis compares luciferase expression for each treatment versus (*) control or versus (**) *at*RA. These data represent one run in triplicate. $p < 0.05$.

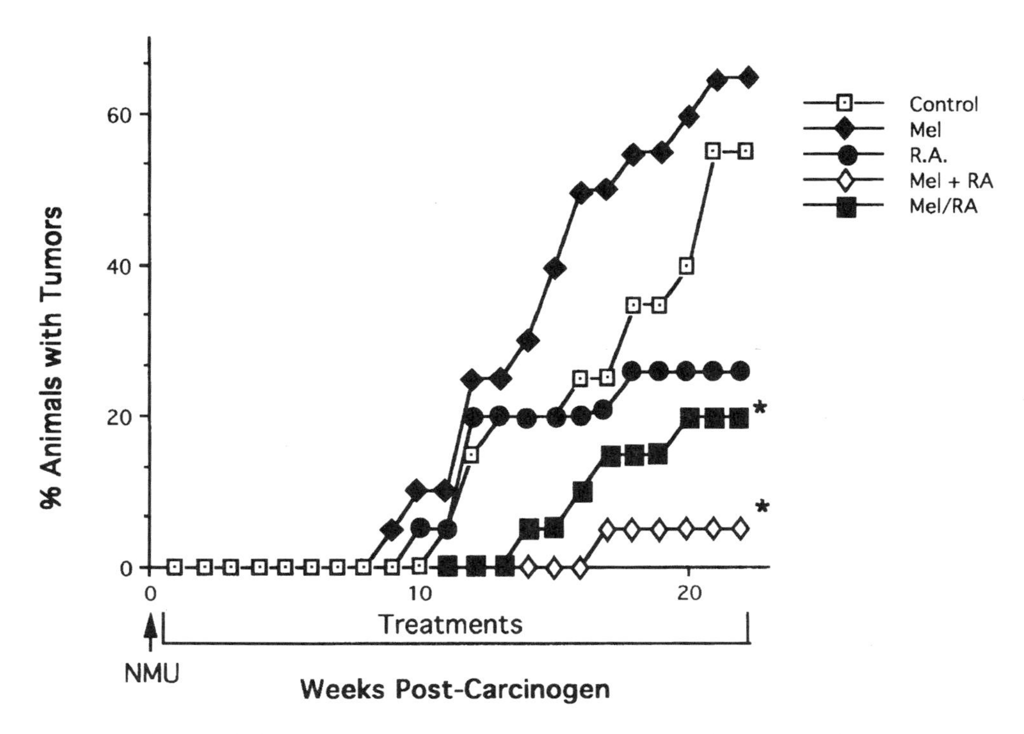

Figure 6. Incidence of NMU-induced mammary tumors in rats treated with 9*c*RA, MLT or a combination of MLT and 9*c*RA. N-nitroso-N-methylurea (50 mg/kg body weight) was administered i.p. to Sprague-Dawley rats at 50 days of age. Treatment with diluent (saline with 0.01% ethanol), MLT (500 µg/day), 9*c*RA (RA) [30 mg/kg/chow], MLT and 9*c*RA on alternating days (Mel/RA), or MLT and 9*c*RA every day (Mel + RA) were initiated on Day 1 following NMU administration, and continued through Week 22 of the experiment. MLT was administered s.c. daily at 3 p.m. Twenty rats were in each treatment group and were housed under long photoperiod conditions (12 h light/12 h dark). This graph also depicts latency to tumor onset. * $p < 0.01$.

suppressing tumor formation was the daily administration of both MLT and 9*c*RA, which almost completely prevented tumor formation, the animals receiving MLT and 9*c*RA every other day, and thus, only half the total retinoid dosage of those animals treated daily with 9*c*RA, also showed a lower tumor incidence than those treated daily with 9cRA alone. These data, however, must be interpreted with care because the decrease in tumor incidence from 26% to 20% is not statistically significant. From these studies it appears as though the addition of MLT sensitizes the animals to the anti-tumor effects of 9*c*RA, so that the dosage of 9*c*RA can be cut in half while maintaining an equivalent or even greater tumor suppressive effect. Thus, the additive or synergistic interaction between these two signaling pathways could allow non-toxic doses of retinoids to be used as a chemotherapeutic agent, thus providing a more effective, less toxic treatment regimen for breast cancer.

5. CONCLUSION

Previous work by our laboratory (75), as well as that of others (22), has clearly demonstrated the growth-inhibitory effects of MLT on human breast tumor cells. In addition, numerous laboratories have reported that both *at*RA and *9c*RA are effective inhibitors of breast tumor cell proliferation (55–57). Although each hormone has been shown to slow tumor proliferation, neither hormone alone has been shown to produce cytocidal effects in breast cancer cells at physiological concentrations. However, the results presented here show that, when used in a sequential regimen (MLT followed 24h later by *at*RA), these hormones are able to act in an additive or synergistic manner to induce a cytocidal response and apoptosis in hormone-responsive breast tumor cells. The cytocidal effects produced by the sequential regimen do not appear to result from non-specific cytotoxicity, but rather, are cell- and treatment-specific. This is evident by the lack of response of ER-negative (hormone-insensitive) MDA-MB-231 and BT-20 cells to the sequential regimen.

These data suggest that there is some level of cross-talk between the MLT and RA signaling pathways, and that the signaling pathways between the MCF-7 and T47D cell lines may be somewhat different. Most evidence indicates that the effects of MLT are mediated via membrane-associated receptors, two of which (mt_1 and MT_2) have recently been cloned (29,31). Though controversial, some reports have also suggested that MLT is able to bind and activate nuclear RORα receptors (32). Transcripts for both the mt_1 and RORα receptors are expressed on MCF-7 cells (34). If MLT's effects are mediated via the membrane receptors, it is possible that cross-talk occurs between the MLT and RA signaling pathways via phosphorylation of RAR or RXR receptors. Cross-talk between the RORα and RAR/RXR receptors has already been demonstrated at the level of the hormone response element (70).

Expression of the "death suppressor", Bcl-2, has previously been shown to be up-regulated by estrogens in MCF-7 cells (74), a process clearly mediated via the ER. The significant diminution of ER mRNA levels in response to the sequential regimen of MLT and *at*RA raises the possibility that the down-regulation of Bcl-2 expression by this treatment is mediated directly via the reduction in ER expression. However, the lack of a 1:1 correlation between the percent reduction in the levels of ER and Bcl-2 suggests that this treatment regimen may involve additional pathways that modulate Bcl-2 expression. Another potential contributor to the apoptotic effects could be the up-regulation of the "death inducer", Bak, seen in response to the sequential treatment with MLT and *at*RA. However, it is probable that the overall ratio of Bcl-2 to Bak is more important in mediating apoptosis than is either gene alone.

It is also possible that Bcl-2- and Bak-associated pathways play a secondary role, and that the over-expression of TGF-β1 induced by this sequential MLT and *at*RA treatment is the critical event leading to apoptosis. TGF-β1 has been shown to be a potent growth-inhibitor of breast epithelium and breast tumor cells, particularly MCF-7 cells (76). We have previously demonstrated that MLT can up-regulate TGF-β1 mRNA expression in a time-course independent of estrogen's effects (72). Thus, with the sequential treatment where super-induction of TGF-β1 expression occurs, two mechanisms are possible. First, the effects of this regimen may be mediated primarily through the regulation of Bcl-2 and Bak expression, and that up-regulation of TGF-β1 is only a secondary event. Second, the modulation of TGF-β1 may be the primary event leading to apoptosis, while changes in Bcl-2 and Bak are secondary. Finally, the changes in TGF-β1 and the Bcl-2 family members may be of relatively equal importance in the apoptotic pathway utilized by MLT and RA.

Both MLT and RA can inhibit the proliferation of various malignant cell types, including breast cancer. However, only supraphysiological concentrations of retinoids have been shown to induce cell death (65). Thus, at least for RA, the major drawback to its use as a therapeutic agent is the toxicity induced at pharmacological doses. For this reason, the development of combinatorial therapies that would reduce the concentrations needed for clinical efficacy, yet still enhance anti-tumorigenic activity, would be of great benefit. Our data indicate that the antiproliferative effects of MLT and RA on human breast cancer cells may be additive or synergistic when administered in the appropriate order, time, and dose. In fact, when used in a given paradigm, these two agents may be able to specifically induce apoptosis of breast tumor cells and the regression of breast tumors. This concept is supported by the results from our studies in the NMU rat mammary tumor model. From these studies, it is clear that the *in vitro* cross-talk between the MLT and RA signaling pathways also occurs in the *in vivo* setting. Clearly, reducing the dosage of retinoids by half should alleviate much of the induced toxicities observed clinically. Finally, our preliminary studies examining the effects of MLT on RAR/RXR transactivation suggest that MLT sensitizes the RA response pathway at the receptor level, allowing sub-optimal doses of RA to generate enhanced levels of gene transcription. If this observation is substantiated, then MLT treatment could be used to prime patients for RA treatment. This would allow significantly lower dosages of RA to be used in the management of breast cancer, thus, reducing RA-induced toxicity, while still maintaining optimal efficacy. These studies demonstrate that MLT functions as a potent biological modifier, and that it is capable of cross-talking with other cell signaling pathways to generate an endocrine environment that suppresses breast tumor cell growth and development.

REFERENCES

1. Davis, D.E., Hoel, D., Fox, J., and Lopez, A. International trends in cancer mortality in France, West Germany, Italy, Japan, England, and Wales, and the USA. Lancet 336:474–481, 1990.
2. Parker, S.L., Tong, T., Bolden, S., and Wingo, P.A. Cancer Statistics. CA Cancer J Clin 65:5–27, 1996.
3. Jordan, V.C. and Murphy, C.S. Endocrine pharmacology of antiestrogens as antitumor agents. Endocr Rev 11:578–610, 1991.
4. Jordan, V.C. Molecular mechanisms of antiestrogen action in breast cancer. Breast Cancer Res Treat 31:41–52, 1994.
5. Freiss, G., Pretois, C., and Vignon, F. Control of breast cancer growth by steroids and growth factors: Interactions and mechanisms. Breast Cancer Res Treat 27:57–68, 1993.
6. Knabbe, C., Lippman, M.E., Wadefield, L.M., Flanders, K.C., Kasid, A., Derynck, R., and Dickson, R.B. Evidence that transforming growth factor-beta is a hormonally regulated negative growth factor in human breast cancer cells. Cell 48:417–428, 1987.
7. Blask, D.E. The emerging role of the pineal gland and melatonin in oncogenesis. In: Extremely Low Frequency Eletromagnetic Fields; The Question of Cancer. B.W. Wilson, R.G. Stevens, L.E. Anderson, eds., Battele Press, Columbus, OH, 1990, pp. 319–335.
8. Blask, D.E. and Hill, S.M. Melatonin and cancer. Basic and clinical perspectives. In: Melatonin-Clinical Persectives. A. Miles, D.R.S. Philbrick, C. Thompson, eds., Oxford Press, New York, 1988, pp. 128–173.
9. Blask, D.E. Melatonin in oncology. In: Melatonin. Biosynthesis, Physiological Effects, and Clinical Applications. H-S Yu, R.J. Reiter, eds., CRC Press, Boca Raton, Florida, 1993, pp. 448–475.
10. Blask, D.E., Hill, S.M., Orstead, K.M., and Massa, J. Inhibitory effects of the pineal hormone melatonin and underfeeding during the promotional phase of 7,12-dimethylbenzanthracene (DMBA)-induced mammary tumorigenesis. J Neural Trans 67:125–138, 1986.
11. Blask, D.E., Pelletier, D.B., Hill, S.M., Lemus-Wilson, A., and Grosso, D. Pineal melatonin inhibition of tumor promotion in the N-nitroso-N-methylurea of mammary carcinogenesis: Potential involvement of antiestrogenic mechanisms in vivo. J Cancer Clin Oncol 117:526–532, 1991.
12. Kothari, L. and Subramanian A. A possible modulatory influence of melatonin on representative phase I and phase II drug metabolizing enzymes of 9,10-dimethyl-1,2-benzanthracene induced rat mammary tumorigenesis. Anti-Cancer Drugs 3:623–628, 1992.

13. Hill, S.M., Spriggs, L.L., Simon, M.A., Muraoka, H., and Blask, D.E. The growth inhibitory action of melatonin on human breast cancer cells in linked to the estrogen response system. Cancer Lett 64:249–256, 1992.
14. Soule, M.D., Vazquez, J., Long, A., Albert, S., and Breanan, M. A human cell line from a pleural effusion derived from a breast carcinoma. J Natl Cancer Inst 51:1409–1413, 1973.
15. Eckert, R.L. and Katzenellenbogen, B.S. Effects of estrogens and antiestrogens on estrogen receptor dynamics and the induction of progesterone receptor in MCF-7 human breast cancer cells. Cancer Res 42:139–144, 1982.
16. Horwitz, K.B., Costlow, M.E., and McGuire, W.L. MCF-7: a human breast cancer cell line with estrogen, androgen, progesterone, and glucocorticoid receptors. Steroids 26:785–795, 1975.
17. Weinberg, U., E'Elletto, R.D., Weitzman, E.D., Erlich, S., and Hallander, C.S. Circulating melatonin in man: secretion through the light-dark cycle. J Clin Endocrinol and Metab 48:114–118, 1979.
18. Molis, T.M., Spriggs, L.L., and Hill, S.M. Modulation of estrogen receptor mRNA expression by melatonin in MCF-7 human breast cancer cells. Mol Endocrin 8:1681–1690, 1994.
19. Molis, T.M., Walters, M.R., and Hill, S.M. Melatonin modulation of estrogen receptor mRNA in MCF-7 breast cancer cells. Int J Oncol 3:687–694, 1993.
20. Blask, D.E., Cos, S., Hill, S.M., Burns, D.M., Lemus-Wilson, A., Pelletier, D.B., Liaw, L., and Hill, A. Breast cancer: a target for the oncostatic actions of pineal hormones. Adv Pineal Res 4:267–274, 1990.
21. Cos, S., Blask, D.E., Lemus-Wilson, A., and Hill, S.M. Effects of melatonin on the cell cycle kinetics and "estrogen-rescue" of MCF-7 human breast cancer cells in culture. J Pineal Res 8:21–27, 1991.
22. Cos, S., Recio, J., and Sanchez-Barcelo, E.J. Modulation of the length of the cell cycle time of MCF-7 human breast cancer cells by melatonin. Life Sciences 58:811–816, 1996.
23. Cos, S., Fernandez, F., and Sanchez-Barcelo, E.J. Melatonin inhibits DNA synthesis in MCF-7 breast cancer cells *in vitro*. Life Sciences 58:2447–2453, 1996.
24. Tamarkin, L., Danforth, D., Lichter, A., Demoss, E., Cohen, M., Chabner, B., and Lippman, M. Decreased nocturnal plasma melatonin peak in patients with estrogen receptor-positive breast cancer. Science 216:1003–1005, 1982.
25. Bartsch, C., Bartsch, H., Jain, A.K., Laumas, K.R., and Wetterberg, L. Urinary melatonin levels in human breast cancer patients. J Neural Trans 52:281–294, 1981.
26. Stanberry, L.R., Das Gupta, T.K., and Beattie, C.W. Photoperiodic control of melanoma growth in hamsters: influence of pinealectomy and melatonin. Endocrinology 113:469–475, 1993.
27. Lissoni, P., Crispino, S., and Barni, S. Pineal gland and tumor cell kinetics: serum levels of melatonin in relation to Ki-67 labeling rate in breast cancer. Oncology 47:275–277, 1990.
28. Lissoni, P., Barni, S., Meregalli, S., Fossati, V., Cazzaniga, M., Esposti, D., and Tancini, G. Modulation of cancer endocrine therapy by melatonin: a phase II study of tamoxifen plus melatonin in metastatic breast cancer patients progressing on tamoxifen alone. Br J Cancer 71:854–856, 1995.
29. Reppert, S.M., Weaver, D.R., and Ebisawa, T. Cloning and characterization of a melatonin receptor that mediates reproductive and circadian responses. Neuron 13:1177–1185, 1994.
30. Birnbaumer, L.G. Proteins in signal transcduction. Ann Rev Pharmac Toxic 30:675–705, 1990.
31. Ebisawa, T., Karne, S., Lerner, M.R., and Reppert, S.M. (1994) Expression cloning of a high-affinity melatonin receptor from Xenopus dermal melanophores. Proc Natl Acad Sci USA 91:6133–6137, 1994.
32. Becker-Andre, M., Wiesenberg, I., Schaeren-Wiemers, N., Andre, E., Missbach, M., Saurat, J.-H., and Carlberg, C. Pineal gland hormone melatonin binds and activates an orphan of the nuclear receptor superfamily. J Bio Chem 269:28531–2853, 1994.
33. Giguere, V., Tini, M., Flock, G., Ong, E., Evans, R.M., and Otulakowski, G. Isoform-specific amino-terminal domains dictate DNA-binding properties of RORα, a novel family of orphan hormone nuclear receptors. Genes and Develop 8:538–553, 1994.
34. Ram, Prahlad. Identification and characterization of the melatonin receptor and its modulation of estrogen receptor phosphorylation and transactivation in MCF-7 human breast tumor cells. *Dissertation*, pp. 34–42, 1997.
35. Lotan, R. Effects of vitamin A and its analogs (retinoids) on normal and neoplastic cells. Biochem Biophys Acta 60:33–41, 1981.
36. Roberts, A.B. and Sporn, M.B. Cellular biology and biochemistry of the retinoids. In: The Retinoids, Sporn, M.B., Roberts, A.B., Goodman, D.S., eds., The Academic Press, Orlando, 1984.
37. Morris-Kay, G. ed. Retinoids in Normal Development and Teratogenesis. Oxford Science Publications, Oxford, U.K., 1992.
38. Giguere, V., Ong, E.S., Seigi, P., and Evans, R.M. Identification of a receptor for the morphogen retinoic acid. Nature 330:624–629, 1987.

39. Pfahl, M. Vertetrate receptors: Molecular biology, dimerization and response elements. Semn in Cell Biol 5:95–103, 1994.
40. Yu, V.C., Delsert, C., Andersen, B., Holloway, M.M., Devary, O.V., Nr, A.M., Kim, S.Y., Boutin, J-M., Glass, C.K., and Rosenfeld, M.G. RXRβ: a coregulator that enhances binding of retinoic acid, thyroid hormone, and vitamin D receptors to their cognate response elements. Cell 67:1251–1266, 1991.
41. Bugge, T.H., Pohl, J., Lonnoy, O., and Stunnenberg, H.G. RXRα, a promiscuous partner of retinoic acid and thyroid hormone receptors. EMBO J 11:1409–1410, 1992.
42. Zhang, X-K., Hoffman, B., Pran, P., Graupner, G., and Pfahl, M. Retinoid X receptor is an auxillary protein for thyroid hormone and retinoic acid receptors. Nature 355:441–446, 1992.
43. Lohnes, D., Mark, M., Mendelsohn, C., Dolle, P., Dierich, A., Gorry, P., Gansmuller, A., and Chambon, P. Function of the retinoic acid receptors (RARs) during development (I) Craniofacial and skeletal abnormalities in RAR double mutants. Development 120:2723–2748, 1994.
44. Mehndelsohn, C., Lohnes, D., Decimo, D., Lufkin, T., LeMeur, M., Chambon, P., and Mark, M. Function of the retinoic acid receptors (RARs) during development (II) Multiple abnormalities at various stages of organogenesis in RAR double mutants. Development 120:2749–277, 1994.
45. Graupner, G., Wills, K.N., Tzukerman, M., Zhang, X-K., and Pfahl, M. Dual regulatory role for thyroid hormone receptors allows control of retinoic acid receptor activity. Nature 340:653–656, 1989.
46. Hermann, T., Hoffmann, B., Piedrafita, J.F., Zhang, X-K., and Pfahl, M. V-erbA requires auxillary proteins for dominant negative activity. Oncogene 8:55–65, 1993.
47. Apfel, R., Benbrook, D., Lernhard, E., Ortiz-Caseda, M.A., and Pfahl, M. A novel orphan receptor with a unique ligand binding domain and its interaction with the retinoid/thyroid hormone receptor superfamily. Mol Cell Biol 14:7025–7035, 1992.
48. Mangelsdorf, D.J., Umesono, K., Kliewer, S.A., Borgmeyer, U., Ong, E.S., and Evans, R.M. A direct repeat in the cellular retinol-binding protein type II gene confers differential regulation by RXR and RAR. Cell 66:555–561, 1991.
49. Damm, K., Thompson, C.C., and Evans R.M. Protein encode by v-erbA functions as a thyroid-hormone receptor antagonist. Nature 339:593–597, 1989.
50. Zhang, X-K. and Pfahl, M. Regulation of retinoid and thyroid hormone action through homodimeric and heterodimeric receptors. Trends Endocrinol Metab 4:156–162, 1993.
51. Lehmann, M.M., Zhang, X-K., Graupner, G., Lee, M-O., Hermann, T., Foffmann, B., and Pfahl, M. Formation of retinoid X receptor homodimers leads to repression of T_3 response: hormonal cross-talk by ligand-induced squelching. Mol Cell Biol 13:7698–7707, 1993.
52. MacDonald, P.N., Dowd, DR., Nakajima, S., Galligan, M.A., Reeder, M.C., Haussler, C.A., Ozato, K., and Hausler, M.R. Retinoid X receptors stimulate and 9-cis retinoic acid inhibits 1,25-dihydroxyvitamin D_3-activated expression of the rat osteocalcin gene. Mol Cel Biol 13:5907–5917, 1993.
53. Schule, R., Umessono, K., Mangelsdorf, D.M., Bolado, J., Pike, J.W., and Evans, R.M. Jon-Fos and receptors for vitamin A and D recognize a common response element in the human osteocalcin gene. Cell 61:497–504, 1990.
54. Lotan, R. Different susceptibilities of human melanoma and breast carcinoma cell lines to retinoic acid-induced growth inhibition. Cancer Res 39:1014–1019, 1979.
55. Fontana, J.A., Mezu, A.B., Cooper, B.N., and Miranda D. Retinoid modulation of estradiol-stimulated growth and of protein synthesis and secretion in human breast carcinoma cells. Cancer Res 50:1997–2002, 1990.
56. van der Burg, B., van der Leede, B-j.M., Kwkkenbos-Isbrucker, L., Salverda, S., de Laat, S.W., and van der Saag, P.T. Retinoic acid resistance of estradiol-independent breast cancer cells coincides with diminished retinoic acid receptor function. Mol Cell Endocrinol 91:149–147, 1993.
57. Fontana, J.A., Burrows-Mezu, A., Clemmons, D.R., and LeRoith, D. Retinoid modulation of insulin-like growth factor binding proteins and inhibition of breast carcinoma proliferation. Endocrinology 128:1115–1122, 1991.
58. Clarke, C.L., Roman, S.D., Graham, J., Koga, M., and Sutherland, R.L. Progesterone receptor regulation by retinoic acid in the human breast cancer cell line T-47D. J Biol Chem 265:12694–12700, 1990.
59. Rubin, M., Fenig, E., Rosenauer, A., Menendez-Botet, C., Achkar, C., Bentel, J.M., Yahalom, J., Mendelsohn, M., and Miller, W.H. 9-cis Retinoic acid inhibits growth of breast cancer cells and down-regulates estrogen receptor RNA and protein. Cancer Res 54:6549–6556, 1994.
60. Bacus, S.S., Kiguchi, K., Chin, D., King, C.R., and Huberman, E. Differentiation of cultured human breast cancer cells (AU-565 and MCF-7) associated with loss of cell surface Her-2/neu antigen. Mol Carcinog 3:350–362, 1990.

61. Moon, R.C., McCormick, D.L., and Mehta, R.G. Inhibition of carcinogenesis by retinoids. Cancer Res 43(Suppl): 2469–2475, 1983.
62. Lecrox, A., Koskas, C., and Bhat, P.V. Inhibition of growth of established N-methyl-N-nitrosourea-induced mammary cancer in rats by retinoic acid and ovariectomy. Cancer Res 59:5731–5735, 1990.
63. Gullino, P.M., Pettigres, H.M., and Grahtham, F.M. N-nitrosomethylurea as mammary gand carcinogen in rats. J Natl Cancer Inst 54:401–414, 1975.
64. Moon, R.C., Hehta, R.G., and Rao, K.V.N. Retinoids and cancer in experimental animals. In: The Retinoids: Biology, Chemistry, and Medicine. Sporn, M.B., Roberts, A.B., and Goodman, D.S., eds., pp. 573–595, Raven Press, New York, 1994.
65. Gottardis, M., Lamph, M., Shalinsky, W.W., Wellstein, D.R., and Heyman, R.A. The efficacy of 9-cis retinooic acid in experimental models of cancer. Breast Cancer Res Treat 38:85–96, 1996.
66. Anzano, M.A., Byers, S.W., Smith, J.M., Peer, C.W., Mullen, L.T., Brown, C.C., Roberts, A.B., and Sporn, M.B. Prevention of breast cancer in the rat with 9-cis-retinoic acid as a single agent in combination with tamoxifen. Cancer Res 54:4614–4617, 1994.
67. Costa, A., Formelli, F., Chiesa, F., Decesi, A., De Palo, G., and Veroesi, U. Prospects of chemoprevention of human cancers with the synthetic retinoid feretinide. Cancer Res 54:2032s-2037s, 1994.
68. Gottardis, M.M., Bishoff, E.D., Shirley, M.A., Wagoner, M.A., Lamph, W.W., and Heyman, R.A. Chemoprevention of mammary carcinoma by LGD1069 (Targretin): and RXR-selective ligand. Cancer Res 56:5566–5570, 1996.
69. Bischoff, E.D., Gottardis, M.M., Moon, T.E., Heyman, R.A., and Lamph, W.W. Beyond tamoxifen: The retionid X receptor-selective ligand LGD1069 (Targretin) causes complete regression of mammary carcinoma. Cancer Res 58:479–484, 1998.
70. Tini, M., Fraser, R.A., and Gigùere, V. Functional interactions between retinoic acid receptors-related orphan nuclear receptor (RORα) and the retinoic acid receptors in the regulation of the γF-crystalline promoter. J Biol Chem 270:20156–20161, 1995.
71. Eck, K.M., Yuan, L., Duffy, L., Ram, P.T., Ayettey, S., Chen, I., Cohn, C.S., Reed, J.C., and Hill, S.M. A sequential treatment regimen with melatonin and all-trans retinoic acid induces apoptosis in MCF-7 tumor cells. Br J Cancer 77:2129–2137, 1998.
72. Molis, T.M., Spriggs, L.L., Jupiter, Y., and Hill, S.M. Melatonin modulation of estrogen-regulated proteins, growth factors, and proto-oncogenes in human breast cancer. J Pineal Res., 18:93–103, 1995.
73. Krajewski, S., Blomqvist, C., Franssila, K., Krajewsi, M., Wasenius, V.M., Niskanen, E., Nording, S., and Reed, J.C. Reduced expression of proapoptotic gene Bax is associated with poor response rates to combination chemotherapy and shorter survival in women with metastatic breast adenocarcinoma. Cancer Res 55:4471–4478, 1995.
74. Teixeira, C., Reed, J.C., and Pratt, M.A.C. Estrogen promotes chemotherapeutic drug resistance by a mechanism involving Bcl-2 proto-oncogene expression in human breast cancer cells. Cancer Res 55:3902–3907, 1995.
75. Hill, S.M. and Blask, D.E. Effects of the pineal hormone melatonin on the proliferation and morphological characteristics of human breast cancer cells (MCF-7) in culture. Cancer Res 48:6121–6129, 1988.
76. Arteaga, C.L., Coffey, R.J., Dugger, T.C., McCutchen, C.M., Moses, H.L., and Lyons, R.M. Growth stimulation of human breast cancer cells with anti-transforming growth factor β antibodies; evidence for negative autocrine growth regulation by transforming growth factor β. Cell Growth Differ 1:367–374, 1990.
77. Nicoletti, I., Migliorati, G., Paglancci, M., Grignani, F., and Riccardi, C. A rapid and simple method for measuring thymocyte apoptosis by propidium iodide staining and flow cytometry. J Immunol Methods 139:271–279, 1991.

MELATONIN AND 9-*cis*-RETINOIC ACID IN THE CHEMOPREVENTION OF NMU-INDUCED RAT MAMMARY CARCINOMA

S. R. Teplitzky,[1] D. E. Blask,[4] Q. Cheng,[3] L. Myers,[3] and S. M. Hill[2,3]

[1] Department of Surgery
[2] Department of Anatomy
[3] Tulane Cancer Center
Tulane University School of Medicine
1430 Tulane Avenue
New Orleans, Louisiana 70112
[4] Mary Imogene-Bassett Research Institute
Cooperstown, New York 13326

1. INTRODUCTION

The epithelium of the breast requires a variety of hormones, including estrogen, progestins, insulin and growth hormone for normal develoment, growth and function. Estrogen plays a central role in the growth of normal as well as neoplastic breast tissue. As such, breast cancer is a hormone responsive neoplasm. Approximately 60% of primary breast tumors are estrogen receptor (ER)-positive and maintain estrogen dependence. The estrogen-ER complex binds to the estrogen response element, which modulates the transcription of specific sets of genes involved in the regulation of cell growth. This pathway has provided a target for the hormonal treatment of breast cancer by anti-estrogens such as tamoxifen. However, a certain percentage of breast tumors are unresponsive to anti-estrogen thereapy. Therefore, it is imperative to discover other mechanisms by which neoplastic growth can be manipulated and against which therapeutic treatments can be developed.

A number of studies have examined the role of the pineal gland and its hormone, melatonin, in tumorigenesis. Substantial evidence has demonstrated that the pineal gland is an important regulator of neuroendocrine mechanisms controlling the growth of endocrine responsive neoplasias (1,2). The pineal, via melatonin, has been shown to exert an inhibitory influence on the development and growth of carcinogen-induced

Melatonin after Four Decades, edited by James Olcese.
Kluwer Academic / Plenum Publishers, New York, 2000.

rat mammary tumors (3). Maximal stimulation of pineal melatonin secretion, through light deprivation or blinding, as well as the chronic daily late afternoon injection of melatonin have both been shown to inhibit the development and growth of carcinogen-induced rat mammary tumors (1,4–7). In contrast to animal studies, only a marginal amount of work has been done with respect to the role of melatonin in human breast cancer. Our laboratory has demonstrated that physiological concentrations of melatonin, corresponding to peak nighttime plasma levels in humans, are able to supress the growth and alter the morphology of estrogen-responsive MCF-7 human breast cancer cells *in vitro* (8). This direct growth inhibitory effect of melatonin on breast cancer cell lines has been confirmed by other groups (9–10). Growth inhibition is limited to ER-positive breast cancer cell lines (11) and is cell cycle-specific, inducing a G_0/G_1 transition delay in MCF-7 cells (10). Furthermore, melatonin is able to block the estrogen rescue of cells inhibited by tamoxifen (9). These observations suggest that melatonin's actions may be linked to the cell's estrogen-response pathway.

Retinoids are vitamin A derived hormones whose nuclear receptors are members of the steroid hormone receptor superfamily. Retinoids have ben shown to inhibit cancer development at a number of organ sites, including the mammary gland. In the rat, the induction of mammary tumors by a variety of carcinogens is suppressed by retinoids (12,13). Like melatonin, retinoids have also been shown to suppress *in vitro* proliferation of ER-positive but not ER-negative human breast cancer cell lines (13,14). Retinoids have been shown to exert their growth inhibitory effects by binding to and activating nuclear retinoic acid receptors (RARs and RXRs) (15–17). 9-*cis*-Retinoic acid (9cRA) can bind to both RAR and RXR receptors while *all-trans* retinoic acid (atRA) binds only the RAR receptor. Our laboratory has shown that treatment of MCF-7 cells with both melatonin and retinoids in a specific fashion, melatonin followed 24 hours later with atRA, results not only in growth inhibition but more importantly in apoptosis, as evidenced by morphologic and DNA fragmentation studies (18).

Mammary adenocarcinomas can be induced in the rat using either the carcinogen 7,12-dimethylbenzanthracene (DMBA) or *N*-nitroso*N*-methylurea (NMU) (19–20). DMBA-induced mammary tumors are primarily prolactin-dependent, whereas NMU tumors are estrogen-dependent and more closely resemble human breast cancer. Based on the above observation that a timed treatment of MCF-7 cells with melatonin followed by atRA induces apoptosis *in vitro*, preclinical trials were conducted to investigate the hypothesis that a timed treatment regimen of melatonin followed 24 hours later by retinoic acid will inhibit *in vivo* mammary tumorigenesis in the rat. A preclinical trial was performed using the NMU-induced rat mammary tumor model.

2. MATERIALS AND METHODS

The day following tumor induction with NMU (50 mg/kg, IP), twenty female Sprague-Dawley rats were assigned to five treatment groups. Animals were housed two per cage under a long (12 h light: 12 h dark) photoperiod. Control animals received 0.1% ethanol saline s.c. daily at 3 p.m.. Melatonin (mlt) treated animals received 500 μg mlt s.c. daily at 3 p.m. The retinoid treated group received 9-*cis*-retinoic acid (9cRA) 30 mg/kg chow daily at 11 a.m. Two combinatorial treatment groups were included, one group which received both mlt and 9cRA (mlt+9cRA) everyday and one which received mlt alternating every other day with 9cRA (mlt/9cRA). Treatments were administered for 22 weeks. Animals were weighed weekly. Tumors were palpated weekly and measured with calipers. Tumor volume was calculated as the product of the two greatest dimensions (LxW = mm^2). At the conclusion of the study, animals were sacrificed by rapid decapitation. Uterine wet weight was determined for each treatment group. Tumors were harvested, snap frozen in liquid nitrogen and stored at −70°C.

3. RESULTS

Histopathological studies confirmed the presence of mammary adenocarcinoma in each of the tumors. At the conclusion of the study, tumor incidence was 55% in the control animals and 65% in the melatonin treated animals, which was not statistically different. Treatment with 9cRA alone resulted in a tumor incidence of 26% which also was not statistically different from controls. However, both combinatorial treatment groups had a decreased tumor incidence as compared to the control group. Tumor incidence for the group receiving the timed treatment regimen of melatonin alternating every other day with 9cRA was 20% ($p < 0.05$ compared to control). Tumor incidence in the group receiving both mlt and 9cRA daily was only 5% ($p < 0.01$ compared to control), which represents one animal developing one tumor. Many of the animals developed more than one tumor. Therefore, tumor multiplicity, or the average number of tumors per rat, was analyzed. The combinatorial treatments significantly decreased tumor multiplicity ($\alpha = 0.05$) and were more effective than 9cRA alone in decreasing the number of tumors which developed per animal. Combination treatment with mlt and retinoids also significantly increased the latency to onset of tumor development. Tumors developed much later in these animals, at 13–17 weeks, as compared to the other treatment groups which developed tumors by 8–9 weeks. The cumulative tumor burden was calculated weekly as the summation of tumor volumes within a given treatment group. For animals requiring termination prior to 22 weeks, the final tumor volume obtained was included in subsequent calculations. The combination of melatonin and 9cRA was very effective in decreasing tumor burden. This analysis also reveals that melatonin treated animals had a lower tumor burden than controls. So, although melatonin treatment did not decrease tumor incidence in this study as expected, it was associated with smaller tumors in these animals. There was no statistical difference in uterine wet weight among treatment groups (control 273 gm, mlt 273 gm, 9cRA 267 gm, mlt+9cRA 261 gm, mlt/9cRA 264 gm).

4. CONCLUSIONS

This study demonstrated that the combinatorial use of melatonin and retinoids is effective in the chemoprevention of carcinogen-induced rat mammary adenocarcinoma. This therapy, especially when both treatments were administered daily, was more effective than either agent used alone in preventing the promotion and progression of mammary tumors. This result varied from the *in vitro* setting in which a timed treatment regimen of melatonin followed 24 hours later by retinoic acid induces the greatest effect. This may reflect the more complex nature of the intact animal. Surprisingly, this study did not demonstrate a decrease in tumor incidence with melatonin treatment alone. This result varies from those seen by this and other laboratories. This may have resulted from melatonin administration too early in the afternoon to acheive this effect on its own. This is reflected by the fact that the uterine wet weight of the melatonin treated animals was not decreased as would be expected. However, there was clearly an additive or synergistic effect between melatonin and retinoic acid. This interaction between melatonin and retinoids could potentially allow nontoxic doses of retinoids to be used in the therapy of hormone responsive cancer. This could provide a useful clinical treatment regimen for breast cancer and other malignancies.

REFERENCES

1. Blask, D.E. The pineal: An oncostatic gland? In: The Pineal Gland. RJ Reiter, ed. Raven Press, New York, N.Y., 1984, pp. 253.
2. Cohen, M., Lippman, M., and Chabner, B. Role of pineal gland in aetiology and treatment of breast cancer. Lancet 2:814, 1978.
3. Lapin, V. In: The Pineal Gland and Cancer. D. Gupta, A Attanasci, RJ Reiter, eds. Brain Research Promotion, Tubingen, 1988, pp. 1.
4. Chang, N., Spaulding, T.S., and Tseng, M.T. Inhibitory effects of superior cervical ganglionectomy on dimethylbenz(a)anthracene-induced mammary tumors in the rat. J. Pineal Res. 2:331, 1985.
5. Aubert, C.H., Janiaud, P., and Leclavez, J. Effect of pinealectomy and melatonin on mammary tumor growth in Sprague-Dawley rats under different condition of lighting. J. neural Transm. 47:121, 1980.
6. Blask, D.E., Pelletier, D.B., Hill, S.M., Lemus-Wilson, A., Grosso, D.S., Wilson, S.T., and Wise, M.E. Pineal melatonin inhibition of tumor promotion in the N-nitroso-N- methylurea model of mammary carcinogenesis: potential involvement of antiestrogenic mechanisms in vivo. J. Cancer Res. Clin Oncol. 117:526, 1991.
7. Tamarkin, L., Cohen, M., Roselle, D., Reichert, C., Lippman, M., and Chabner, B. Melatonin inhibition and pinealectomy enhancement of 7–12 dimethylbenz-anthracene-induced mammary tumors in the rat. Cancer Res. 41:4432, 1982.
8. Hill, S.M. and Blask, D.E. Effects of the pineal hormone melatonin on the proliferation and morphological characteristics of human breast cancer cells (MCF-7) in culture. Cancer Res. 48:61210, 1988.
9. Cos, S. and Blask, D.E. Effects of the pineal hormone melatonin on the anchorage-independent growth of human breast cancer cells (MCF-7) in a clonogenic culture system. Cancer Lett. 64:249, 1992.
10. Cos, S., Blask, D.E., Lemus-Wilson, A., and Hill, A.B. Effects of melatonin on the cell cycle kinetics and "estrogen-rescue" of MCF-7 human breast cancer cells in culture. J. Pineal Res. 10:36, 1991.
11. Hill, S.M., Spriggs, L.L., Simon, M.A., Muraoka, D.E., and Blask, D.E. The growth inhibitory action of melatonin on human breast cancer cells is linked to the estrogen response system. Cancer Lett. 64:249, 1992.
12. Zhang, X. and Pfahl, M. Regulation of retinoid and thyroid hormone action through homodimeric and heterodimeric receptors. Trends Endocrinol. Metab. 4:156, 1993.
13. Clarke, C.L., Graham, J., Roman, S.D., and Sutherland, R.L. Direct transcriptional regulation of the progesterone receptor by retinoic acid diminishes progestin responsiveness in the breast cancer cell line T-47D. J. Biol. Chem. 266:18969, 1991.
14. Rubin, M., Fenig, E., Gosenauer, A., Menendez-Botet, C., Achkar, C., Bentel, J.M., Yahalom, J., Mendelsohn, M., and Miller, W.H. 9-cis Retinoic acid inhibits growth of breast cancer cells and down-regulates estrogen receptor mRNA and protein. Cancer Res. 54:6549, 1994.
15. Strasser, A., Harris, A.W., Jacks, T., and Cory, S. DNA damage can induce apoptosis in proliferating lymphoid cells via p53-independent mechanisms inhibitable by *bcl*-2. Cell 9:329, 1994.
16. Clarke, A.R., Purdie, C.A., Harrison, D.J., Morris, R.G., Bird, C.C., Hooper, M.L., and Wyllie, A.H. Thymocyte apoptosis induced by p53-dependent and independent pathways. Nature 362:849, 1993.

17. Berges, R.R., Furuya, Y., Remington, L., English, H.F., Jacks, T., and Isaacs, J.T. Cell proliferation, DNA repair, and p53 function are not required for programmed cell death of prostatic glandular cells induced by androgen ablation. Proc. Natl. Acad. Sci. USA 90:8910, 1995.
18. Eck, K.M., Yuan, L., Duffy, L., Ram, P.T., Ayettey, S., Chen, I., Cohn, C.S., Reed, J.C., and Hill, S.M. A sequential treatment regimen with melatonin and *all-trans* retinoic acid induces apoptosis in MCF-7 tumour cells. Br. J. Cancer 77(12):2129–2137, 1998.

41

THE ANTIPROLIFERATIVE EFFECTS OF MELATONIN ON EXPERIMENTAL PITUITARY AND COLONIC TUMORS

Possible Involvement of the Putative Nuclear Binding Site?

Marek Pawlikowski, Jolanta Kunert-Radek, Katarzyna Winczyk, Gabriela Melen-Mucha, Anna Gruszka, and Michal Karasek*

Institute of Endocrinology and * Laboratory of Electron Microscopy
Chair of Pathomorphology
Medical University of Lodz
91-425 Lodz, Poland

1. INTRODUCTION

It has been repeatedly shown that the pineal hormone melatonin (MLT) inhibits the growth of various experimental tumors in rodents. The oncostatic action of MLT is connected, at least in part, with its antiproliferative effects on the tumor cells. However, the molecular mechanisms by which MLT suppresses cell proliferation still remains unclear. It was suggested that MLT acts not only via membrane receptors but also via nuclear binding sites and the latter are identical with so-called RZR/ROR α-receptors (2,6). Since a synthetic thiazolidinedione compound CGP 52608 has been recently shown to be an exogenous ligand for RZR/ROR receptors it seemed to us worthwhile to see whether it exerts also the antiproliferative effects. We have studied the effects of MLT and CGP 52608 on cell proliferation of the diethylstilbestrol (DES)-induced rat pituitary tumors and of the murine colonic adenocarcinoma Colon 38.

2. MATERIALS AND METHODS

2.1. DES-Induced Pituitary Tumors in Rats

Four week old male Wistar rats were submitted to subcutaneous implantation of silastic capsules containing DES (10 mg/capsule). Four months later the animals were

Melatonin after Four Decades, edited by James Olcese.
Kluwer Academic / Plenum Publishers, New York, 2000.

divided into experimental groups (each of 7 rats) receiving daily evening injections of MLT (25 μg), CGP 52608 (25 μg) or solvent (controls), during 10 days. Two hours before the end of experiment the animals were injected intraperitoneally with bromodeoxyuridine (BrDU, 50 mg/kg b.w.). The incorporated BrDU was revealed by the immunocytochemical method and the number of BrDU-positive cell nuclei/1000 randomly scored pituitary cells was used as an index of cell proliferation.

2.2. Colon 38 Tumors *in Vivo*

Male mice, first generation of the cross-bred between C57BL/6 and DBA/2 strains were used in the experiment. The animals were subcutaneously implanted with Colon 38 cells suspension. Ten days after tumor implantation, the animals were divided into six experimental groups, and then received subcutaneously the following substances in the daily evening injections: MLT, 10 μg or 100 μg, CGP5 2608 10 μg or 100 μg, saline or 10% ethanol in saline, during 6 days. The animals were sacrificed by spinal cord dislocation 12 hours after the last injection. Two hours before the sacrifice all the animals were injected i.p. with BrDU in a dose as above. The assessment of cell proliferation was done as in the previous experiment.

2.3. Colon 38 Cells *in Vitro*

The tumor cells were isolated from Colon 38 tumors and cultured according the procedure described in details elsewhere (5). In brief, the tumor cells were incubated 24 hours in RPMI medium supplemented with 15% fetal calf serum in the presence of [^{3}H]thymidine and in the presence or absence of the investigated compounds. The incorporation of [^{3}H]thymidine into DNA was considered as an index of cell proliferation.

2.4. Statistical Analysis

The statistical analysis of the data was performed using the Mann-Whitney test.

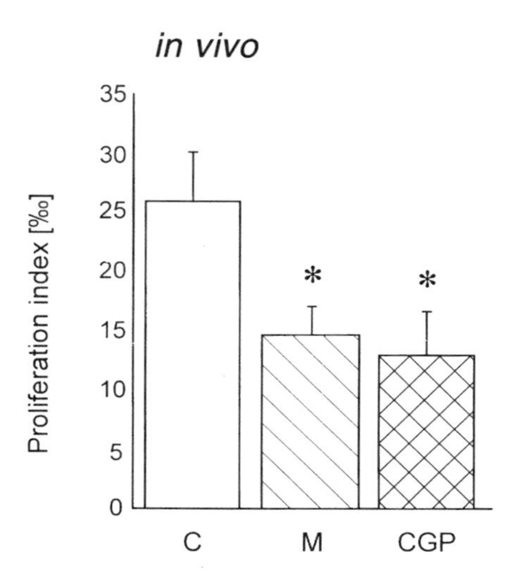

Figure 1. Proliferation indices of DES-induced prolactinomas in control (C), melatonin-treated (M) and CGP 52608-treated (CGP) rats. * $p < 0.05$.

3. RESULTS

3.1. Rat Pituitary Tumors

In groups treated with MLT or CGP 52608 the BrDU mean labelling index was approximately 50% lower than in the respective control (Figure 1).

3.2. Colon 38 Murine Tumors *in Vivo*

MLT and CGP 52608 induced a significant decrease of the BrDU labelling indices vs controls. The effect was the strongest with 100 μg of MLT and with 10 μg of CGP 52608 (Figure 2).

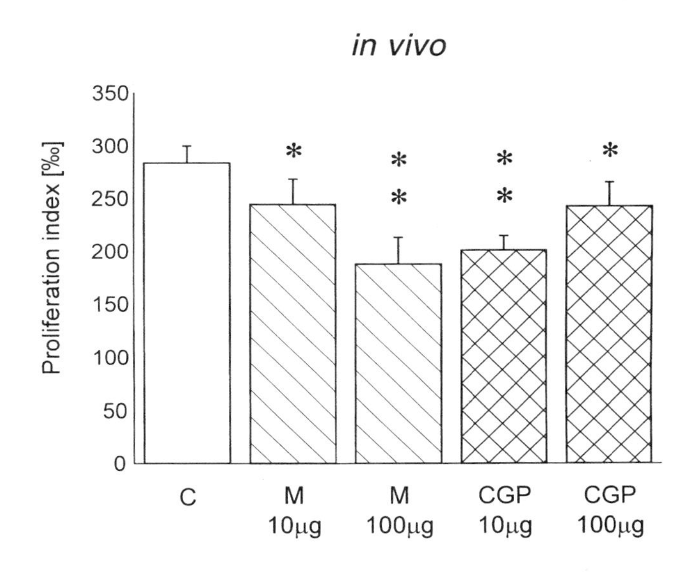

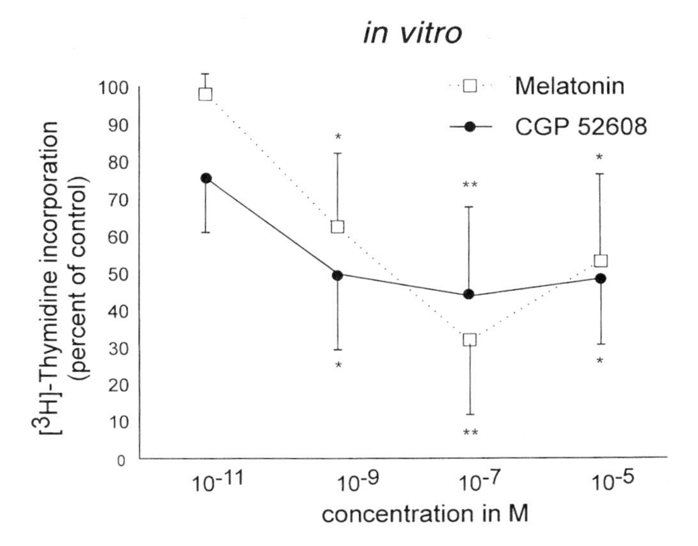

Figure 2. Proliferation indices of Colon 38 adenocarcinomas in control (C), melatonin-treated (M) and CGP 52608-treated (CGP) mice. $*p < 0.05$, $**p < 0.01$, (according to (3)).

3.3. Colon 38 Tumor Cells *in Vitro*

Both compounds inhibited the [^{3}H]thymidine incorporation into DNA of Colon 38 cells. The effect of MLT was present in the concentration of 10^{-9} M and reaches the maximum at 10^{-7} M. CGP 5208 was slightly more potent than MLT but followed in principle the same concentration/effect curve as MLT, reaching a maximal inhibitory effect in concentration of 10^{-7} M.

4. DISCUSSION AND CONCLUSIONS

The data presented in this paper show that both melatonin and RZR/ROR receptor ligand exerts the antiproliferative effects on two experimental tumors in rodents, acting with roughly the same potency. These findings corroborate with the recent observation of Moretti et al. (4), that MLT and CGP 52608 inhibited the proliferation of human prostate cancer cells DU145. The quoted authors suggest that both substances act via RZR/ROR α-receptors, which are expressed by DU 145 cells. In contrast, DU 145 cells do not express the membrane MLT receptors. It remains to establish whether the pituitary and colonic tumors investigated by us express either MLT membrane receptors or RZR/ROR receptors. The interaction of MLT with RZR/ROR receptors has been recently denied (1). Apparently our data support the hypothesis of the involvement of the RZR/ROR nuclear sites in the oncostatic action of melatonin. However, we cannot exclude that the antiproliferative effects of MLT and CGP 52608 are unrelated and occur by different intracellular mechanisms.

REFERENCES

1. Becker-Andre, Schaeren-Wiemers, and Andre. Correction to the article: [Becker-Andre M. et al., J Biol Chem 269:28531–28534, 1994] J Biol Chem 272:16707, 1997.
2. Carlberg C., Wiesenberg I., and Schraeder M. Nuclear signalling of melatonin Front Horm Res 23:25–35, 1997.
3. Karasek M., Winczyk K., Kunert-Radek J., Wiesenberg I., and Pawlikowski M. Antiproliferative effects of melatonin and CGP 52608 on the murine Colon 38 adenocarcinoma in vitro and in vivo. Neuroendocrinol Lett 19:71–78, 1998.
4. Moretti R.M., Montagnani Marelli M., Dondi D., Motta M., and Limonta P. Antiproliferative action of melatonin on androgen-independent prostate cancer cells: possible mechanism of action Abstracts IV Eur Congress of Endocrinology, Sevilla, 9–13 May 1998, abstr P-2-244.
5. Pawlikowski M. and Kunert-Radek J. Differential effects of somatostatin analogues on proliferation of murine colonic cancer cells in vitro Cytobios 89:183–187, 1997.
6. Wiesenberg I., Missbach M., Kahlen J.P., Schraeder M., and Carlberg C. Transcriptional activation of the nuclear receptor ROR gamma by the pineal hormone melatonin and identification of CGP 52608 as a synthetic ligand Nucleic Acid Res 23:327–333, 1995.

42

CYTOCHALASIN B INFLUENCE ON MEGAKARYOCYTE PATCH-CLAMP

L. Di Bella, L. Gualano, C. Bruschi, S. Minuscoli, and G. Tarozzi

Private Laboratory of Physiology—Via Marianini
45 Modena, Italy

RESEARCH SUBJECT

Cytochalasins have been found in megakaryocytes and platelets (1), in intestines and other organs with smooth muscle cells (2), in kidney and other tight epithelia (3) where they modify numerous cellular functions related to DNA synthesis or fragmentation, contractile actin microfilaments, delivery of newly synthesized membrane proteins, activation of apical K^+ channels, control of exocytic events, cell volume regulation. Many such functions could carry a great weight, both in the etiopathogenetic and therapeutic aspects of cancer.

We have already prudently applied very low doses of Cytochalasin B (Cyt. B) on human spontaneous and animal experimental cancer.

Therefore we further studied some effects of Cytochalasins on the excellent target cells of megekaryocytes (4).

METHODS

Male Wistar rats, weighing 300–400 g, were anesthetized and bled to death. Femurs were removed and bone marrow driven out into a plastic dish, containing the following Ca-free saline solution (mM): 135 NaCl, 5 KCl, 1 $MgCl_2$, 10 D-glucose, 10 HEPES, and stored at 4°C until use (5). A petty fragment of the split bone marrow was transferred to the recording chamber, containing about 1 ml of the following external solution (5) (mM): 120 NaCl, 5 KCl, 10 $CaCl_2$, 1 $MgCl_2$, 10 D-glucose, 10 HEPES to final pH 7.3 at 25°C. The internal saline solution contained (5) (mM): 150 KCl, 1 $MgCl_2$, 10 HEPES to final pH 7.2 with KOH 1 M. The base of the recording chamber was coated in advance with adhesive polypeptide (cell-tak®, Collaborative Biomedical Products; Two Oak Park, Bedford, MA). After megakaryocyte settlement on the bottom, the chamber was mounted on the movable stage of an inverted microscope (Leitz DM IL),

Melatonin after Four Decades, edited by James Olcese.
Kluwer Academic / Plenum Publishers, New York, 2000.

and permanently perfused with external saline solution at room temperature. Cytochalasin B (Sigma-Aldrich, Milano, Italy) 0.4 mM was added to the external saline solution to yield a 100 μM concentration. Patch pipettes were pulled from 1.5 mm capillary glass (Baxter, USA) in a two-stage vertical puller (Narishige, Japan); when filled with standard internal solution, the resistance was 2–6 Mς. The current and voltage were measure with a patch-clamp amplifier (Axopatch model 1D, Axon Instrument, USA), and were monitored with a storage oscilloscope (HM204-2, Germany). Signals were recorded and analyzed using a Pentium computer equipped with a Digidata 1200 data acquisition system and pCLAMP software (Axon Instrument, USA).

RESULTS

1) Megakaryocytes spontaneously settled were normally 23 ± 9 μm diameter;
2) V_m of megakaryocyte slowly declines following perfusion with Cyt. B (Figure 1);
3) When Cyt. B solution is injected very near megakaryocytes, V_m declines rapidly within 6 sec. after a short latency of 1 sec. A new, slow, similar decline begins within 11.25 sec., after which potential remains comfortably constant (Figure 2);
4) With a patch-pipette filled with Cyt. B solution, very large and irregular V-waves appear that soon and spontaneously fade (Figure 3).

CONCLUSIONS

Patch-clamp technique is a recent laboratory technique that gives precise intimate understanding of many cellular mechanisms.

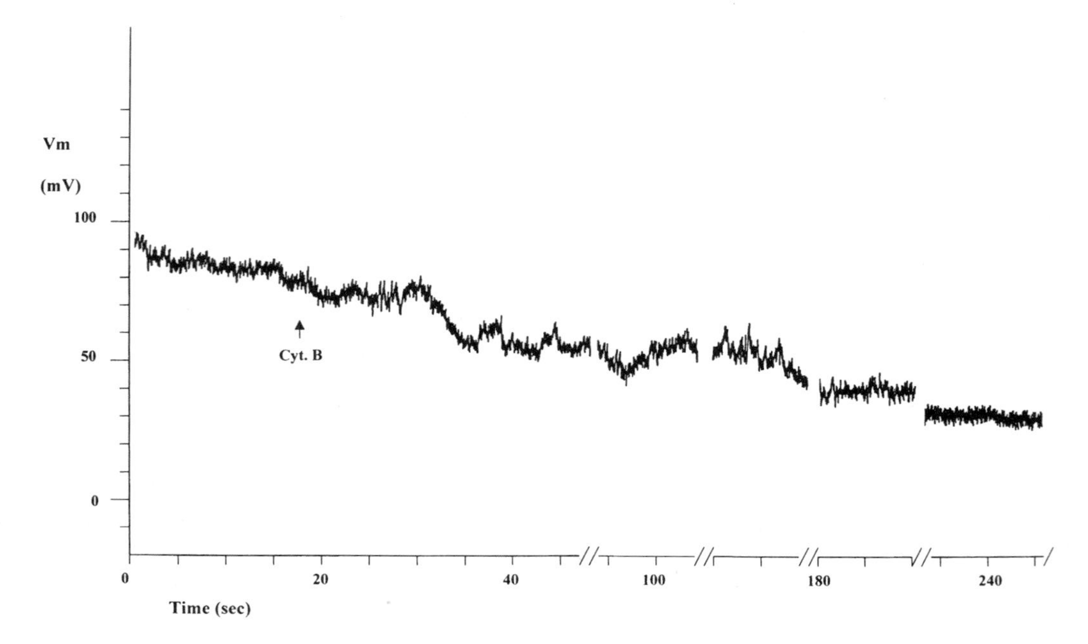

Figure 1. Megakaryocyte in cell-attached configuration. Following Cyt. B perfusing the Vm value declines.

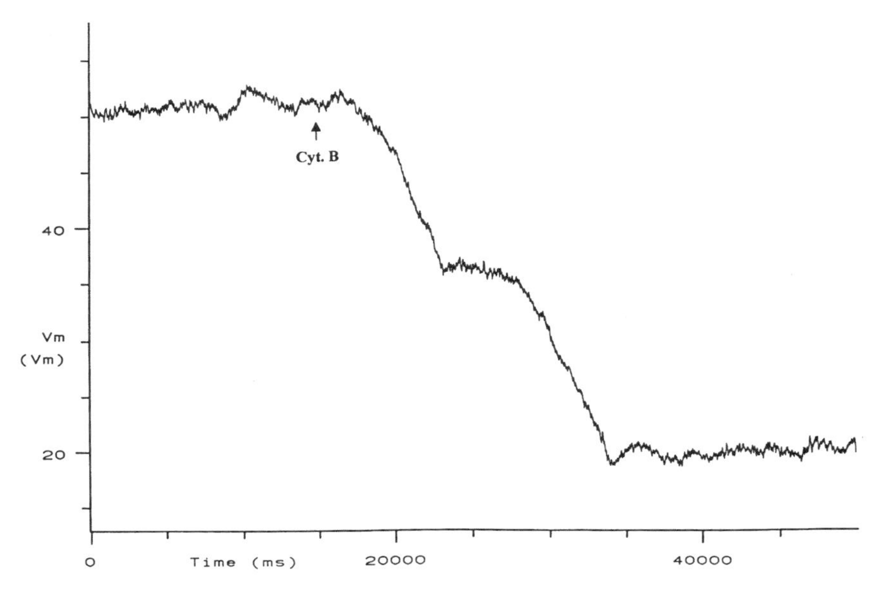

Figure 2. Megakaryocyte in cell-attached configuration. Cytochalasin B solution is injected near megakaryocytes at 20 sec. After a later time of about 1 sec. Vm declines by 50% within 6 sec. The decline stops after 8.5 sec. and is followed by a second almost identical decline by 50% (from +36mv to +18mV) after 11.25 sec. Vm then proceeds nearly stable.

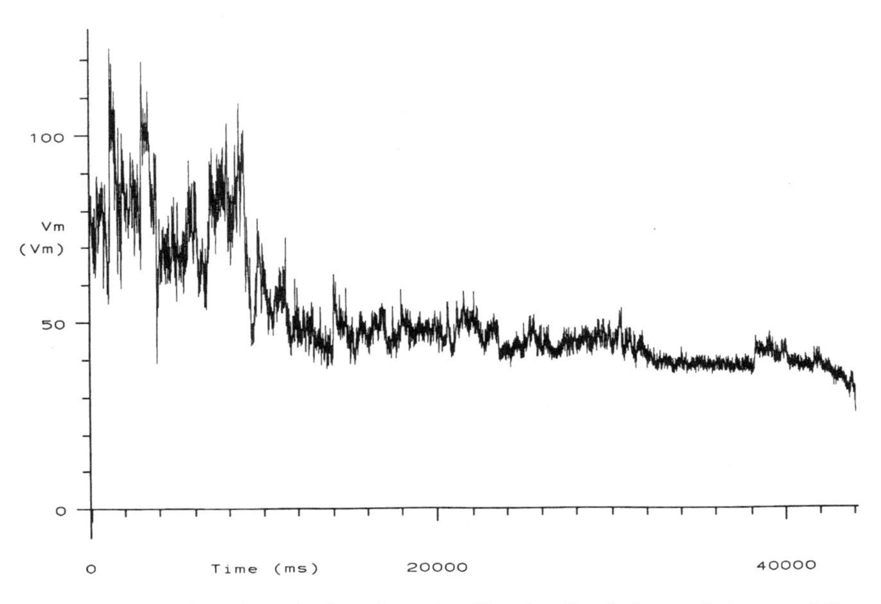

Figure 3. Megakaryocyte in cell-attached configuration. The Cyt. B solution perfusion was followed by a sprightly 7.6 sec. duration (about 50% intense decline) before gradually stabilizing at the lower level.

Cytochalasins dramatically influence membrane properties of megakaryocytes: in some conditions complete block of many channels can be achieved.

These events could probably acquire some importance when applied to cancer cells, particularly some cells such as smooth cell lung carcinoma.

Future investigations will suggest the practical importance of Cytochalasins in the treatment of cancer.

REFERENCES

1. White J.G. in Caen J.: Platelet aggregation. Paris 1971, 15; *Am. J. Pathol.* 1971, *63*, 403 e 1972, *66*, 295.
2. Obara A.U. in H. Yabu: *Eur. J. Pharmacol.* 1994, 139–147.
3. ELS W.J., CHOU K.Y.: *J. Physiol.* 1993, *462*, 447–464.
4. Fedorko M.E.: *Lab. Inv.* 1977, *36*, 321–8.
5. Kawa K.: Voltage-gate calcium and potassium currents in megakaryocytes dissociated from guinea-pig bone marrow. *J. Physiol.* (Lond.) 431:187–206, 1990.

RELATIONSHIPS BETWEEN MELATONIN, GLUTATHIONE PEROXIDASE, GLUTATHIONE REDUCTASE, AND CATALASE

Endogenous Rhythms on Cerebral Cortex in *Gallus domesticus*

M. T. Agapito,* I. Redondo, R. Plaza, S. Lopez-Burillo, J. M. Recio, and M. I. Pablos

Dept. Biochemistry and Molecular Biology
Faculty of Sciences
Univ. of Valladolid
47005-Valladolid, Spain

In this report, we studied the endogenous rhythms of three antioxidant enzymes: glutathione peroxidase (E.C.1.11.1.9), glutathione reductase (E.C.1.6.4.2) and catalase (E.C.1.11.1.6) in cortex of chick brain and correlate them with physiological blood melatonin concentrations.

1. INTRODUCTION

Melatonin was recently shown to be a component of the antioxidative defence system of organism due to its free radical scavenging and antioxidant activities. It was previously found that pharmacologically melatonin increases glutathione peroxidase (E.C. 1.11.1.9.) (GSH-Px) activity in rat brain (1). Also, GSH-Px activity is stimulated in several tissues of chicks (2) when pharmacological amounts of melatonin are administered. The melatonin-induced increase in GSH-Px activity, which peaks about 90 min

* Whom the correspondence to be send: teresaa@wamba.cpd.uva.es

Melatonin after Four Decades, edited by James Olcese.
Kluwer Academic / Plenum Publishers, New York, 2000.

after melatonin injection, was seen in every tissue (liver, kidney, gut, erythrocytes, brain and pineal gland) studied in the chick (3). Of special interest is that melatonin accumulates in the nucleus of the cells (4) (5) where binding sites for the indole seem to exist (6) (7); at least some of these may be receptors which belong to the superfamily of orphan nuclear receptors (8). It is possible that the effects of melatonin on GSH-Px activity may be mediated via a genomic mechanism (9) which involves the nuclear binding sites that have been identified, although the possibility of membrane receptor involvement cannot be excluded.

Considering the stimulatory action of exogenously administered melatonin on GSH-Px activity, it seemed possible that endogenous melatonin cycle may normally induce a circadian rhythm of neural GSH-Px activity. This assumption served as the rationale for the present study. At the same time, we measured glutathione reductase (1.6.4.2) (GSSG-Rd) and catalase (1.11.1.6) activities in chick cerebral cortex.

2. MATERIAL AND METHODS

Ten-day-old chicks were acclimated to a 12:12 light:dark cycle and maintained in a temperature controlled room (22°C ± 2) with water and food *ad libitum.*

Chicks were decapitated every 2h over a single light:dark cycle. the blood and brain from each bird was collected and all tissues were frozen at −80°C until the analyses were performed.

To test whether abolition of the endogenous melatonin rhythm would alter the GSH-Px, GSSG-Rd and Catalase activity levels, another group of chicks (10 each) was exposed to constant light for six days. A control group of chicks was simultaneously maintained under 12L:12D photoperiod. At 9h after darkness onset of the 6th day chicks were killed by decapitation and trunk blood and brains were collected.

Melatonin levels in sera were measured by radioimmunoassay using ^{3}H-melatonin as a tracer and a Guildford antibody as described previously (10).

GSH-Px activity was measured by a coupled reaction with GSSG-Rd using cumene hydroperoxide as substrate and measuring the decrease in the NADPH absorbance at 340nm as reported (11) (12). GSSG-Rd was assayed following the decrease in the NADPH absorbance at 340nm in presence of oxidized glutathione (13). Catalase was determined by a UV spectrophotometric method following the decrease absobance of H_2O_2 at 240nm (14). Proteins were measured by Lowry et al., methods using bovine serum albumin as standard.

In order to estimated the circadian rhythms of each antioxidant enzymes and melatonin, a cosinor analysis was performed with individual data with software developed by Revilla and Ardura (15).

3. RESULTS AND DISCUSSION

Glutathione peroxidase exhibited a marked circadian rhythm with a peak activity which had an acrophase of 0781 ± 0073h. after lights off (mesor: 33.8 ± 1.2m units./mg prot; amplitude: 9.4 ± 0.6m units/mg prot.) and about 4h after serum melatonin peak. Glutathione reductase and Catalase exhibited similar robust rhythms with the peak occurring respectively 1540 ± 0142h (mesor: 11.67 ± 0.58m units/mg prot.;

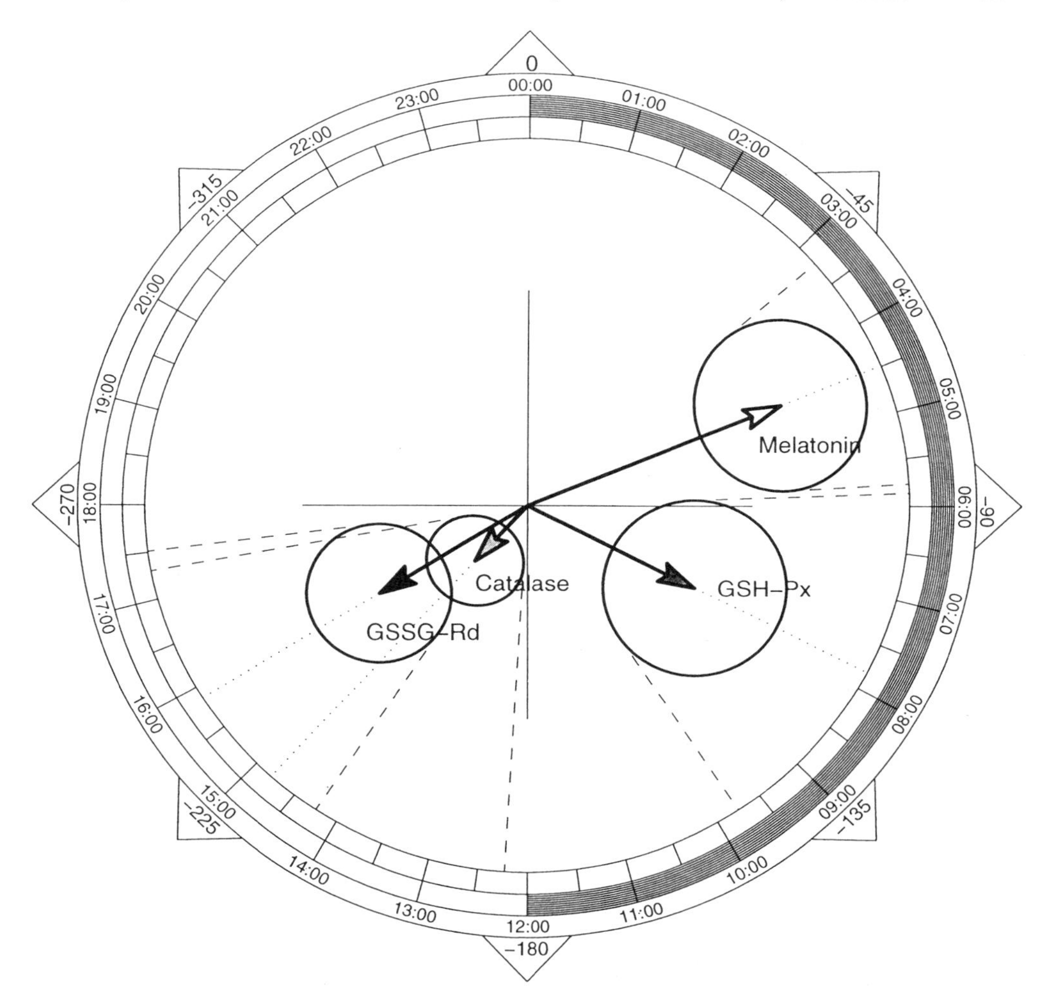

Figure 1. Population cosinor. Comparison between: Cerebral cortex GSH-Px, GRRG-Rd, Catalase activities and serum Melatonin acrophases.

amplitude: 2.42 ± 1.03 m units/mg prot.) and 1526 ± 0041 h (mesor: 4.42 ± 0.25 U.I/mg prot; amplitude: 1.28 ± 0.39/mg prot.) after lights off about 8 h after glutathione peroxidase peak and after 11 h after melatonin peak (Figure 1).

In the Figure 1 we show the comparison between: Cerebral cortex GSH-Px, GRRG-Rd, Catalase activities and serum melatonin acrophases. We suggest that neural glutathione peroxidase activity increase due to the rise of nocturnal melatonin levels while glutathione reductase and catalase activities rise slightly later, possibly due to an increase of their substrates.

Exposure of chicks to constant light for 6 days eliminated the melatonin rhythm as well as the peaks in glutathione peroxidase, glutathione reductase and catalase. Exposure of chicks to constant light for 6 days eliminated the melatonin rhythm as well as the peaks in glutathione peroxidase, glutathione reductase and catalase (Figure 2).

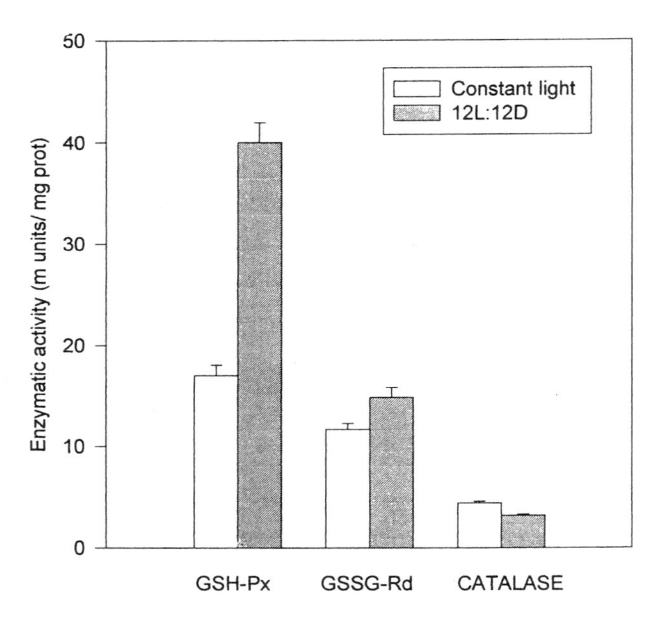

Figure 2. Inhibition of GSH-Px and GSSG-Rd activities and variation of catalase activity, in cortex cerebral of chicks exposed to constant light and killed at 0600 h, control chicks were kept in 12L:12D and killed in darkness at 0600 h. Data are mean ± SEM.

In this paper we demonstrate that three major antioxidant enzymes, GSH-Px, GSSG-Rd and Catalase, exhibit a circadian rhythm in chick cerebral cortex. Glutathione peroxidase exhibited a marked circadian rhythm with a peak activity which had an acrophase of 0781 ± 0073 h after lights off (mesor: 33.8 ± 1.2 U.I./mg prot; amplitude: 9.4 ± 0.6 U.I./mg prot.) and about 4 h after serum melatonin peak. Glutathione reductase and Catalase exhibited similar robust rhythms with the peak occurring respectively 1540 ± 0142 h (mesor: 11.67 ± 0.58 U.I./mg prot.; amplitude: 2.42 ± 1.03 U.I./mg prot.) and 1526 ± 0041 h (mesor: 4.42 ± 0.25 U.I./mg prot; amplitude: 1.28 ± 0.39 U.I./mg prot.) after lights off about 8 h after glutathione peroxidase peak and after 11 h after melatonin peak. Exposure of chicks to constant light for 6 days eliminated the melatonin and Catalase rhythms as well as the peaks in Glutathione peroxidase and Glutathione reductase activities.

This work was supported by Junta de Castilla y Leon reference project VA45/98. (B.O.C. y L. January 29th 1998).

REFERENCES

1. Barlow-Walden, L.R., Reiter R.J., Abe, M., Pablos, M.I., Menéndez-Peláez, A., Chen, L.D., and Poeggeler, B. (1995). Melatonin stimulates brain glutathione peroxidase activity. Neurochem. Int., 26, 497–502.
2. Pablos, M.I., Agapito, M.T., Gutierrez, R., Recio J.M., Reiter, R.J., Barlow-Walden, L.R., Acuña-Castroviejo, D., and Menéndez-Peláez, A. (1995). Melatonin stimulates the activity of the detoxifying enzime glutathione peroxidase in severall tissues of the chicks. J. Pineal Res., 19, 111–115.
3. Pablos, M.I., Chuang, J.I., Reiter, R.J., Ortiz, G.G., Daniels, W.M.U., Sewerynek, E., Melchiorri, D., and Poeggeler, B. (1995). Time course of melatonin-induced increase in glutathione peroxidase activity in chicks tissues. Biol. Signals, 4, 325–330.

4. Menéndez-Peláez, A. and Reiter, R.J. (1993). Distribution of melatonin in mammalian tissues. The importance of nuclear versus cytosolic localization. J. Pineal Res., 15, 59–67.
5. Menéndez-Peláez, A., Poeggeler, B., Reiter, R.J., Barlow-Walden, L.R., Pablos, M.I., and Tan, D.X. (1994). Nuclear localization onf melatonin in different mammalian tissues: Inmunocytochemical and radioinmunoassay evidence. J. Cell. Biochem., 53, 373–382.
6. Acuña-Castroviejo, D., Pablos, M.I., Menendez Pelaez, A., and Reiter, R.J. (1993). Melatonin receptor in purified cell nuclei of liver. Res. Commun. Chem. Pathol. Pharmacol., 82, 253–256.
7. Acuña-Castroviejo, D., Reiter, R.J., Menendez, Pelaez, A., Pablos, M.I., and Burgos, A. (1994). Characterization of high affinity melatonin binding sites in purified cell nuclei of rat liver. J. Pineal Res., 16, 100–112.
8. Carlberg, C. and Wiesenberg, I. (1995). The orphan receptor family RZR/ROR, melatonin and 5-lipoxygenase: An unexpected relationship. J. Pineal Res., 18, 171–178.
9. Kotler, M., Rodriguez, C., Sainz, R.S., Antolín, I., and Menéndez-Peláez, A. (1998). Melatonin increase gene expression for antioxidant enzymes in rat brain cortex. J. Pineal. Res. 24, 83–89.
10. Fraser, S., Cowen, P., Franklin, M., Franey, C., and Arendt, J. (1983). Direct radioinmunoassay for melatonin in plasma. Clin. Chem., 29, 396–397.
11. Paglia, D.E. and Valentine, W.N. (1967). Studies of quantitative and qualitative characterization of erythrocyte glutathione peroxidase. J. Lab. Clin. Med., 70, 158–169.
12. Bompart, B.J., Prevot, D.S., and Bascands, J.L. (1990). Rapid automated analysis of glutathione reductase, peroxidase and S-transferase activities: Aplication to cisplatin-induced toxicity. Clin. Biochem., 23, 501–504.
13. Goldberg, D.M. and Spooner, R.J. (1985). Glutathione reductase. In methods in enzymatic analysis, de. H.U. Bergmeyer, 3rd ed. Vol 3 pp. 258–265.
14. Chance, B. and Herbert, D. (1950). The Enzyme-substrate compounds of bacterial catalase and peroxides. Biochem. J., 46, 402–414.
15. Revilla, M.A. and Ardura, J. (1993). Rhythmometric: A multiple platform for biological rhythms analysis. Abstract of 21st Conference of the International Society of Chronobiology. Quebec, Canada, 11–15. July. p. 85.

44

EFFECT OF PINEALECTOMY ON MELATONIN LEVELS IN THE GASTROINTESTINAL TRACT OF BIRDS

Iveta Herichová[1] and Michal Zeman[2]

[1] Department of Animal Physiology and Ethology
Comenius University
Bratislava, Slovakia
[2] Institute of Animal Biochemistry and Genetics
SASci, Ivanka pri Dunaji, Slovakia

1. INTRODUCTION

The presence of melatonin and enzymes in its forming pathway was proven in the gastrointestinal tract (GIT) of several species of higher vertebrates, including birds. High levels of melatonin were found in the GIT of pigeon (12), duck (9), zebra finch (13) and chicken (7). Both enzymes of the melatonin forming pathway, N-acetyltransferase (NAT) and hydroxyindole-O-methyltransferase, are present in the gut of rat (5,10) and NAT was found also in the GIT of quail (8).

In rats, immunoreactive melatonin in the GIT exhibits a regional distribution very similar to the density of enterochromaffin cells of intestinal mucosa (3). These cells are in higher vertebrates an important site of synthesis of melatonin precursor—serotonin. Since the gut contains the substrate as well as the biochemical apparatus for synthesis of melatonin, a local production of melatonin in enterochromaffin cells was suggested (11,3).

To evaluate the origin of melatonin in the GIT of chicken we studied effects of pinealectomy (PINX) on the melatonin content in the plasma, duodenum, jejunum, pancreas and kidney.

2. MATERIALS AND METHODS

Pinealectomy and sham-operations were performed in 2-day-old broiler chicks and then all chicks were kept under the cycle 12L : 12D with food and water ad libitum.

Melatonin after Four Decades, edited by James Olcese.
Kluwer Academic / Plenum Publishers, New York, 2000.

Light was provided by fluorescent tubes that produced illumination 90–100 lux at the center of the room. One week after surgery birds were decapitated in the middle of the light and dark period and melatonin levels in plasma and tissues were assayed.

Melatonin was measured by radioimmunoassay (RIA) (6) directly in plasma or after solvent extraction of pineal glands (methanol) and gut tissue (chloroform). Statistical analysis of data was performed using t-test and ANOVA.

3. RESULTS

A clear daily melatonin rhythm with high levels during the nighttime and low levels during the daytime was observed in the plasma, duodenum, jejunum and pancreas of intact and sham-operated birds ($P < 0.001$) and there were no differences between melatonin levels in intact and sham-operated animals. In kidney, concentrations of melatonin were low and PINX did not affect them.

Pinealectomy considerably decreased ($P < 0.05$) high nighttime melatonin levels in the plasma, duodenum and pancreas but the residual melatonin rhythm was still preserved ($P < 0.05$). Removing of the pineal gland abolished completely the melatonin rhythm in the jejunum.

Our results suggest that majority of melatonin found in the GIT of chicken during the nighttime comes from the pineal gland. However, some amount of melatonin was present in the plasma and tissues of the GIT after pinealectomy.

4. DISCUSSION

Our results confirmed presence of melatonin in the gut of chickens. Melatonin levels in the GIT exhibited a clear daily rhythm. These results are in accordance with data of Vakkuri et al. (12) who demonstrated the daily melatonin rhythm in the duodenum of pigeons and results of Van't Hof and Gwinner (13) who observed the daily rhythm of melatonin levels in the GIT of zebra finch. The range of plasma, pineal and GIT concentrations of melatonin in our experiments is consistent with published data (12,9).

Results on effects of PINX on melatonin levels in the GIT of birds are scarce. Removing of the pineal gland did not influence melatonin levels in the GIT of ducks during the light and dark period (9) and pigeons during the dark period (12). In our study a significant decrease of nighttime melatonin concentrations in the GIT tissues of PINX chickens was observed while daytime levels were unaffected. In spite of this, pinealectomy did not result in the complete elimination of melatonin from the gut.

On the basis of these results we can conclude that in chickens the majority of melatonin in the GIT during the nightime comes from the pineal gland. Differences between residual levels of melatonin in the GIT after PINX may result from interspecies variability. Melatonin present in the GIT after pinealectomy can be derived from an extrapineal source of this indolamine including eyes and the Harderian gland, but enterochromaffin cells of the gut should be also taken into consideration.

High levels of melatonin in the gut as well as the pronounced daily rhythm that is present already during the embryonic life of chicken (7) indicate a possible role of melatonin in modulation of gastrointestinal functions. Several studies performed till now with mammalian experimental models show that melatonin in the gastrointestinal tract can contribute to paracrine regulation of gut motility (2), modulation of gut nervous system activity (1) and to protection of tissues against reactive oxygen species (4). Therefore evaluation of melatonin functions in the GIT of birds would be of a great interest.

ACKNOWLEDGMENT

This study was supported by grant of the Scientific Agency of Slovak Republic (2/5044/98).

REFERENCES

1. Barajas-López, C., Peres, A.L., Espinosa-Luna, R., Reyes-Vázquez, C., and Prieto-Gómez, B. Melatonin modulates cholinergic transmission by blocking nicotinic channels in the guinea-pig submucous plexus. Eur J Pharmacol 312:319–325, 1996.
2. Bubenik, G.A. and Dhanvantari, S. Influence of serotonin and melatonin on some parameters of gastrointestinal activity. J Pineal Res 7:333–334, 1989.
3. Bubenik, G.A., Brown, G.M., and Grota, L.J. Immunohistological localization of melatonin in the rat digestive system. Experientia 33:662–663, 1977.
4. De La Lastra, C.A., Cabeza, J., Motilva, V., and Martin, M.J. Melatonin protects against gastric ischemia-reperfusion injury in rats. J Pineal Res 23:47–52, 1997.
5. Ellison, N., Weller, J., and Klein, D.C. Development of a circadian rhythm in the activity of pineal serotonin N-acetyltransferase. J Neurochem 19:1335–1340, 1972.
6. Fraser, S., Cowen, P., Franklin, M., Franey, C., and Arendt, J. Direct radioimmunoassay for melatonin in plasma. Clin Chem 29:396–397, 1983.
7. Herichová, I. and Zeman, M. Perinatal development of melatonin production in gastrointestinal tract of domestic chicken. In: Tönhardt, H. and Lewin, R. (eds): Investigations of Perinatal Development of Birds. Berlin, Free University, pp. 117–122, 1996.
8. Hong, G.X. and Pang, S.F. N-acetyltransferase activity in the quail (Coturnix coturnix jap.) duodenum. Comp Biochem Physiol 112B: 251–255, 1995.
9. Lee, P.N., Shiu, S.Y.W., Chow, P.H., and Pang, S.F. Regional and diurnal studies of melatonin and melatonin binding sites in the duck gastrointestinal tract. Biol Signals 4:212–224, 1995.
10. Quay, W.B. and Ma, Y.H. Demonstration of gastrointestinal hydroxyindole-O-methyltransferase. Int Rep Clin Sci, Med Sci Lib Compend 4:563, 1976.
11. Raikhlin, N.T., Kvetnoy, I.M., and Tolkachev, V.M. Melatonin may be synthesised in enterochromaffin cells. Nature 255:344–345, 1975.
12. Vakkuri, O., Rintamaki, H., and Leppaluoto, J. Plasma and tissue concentrations of melatonin after midnight light exposure and pinealectomy in the pigeon. J Endocrinol 105:263–259, 1985.
13. Van't Hof, T.J. and Gwinner, E. Circadian variation in zebra finch (Poephila gutata) gastrointestinal tract melatonin. Ital J Anat Embryol 101:78, 1996.

MELATONIN AND 5-METHOXYTRYPTAMINE IN THE BIOLUMINESCENT DINOFLAGELLATE *Gonyaulax polyedra*

Restoration of the Circadian Glow Peak after Suppression of Indoleamine Biosynthesis or Oxidative Stress

R. Hardeland, S. Burkhardt, I. Antolín, B. Fuhrberg, and A. Coto-Montes

Institute of Zoology and Anthropology
University of Göttingen
Berliner Str. 28, D-37073 Göttingen
Germany

1. INTRODUCTION

The bioluminescent dinoflagellate *Gonyaulax polyedra* produces melatonin (MLT) and 5-methoxytryptamine (5MT) in relatively high concentrations, which depend on circadian phase and ambient temperature (review: ref. 1): At normal rearing temperature of 20°C, the MLT peak (ca. 1 μM) is found shortly after the onset of darkness, whereas 5MT reaches its maximum (ca. 0.7 μM) during late scotophase; at lower temperatures, e.g. 15°C, these indoleamines rise by many fold (2). The 5MT peak coincides with the maximum of spontaneous bioluminescence, which is caused by a strong increase in one mode of light emission, glow. 5MT was shown to be a potent stimulator of bioluminescence, acting at physiological concentrations, in contrast to other indolic compounds including MLT (3). Therefore, we investigated the possibility that 5MT might be required for the expression of the nocturnal glow peak.

2. MATERIALS AND METHODS

Gonyaulax polyedra was grown at 20°C, in LD 12:12 (2600:0 lx), using a modified f/2 seawater medium according to Hoffmann and Hardeland (4). Experiments were

Melatonin after Four Decades, edited by James Olcese.
Kluwer Academic / Plenum Publishers, New York, 2000.

carried out either in LD 12:12 or in DD (cf. figure legend). Melatonin-deacetylating aryl acylamidase activity was measured as described earlier (5). 5MT was determined by HPLC with electrochemical detection in extracts from shock-frozen cells (2). Bioluminescence was recorded in DD, using a temperature-controlled scintillation spectrometer operated under conditions optimized for dinoflagellate bioluminescence (6). In experiments on suppression of tryptophan hydroxylase, cells were transferred to the scintillation spectrometer at ZT 12; in experiments on effects of paraquat and buthionine sulfoximine, cells were transferred already at ZT 6, in order to distinguish the physiological glow peak from bioluminescence caused by possibly occurring cell damage (7).

3. RESULTS AND DISCUSSION

The nocturnal maxima of melatonin-deacetylating aryl acylamidase activity, reflecting 5MT formation, and of 5MT concentration are found in identical phase positions; moreover, the circadian glow peak appears at the time of highest 5MT levels (Figure 1). Therefore, a substantial precondition is fulfilled for an involvement of 5MT in the generation of the glow peak. Inhibition of indoleamine biosynthesis by blocking tryptophan hydroxylase leads to a marked suppression of the glow peak, as demonstrated by three different inhibitors, p-chlorophenylalanine, p-fluorophenylalanine and 5-fluorotryptophan (Figure 1). We had previously shown that a glow maximum suppressed by p-chlorophenylalanine can be restored by 5-hydroxytryptophan (8). Here we demonstrate that such a restoration is also possible with MLT and 5MT (Figure 1). In the case of 5MT, near-physiological concentrations are sufficient for this effect. Higher doses of MLT are required, in order to generate stimulatory amounts of 5MT, which is readily catabolized to 5-methoxytryptophol (1). More extended studies using all three inhibitors and all intermediates (not shown) revealed that any indolic metabolite which can be converted to 5MT is capable of restoring the glow peak; however, other products such as substituted tryptophols, not leading to 5MT formation, are ineffective.

We have recently shown that both MLT and 5MT strongly decline in *Gonyaulax* exposed to oxidative stress, as induced by differently acting substances, such as (a) hydrogen peroxide, a source of hydroxyl radicals, (b) paraquat (PQ), a herbicide generating superoxide anions, and (c) buthionine sulfoximine (BSO), an agent decreasing reduced glutathione (9). While hydrogen peroxide is highly toxic to *Gonyaulax* (7), the weaker stressors PQ and BSO are tolerated by the dinoflagellate up to remarkably high concentrations. Here we demonstrate that both PQ and BSO also suppress the nocturnal glow peak (Figure 1). Again, the maximum can be restored by MLT, although this is not a stimulator of bioluminescence per se. The restoration is more clearly expressed in the case of PQ, as compared to BSO treatment. Also 5MT is capable of enhancing bioluminescence in PQ- or BSO-treated cells (Figure 1); however, concentrations higher than physiological are required (30 or 10 μM, respectively), presumably because of the continuous presence of the stressors leading to permanent intracellular oxidative destruction of the indoleamine. For this reason, increases of bioluminescence in a phase position similar to that of the normal glow peak are only seen when 5MT is given shortly before this phase. On the other hand, its precursor MLT, which does not stimulate bioluminescence directly and, therefore, can be given at higher concentrations (250 or 100 μM, respectively), is also effective when added much earlier, at the

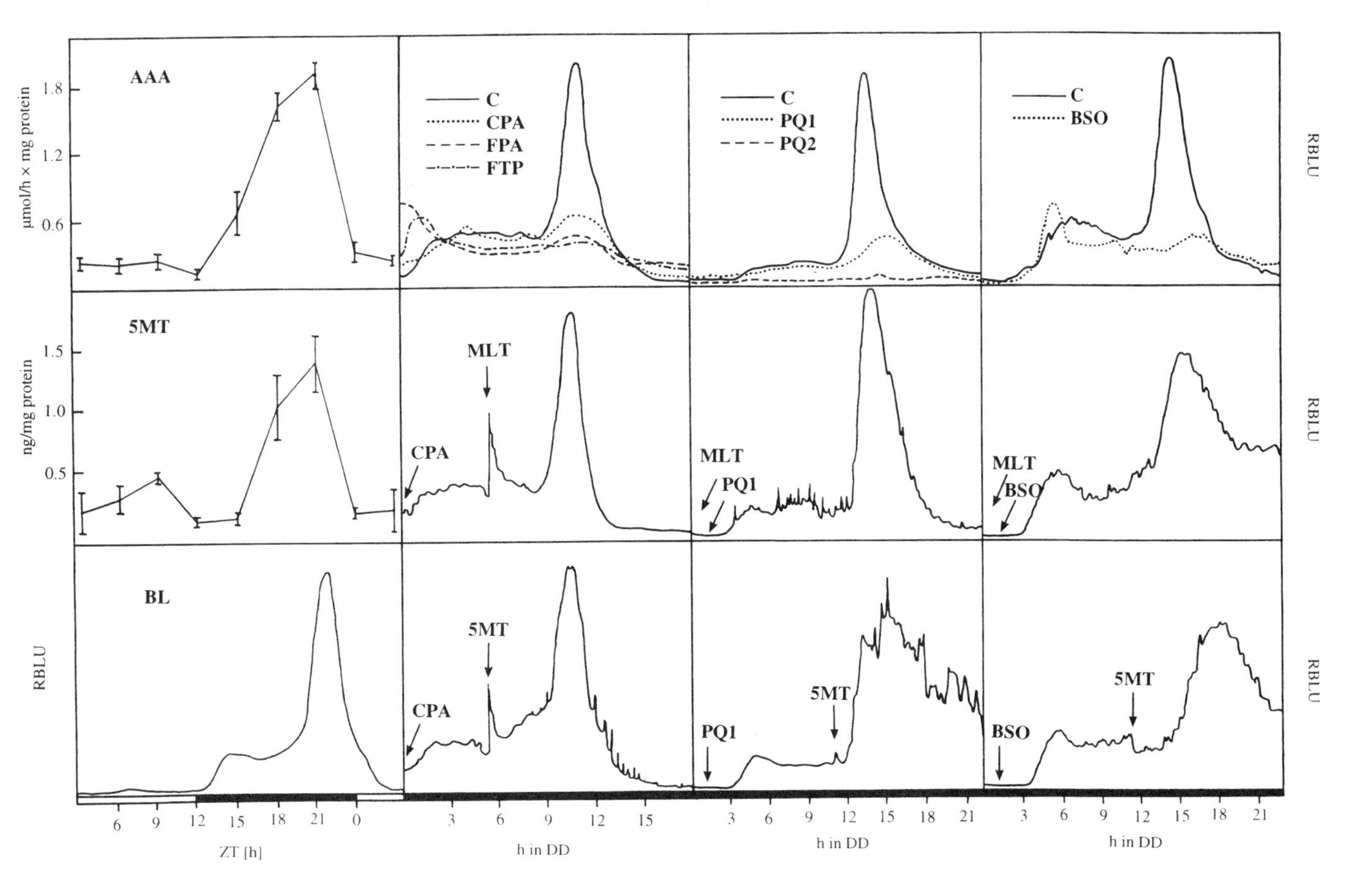

Figure 1. 5-Methoxyindoles and the expression of the glow peak. Left column: Coincidence of maxima of aryl acylamidase (AAA), 5MT and spontaneous bioluminescence (BL); experiments in LD. 2nd to 4th columns: Suppression of the glow peak by inhibitors of tryptophan hydroxylase, by paraquat and by buthionine sulfoximine, and its restoration by MLT and 5MT; experiments in DD. RBLU—relative bioluminescence units. Black bars: dark time. Vertical lines: s.e.m. Other abbreviations: C—control; CPA—p-chlorophenylalanine 0.1 mM; FPA—p-fluorophenylalanine 0.1 mM; FTP—5-fluorotryptophan 1 mM; PQ1—paraquat 1 mM; PQ2—paraquat 2 mM; BSO—buthionine sulfoximine 10 mM; doses of MLT—2nd and 4th columns: 0.1 mM, 3rd column: 0.25 mM; doses of 5MT—2nd column: 1 μM, 3rd column: 30 μM, 4th column: 10 μM.

time of transfer to DD (Figure 1). In this case, MLT will be preferentially converted to 5MT not before the second half of subjective night, due to the phasing of aryl acylamidase, and therefore lead to a glow peak in approximately correct phase position.

REFERENCES

1. Hardeland, R. and Fuhrberg, B. Ubiquitous melatonin—Presence and effects in unicells, plants and animals. Trends Comp Biochem Physiol 2:25–45, 1996.
2. Fuhrberg, B., Hardeland, R., Poeggeler, B., and Behrmann, G. Dramatic rises of melatonin and 5-methoxytryptamine in *Gonyaulax* exposed to decreased temperature. Biol Rhythm Res 28:144–150, 1997.
3. Balzer, I. and Hardeland, R. Stimulation of bioluminescence by 5-methoxylated indoleamines in the dinoflagellate, *Gonyaulax polyedra*. Comp Biochem Physiol 98C:395–397, 1991.
4. Hoffmann, B. and Hardeland, R. Membrane fluidization by propranolol, tetracaine, and 1-aminoadamantane in the dinoflagellate, *Gonyaulax polyedra*. Comp Biochem Physiol 81C:39–43, 1985.
5. Hardeland, R., Fuhrberg, B., and Lax, P. Aryl acylamidase of the dinoflagellate *Gonyaulax polyedra*. In: Metabolism and Cellular Dynamics of Indoles (R. Hardeland, ed.), Univ. of Göttingen, Göttingen 1996, pp. 47–51.
6. Hardeland, R. Effects of catecholamines on bioluminescence in *Gonyaulax polyedra* (Dinoflagellata). Comp Biochem Physiol 66C:53–58, 1980.
7. Antolín, I., Obst, B., Burkhardt, S., and Hardeland, R. Antioxidative protection in a high-melatonin organism: The dinoflagellate *Gonyaulax polyedra* is rescued from lethal oxidative stress by strongly elevated, but physiologically possible concentrations of melatonin. J Pineal Res 23:182–190, 1997.
8. Burkhardt, S., Meyer, T.J., Hardeland, R., Poeggeler, B., Fuhrberg, B., and Balzer, I. Requirement of indoleamines and a V-type proton ATPase for the expression of the circadian glow rhythm in *Gonyaulax polyedra*. Biol Rhythm Res 28:151–159, 1997.
9. Burkhardt, S., Hardeland, R., and Poeggeler, B. Various forms of oxidative stress strongly diminish 5-methoxylated indoleamines. In: Biological Rhythms and Antioxidative Protection (R. Hardeland, ed.), Cuvillier, Göttingen 1997, pp. 98–102.

PRESENCE AND POSSIBLE ROLE OF MELATONIN IN A SHORT-DAY FLOWERING PLANT, *Chenopodium rubrum*

Jan Kolář,[1] Carl H. Johnson,[2] and Ivana Macháčková[3]

[1] Department of Plant Physiology
Faculty of Sciences
Charles University
Praha, and Institute of Experimental Botany
Academy of Sciences of the Czech Republic
Praha, Czech Republic
[2] Department of Biology
Vanderbilt University
Nashville, Tennessee
[3] De Montfort University
Norman Borlaug Institute for Plant Science Research
Institute of Experimental Botany
Academy of Sciences of the Czech Republic
Praha, Czech Republic

1. INTRODUCTION

Melatonin plays a role in regulating vertebrate circadian rhythms and photoperiodic reactions (11). Melatonin is also produced by many groups of invertebrates, including insects (7), molluscs (1), and even planarians (9). During the last decade, melatonin was found in the unicellular alga *Gonyaulax polyedra*. Melatonin levels in *Gonyaulax* oscillate in diurnal (10) and circadian patterns (3). A recent model (5) proposes that melatonin induces its own conversion to 5-methoxytryptamine which affects the circadian rhythms in bioluminescence and also induces photoperiodically controlled encystment of this alga (2). Melatonin was also detected in many dicot and monocot plant species (4,6).

This widespread occurence of melatonin led us to investigate its presence, daily rhythm, and functions in the dicot plant *Chenopodium rubrum*. *C. rubrum* is

Melatonin after Four Decades, edited by James Olcese.
Kluwer Academic / Plenum Publishers, New York, 2000.

a qualitative short-day flowering plant used as a model for studies of plant photoperiodism (12).

2. MELATONIN IN *Chenopodium rubrum*

We have identified melatonin in *C. rubrum* shoots both by radioimmunoassay and liquid chromatography-tandem mass spectrometry (LC-MS/MS). The daughter mass spectra of melatonin standard and the purified shoot extract were identical; this confirmed the identity of the compound (8). Melatonin oscillations were studied in the shoots of 25-day-old plants that were grown in constant light, then transferred to the desired photoperiodic regime and sampled during the 5th light/dark cycle. Melatonin (measured by LC-MS/MS) exhibited a strong diurnal rhythm with low or undetectable levels during the light and a maximum of several hundred pg/g fresh weight in the first half of the night. The timing, not the duration, of the melatonin peak is probably determined by photoperiod. The melatonin maximum occured 4–6 h after lights off in 12 h light : 12 h dark cycle, but already 2 h after lights off in 16 h light : 8 h dark cycle. Melatonin rhythms persisted in *C. rubrum* plants transferred from 12 h light : 12 h dark regime to constant 72-h darkness.

3. THE FUNCTION OF MELATONIN IN *Chenopodium rubrum*

The data mentioned above indicate that the melatonin rhythm in *C. rubrum* is similar to the rhythms described in vertebrates. Therefore, we investigated possible effects of melatonin application on photoperiodic flower induction of *C. rubrum*. 4.5-day-old plants were induced to flowering by a single 12-h darkness and the chemicals, dissolved in DMSO and diluted with water, were applied onto the cotyledons and apical parts at different times. The percentage of flowering plants was assessed after 7 days of futher growth in constant light.

Melatonin, several putative precursors, melatonin agonists and antagonists and inhibitors of tryptophan decarboxylase were tested. Of these compounds, only melatonin and the agonist CGP 52608 showed significant effects. CGP 52608 (a thiazolidine dione) was shown to bind specifically to putative human nuclear melatonin receptors RZRα (13). CGP 52608 in concentrations from 2×10^{-5} M to 3×10^{-4} M reduced flowering when applied 3 or 1 h before, or 2 h after the beginning of 12-h darkness. The effect was dose-dependent. Another experiment showed that 2×10^{-4} M CGP 52608 decreased flowering if applied in the interval from 3 h before to 6 h after the beginning of darkness. Later application was ineffective. The treatment with 2×10^{-5} M to 5×10^{-4} M melatonin 1 h before the beginning of darkness also reduced flowering. The application of both melatonin and CGP 52608 together was even more efficient. No signs of toxicity were observed even with the highest concentrations employed.

Our results support the notion that melatonin might actually have similar functions in plants as it has in animals. The aim of our future research is either to confirm, or reject this hypothesis.

ACKNOWLEDGMENT

We thank Novartis and especially Dr. I. Wiesenberg for generously providing CGP 52608. This work was supported by grants # ME 056 of program "KONTAKT",

INT-9605193 from the USA's NSF and 204/95/1576 from the Grant Agency of the Czech Republic.

REFERENCES

1. Abran, D., Anctil, M., and Ali, M.A. Melatonin activity rhythms in eyes and cerebral ganglia of *Aplysia californica*. Gen Comp Endocrinol 96:215–222, 1994.
2. Balzer, I. and Hardeland, R. Photoperiodism and effects of indoleamines in a unicellular alga, *Gonyaulax polyedra*. Science 253:795–797, 1991.
3. Balzer, I., Poeggeler, B., and Hardeland, R. Circadian rhythms of indolamines in a dinoflagellate, *Gonyaulax polyedra*: Persistence of melatonin rhythm in constant darkness and relationship to 5-methoxytryptamine. In: Touitou, Y., Arendt, J., and Pevet, P. Melatonin and the pineal gland. From basic science to clinical application. 83–186, 1993. Amsterdam, Excerpta Medica.
4. Dubbels, R., Reiter, R.J., Klenke, E., Goebel, A., Schnakenberg, E., Ehlers, C., Schiwara, H.W., and Schloot, W. Melatonin in edible plants identified by radioimmunoassay and by high performance liquid chromatography-mass spectrometry. J Pineal Res 18:28–31, 1995.
5. Hardeland, R. and Fuhrberg, B. Ubiquitous melatonin—presence and effects in unicells, plants and animals. Trends Compar Biochem Physiol 2:25–45, 1996.
6. Hattori, A., Migitaka, H., Iigo, M., Itoh, M., Yamamoto, K., Ohtanikaneko, R., Hara, M., Suzuki, T., and Reiter, R.J. Identification of melatonin in plants and its effects on plasma melatonin levels and binding to melatonin receptors in vertebrates. Biochem Mol Biol Int 35:627–634, 1995.
7. Itoh, M.T., Hattori, A., Nomura, T., Sumi, Y., and Suzuki, T. Melatonin and arylalkylamine N-acetyltransferase activity in the silkworm, *Bombyx mori*. Mol Cell Endocrinol 115:59–64, 1995.
8. Kolář, J., Macháčková, I., Eder, J., Prinsen, E., Van Dongen, W., van Onckelen, H., and Illnerová, H. Melatonin: occurrence and daily rhythm in *Chenopodium rubrum*. Phytochemistry 44:1407–1413, 1997.
9. Morita, M. and Best, J.B. The occurence and physiological functions of melatonin in the most primitive eumetazoans, the planarians. Experientia 49:623–626, 1993.
10. Poeggeler, B., Balzer, I., Hardeland, R., and Lerchl, A. Pineal hormone melatonin oscillates also in the dinoflagellate *Gonyaulax polyedra*. Naturwissenschaft 78:268–269, 1991.
11. Reiter, R.J. The melatonin rhythm: both a clock and a calendar. Experientia 49:654–664, 1993.
12. Ullmann, J., Seidlová, F., Krekule, J., and Pavlová, L. *Chenopodium rubrum* as a model plant for testing the flowering effects of PGRs. Biol Plant 27:367–372, 1985.
13. Wiesenberg, I., Missbach, M., Kahlen, J.P., Schrader, M., and Carlberg, C. Transcriptional activation of the nuclear receptor RZRa by the pineal gland hormone melatonin and identification of CGP 52608 as a synthetic ligand. Nucleic Acids Res 23:327–333, 1995.

47

MLT AND THE IMMUNE-HEMATOPOIETIC SYSTEM

Georges J. M. Maestroni

Istituto Cantonale di Patologia, Center for Experimental Pathology, 6601 Locarno, Switzerland

1. ABSTRACT

It is now well recognized that a main actor in the continous interaction between the nervous and immune systems is the pineal hormone MLT. T-helper cells bear G-protein coupled MLT cell membrane receptors and, perhaps, MLT nuclear receptors. Activation of MLT receptors enhances the release of T-helper cell type 1 (Th1) cytokines, such as gamma-interferon and interleukin-2 (IL-2), as well as of novel opioid cytokines which crossreact immunologically with both interleukin-4 (IL-4) and dynorphin B. MLT has been reported also to enhance the production of interleukin-1 (IL-1), interleukin-6 (IL-6) and interleukin-12 (IL-12) in human monocytes. These mediators may counteract stress-induced immunodepression and other secondary immunodeficiences, protect mice against lethal viral and bacterial diseases, synergize with IL-2 in cancer patients and influence hematopoiesis. In cancer patients, MLT seems to be required for the effectiveness of low dose IL-2 in those neoplasias that are generally resistant to IL-2 alone. Hematopoiesis is apparently influenced by the action of the MLT-induced-opioids (MIO) on kappa-opioid receptors present on stromal bone marrow macrophages. Most interestingly, gamma-interferon and colony stimulating factors (CSFs) may modulate the production of MLT in the pineal gland. A hypothetical pineal-immune-hematopoietic network is, therefore, taking shape. From the immunopharmacological and ethical point of view, clinical studies on the effect of MLT in combination with IL-2 or other cytokines in viral disease including human immunodeficiency virus-infected patients and cancer patients are needed. In conclusion, MLT seems to play a crucial role in the homeostatic interactions between the brain and the immune-hematopoietic system and deserves to be further studied to identify its therapeutic indications and its adverse effects.

Melatonin after Four Decades, edited by James Olcese.
Kluwer Academic / Plenum Publishers, New York, 2000.

2. NEUROIMMUNOMODULATION

Maintenance of health depends to a significant extent on the ability of the exposed host to respond appropriately and, eventually, to adapt to environmental stressors. It is now well established that inappropriate or maladaptative response to such stressors weaken the body's resistance to other stimuli from the environment such as pathogenic organisms or cancer cells. It is fair to consider the social environment as part of the general environment which has an impact on the body via redundant and reciprocal interactions between the body and the brain. These are linked by the nervous, endocrine, and immune systems and utilize a large array of chemical messengers including hormones, cytokines and neurotransmitters (72). There is abundant evidence that there are functional afferent nerve endings in the tissue of the immune and hematopoietic systems arising from both the sympathetic and parasympathetic systems (1,2). It is also clear that many neurotransmitters, neuroendocrine factors and hormones can drastically change immune functions and that, on the contrary, cytokines derived from immunocompetent cells can profoundly affect the central nervous system (1,2). As a consequence, any environmental stimulus to the nervous system will affect the immune system and viceversa essentially via the endocrine system. On this conceptual basis, it should not be surprising that the day/night photoperiod, which constitutes a basic environmental cue for any organism, can also influence the immune-hematopoietic system. As for many other adaptative responses, a major mediator of this influence seems to be the pineal gland which tranduces the light/dark rhythm into the circadian synthesis and release of MLT (78).

3. MLT EFFECTS

Early studies about a possible link between the pineal gland and the immune system claimed that absence of the pineal gland stimulated the proliferation of immunocompetent cells (11,21,70) and others just the opposite (6,32,73). However, most studies agreed that pinealectomy is associated with a precocious involution and histological disorganization of the thymus (6,20,21,73). The mechanism of this effect was postulated to depend on increased steroid gonadal hormones. By various pharmacological interventions aimed at inhibiting MLT synthesis, we provided a first evidence about a possible involvement of endogenous MLT on humoral and T cell immune reactions, as well as on spleen and thymus cellularity in mice (59,63). In another report we show that pinealectomy inhibits leukemogenesis in a radiation leukemia virus murine model and that MLT has a promoting effect on the disease (18). A number of other authors have then further extended this type of evidence. Pinealectomized mice were reported to have depressed humoral immunity (7). In another report, inhibition of endogenous MLT in hamster produced a decrease of spleen weight and reduced T cell blastogenesis. MLT administration counteracted this effect (16). IL-2 production and antibody-dependent cellular cytotoxicity were inhibited in pinealectomized mice and exogenous MLT restored these important functions (22,67). Endogenous MLT has been also reported to influence the concentration of bone marrow granulocyte/macrophage colony-forming unity (GM-CFU) (28). An interesting finding which might be associated with and explained by the immunoenhancing action of endogenous MLT, is the widely documented oncostatic role of the pineal gland and of MLT (12). From the pharmacological point of view, MLT can augment the immune response and correct immunodeficiency states which may follow acute stress, viral diseases, or drug treatment (51,53,54,56,59–61). This finding has been then confirmed and extended either in mice or in humans to a variety of immune parameters (15,16,22,26,49,64,67,68). On this line, a very significant biological effect of MLT is the protection of mice against encephalitis viruses and lethal bacterial (8,9,13). In general, the immunoenhancing action of MLT seems restricted to T-dependent antigens and to be most pronounced in immunodepressed situations. For example, MLT may completely counteract thymus involution and the immunological depression induced by stress events or glucocorticoid treatment (54). MLT is active only when injected in the afternoon or in the evening, i.e. with a schedule consonant with its physiological rhythm (59,60). In addition, MLT is most active on antigen or cytokine activated immunocompetent cells (59,60). Consistent with these requirements, a recent report shows that MLT may also restore depressed immunological functions after soft-tissue trauma and hemorrhagic shock (74). Beside acquired immunity, natural immune parameters seem also influenced by MLT (3,22,35). In a tumor model of established lung metastases we found that MLT could synergize with the anti-cancer effect of IL-2 (56). More recently, we reported that MLT may rescue hematopoiesis in mice transplanted with Lewis Lung Carcinoma (LLC) and treated with cancer chemotherapeutic compounds (58,61). Interestingly, a most recent reports confirms the protective effect of melatonin in rats treated with cytotoxic drugs (4). However, when the mice were tumor-free MLT augmented the chemotherapy-induced myelotoxicity (52). This indicates that MLT does not have only beneficial or therapeutic effects. As a matter of fact, MLT has been reported to exaggerate collagen-induced arthritis (29) and to promote T-cell leukemia (18).

4. CYTOKINES WHICH MEDIATE THE EFFECTS OF MLT

IL-2 and gamma-interferon and opioid peptides released by activated Th cells seem to mediate, at least in part, the immunopharmacological action of MLT (15,17,22,24,31,54,68). MLT activates also human monocytes and stimulate IL-1, IL-6 and IL-12 production (24,47,64). However, T lymphocytes seems to be the main target of MLT in mice (31,54) and humans in which physiological concentrations of MLT stimulate IL-2 and gamma-interferon production (24,27). We have reported that the immunoenhancing and anti-stress effect of MLT is neutralized by the opioid antagonist naltrexone (24,27,54,55). Known opioid peptides could mimic the effects of MLT, with the kappa-agonist dynorphin being the most potent agent (53). The hematopoietic protection involved the release of granulocyte/macrophage colony-stimulating factor (GM-CSF) from bone marrow stroma upon stimulation by a Th cells factor induced by MLT (61). This factor was immunologically and biologically indistinguishable from IL-4 (58). Nevertheless, further investigations revealed that this Th cell factor was constituted by 2 cytokines of 15 and 67 kDa MW with the common opioid sequence (Tyr-Gly-Gly-Phe) at their amino terminal and a carboxy-terminal estension which resemble both IL-4 and dynorphin (50). Both activated lymph nodes Th cells and bone marrow Th cells released these opioid cytokines which were named MIO (50). Due to their size and unusual immunological characterization, the MIO might represent novel opioid cytokines. The lower molecular weight MIO (MIO-15) seems to mediate both the anti-stress and hematopoietic effects of MLT (50). Our data may reflect a physiological requirement for sustained MLT regulation of hematopoiesis. Most recently, we performed experiments in which we compared the ability of MLT to protect hematopoiesis in LLC-bearing mice and in tumor-free normal mice treated with the cytotoxic drug cyclophosphamide. This experiment was suggested by the fact that MLT added in GM-CFU cultures could directly enhance the number of GM-CFU but only in presence of suboptimal concentration of colony stimulating factors (CSF), i.e. in presence of activated bone marrow adherent cells (58,61). In addition, LLC is known to produce CSF and exert myelopoietic activity in vivo (77). MLT did not exert any hematopoietic protection in tumor-free mice but rather increased the bone marrow toxicity of cyclophosphamide. However, both in tumor-free and LLC-bearing mice the effect of MLT was neutralized by naltrexone which suggested the involvement of MIO. Most recently, the MIO were found to be acting on a single opioid binding site present in adherent bone marrow cells. Opiod agonists could mimic the colony stimulating activity of MLT in murine adherent bone marrow cells with an order of potency (dynorphin A > ICI 199,441 ≥ U50488H > ICI204,449 ≥ DPDPE ≥ DAMGO) which suggested the presence of a type 1 k-opioid receptor. Consistently, the specific k-opioid receptor antagonist nor-binaltorphimine neutralized the in vitro effect of MLT and, most relevant, inhibited regeneration of hematopoiesis in mice treated with carboplatin. Like the MIO, dynorphin A could exerts a colony stimulating activity on adherent bone marrow cells but not on non-adherent cells and only in presence of GM-CFS. The effect of dynorphin A was abolished by overnight incubation of adherent cells with antisense oligodeoxynucleotide to k-opioid receptor or by addition of anti-IL-1 monoclonal antibody. The adherent bone marrow cells which express kORs were identified by a specific anti-kOR mAb and found to be macrophages (Maestroni, submitted for publication).

5. MLT RECEPTORS

As far as it concerns MLT receptors in immunocompetent cells, high affinity binding sites for MLT have been described in the membrane homogenates of thymus, bursa of Fabricius and spleen of a number of birds and mammals (69). We have described an high affinity binding site in bone marrow Th cells (62). Another study showed that MLT binds to human lymphoid cells modulating their proliferative response. Consistent with our findings, T cells activation significantly increased MLT binding (34). MLT binding sites and mt_1 mRNA was mostly found in humanTh cells, but also in CD8+ T cells and B cells (24,25,27,34). Beside membrane receptors, nuclear receptors for MLT has been described in human myeloid cells. MLT seems to be the natural ligand for nuclear orphan receptors RZR/ROR. It appears that MLT down-regulates the expression of the RZR/ROR responsive gene 5-lipoxygenase, a key enzyme in allergic and inflammatory disease (14).

6. CLINICAL STUDIES

We showed that MLT may synergize with IL-2 in controlling tumor growth (56). On this basis, Dr. Paolo Lissoni and coworkers in Italy have conducted an impressive series of clinical studies in cancer patients. Over 200 advanced solid tumor patients in which the standard anticancer therapies were not tolerated or not effective were treated with IL-2 and MLT. The results obtained show that this neuroimmunotherapeutic strategy may amplify the anti-tumoral activity of low dose IL-2, induces objective tumor regression, prolongs progression-free time and overall survival and, moreover, the treatment was very well tolerated. It should be stressed that MLT seems to be required for the effectiveness of low dose IL-2 in those neoplasias that are generally resistant to IL-2 alone (reviewed in 19). Similar findings were obtained in a smaller study in which MLT was combined with gamma-IFN in metastatic renal cell carcinoma (66). Similarly, MLT in combination with low-dose IL-2 was able to neutralize the surgery-induced lymphocytopenia in cancer patients. In another recent study, MLT has been shown to activate the cytokine system which exerted tumor growth inhibitory effects in cancer patients (65). On the contrary, a most recent double blind study investigating the myeloprotective effect of MLT given in combination with carboplatin and etoposide to lung cancer patients shows that MLT did not influence the chemotherapy-induced hematopoietic toxicity [Ghielmini, submitted for publication].

7. MECHANISM OF ACTION

MLT binds to specific MLT receptors in Th cells and/or monocytes stimulating the production of gamma-IFN, IL-2, MIO, IL-1, IL-6 and IL-12 which in turn upregulate the immune response. Second messengers are not completely understood but include G-protein and inhibition of cAMP (27). The immunotherapeutic effect of MLT against encephalitis viruses or bacteriall infections (8,9,13) might be explained by the increased production of IL-1, IL-12, gamma-IFN and/or IL-2 as well as by an increased myelopoiesis due to the hematopoietic action of the MIO. A mechanism involving Th type 1 cytokines might also account for the capacity of MLT to restore immunodeficiency states secondary to aging (15,68), trauma-hemorrhage (74,75) or to synergize with IL-2 in cancer patients (36–46,49). In this regard, it is noteworthy to recall that MLT is most active on antigen or cytokine activated cells (34,59,60). Consistently, IL-2 treatment in patients results in activation of the whole immune system and creates the most suitable biological background for MLT. The finding that MLT can stimulate IL-12 production from human monocytes only if incubated in presence of IL-2 further support this concept (47). In fact, the plasma concentration of IL-12 was increased in patients who showed a partial response to the MLT/IL-2 combination (48). This finding seems important because IL-12 which is mainly produced by monocyte/macrophages plays a relevant role in cytokine therapy (5). The ability of MLT to counteract the thymus involution and immunodepression caused by stress or corticosteroid treatment seems to be mediated by the MIO. We do not yet know whether the MIO action is exerted on peripheral immunocompetent cells and in the thymus or whether the hematopoietic effects of these novel opioid cytokines are also involved. On the contrary, it seems clear that the hematopoietic effects of MLT depend on a complex series of events which involve the effect of the MIO on kappa-opioid receptors expressed on bone marrow macrophages which, in turn, results in an increased of IL-1 gene expression (Maestroni, unpublished results). This seems of considerable relevance. for understanding the MLT effects and, in general, the physiology of hematopoiesis. MIO might belong to a new family of endogenous kappa-opioid agonists which, in the case of hematopoietic protection, seem to synergize with GM-CSF on stromal cells kappa-receptors (52). This would explain why in LLC-bearing mice, MIO rescue hematopoiesis against the toxic action of cancer chemotherapy. LLC is, in fact, known to release GM-CSF (77). Activated Th cells may produce also GM-CSF. This might account for the therapeutic and positive hematopoietic effects of MLT when administered together with IL-2 in cancer patients (19). The cytokines involved in the immune-hematopoietic action of MLT may exert an influence on the production of MLT by the pineal gland. The pineal gland is, in fact, located outside the blood-brain barrier and some reports show that gamma-IFN may directly affect the synthesis of MLT in the pineal gland (76).

MLT whose synthesis is regulated by the photoperiod, activates its specific membrane receptor (mt_1) in T-helper type 1 cells (Th1) or the nuclear receptor RZR/ROR (NR) in macrophages (M). This stimulates the secretion of cytokines such as gamma-IFN, IL-2, MIO and IL-1, IL-6 respectively. In turn, these cytokines may upregulate immune effectors which may counteract secondary immunodeficiencies and protect against viral and bacterial diseases and cancer. Besides the immune response, MIO may also exert hematopoietic effects. MIO bind to kappa 1 opioid receptors (kOR1) on bone marrow stromal cells (SC). Depending on the affinity state of the kOR1 receptor the effect may be an increased production of CSF which rescue GM-CFU from myelotoxic drugs or, by a still obscure mechanism, an increased toxicity. Gamma-IFN, IL-2, MIO, IL-1, IL-6 and CSF might modulate MLT production in the pineal gland.

8. CONCLUSION AND PERSPECTIVES

MLT may be considered as a potential immunotherapeutic agent and an important endogenous neuroimmunomodulator. An hypothetical pineal gland-immune system physiological network might, therefore, take shape (Figure 1). The proper functioning of such network might be crucial in the adaptative response of the organism to environmental demands and, thus, in the maintenance of health. However, we are still far from a complete understanding of the mechanism underlying the immunological and hematopoietic action of MLT. For example, it is not clear whether MLT acts on Th1 or Th2 cells or on both. In addition, it is not known whether MLT may induce cytokine gene expression or whether its action is posttranslational only. These seem important questions as the Th1/Th2 balance and the resulting cytokines production are crucial for a successful immune response and may be relevant in immune-based pathologies (23). The stimulatory effect of MLT on IL-2 and gamma-IFN and the lack of influence on IL-4 suggests the involvement of Th1 cells. Perhaps, the same Th cell type may also produce MIO which are radically different from the enkephalin-containing molecules reported to be produced by Th2 cells (30). On the other hand, the dramatic protection exerted by MLT in experimental models of viral encephalitis and lethal bacterial infections as well as its capacity to restore depressed immune functions is consonant with a Th1 cells involvement as well as with the action of MLT on monocyte/macrophage cytokines (23). Physiologically, it seems possible to distinguish two different roles. The first one occurs in acute conditions during a viral or bacterial infection which produces a substantial activation of the immune system. In that condition, endogenous and/or exogenous MLT may optimize the immune response by sustaining Th cell and macrophage functions and production of cytokines, part of which (MIO, IL-6, IL-1) may also affect hematopoiesis. A second, more general role may be

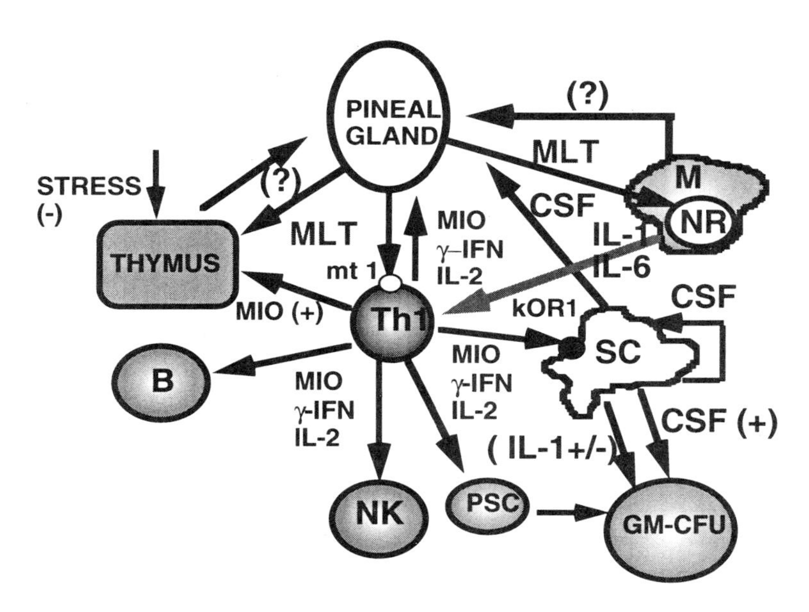

Figure 1. The MLT-immune network. (PSC: pluripotent stem cells; NK: natural killer cells, B: B cells.)

exerted at the hematopoietic-immune level by a chronic circadian resetting of the immunological machinery to maintain the immune homeostasis. This is suggested by the observation that in healthy mice, i.e. in absence of any infection and immunological activation, only the Th cells which sit in the bone marrow are sensitive to MLT (50,51). Products of this MLT-bone marrow Th cells interaction are the MIO which may affect hematopoiesis and thymocytes proliferation (50,51). Both the acute and chronic mechanisms might be exploited in the use of MLT as immunotherapeutic agent to correct secondary immunodeficiency or fight viral diseases. As we already stated in preceding reviews (56,57), we would like to stress the need for a large double blind study in human immunodeficiency virus (HIV)-positive patients. In presence of normal Th cell counts, the apparent ability of MLT to sustain Th cell functions and IL-2 and gamma-IFN production might result in delayed development or occurrence of AIDS. Reduction of plasma viremia was associated with an increased IL-2 mRNA expression in lymph nodes of HIV-infected patients (71). IL-2 is the most potent cytokine capable of inducing the CD8+ T cell -mediated inhibition of HIV replication which seems to override the ability of IL-2 to stimulate HIV expression (33). If effective, MLT administration would be a relatively cheap and safe prevention of this devastating disease. Alternatively, MLT might be associated with low-dose IL-2 which seems to be beneficial in HIV-associated malignancies (10) or with HIV protease inhibitors. The use of MLT in combination with IL-2 in cancer neuroimmunotherapy might prolong survival and improve the patients' quality of life (19). These encouraging results obtained by Lissoni and coworkers deserve, therefore, to be expanded and challenged in other studies. In addition, MLT might be useful in enhancing the immune response against tumor antigens, a promising therapeutic strategy for cancer vaccines. In regard to the use of MLT in combination with cancer chemotherapeutic drugs the results obtained so far are disappointing. MLT administered alone seems to worsen the bone marrow toxicity of common cancer chemotherapeutic regimens or, at best, be ineffective. This fact calls for further studies to understand the role of MLT in hematopoiesis and indicates that, in certain conditions, MLT may have adverse effects.

ACKNOWLEDGMENTS

The work performed in Locarno has been supported by Swiss Nationalfonds grants no. 3.267.0.85; 31.25350.88; 31.36128.92, 31.45532.95 and by the Helmut Horten Foundation.

REFERENCES

1. Ader, R., Cohen, N., and Felten, D.: Psychoneuroimmunology: interactions between the nervous system and the immune system. Lancet 345 (8942):99–103, 1995.
2. Ader, R., Felten, D.L., and Cohen, N.: Psychoneuroimmunology II. San Diego, Academic Press, 1991.
3. Angeli, A., Gatti, G., Sartori, M.L., Del Ponte, D., and Cerignola, R.: Effect of exogenous melatonin on human natural killer (NK) cell activity. An approach to the immunomodulatory role of the pineal gland, in Gupta, D., Attanasio, A., and Reiter, R.J. (eds): The Pineal Gland and Cancer. Tübingen, Müller and Bass, 1988, pp. 145–157.
4. Anwar, M.M., Mahfouz, H.A., and Sayed, A.S.: Potential protective effects of melatonin on bone marrow of rats exposed to cytotoxic drugs. Comp. Biochem. Physiol. 119:493–591, 1998.
5. Banks, R.E., Patel, P.M., and Selby, P.J.: Interleukin-12: a new player in cytokine therapy. Br. J. Cancer 71:655–659, 1995.
6. Barath, P. and Csaba, G.: Histological changes in the lung, thymus and adrenal one and half year after pinealectomy. Acta Biol. Acad. Sci. Hung. 25:123–125, 1974.
7. Becker, J., Veit, G., Handgretinger, R., Attanasio, G., Bruchett, G., Trenner, I., Niethammer, D., and Gupta, D.: Circadian variations in the immunomodulatory role of the pineal gland. Neuroendocrinology Letters 10:65–80, 1988.
8. Ben-Nathan, D., Maestroni, G.J., Lustig, S., and Conti, A.: Protective effects of melatonin in mice infected with encephalitis viruses. Arch. Virol. 140:223–230, 1995.
9. Ben-Nathan, D., Maestroni, G.J.M., and Conti, A.: Protective effect of melatonin in viral and bacterial infections, in Maestroni, G.J.M., Conti, A., and Reiter, J.R. (eds): Frontiers in Hormone Research, vol. 23. Basel, Karger, 1997, pp. 72–81.
10. Bernstein, Z.P., Porter, M.M., Gould, M., Lipman, B., Bluman, E.M., Stewart, C.C., Hewitt, R.G., Fyfe, G., Poiesz, B., and Caligiuri, M.A.: Prolonged administration of low-dose interleukin-2 in human immunodeficiency virus-associated malignancy results in selective expansion on innate immune effectors without significant clinical toxicity. Blood 86:3287–3294, 1995.
11. Bindoni, M. and Cambria, A.: Effects of pinealectomy on the in vivo and in vitro biosynthesis of nucleic acids on the mitotic rate in some organs of the rat. Arch. Sci. Biol. 52:271–283, 1968.
12. Blask, D.E.: Melatonin in oncology, in Yu, H.-S. and Reiter, R.J. (eds): Melatonin. Biosynthesis, Physiological Effects, and Clinical Applications. Boca Raton, CRC Press, 1993, pp. 447–477.
13. Bonilla, E., Valerofuenmajor, N., Pons, H., and Chacin-Bonilla, L.: Melatonin protects mice infected with Venezuelan equine encephalomyelitis virus. Cell Mol. Life Sci. 53:430–434, 1997.
14. Carlberg, C. and Wiesenberg, I.: The orphan receptor family RZR/ROR, melatonin and 5-lipoxygenase: an unexpected relationship. J. Pineal Res. 18:171–178, 1995.
15. Caroleo, M.C., Frasca, D., Nistico, G., and Doria, G.: Melatonin as immunomodulator in immunodeficient mice. Immunopharmacology 23:81–89, 1992.
16. Champney, T.H. and McMurray, D.N.: Spleen morphology and lymphoproliferative activity in short photoperiod exposed hamsters, in Fraschini, F. and Reiter, R.J. (eds): Role of melatonin and pineal peptides in neuroimmunomodulation. London, Plenum Publ Co, 1991, pp. 219–225.
17. Colombo, L.L., Chen, G.-J., Lopez, M.C., and Watson, R.R.: Melatonin induced increase in gamma-interferon production by murine splenocytes. Immunol. Lett. 33:123–126, 1992.
18. Conti, A., Haran-Ghera, N., and Maestroni, G.J.M.: Role of pineal melatonin and melatonin-induced-immuno-opioids in murine leukemogenesis. Med. Oncol. Tumor Pharmacoth. 9:87–92, 1992.

19. Conti, A. and Maestroni, G.J.M.: The clinical neuroimmunotherapeutic role of melatonin in oncology. J. Pineal Res. 19:103–110, 1995.
20. Csaba, G. and Barath, P.: Morphological changes of thymus and the thyroid gland after postnatal extirpation of pineal body. Endocrinol. exp. 9:59–67, 1975.
21. Csaba, G., Bodocky, M., Fischer, J., and Acs, T.: The effect of pinealectomy and thymectomy on the immune capacity of the rat. Experientia 22:168–175, 1965.
22. Del Gobbo, V., Libri, V., Villani, N., Caliò, R., and Nstico, G.: Pinealectomy inhibits interleukin-2 production and natural killer activity in mice. Int. J. Immunopharmacol. 11:567–577, 1989.
23. Del Prete, G., Maggi, E., and Romagnani, S.: Human Th1 and Th2 cells: Functional properties, mechanisms of regulation and role in disease. Lab. Invest. 70:299–307, 1994.
24. Garcia-Mauriño, S., Gonzales-Haba, M.G., Calvo, J.R., Rafii-El-Idrissi, M., Sanchez-Margalet, V., Goberna, R., and Guerrero, J.M.: Melatonin enhances IL-2, Il-6, and IFN-γ production by human circulating CD4+ cells. J. Immunol. 159:574–581, 1997.
25. Garcia-Perganeda, A., Pozo, D., Guerrero, J.M., and Calvo, J.R.: Signal transduction for melatonin in human lymphocytes. Involvement of a pertussis toxin-sensitive G protein. J.Immunol. 159:3774–3781, 1997.
26. Giordano, M. and Palermo, M.S.: Melatonin-induced enhancement of antibody dependent cellular cytotoxicity. J. Pineal Res. 10:117–121, 1991.
27. Guerrero, J.-M., Garcia-Maurino, S., Gil-Haba, M., Rafii-El-Idrissi, M., Pozo, D., Garcia-Perganeda, A., and Calvo, J.R.: Mechanisms of action on the human immune system: Membrane receptors versus nuclear receptors., in Maestroni, G.J.M., Conti, A., and Reiter, R.J. (eds): Therapeutic potential of the Pineal Hormone Melatonin, vol. 23, Frontiers of Hormone Research. Basel, Karger, 1996, pp. 43–52.
28. Haldar, C., Haussler, D., and Gupta, D.: Response of CFU-GM (colony forming units for granulocytes and macrophages) from intact and pinealectomized rat bone marrow to macrophage colony stimulating factor (rGM-CSF) and human recombinant erythropoietin (rEPO). Prog. Brain Res. 91:323–325, 1992.
29. Hansson, I., Holmdahl, R., and Mattsson, R.: The pineal hormone melatonin exaggerates development of collagen-induced arthritis in mice. Journal of Neuroimmunology 39:23–31, 1992.
30. Hiddinga, H.J., Isaak, D.D., and Lewis, R.V.: Enkephalin-containing peptides processed from proenkephalin significantly enhance the antibody-forming cell responses to antigens. J. Immunol. 152:3748–3758, 1994.
31. Hofbauer, L.C. and Heufelder, A.E.: Endocrinology meets immunology: T lymphocytes as novel targets for melatonin. Eur. J. Endocrinol. 134:424–425, 1996.
32. Jankovic, B.D., Isakovic, S., and Petrovic, S.: Effect of pinealectomy on immune reactions in the rat. Immunology 8:1–6, 1970.
33. Kinter, A.L., Bende, S.M., Hardy, E.C., Jackson, R., and Fauci, A.S.: Interleukin 2 induces CD8+ T cell-mediated suppression of human immunodeficiency virus replication in CD4+ T cells and this effect overrides its ability to stimulate virus expression. Proc. Natl. Acad. Sci. USA 92:19985–10989, 1995.
34. Konakchieva, R., Kyurkchiev, S., Kehayov, I., Taushanova, P., and Kanchev, L.: Selective effect of methoxyndoles on the lymphocyte proliferation and melatonin binding to activated human lymphoid cells. J. Neuroimmunol. 63:125–132, 1995.
35. Lewinsky, A., Zelazowsky, P., Sewerynek, E., Zerek-Melen, G., and Szudlinsky, M.: Melatonin-induced immunosuppression of humal lymphocyte natural killer activity in vitro. J. Pineal Res. 7:153–164, 1989.
36. Lissoni, P., Ardizzoia, A., Barni, S., Brivio, F., Tisi, E., Rovelli, F., Tancini, G., and Maestroni, G.J.M.: Efficacy and tolerability of cancer neuroimmunotherapy with subcutaneous low-dose interleukin-2 and the pineal hormone melatonin: A progress report of 200 patients with advanced solid neoplasms. Oncology Reports 2:1063–1068, 1995.
37. Lissoni, P., Ardizzoia, A., Tisi, E., Rossini, F., Barni, S., Tancini, G., Conti, A., and Maestroni, G.J.M.: Amplification of eosinophilia by melatonin during the immunotherapy of cancer with interleukin-2. J. Biol. Reg. Homeos. Agents 7:34–36, 1993.
38. Lissoni, P., Ardizzoia, A., Tisi, E., Rossini, F., Barni, S., Tancini, G., Conti, A., and Maestroni, G.J.M.: Amplification of eosinophilia by melatonin during the immunotherapy of cancer with interleukin-2. J. Biol. Reg. Homeos. Agents 7:34–36, 1993.
39. Lissoni, P., Barni, S., Ardizzoia, A., Brivio, F., Tancini, G., Conti, A., and G.J.M., M.: Immunological effects of a single evening subcutaneous injection of low-dose interleukin-2 in association with the pineal hormone melatonin in advanced cancer patients. J. Biol. Reg. Homeos. Agents 6:132–136, 1992.

40. Lissoni, P., Barni, S., Rovelli, F., Brivio, F., Ardizzoia, A., Tancini, G., Conti, A., and Maestroni, G.J.M.: Neuroimmunotherapy of advanced solid neoplasms with single evening subcutaneous injection of low-dose interleukin-2 and melatonin: preliminary results. Eur. J. Cancer 29A:185–189, 1993.
41. Lissoni, P., Barni, S., Tancini, G., Ardizzoia, A., Ricci, G., Aldeghi, R., Brivio, F., Tisi, E., Rescaldani, R., Quadro, G., and Maestroni, G.J.M.: A randomised study with subcutaneous low-dose interleukin 2 alone vs interleukin 2 plus the pineal neurohormone melatonin in advanced solid neoplasms other than renal cancer and melanoma. Br. J. Cancer 69:196–199, 1994.
42. Lissoni, P., Barni, S., Tancini, G., Ardizzoia, A., Rovelli, F., Cazzaniga, M., Brivio, F., Piperno, A., Aldeghi, R., Fossati, D., Characjeius, D., Kothari, L., Conti, A., and Maestroni, G.J.M.: Immunotherapy with subcutaneous low-dose interleukin-2 and the pineal indole melatonin as a new effective therapy in advanced cancers of the digestive tract. Br. J. Cancer 67:1404–1407, 1993.
43. Lissoni, P., Barni, S., Tancini, G., Ardizzoia, A., Rovelli, F., Cazzaniga, M., Brivio, F., Piperno, A., Aldeghi, R., Fossati, D., Characjeius, D., Kothari, L., Conti, A., and Maestroni, G.J.M.: Immunotherapy with subcutaneous low-dose interleukin-2 and the pineal indole melatonin as a new effective therapy in advanced cancers of the digestive tract. Br. J. Cancer 67:1404–1407, 1993.
44. Lissoni, P., Barni, S., Tancini, G., Rovelli, F., Ardizzoia, A., Conti, A., and Maestroni, G.J.M.: A study of the mechanisms involved in the immunostimulatory action of the pineal hormone in cancer patients. Oncology 50:399–402, 1993.
45. Lissoni, P., Barni, S., Tancini, G., Rovelli, F., Ardizzoia, A., Conti, A., and Maestroni, G.J.M.: A study of the mechanisms involved in the immunostimulatory action of the pineal hormone in cancer patients. Oncology 50:399–402, 1993.
46. Lissoni, P., Brivio, F., Brivio, O., Fumagalli, L., Gramazio, F., Rossi, M., Emanuelli, G., Alderi, G., and Lavorato, F.: Immune effects of preoperative immunotherapy with high-dose subcutaneous interleukin-2 versus neuroimmunotherapy with low-dose interleukin-2 plus the neurohormone melatonin in gastrointestinal tract tumor patients. J. Biol. Regul. Homeost. Agents 9:31–33, 1995.
47. Lissoni, P., Pittalis, S., Barni, S., Rovelli, F., Fumagalli, L., and Maestroni, G.J.-M.: Regulation of interleukin-2/interleukin-12 interactions by the pineal gland. Int. J. Thymology 5:443–447, 1997.
48. Lissoni, P., Rovelli, F., Brivio, F., Pittalis, S., Maestroni, G.J.M., and Fumagalli, L.: In vivo study of the modulatory effect of the pineal hormone melatonin on interleukin-12 secretion in response to interleukin-2. Int. J. Thymology 5:482–486, 1997.
49. Lissoni, P., Tisi, E., Barni, S., Ardizzoia, A., Rovelli, F., Rescaldani, R., Ballabio, D., Benenti, C., Angeli, M., Tancini, G., Conti, A., and Maestroni, G.J.M.: Biological and clinical results of a neuroimmunotherapy with interleukin-2 and the pineal hormone melatonin as a first line treatment in advanced non-small cell lung cancer. Br. J. Cancer 66:155–158, 1992.
50. Maestroni, G., Hertens, E., Galli, P., Conti, A., and Pedrinis, E.: Melatonin-induced T-helper cell hematopoietic cytokines resembling both interleukin-4 and dynorphin. J. Pineal Res. 21:131–139, 1996.
51. Maestroni, G.J.M.: The immunoneuroendocrine role of melatonin. J. Pineal Res. 14:1–10, 1993.
52. Maestroni, G.J.M.: Kappa-opioid receptors in marrow stroma mediate the hematopoietic effects of melatonin-induced opioid cytokines. Neuroimmunomodulation: Molecular Aspects, Integrative Systems, and Clinical Advances, 1998, pp. 411–420.
53. Maestroni, G.J.M. and Conti, A.: Beta-endorphin and dynorphin mimic the circadian immunoenhancing and anti-stress effects of melatonin. Int. J. Immunopharmacol. 11:333–340, 1989.
54. Maestroni, G.J.M. and Conti, A.: The pineal neurohormone melatonin stimulates activated CD4+, Thy-1+ cells to release opioid agonist(s) with immunoenhancing and anti-stress properties. J. Neuroimmunol. 28:167–176, 1990.
55. Maestroni, G.J.M. and Conti, A.: Anti-stress role of the melatonin-immuno-opioid network. Evidence for a physiological mechanism involving T cell-derived, immunoreactive β-endorphin and met.enkephalin binding to thymic opioid receptors. Int. J. Neurosci. 61:289–298, 1991.
56. Maestroni, G.J.M. and Conti, A.: Melatonin in relation to the immune system, in Yu, H.-S. and Reiter, R.J. (eds): Melatonin. Biosynthesis, Physiological effects and Clinical Applications. Boca Raton, CRC Press, 1993, pp. 290–306.
57. Maestroni, G.J.M. and Conti, A.: Melatonin and the immune-hematopoietic system. Therapeutic and adverse pharmacological correlates. Neuroimmunomodulation 3:325–332, 1996.
58. Maestroni, G.J.M., Conti, A., and Lissoni, P.: Colony-stimulating activity and hematopoietic rescue from cancer chemothereapy compounds are induced by melatonin via endogenous interleukin 4. Cancer Res. 54:4740–4743, 1994.
59. Maestroni, G.J.M., Conti, A., and Pierpaoli, W.: Role of the pineal gland in immunity. Circadian

synthesis and release of melatonin modulates the antibody response and antagonize the imunosuppressive effect of corticosterone. J. Neuroimmunol. 13:19–30, 1986.
60. Maestroni, G.J.M., Conti, A., and Pierpaoli, W.: Role of the pineal gland in immunity: II. Melatonin enhances the antibody response via an opiatergic mechanism. Clin. Exp. Immunol. 68:384–391, 1987.
61. Maestroni, G.J.M., Covacci, V., and Conti, A.: Hematopoietic rescue via T-cell-dependent, endogenous GM-CSF by the pineal neurohormone melatonin in tumor bearing mice. Cancer Res. 54:2429–2432, 1994.
62. Maestroni, G.J.M., Flamigni, L., Hertens, E., and Conti, A.: Biochemical and functional characterization of melatonin-induced opioids in spleen and bone marrow T-helper cells. Neuroendocrinol. Lett. 17:145–153, 1995.
63. Maestroni, G.J.M. and Pierpaoli, W.: Pharmacological control of the hormonally modulated immune response, in Ader, R. (ed): Psychoneuroimmunology, vol. I. New York, Academic Press, 1981, pp. 405–425.
64. Morrey, M.K., McLachlan, J.A., Serkin, C.D., and Bakouche, O.: Activation of human monocytes by the pineal neurohormone melatonin. J. Immunol. 153:2671–2680, 1994.
65. Neri, B., DeLeonardis, V., Gemelli, M.T., DiLoro, M., Mottola, A., Ponchietti, R., Raugei, A., and Cini, G.: Melatonin as biological response modifier in cancer patients. Anticancer Res. 18:1329–1332, 1998.
66. Neri, B., Fiorelli, C., Moroni, F., Nicita, G., Paoletti, M.C., Ponchietti, R., Raugei, A., Santoni, G., Trippitelli, A., and Grechi, G.: Modulation of human lymphoblastoid interferon activity by melatonin in metastatic renal cell carcinoma. A phase II study. Cancer 73:3015–3019, 1994.
67. Palermo, M.S., Vermeulen, M., and Giordano, M.: Modulation of antibody-dependent cellular cytotoxicity by pineal gland, in Maestroni, G.J.M., Conti, A., and Reiter, R.J. (eds): Advances in Pineal Research, vol. 7. London, John Libbey, 1994, pp. 143–147.
68. Pioli, C., Caroleo, M.C., Nistico, G., and Doria, G.: Melatonin increases antigen presentation and amplifies specific and non specific signals for T-cell proliferation. Int. J. Immunopharmacol. 15:463–469, 1993.
69. Poon, A.M., Liu, Z.M., Pang, C.S., Brown, G.M., and Pang, S.F.: Evidence for a direct action of melatonin on the immune system. Biol. Signals 3:107–117, 1994.
70. Rella, W. and Lapin, V.: Immunocompetence of pinealectomized and simultaneously pinealectomized and thymectomized rats. Oncology 33:3–6, 1978.
71. Sei, S., Akiyoshi, H., Bernard, J., Venzon, D.J., Fox, C.H., Schwartzentruber, D.J., Anderson, B.D., Kopp, J.B., Mueller, B.U., and Pizzo, P.A.: Dynamics of virus versus host intercation in children with human immunodeficiency virus type 1 infection. J. Infect. Dis. 173:1485–1490, 1996.
72. Spector, N.H., Dolina, S., Cornelissen, G., Halberg, F., Markovic, B.M., and Jankovic, B.D.: Neuroimmunomodulation: neuroimmune interactions with the environment, in Blatteis, C.M. (ed): Handbook of Physiology: The Environment. Bethesda, Amer. Physiol. Society, 1996, pp. 1437–1549.
73. Vaughan, M.K. and Reiter, R.J.: Transient hypertrophy of the ventral prostate and coagulating glands and accelerated thymic involution following pinealectomy in the mouse. Texas Rep. Biol. Med. 29:579–586, 1971.
74. Wichmann, M., Zellweger, R., DeMaso, A., Ayala, A., and Chaudry, I.: Melatonin administration attenuates depressed immune functions trauma-hemorrhage. J. Surg. Res. 63:256–262, 1996.
75. Wichmann, M.W., haisken, J.M., Ayala, A., and Chaudry, I.H.: Melatonin administration following hemorrhagic shock decreases mortality from subsequent septic challenge. J. Surg. Res. 65:109–114, 1996.
76. Withyachumnarnkul, B., Nonaka, K.O., Santana, C., Attia, A.M., and Reiter, R.J.: Interferon-gamma modulates melatonin production in rat pineal gland in organ culture. J. Interferon Res. 10:403–411, 1990.
77. Young, M.R.I., Young, M.E., and Kim, K.: Regulation of tumor-induced myelopoiesis and the associated immune suppressor cells in mice bearing metastatic Lewis lung carcinoma by prostaglandin E2. Cancer Res. 48:6826–6831, 1988.
78. Yu, H.-S. and Reiter, R.J.: Melatonin. Biosynthesis, Physiological Effects, and Clinical Applications. Boca Raton, CRC Press, 1993.

THE USE OF MELATONIN AND CO-TREATMENT WITH AUTOLOGOUS OR ALLOGENEIC CELLS AS A MODEL FOR CONTROL OF MALIGNANT β-CELL LEUKEMIA

I. Nir,[1] L. Weiss,[2] and S. Slavin[2]

[1] Hebrew University
Faculty of Medicine
[2] Hadassah University Hospital
Jerusalem, Israel

The purpose of the present study was to evaluate the possible effects of pharmacological doses of melatonin upon the survival of leukemic mice using a murine model of originally spontaneous, subsequently transplantable and non-immunogenic β-cell leukemia (BCL_1). (BALB/C × C57BL/6) F_1 mice inoculated with BCL_1 cells were conditioned by total irradiation (750cGy) and reconstituted with either bone marrow (BM) cells, a syngeneic transplantation, as a model of autologous transplantation, or BMC 57BL/6(C57) allogeneic transplantation (1). The leukemia produced is believed to exert damaging effects on cellular DNA, causing disruption of chemical bonds in the molecular structure of DNA, the destruction by the active free radicals interacting with cellular DNA. Melatonin, the compound tested, is an antioxidant believed to detoxify the devastating OH radicals, thus providing protection against the destructive biological effect and genetic damage of malignancy.

Our aim was to establish whether protection is provided by melatonin against the destructive effect of leukemia, and if so, whether it is due to its antioxidative capacity. Four groups of 12 mice each (2–3 months old, purchased from Harlem, Jerusalem) were irradiated according to the outlined procedure and on the following day injected with BCL_1 cells and reconstituted with either BMC57 or BMF_1 cells. Melatonin was given by s.c. injection 10 mg/kg for 4 consecutive days and the animals left for observation until the development of leukemia or death. Leukemia and consequently death developed twice as quickly in the mice reconstituted with syngeneic cells [Figure 1

Melatonin after Four Decades, edited by James Olcese.
Kluwer Academic / Plenum Publishers, New York, 2000.

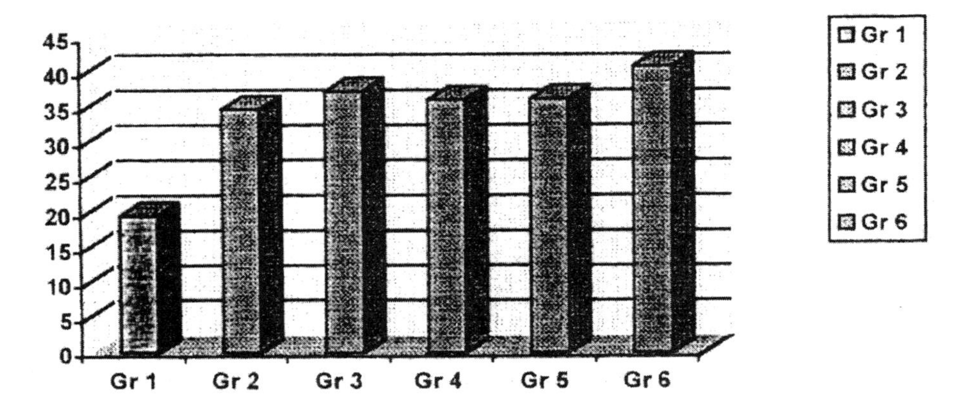

Figure 1. Development of leukemia (mean number of days) in mice receiving syngeneic cells (Gr1) syngeneic cells and melatonin (Gr2), allogeneic cells (Gr3), or allogeneic cells and melatonin (Gr4). Groups 5 and 6 equal Groups 3 and 4 respectively with some variation in the mode of application of melatonin. Groups 2–6 are statistically different from group 1 ($p < 0.05$).

(Gr1)] compared with those reconstituted with allogeneic cells (Gr. 3). The simultaneous application of melatonin, however, could reduce the destruction in animals given the syngeneic cells to the same level as those of the mice reconstituted with allogeneic cells [Figure 1 (Gr2)]. No difference was found in the pace of development of leukemia in the mice with allogeneic cells, irrespective of whether or not the animals received melatonin [Figure 1 (Gr3 & 4)].

This finding indicates the absence of a leukemia defensive factor in the syngeneic cells such as is found in allogeneic cells. Melatonin administration to the syngeneic cell recipients raised the defensive function to that recorded in the mice reconstituted with allogeneic cells. The question arising, furthermore, was whether the additional protection against leukemic malignancy provided by melatonin in the presence of syngeneic cells is due to an antioxidative effect against free radicals. This possibility was tested by determining the antioxidative protective capacity and actual erythrocyte sensitivity of the various groups of experimental animals. The erythrocyte sensitivity and oxidative stress was determined with 2′2 azobis-amidinopropane (AAPH), a radical inhibitor which brings about membrane alteration and hemolysis (2). The toxicity related to oxygen free radicals appeared not to differ in the plasma and erythrocytes of the animals under the various experimental conditions, indicating a similar deficit of protection against radicals in the various instances (Figure 2). *In vitro* melatonin significantly inhibited hemolysis of red blood corpuscles, thus indicating a protective capacity against oxygen free radicals. Furthermore, an additional screening test to establish the anti-radical elements present in the plasma of the variously treated mice was undertaken using a cyclic voltametry procedure to determine low molecular weight antioxidative capacity (3). The major contributors of the antioxidative capacity of biological fluids and tissues were established to be ascorbic-dehydro ascorbic acid and uric acid, acting as direct chemical scavengers. Here too, the values determined appear to be similar in the animals under the various experimental conditions.

It may be concluded, based on the results recorded in our leukemic model of mice, that melatonin could, under certain experimental conditions, bring about a restraint

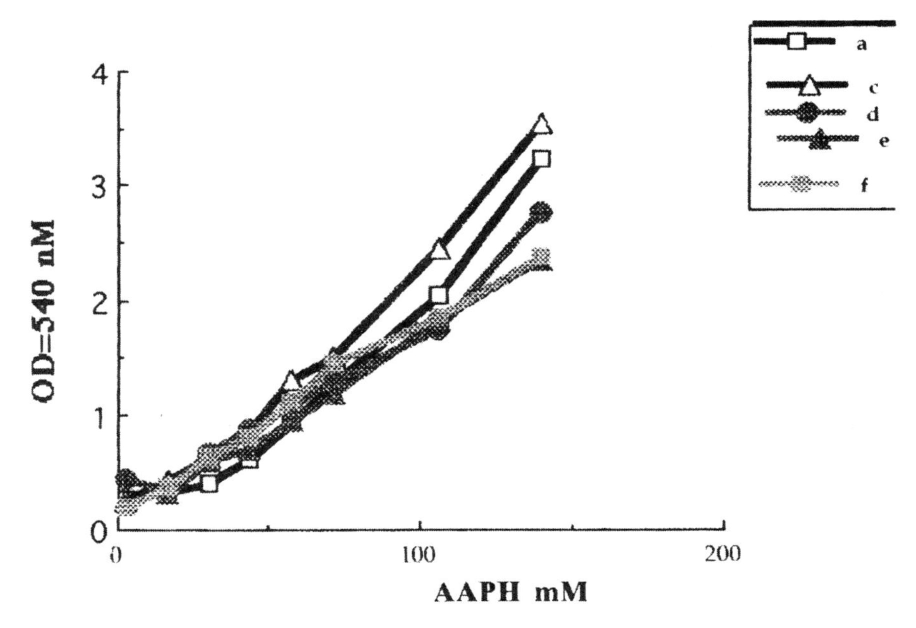

Figure 2. The antioxidative capacity of the plasma obtained from the six variously treated groups of mice (a,b,c,d,e,f) as indicated from the degree of hemolysis determined in vitro (inhibitory 50% plasma volume). There was no statistical difference between any of the groups determined. (a,b,c,d,e,f correspond to the groups 1,2,3,4,5,6 described in Figure 1).

in oncogenicity, this effect not being dependent on its antioxidative status. Moreover, the positive effect of melatonin cotreatment with syngeneic cells established in BCL_1 leukemic sygeneic mice, if found applicable for human bone marrow transplantation, would be of prime importance in clinical practice, allowing use of cells taken from the patient himself instead of the accepted use of cells from donors.

REFERENCES

1. Weiss, L., Reich, S., and Slavin, S. Use of recombinant human interleukin 2 in conjunction with bone marrow transplantation as a model for control of minimal residual disease in malignant hematological disorders: I. Treatment of murine leukemia in conjunction with allogeneic bone marrow transplantation and IL-2 activated cell-mediated immunotherapy. *Cancer Investigation*, **10**, 19–26 (1992).
2. Abella, A., Messaoudi, C., Laurent, D., Marot, D., Chalas, J., Breux, J., Claise, C., and Lindenbaum, A. A method for simulataneous determination of plasma and erythrocyte antioxidant status. Evaluation of the antioxidant activity of vitamin E in healthy volunteers. *Br J Clin. Pharmacol*, **42**, 737–741 (1996).
3. Chevion, S., Berry, E.M., Ktrowssky, N., and Kohen, R. Evaluation of plasma low molecular weight antioxidants by voltametry. *Free Rad. Biol. Med.*, **22**, 411–471 (1997).

49

AUTORADIOGRAPHIC DETECTION OF 2-(^{125}I)-IODOMELATONIN BINDING SITES IN IMMUNE TISSUE OF RATS

R. Konakchieva,[1,2] S. Manchev,[2] P. Pevét,[1] and M. Masson-Pevét[1]

[1] Laboratoire de Neurobiologie des Fonctions Rythmiques et Saisonnières
CNRS UMR 7518
Université Louis Pasteur
Strasbourg, France
[2] Department of Immunoneuroendocrinology
IBIR, Bulgarian Academy of Sciences
Sofia, Bulgaria

1. INTRODUCTION

The principal secretory product of the mammalian pineal gland—the hormone melatonin (MLT) has been implicated in the circadian regulation of a variety of physiological processes (1,2,3) as well as in neuroendocrine-immune circuits (4). Findings of immunomodulatory activity of MLT *in vivo* and *in vitro* have been substantiated by reports of membrane specific binding sites for the hormone in immune tissue using 2-(^{125}I)-iodomelatonin (4,5,6,7). However, there has not been yet unambiguous indication of the cell type/s which express the binding. The reported characteristics of the putative receptors in lymphoid cell preparations vary in a broad range, with an apparent affinity for the 2-(^{125}I)-iodomelatonin ligand beween 0.1 to 2000 nM. Peculiar about melatonin action in the immune system seems to be its inability to influence directly in physiological concentrations humoral and cellular immunity under resting conditions (6,8,9), but to counteract immunodepressive states (4).

In previous studies we have shown that circadian timed application of MLT in rats can alleviate several symptoms of dysregulation of the limbic-hypothalamic-pituitary-adrenal (LHPA) axis, resulting from transient exposure to excessive glucocorticoid plasma levels, which parallel age- and chronic stress-induced changes (10). These results illustrate a novel aspect of melatonin's potential to influence neuroen-

Melatonin after Four Decades, edited by James Olcese.
Kluwer Academic / Plenum Publishers, New York, 2000.

docrine adaptive processes in a protective fashion and suggest that the hormone might counteract stress-induced immunosuppression by affecting glucocorticoid-mediated control on immune function.

In the present study we aimed to examine the ability of MLT, administered *in vivo* to influence glucocorticoid-provoked changes of corticosteroid receptor characteristics in primary (thymus) and secondary (spleen) lymphoid organs of rats. Using 2-(^{125}-I)-iodomelatonin autoradiography we have investigated the distribution pattern of melatonin receptor binding in both tissues.

2. MATERIALS AND METHODS

2.1. Experimental Conditions

In all experiments male Wistar rats weighing 180–200 g were used. The animals were given subcutaneous evening injections of MLT (80 μg/kg b.w.) or vehicle for 10 days. Three days after start of the treatment some of the animals were administered a daily dose of 500 μg/kg b.w. dexamethasone (Dex) which continued for 5 days. Two days after cessation of the treatments spleen tissue was collected on ice, splenocytes were isolated in a primary culture and subjected to (^{3}H)-Thymidine incorporation assay upon stimulation with Concanavalin A as described elsewhere (6). Glucocorticoid receptor (GR) binding characteristics were determined in cytosol fractions prepared from spleen and thymus using (^{3}H)-dexamethasone (NEN, Austria) as radioligand and an established protocol (11).

2.2. Autoradiography

Male Wistar rats were entrained to a 12L/12D regime for two weeks. Serial 20 μm thick cross-sections of spleen and thymus were cut on a cryostat, thaw-mounted onto gelatin coated slides and kept at –80°C untill processed for 2-(^{125}I)-iodomelatonin autoradiography. The radioligand was synthesized by the method of Vakkuri et al. (1984) and purified by HPLC yelding a specific activity of 1600–2300 Ci/mmol. The autoradiographic procedure was as described elsewhere (12). Non-specific binding was determined in presence of 1 μM excess of melatonin (Sigma, USA).

3. RESULTS

Transient exposure of rats to high doses of dexamethasone (500 μg/kg b.w.) causes down-regulation of glucocorticoid receptors and a decrease of the binding capacity (Bmax) in the cytosol from spleen to undetectable levels (Table 1). Concomitant treatment with moderate doses of melatonin restores the receptor binding characteristics to values which are not significantly different from those obtained in intact animals. In the thymus the effect of melatonin administration (MLT/DEX) on glucocorticoid receptor characteristics is not statistically significant as compared to the effect of dexamethasone (placebo/DEX) (Table 2). In both groups the glucocorticoid receptor

Table 1. (^{3}H)-dexamethasone binding in rat spleen cytosol fractions following melatonin (MLT) and/or dexamethasone (DEX) treatment *in vivo*

Group	Kd (nM)	Bmax (fmol/mg protein)
Placebo/DEX	— undetectable —	
MLT/DEX	4.4 ± 2.5	124.7 ± 35.4
Naive	3.0 ± 0.4	193.1 ± 44.3

Table 2. (^{3}H)-dexamethasone binding in rat thymic cytosol fractions following melatonin (MLT) and/or dexamethasone (DEX) treatment *in vivo*

Group	Kd (nM)	Bmax (fmol/mg protein)
Placebo/DEX	1.8 ± 0.5	32.3 ± 7.2*
MLT/DEX	3.6 ± 0.9	56.3 ± 24.6*
Naive	2.5 ± 1.6	110.0 ± 14.1

Glucocorticoid receptor assay was performed with hot saturation up to 20 nM (^{3}H)—dexamethasone in presence or abscence of 100-fold excess of cold ligand. Data represent mean ± SEM from duplicates in three independent determinations. * -significant difference to controls at $P < 0.05$.

binding capacity in the cytosol fractions remains statistically lower in comparison to the receptor values in the intact animals.

The autoradiography allowed localization of the expression of the specifc 2-(^{125}I)-iodomelatonin binding in discrete areas of lymphocyte subpopulations in the rat spleen and thymus (Figure 1). The localization was confirmed by staining of the same sections in Cresyl violet (Figure 1, B). In the spleen the signal appeared in the area surrounding the germinal centres, comprising of T- and B-lymphocytes and was completely displaced by the excess of non-labelled ligand (data not shown). In the thymus the highest density of 2-(^{125}-I)-iodomelatonin binding was obtained in the cortical zones of relatively immature proliferating lymphocytes.

4. CONCLUSIONS

Data presented here define morphologically the localization of 2-(^{125}I)-iodomelatonin binding in the rat spleen and thymus. They demonstrate that the melatonin receptors are confined to subpopulations of lymphocytes at a different stage of maturity in the two structures studied. This finding suggests a selective mode of action of MLT in primary and secondary immune tissue and most probably reflects the difference in the observed changes on the glucocorticoid receptor characteristics in these organs. The demonstration of a glucocorticosteroid-dependent modulation of the glucocorticoid receptor binding capacity in the spleen, in reach of mature immunocompetent cells, suggest that the pineal hormone might interact with corticosteroids to regulate immune function.

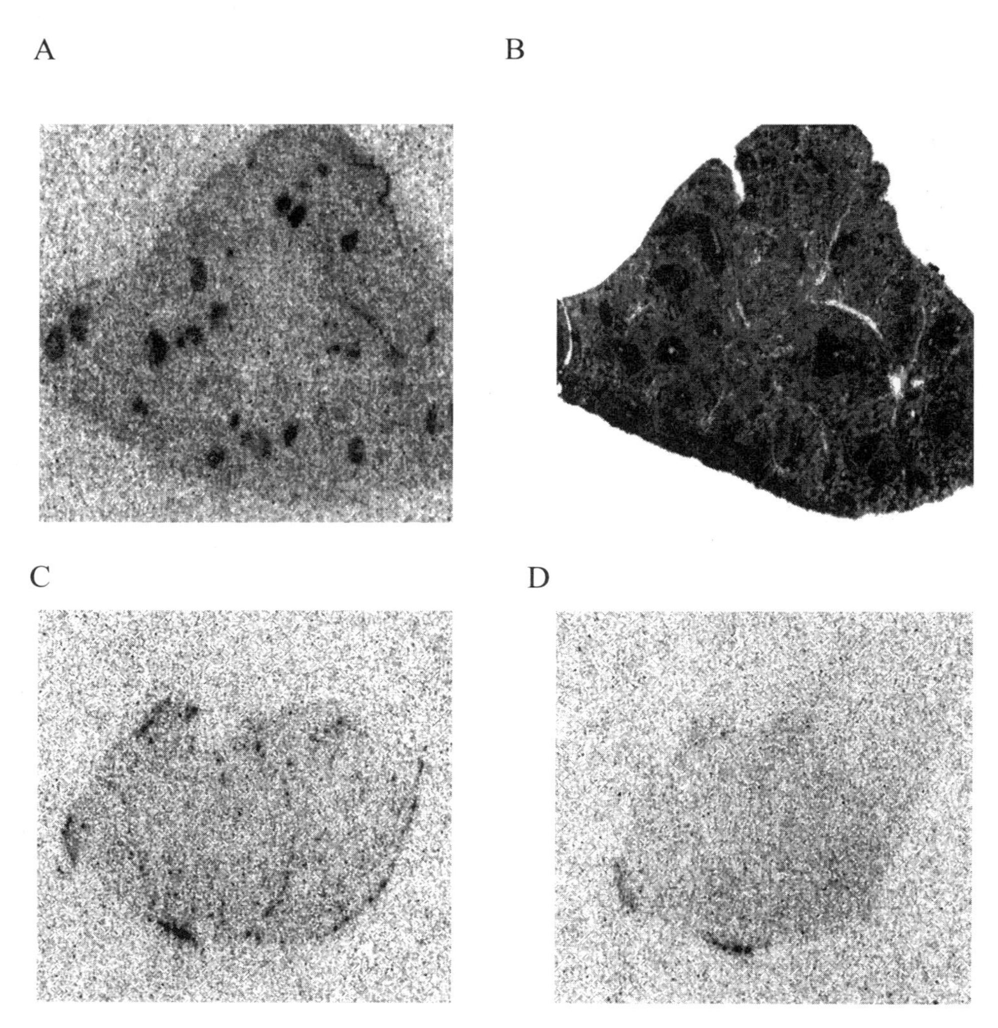

Figure 1. Autoradiograms showing the 2-(^{125}I)-iodomelatonin binding in cryostat spleen (A) and thymic (C) sections. (B)-Cresyl-violet staining of the spleen section labelled by autoradiography. (D)-Non-specific binding in thymus in presence of 1 μM excess of melatonin.

REFERENCES

1. Arendt, J. Melatonin. Clin Endocrinol 29:205–229, 1988.
2. Pévet, P. Environmental control of the annual reproductive cycle in mammals. In: Comparative Physiology of Environmental Adaptations (Pévet, P. ed) Basel, Karger, 1987, vol 3, pp 82–100.
3. Cassone, V.M. Effect of melatonin on vertebrate systems. Trends Neurosci 13:457–464, 1990.
4. Maestroni, G.J.M. The immunoneuroendocrine role of melatonin. J Pineal Res 14:1–10, 1993.
5. Guerrero, J.M., Calvo, J.R., Osuna, C., and Lopez-Gonzalez, M.A. Binding of melatonin by lymphoid cells in humans and rodents. Adv Pineal Res 7:109–117, 1994.
6. Konakchieva, R., Kyurkchiev, S., Kehayov, I., Taushanova, P., and Kanchev, L. Selective effect of methoxyindoles on the lymphocyte proliferation and melatonin binding to activated human lymphoid cells. J Neuroimmunol 63:125–132, 1995.

7. Poon, A.M.S. and Pang, S.F. Pineal melatonin-immune system interaction. In: Melatonin: A Universal Photoperiodic Signal with Diverse Actions. Front Horm Res (Tang, P.L., Pang, S.F., and Reiter, R.J., eds) Basel, Karger, 1996, vol 21, pp 71–83.
8. Hill, S.M. and Blask, D.E. Effects of the pineal hormone melatonin on the proliferation and morphological characteristics of human breast cancer cells (MCF-7) in culture. Cancer Res 48:6121–6126, 1988.
9. Maestroni, G. and Conti, A. The pineal neurohormone melatonin stimulates activated CD4+ Thy1+ cells to release opioid agonists with immunoenhancing and anti-tumor properties. J Neuroimmunol 28:167–176, 1990.
10. Konakchieva, R., Mitev, Y., Almeida, O.F.X., and Patchev, V. Chronic melatonin treatment counteracts glucocorticoid-induced dysregulation of the hypothalamic-pituitary-adrenal axis in the rat. Neuroendocrinol 67:171–180, 1998.
11. Marinova, C., Persengiev, S., Konakchieva, R., Ilieva, A., and Patchev, V. Melatonin effects on glucocorticoid receptors in rat brain and pituitary: significance in adrenocortical regulation. Int J Biochem 23:479–481, 1991.
12. Gauer, F., Masson-Pévet, M., and Pévet, P. Effects of short and long term pinealectomy on the density of melatonin receptors in the suprachiasmatic nuclei and the pars tuberalis of the rat. J Pineal Res 16:73–76, 1994.

50

IS MELATONIN A PHOTOPERIODIC SIGNAL IN HUMANS?

Josephine Arendt

School of Biological Sciences
University of Surrey
Guildford
Surrey
GU2 5XH
UK

1. INTRODUCTION

In view of the sustained interest in melatonin as a chronobiotic (1,2), its possible physiological role in humans has surprisingly not been addressed in detail. It is primarily a photoperiodic messenger molecule in mammalian physiology. The evidence for an important physiological role in circadian physiology is sparse and may be restricted to its effects during development (3), together with reinforcement of "darkness physiology". For example several strains of mice have no detectable melatonin production (4), but do not appear to have circadian abnormalities. Pinealectomy in laboratory rats leads to little change in overt behaviour. Faster reentrainment to a shift in the light dark cycle is seen (rate of adaptation is a photoperiod dependent function (5 and references in 6) and disrupted rhythms are evident in constant light (7). There are many correlative associations of melatonin with sleep (and other variables) in humans, but strictly causal effects of endogenous melatonin have not been reported. The advent of melatonin antagonists will enable this question to be correctly addressed.

Removal or denervation of the pineal does however have a major influence in clearly photoperiodic mammals: it abolishes the ability to perceive changes in daylength, indicated by changes in the duration of melatonin secretion which reflect the length of the night (references in 6). This is the only essential function attributable to melatonin in mammals to date. Daylength is used to time critically important seasonal functions such as reproduction, coat growth, sleep duration and behaviour. In the absence of daylength information, seasonal functions can become desynchronised from

Melatonin after Four Decades, edited by James Olcese.
Kluwer Academic / Plenum Publishers, New York, 2000.

the solar year (8) leading to inappropriate changes in physiology which, in the long term, may be fatal to the individual or to the population concerned.

This review will address some recent data on the physiological role and pharmacological effects of melatonin in humans with particular reference to photoperiodism.

2. ARE HUMANS PHOTOPERIODIC?

Humans are not generally considered to be photoperiodic. However there are seasonal variations in many human functions which may or may not be related to photoperiod. Seasonal affective disorder (SAD) is probably the most well known (9). Extended winter darkness leading to extended melatonin secretion was thought to be the cause. Certainly increasing the amount of bright light exposure was and is a successful treatment. But to date there is little evidence that this is due to a shortening of the melatonin secretion profile. Indeed it is also possible to suggest that this population may in fact require a longer duration melatonin in winter than is permitted by the artificial light environment imposed by social and professional obligations.

The best correlative evidence for an influence of photoperiod on human physiology was a survey by Roenneberg and Aschoff (1990) (10). The authors analysed hundreds of thousands of births spanning the globe over part of the last century up to the time of publication. They found very strong correlations, especially in the older data, of conception rate (Δ conception/Δ time) with deviation of photoperiod from 12L12D. Much of the correlation was lost, or the rhythm changed shape, in the middle of this century, suggesting that the advent of widespread artificial lighting, central heating and air conditioning may have eliminated to some extent this relationship. Their data lead to the conclusion that in the absence of an artificial environment, the maximum human birthrate is found in the spring with sometimes a second peak in autumn. Thus we are, from this correlative evidence, long day breeders. If this is true, then long duration melatonin should be inhibitory to human reproductive activity and vice versa. This theory alone is sufficient to cause concern with regard to the extensive self medication in countries where melatonin is freely available.

Recent data from Northern temperate (11,12) and Polar regions (13), and from simulated long and short photoperiod experiments (14,15,16), suggests that we retain a number of photoperiodic responses, including changes in the duration of melatonin secretion and sleep as a function of daylength. Whether or not these changes are of physiological importance to, for example, urban humans remains debatable. In general our consciously perceived daylength, artificial or natural, changes little during the year. A small delay in the phase of melatonin with or without a small increase in duration of secretion during the winter is seen in healthy volunteers (as well as in SAD patients (17)) in temperate latitudes and polar regions (11,12,13). A skeleton spring photoperiod or morning bright light alone can restore the summer phase position (13,17). It is of interest here to note that SAD patients have more seasonal patterns of reproduction than the general population.

Bright natural (or artificial) light phase shifts the circadian system according to a phase response curve (PRC) whereby phase advances are seen in the early morning shortly after core temperature nadir and phase delays in the late evening shortly before core temperature nadir (18,19). For those working during a "normal" day indoors (0800 h–1700 h), bright natural early morning light is only seen in the summer in the higher latitudes of temperate regions or polar regions, and this early morning light exposure may well be the environmental cause of the phase advance in melatonin seen in summer. Any shortening of the duration of secretion could be attributed to suppression of the evening rise, again by natural or artificial light (20), at a time when phase delays by such exposure are theoretically small or absent (19). These small changes may influence sleep duration, since this is increased in long nights (14), and may of course influence other aspects of physiology.

Much has been made of the actions of melatonin as a possible endogenous circadian zeitgeber in humans (1,6). Since its profile of secretion depends essentially on the light dark cycle it can only function endogenously as an *adjunct* to light in sighted mammals. Notwithstanding, it clearly can override the prevailing light dark cycle as a photoperiodic signal. When, for example, sheep are treated with exogenous long duration melatonin in summer photoperiod their seasonal functions respond to melatonin rather than to the prevailing photoperiod (6). However it is not able to override the environment and simultaneously induce longer sleep in summer in the diurnal sheep (21) or humans (22) although in the latter case it increased "fatigue". Presumably the behavioural effects of light are distinguishable from their systemic effects.

3. PHASE SHIFTING EFFECTS OF MELATONIN

Following an acute dose of melatonin (0.5–10 mg) core body temperature declines and causal links have been suggested between this effect, the induction of sleepiness and, in the case of early evening administration, a subsequent phase advance of melatonin onset and/or core body temperature (23,24,25). Substance has been added to this speculation by the observation that both the temperature decline and the sleepiness are dependent on posture (26). Subjects who remain upright and/or active after the dose do not show either the sleepiness or the temperature drop. However the induced phase shift may not depend on changes in core temperature. In early work, subjects taking 2 mg melatonin daily at 1700 h for 30 days and remaining active only showed significant evening sleepiness after 4 days as a group (22,27) and rarely slept in the early evening. Moreover in conditions where very little acute change in temperature was found, phase shifts still occurred (22,27).

Fast release melatonin (0.5 mg–5 mg, or less in divided doses) phase advances and delays the circadian system (endogenous melatonin, core temperature, sleep timing) according to a PRC (28,29,30). In our experiments this is a dose related phenomenon for advance phase shifts in the range 0.05–5 mg (25). Infusion experiments with controlled plasma levels indicated that melatonin can phase advance and delay the circadian system in physiological concentrations, however the distinction between endogenous (marker rhythm) and exogenous melatonin in parts of this study was mathematical rather than real (29). The duration of endogenous melatonin may be increased by evening oral administration since the onset can be advanced more than the offset (27,31). In this way the circadian and photoperiodic effects become confounded, if indeed they are distinguishable at all.

It is absolutely relevant to the argument for a photoperiodic function of melatonin in humans that (in an entrained environment) phase advances are seen when oral administration is timed such that the exogenous dose adds to the duration of the endogenous dose by advancing the evening rise, that a dead zone exists in the middle of the night when no duration change would be anticipated (with low doses) and that phase delays are seen when the exogenous dose adds to the duration by delaying the morning offset. An exception to this is the phase delay (in acrophase) reported after infusion of night time levels from 12–15 h (29). However the episodic nature of the profiles reported in this study may have unduly influenced the acrophase estimates.

It is the author's opinion that endogenous melatonin indicates dark onset (the rise) and offset (the decline) and reinforces physiological functions associated with darkness, in humans as in other mammals. Pharmacological doses of melatonin may well act differently. What constitutes a physiological dose of melatonin remains problematic. In the authors' experience fast release doses of melatonin in gelatine-lactose or in corn oil/1% ethanol from 0.05–0.2 mg give, on average, "physiological", i.e. night time plasma concentrations of melatonin during the day (25,27,31,32). Individual pharmacokinetics are extremely variable with plasma levels varying up to 25 fold (33) and this may account for some of the variability in the literature.

Whether target tissue concentrations are comparable to blood or not remains debatable and it is possible that higher doses are required to give physiological target tissue concentrations. This is unlikely since when given by infusion to create "physiological profiles" melatonin has clear photoperiodic effects in rodents and sheep and generates phase shifts in humans (6,29).

4. SYNCHRONISATION OF HUMAN CIRCADIAN RHYTHMS BY MELATONIN?

Since melatonin does show the characteristics of a zeitgeber in that a PRC can be generated, it would be expected to entrain fully the circadian clock in suitable circumstances. Human tau is on average 24.3 h or less in constant dim light (34,35,36) and thus the clock needs to be phase advanced on average by 0.3 h or less each day. Acute phase shifts induced by melatonin in an entrained or free running environment are of at least this magnitude when sleep is permitted. However it has proved impossible so far to demonstrate entrainment in humans, with the exception of apparent entrainment or "stabilisation to 24 h" of the sleep wake cycle (30,37). Given to blind subjects free running in a normal environment a pharmacological dose of melatonin induces phase shifts (38) and can stabilise sleep onset in some subjects (39,40). There is no good evidence for entrainment of strongly endogenous rhythms such as core temperature and endogenous melatonin itself (30,39).

It is possible to maintain the sleep-wake cycle of the majority of sighted subjects transferred to a free run in constant dim light (<8 lux) on a cycle indistinguishable from 24 h by daily administration of a pharmacological dose of melatonin (5 mg, 2000 h) at 24 h intervals for periods of 15 days (30). Less successful was an attempt to reentrain subjects after free running with different periodicities for 15 days in constant dim light by daily melatonin at the same clock time (30). The initial melatonin administration at 2000 h occurred at different circadian phases. Both phase advances and phase delays of sleep and core temperature were seen according to a PRC. However the data indicated that the effects observed were complex and variable. Some subjects showed a stabilization of sleep onset with little effect on sleep offset for periods of several days. There was some evidence for splitting of sleep such that some components delayed and others advanced to resynchronise. One subject showed a double phase delay and appeared to entrain to melatonin given at sleep offset for several days. Core temperature indicated that tau was shortened, in one case to significantly less than 24 h, rather than fully entrained, in many subjects. Since the time series was short (15 days) some taus indistinguishable from 24 h may well not have been synchronised. Similarly a longer study time might have shown synchronisation of temperature in more subjects. Only a very long time series would resolve this question.

The lack of entrainment of strongly endogenous rhythms by melatonin in free running blind and sighted humans remains difficult to explain. The fact that an apparent phase shift is produced by evening melatonin in a normal environment may simply be an artifact of the unnatural experimental situation. If melatonin were given naturalistically such as to extend the duration both in the morning and the evening, no overall phase shift should be seen in endogenous melatonin in an entrained environment. The effects of different doses, with pharmacokinetic evaluation, require detailed exploration. The magnitude of the published PRCs may simply be wrong. As yet there is no single pulse PRC for melatonin in free running conditions and this urgently needs to be measured. The stabilisation of sleep to 24 h may simply be a masking effect—but much data is inconsistent with this hypothesis.

Most surprising of all in the free running study mentioned above (30) was a phenomenon of fragmentation of sleep in two subjects taking 5 mg melatonin close to core temperature maximum immediately after transferring to a dim light environment. Cross-over from melatonin to placebo led to consolidated sleep. A subsequent attempt to time melatonin specifically to be close to core temperature maximum provided two further subjects showing fragmented sleep with melatonin compared to placebo (41). Thus, of a total of 16 subjects studied in dim light in this way, 4 showed sleep fragmentation. There was evidence by spectral analysis for the presence of two components with different periodicity's in 3 of the 4 subjects.

5. COMPLEX EFFECTS OF MELATONIN—THEORETICAL INTERPRETATIONS

If humans remain photoperiodic, some of these effects of melatonin may be explicable. Animal data (42,43) and some human evidence (44,45) show that extending the night length can lead to a bimodal pattern of sleep/activity and sometimes of the endogenous melatonin profile itself. Classical photoperiodic theory attributes this bimodality to two theoretical oscillators, the evening (E) and morning (M) components of a rhythm (42,45). Expansion of a rhythm such as sleep or, in rodents, activity in short days (i.e. long nights) is postulated to be due to an advance of the E and a delay of the M oscillators. Melatonin given only in the early evening (advancing the perceived onset) would be expected to differentially affect these theoretical rhythm generators such that the E is influenced more than the M (or vice versa if given in the morning). The possible differential effect of evening melatonin on melatonin onset (27,31) and, to some extent, on the timing of sleepiness/ alertness (22), lends support to this theory. The evidence for this differential effect with regard to light induced phase shifts is very solid (46).

A possible explanation for the fragmentation of sleep in the free running experiments described above is implicit in this theory if indeed sleep consists of two bouts controlled by two coupled oscillators. Given close to core temperature maximum, a maximum phase advance is induced by the first dose of melatonin, with the greatest effect on the E oscillator. The dose used (5 mg) is not necessarily completely cleared by the end of sleep and a phase delaying effect related to the late decline of melatonin may be simultaneously induced on the M oscillator. In these circumstances it is conceivable that the "sleep oscillators" are uncoupled or literally pulled apart, subsequently free running in the very dim light environment with different periodicies as seen in animal experiments. Another possible explanation may involve interactions of the circadian and homeostatic components of sleep (47) whereby a complex combination of sleep induction by out of phase melatonin goes some way towards compensating sleep debt and a subsequent bout of sleep occurs due to the circadian component of sleep (Process C) which itself may then be shifted by the melatonin administration. The effect if confirmed may be dose related as well as circadian time related.

Whatever the explanation it is clear that this phenomenon would be highly undesirable if consolidated sleep and alertness is required. However a strategy of splitting sleep into two components with melatonin may lead to more rapid adaptation to phase shift by advance of the E and delay of the M component.

6. CONCLUSIONS

The majority of published data indicate that melatonin has therapeutic benefits in circadian rhythm-related sleep disorders. However as yet its mechanism of action remains unclear, the appropriate dose for any given condition and individual is uncertain, the contraindications remain to be defined, there is no data on long term safety, use with concomitant medication or organic disease and very little information concerning its most important function as a photoneuroendocrine transducer in humans. Since it does appear to have some photoperiodic effects, and since in principle daylength has the potential to affect many if not all physiological systems, much further research is needed on its physiological role and pharmacological effects in humans.

REFERENCES

1. Armstrong S.M. Melatonin and circadian control in mammals. Experientia 1989;45:932–939.
2. Arendt J., Skene D.J., Middleton B., Lockley S.W., and Deacon S. Efficacy of melatonin treatment in jet lag, shift work, and blindness. J Biol Rhythms 1997;12:604–617.
3. Davis F. Melatonin: role during development. J Biol Rhythms 1997;12:498–508.
4. Reppert S.M. and Weaver D.R. Melatonin madness. Cell 1995;83:1059–1062.
5. Humlova M. and Illnerova H. Resetting of the rat circadian clock after a shift in the light/dark cycle depends on the photoperiod. Neurosci Res 1992 Mar 13:2, 147–153.
6. Arendt J. Melatonin and the mammalian pineal gland. Chapman Hall, London, 1995.
7. Cassone V. The pineal gland influences rat circadian activity rhythms in constant light. J Biol Rhythms 1992;7:27–40.
8. Woodfill C.J.I., Robinson J.E., Malpaux B.M., and Karsch F.J. Synchronisation of the circannual reproductive rhythgm of the ewe by discrete photoperiodic signals. Biology of Reproduction, 1991;45: 110–121.
9. Rosenthal N.E., Sack D.A., Gillin J.C., Lewy A.J., Goodwin F.K., Davenport Y., Mueller P.S., Newsome D.A., and Wehr T.A. Seasonal affective disorder. A description of the syndrome and preliminary findings with light therapy. Arch Gen Psychiatry 1984;41:72–79.
10. Roenneberg T. and Aschoff J. Annual rhythm of human reproduction: II. Environmental Correlations. J Biol Rhythms 1990;5:217–240.
11. Beck-Friis J., von Rosen D., Kjellman B.F., Ljungen J.G., and Wetterberg L. Melatonin in relation to body measures, sex, age, season and the use of drugs in patients with major affective disorders and healthy subjects. Psychoneuroendocrinology 1984;9:261–277.

12. Illnerova H., Zovolsky P., and Vanecek J. The circadian rhythm in plasma melatonin concentration of the urbanised man: the effect of summer and winter time. Brain Research 1985;328:186–189.
13. Broadway J., Arendt J., and Folkard S. Bright light phase shifts the human melatonin rhythm during the Antarctic winter. Neuroscience Letters 1987;79:185–189.
14. Wehr T.A. The durations of human melatonin secretion and sleep respond to changes in daylength (photoperiod). J Clin Endocrinol Metab 1991;73:1276–1280.
15. Buresova M., Dvorekova M., Zovolsky P., and Illnerova H. Human circadian rhythm in serum melatonin in short winter days and in simulated artificial long days. Neurosci Lett 1992;136:2 173–176.
16. Vondrasova D., Hajek I., and Illnerova H. Exposure to long summer days affects the human melatonin and cortisol rhythms. Brain Res 1997;759:1, 166–170.
17. Lewy A.J., Sack R.L., Miller L.S., and Hoban T.M. Anti-depressant and circadian phase-shifting effects of light. Science 1987;235:352–354.
18. Czeisler C.A., Allan J.S., Strogatz J.S., Ronda J.M., Sandrez R., Rios C.D., Freitag W.O., Richardson G.S., and Kronauer R.E. Bright light resets the human circadian pacemaker independent of the timing of the sleep-wake cycle. Science 1986;233:667–671.
19. Minors D., Waterhouse J., and Wirz-Justice A. A human phase-response curve to light. Neuroscience Letters 1991;133:36–40.
20. Lewy A.J., Wehr T.A., Goodwin F.K., Newsome D.A. and Markey S.P. Light suppresses melatonin secretion in humans. Science 1980;210:1267.
22. Arendt J., Borbely A.A., Franey C., and Wright J. The effect of chronic, small doses of melatonin given in the late afternoon on fatigue in man: a preliminary study. Neurosci Lett 1984;45:317–321.
23. Krauchi K., Cajochen C., Moeri D., Graw P., and Wirz-Justice A. Early evening melatonin and S-20098 advance circadian phase and nocturnal regulation of core body temperature. Am J Physiol 1997;272:R1178–1188.
24. Cagnacci A., Elliot J.A., and Yen S.S. Melatonin: A major regulator of the circadian rhythm of core body temperature in humans. J Clin Endocrinol Metab 1992;75:447–452.
25. Deacon S. and Arendt J. Melatonin-induced temperature suppression and its acute phase-shifting effects correlate in a dose-dependent manner in humans. Brain Res 1995;688:77–85.
26. A relationship between heat loss and sleepiness: effects of postural change and melatonin administration. In: Krauchi K., Cajochen C., and Wirz-Justice A. J Appl Physiol 1997;83:134–139.
27. Arendt J., Bojkowski C., Folkard S., Franey C., Minors D., Waterhouse J., Wever R.A., Wildgruber C., Wright J., and Marks V. Some effects of melatonin and the control of its secretion in man. In: Evered D. and Clark S., eds. Ciba Foundation Symposium 117, Photoperiodism, melatonin and the pineal. London, Pitman, 1985:266–283.
28. Lewy A.J., Saeeduddin A., Latham-Jackson J.M., and Sack R. Melatonin shifts human circadian rhythms according to a phase response curve. Chronobiol Internat 1992;9:380–392.
29. Zaidan R., Geoffriau M., Brun J., Taillard J., Bureau C., Chazot G., and Claustrat B. Melatonin is able to influence its secretion in humans: description of a phase-response curve. Neuroendocrinol 1994;60:105–112.
30. Middleton B., Arendt J., and Stone B. Complex effects of melatonin on human circadian rhythms in constant dim light. J Biol Rhythms 1997;12:467–475.
31. Wright J., Aldhous M., Franey C., English J., and Arendt J. The effects of exogenous melatonin on endocrine function in man. Clin Endocr 1986;24:375–382.
32. Aldhous M., Franey C., Wright J., and Arendt J. Plasma concentrations of melatonin in man following oral absorption of different preparations. Brit J Clin Pharmacol 1985;19:517–521.
33. Waldhauser F., Steger H., and Vorkapic P. Melatonin secretion in man and the influence of exogenous melatonin on some physiological and behavioural variables. Adv Pineal Res 1987;2:207–223.
34. When the human circadian system is caught napping: evidence for endogenous rhythms close to 24 hours. In: Campbell S.S., Dawson D., and Zulley J. Sleep 1993;7:638–640.
35. Middleton B., Arendt J., and Stone B. Human circadian rhythms in constant dim light (<8 lux) with knowledge of clock time. J Sleep Res 1996;5:69–76.
36. Czeisler C.A., Duffy J.F., Shanahan T.L., Brown E.N., Mitchell J.F., Dijk D.-J., Rimmer D.W., Ronda J.M., Allan J.S., Emens J.S., and Kronauer R.E. Reassessment of the intrinsic period (t) of the human circadian pacemaker in young and older subjects. Abstract 505, Second International Congress of the World Federation of Sleep Research Societies, Nassau, The Bahamas, 12–16th September, 1995.
37. Arendt J., Aldhous M., and Wright J. Synchronisation of a disturbed sleep-wake cycle in a blind man by melatonin treatment. Lancet 1988;i:772–773.
38. Sack R.L., Lewy A.J., Blood M.L., Stevenson J., and Keith L.D. Melatonin administration to blind people: phase advances and entrainment. J Biol Rhythms 1991;6:249–261.

39. Folkard S., Arendt J., Aldhous M., and Kennett H. Melatonin stabilises sleep onset time in a blind man without entrainment of cortisol or temperature rhythms. Neurosci Lett 1990;113:193–198.
40. Aldhous M.E. and Arendt J. Melatonin rhythms and the sleep wake cycle in blind subjects. Proceedings of the 7th meeting of the European Society for Chronobiology, Marburg, Germany, 1991. J Interdiscip Cycle Res 1991;22:84–85.
41. Middleton B., Stone B., and Arendt J. Melatonin can induce fragmented sleep patterns. Lancet 1996;348:551–552.
42. Pittendrigh C.S. and Daan S. A functional analysis of circadian pacemakers in nocturnal rodents. V. Pacemaker structure: a clock for all seasons. J Comp Physiol 1976;106:333–355.
43. Arendt J., Symons A.M., and Laud C. Pineal function in the sheep: evidence for a possible mechanism mediating seasonal reproductive activity. Experientia 1981;37:584–589.
44. Wehr T.A., Schwartz P.J., Turner E.H., Feldman-Naim S., Drake C.L., and Rosenthal N.E. Bimodal patterns of human melatonin secretion consistent with a two-oscillator model of regulation. Neurosci Lett 1995;194:105–108.
45. Wehr T.A. Melatonin and seasonal rhythms. J Biol Rhythms 1997;12:518–527.
46. Illnerova H. and Vanecek J. Complex control of the circadian rhythm in N-acetyltransferase activity in the rat pineal gland. In: Vertebrate Circadian Systems, Aschoff J., Daan S., and Groos G., eds. Springer Verlag, Heidelberg, Berlin, 1982:285–296.

51

MELATONIN AS A MARKER AND PHASE-RESETTER OF CIRCADIAN RHYTHMS IN HUMANS

A. J. Lewy

Department of Psychiatry
Oregon Health Sciences University
Portland, Oregon 97201

INTRODUCTION

Forty years after Aaron Lerner and his colleagues discovered melatonin (1), much has been learned about its endogenous function in humans. Dr. Lerner was a dermatologist and hoped that this skin-lightening agent would provide a clue to the pathophysiology and treatment of vitiligo. Although melatonin is the most potent skin-lightening agent in certain species, it lacks this effect in humans.

Nevertheless, melatonin production in virtually every species is confined mainly to nighttime darkness and is acutely suppressed by exposure to light (2). In seasonal animals, melatonin is important in the regulation of annual rhythms, notably reproduction (estrus) (3): the longer duration of melatonin during the winter nights communicates a short day signal to the organism (4–5). Some animals may also use the daily rise and fall of melatonin to aid entrainment by the light/dark cycle and perhaps to help synchronize the constituent cells of the organism to the endogenous circadian pacemaker (ECP) in the suprachiasmatic nucleus of the hypothalamus (6–10).

EFFECTS OF LIGHT IN HUMANS

As in other species, human melatonin production is limited to nighttime darkness. Ambient light exposure suppresses nighttime melatonin production; however, the light needs to be brighter than the intensities that are effective in laboratory animals (Figure 1) (11). To date, the most significant finding related to melatonin has been its role in the discovery that light is of importance in the regulation of human biological rhythms

Melatonin after Four Decades, edited by James Olcese.
Kluwer Academic / Plenum Publishers, New York, 2000.

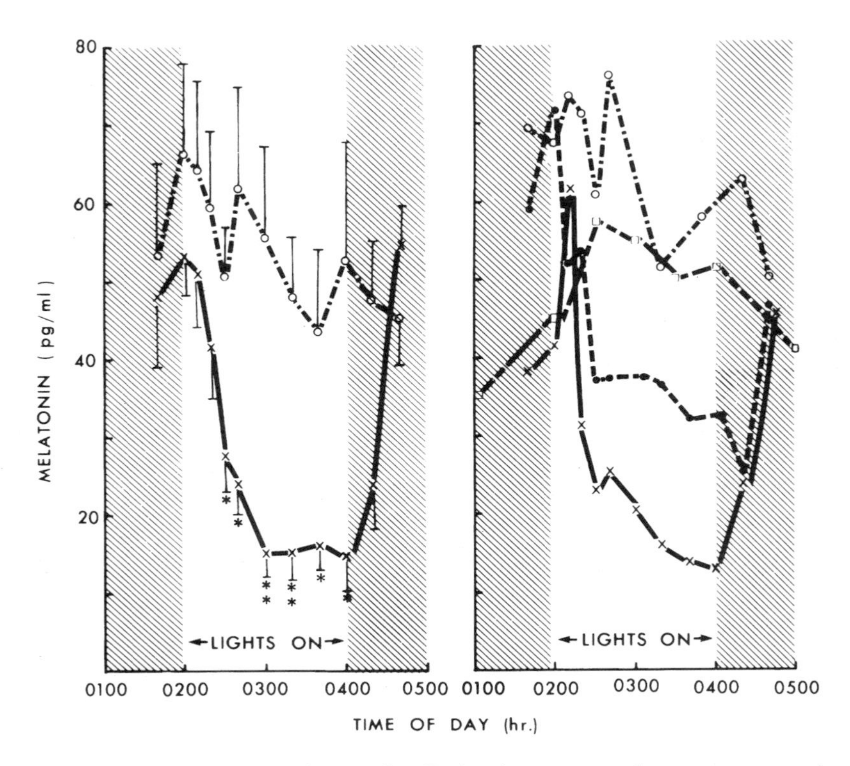

Figure 1 (left). Effect of light on melatonin secretion. Each point represents the mean concentration of melatonin (± standard error) for six subjects. A paired *t*- test, comparing exposure to 500 lux with exposure to 2500 lux, was performed for each data point. A two-way analysis of variance with repeated measures and the Newman-Keuls statistic for the comparison of means showed significant differences between 2:30 a.m. and 4 a.m. (*, $P < .05$; **, $P < .01$). **(right).** Effect of different light intensities on melatonin secretion. The averaged values for two subjects are shown. Symbols: Asleep in the dark (open circles); 500 lux (open squares); 1500 lux (closed circles); and 2500 lux (crosses). From Lewy et al. (11) with permission.

(A. Lerner, personal communication). Light suppression of melatonin is also a useful test for light sensitivity (12–14).

Light exposure resets the ECP according to a phase response curve (PRC) (15). PRCs to light are typically characterized by a "dead zone" of reduced responses in the middle of the day. In the first half of the night, light exposure causes phase delays and in the second half of the night it causes phase advances. The magnitude of these phase shifts increases towards the middle of the night, at which time there is a crossover time that separates delay responses of large magnitude from advance responses of large magnitude. During the middle of the day, there is also a crossover time that separates advance responses of small magnitude from delay responses of small magnitude.

The first suggestion that there was a PRC to bright light in humans came from studies providing or removing bright light in the morning just after awakening or in the evening just before bedtime (16–17); notably, the sleep/wake cycle was held constant. The first evidence for a crossover time in the middle of the night came from a field study of jet lag (18). Subsequently, four complete PRCs have been described (19–22) that are in general agreement with each other and with the one originally hypothesized (23–24).

Along with the early assessments of phase-shifting responses to morning vs. evening light in normal controls, patients with circadian phase disorders were also treated with either morning or evening bright light exposure (24–25). There are two types of circadian phase disorders, the phase-advanced type and the phase-delayed type. In the former category are advanced sleep phase syndrome and east-to-west jet lag. In the latter category are delayed sleep phase syndrome and west-to-east jet lag. Most shift workers also fall into these two categories. Those who are abnormally phase advanced should be exposed to evening light, and those who are abnormally phase delayed should be exposed to morning light.

MELATONIN AS A MARKER FOR CIRCADIAN PHASE POSITION

Determination of circadian phase, normal and pathological, can be done in a number of ways. The easiest but least reliable way is to use the timing of sleep (sleep offset is usually more reliable than sleep onset). Other physiological measures are more reliable but are sometimes difficult to assess, such as core body temperature and cortisol. There are also disagreements over how these data should be collected, specifically whether or not it is necessary or even desirable to use a constant routine in which semi-recumbent subjects are sleep deprived and fed isocalorically for one or two days (26–27).

The circadian rhythm of melatonin production turns out to be perhaps the best marker for circadian phase. Its standard deviation is less than that of the temperature rhythm (28). A constant routine is not necessary for its assessment, although sample collection should be done under dim light (<30 lux). It can be assessed using blood, saliva or urine, although the latter is difficult to collect in intervals shorter than 1–2 hours, and the more frequent the collection intervals the more highly resolved the phase marker.

Frequent sampling around the time of the melatonin onset may turn out to be the most useful of all of the markers for circadian phase. The melatonin onset is a clearly demarcated event (Figure 2). With 30-minute sampling, the standard deviation of the dim light melatonin onset (DLMO) is less than its sampling interval (28). The DLMO usually occurs before bedtime; consequently, there is little or no disturbance of sleep.

Confining measurement of the endogenous melatonin profile to the DLMO has been criticized because assessment of the amplitude of the melatonin profile is not obtained. However, there are few if any reasons to assess amplitude of the melatonin profile. It is questionable that the melatonin peak reflects the amplitude of the ECP (29); there is even some doubt that the temperature rhythm reflects ECP amplitude. Furthermore, amplitude differences in melatonin (or even in the ECP) have yet to be conclusively shown to be important.

Confining measurement of the endogenous melatonin profile to the DLMO has been criticized because it ignores the melatonin offset. However, while it appears that the melatonin offset may be regulated by a separate oscillator in rats, this has not been established in humans (30). Indeed, if there is a separate offset oscillator in humans it appears to be so tightly coupled to the onset oscillator that there appears to be no need to measure both. Given the choice between the onset and the offset, the onset is much preferred, since the offset is confounded by biochemical factors that

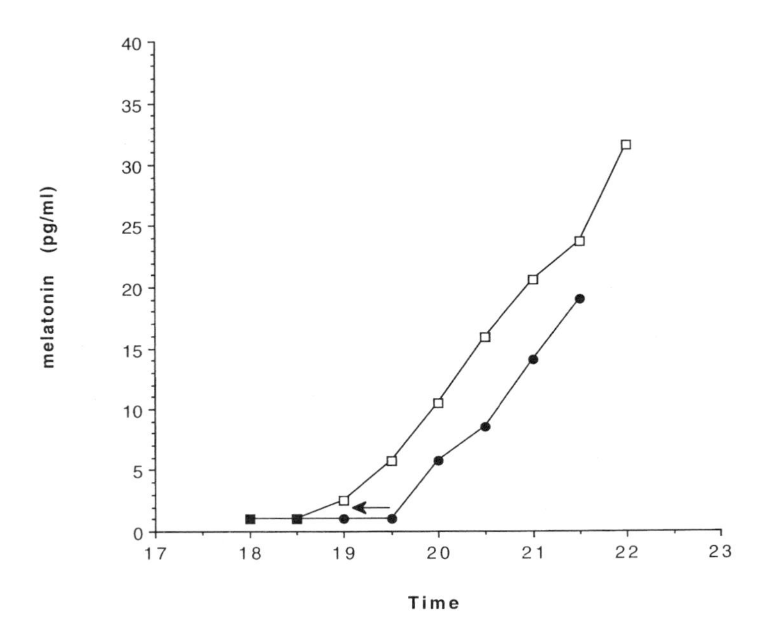

Figure 2. Dim light melatonin onsets (DLMOs) are shown from a patient with winter depression before (closed circles) and after (open squares) successful treatment with afternoon melatonin. An advance in circadian phase accompanied her clinical remission, as expected.

appear to contribute to a higher standard deviation. The actual time that melatonin production ceases is usually a few hours before wake-up time and therefore melatonin half-life will affect the time when circulating levels fall to daytime values (31); in other words, accurate measurement of the true offset requires middle-of-the-night sampling with its attendant disturbance of sleep. It is possible that some pathologies will be evident in the dim light melatonin offset (DLMOff) and not in the DLMO. However, to date no conclusive evidence has been presented suggesting that pathological changes in the DLMOff (not also reflected in the DLMO) predict response to phase-resetting treatment.

Light appears to have identical effects on shifting the phase of the melatonin onset and offset, at least when assessed the day after the last light pulse (32). To be conservative, however, assessments should be controlled for time of the year, in case it turns out that there are significant seasonal differences in melatonin duration (33). In short, there seems to be no reason at the present time not to confine assessment of circadian phase to the DLMO, particularly if practical considerations are important.

This does not mean that more information cannot be obtained by assessing other rhythms. If there are phase differences between the melatonin profile and other circadian rhythms, the melatonin profile in all likelihood most accurately reflects the phase of the ECP, as it is least susceptible to masking and other influences that could distort its waveform. It also appears to be tightly coupled to the ECP (34). Nevertheless, the phase relationships between overt circadian rhythms (particularly between the sleep/wake cycle and the DLMO) are often of interest (35).

WINTER DEPRESSION AND THE PHASE SHIFT HYPOTHESIS

One of the most responsive disorders to bright light is winter depression, which was among the earliest ever treated (36–37). Indeed, winter depression was discovered and identified as a result of the finding that bright light could suppress nighttime melatonin production in humans (38–39). This finding implied that humans had biological rhythms that were cued to natural daylight, relatively unperturbed by ordinary-intensity indoor light, and that biological rhythm disorders could be treated by the use of bright artificial light. Thus, these implications had prepared the minds of researchers when a patient identified himself as having a seasonal rhythm in his affective symptoms.

The DLMOs of winter depressives appear to be delayed in the winter when they are depressed compared to those of normal controls (40–42). Morning light advances the DLMO, while evening light delays it. There is an increasing consensus that bright light scheduled in the morning is more antidepressant than bright light scheduled in the evening (40,42–45). These findings are supportive of the phase shift hypothesis (PSH) of winter depression which states that these patients become depressed in winter at least in part because of a phase delay of their ECP with respect to real time and with respect to sleep (40,46): that is, there is an internal desynchronization between rhythms tightly coupled to the ECP and rhythms associated with the sleep/wake cycle (which is loosely coupled to the ECP) (35,47).

MELATONIN AS A PHASE-RESETTING AGENT

In some animals, exogenous melatonin causes phase shifts and can even provide an entraining signal (see above). Early work by Arendt suggested that melatonin could intermittently have phase-advancing effects (48). Blind people appear to be even more sensitive to these effects than sighted people, perhaps because they are not affected by a competing light/dark cycle (49–51). Eventually, a melatonin PRC (four days of administering 0.5 mg) was described in sighted people that for the first time: 1) showed consistent and robust phase advances; 2) demonstrated phase delays; and 3) described the relationship between the time of melatonin administration and the direction of phase shift (52). These phase shifts were smaller in sighted people, because melatonin had to override the effects of a light/dark cycle which was held constant. The magnitude of these phase shifts increases ten-fold when the sleep/wake and ambient light/dark cycles are shifted 12 hours (53), as is the case with the light PRC study that showed the greatest phase shifts (20). Subsequently, it was shown that these phase shifts can be achieved within one day (54–55) and that these effects are dose-dependent (55).

The original melatonin PRC study has been replicated by another group of investigators (54). Furthermore, an interim report of an additional six subjects given melatonin for 12 trials at times two hours apart showed a PRC similar to the original (28,56). The complete study of 12 subjects (manuscript in preparation) confirms the interim report. In all three PRCs (obtained using a small oral dose of melatonin), the crossover time between advances and delays appears to be at about CT 18 and between delays and advances at about CT 6. [Because of inter-individual variation in internal body clock time, circadian time (CT) is determined by the baseline DLMO (CT 14), since it usually occurs about 14 hours after sleep offset (lights on).]

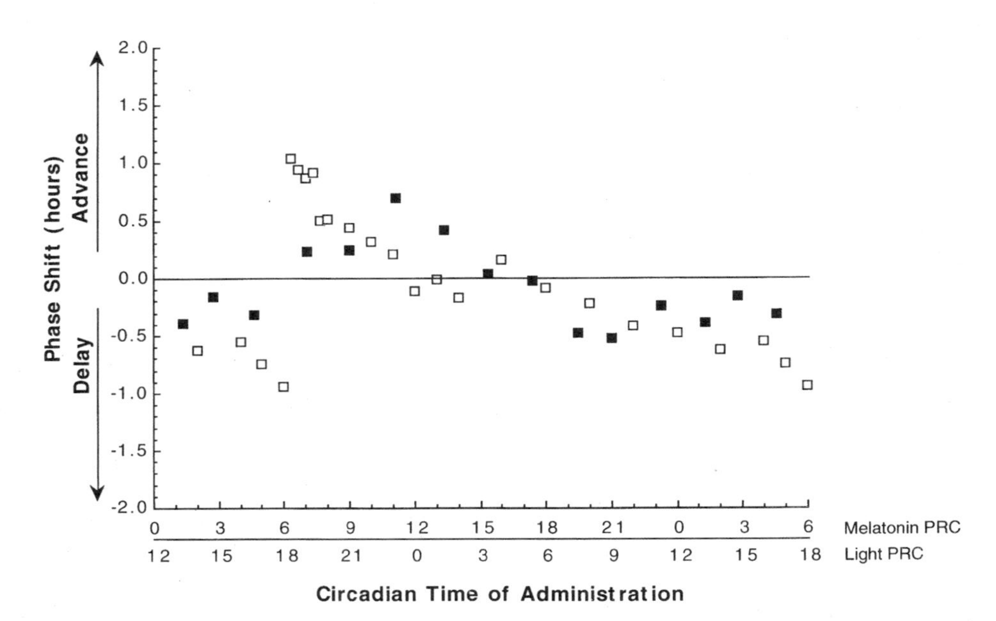

Figure 3. The binned melatonin (closed squares) and light (open squares) PRCs plotted together. The phase-shift magnitudes for the Czeisler's group light PRC (20), which has been adapted from the PRC Atlas (58), have been divided by ten, so that circadian phase positions of the two PRCs can be more easily compared. The scale on the abscissa has been shifted 12 hours for the light PRC. The overlap of the data is quite striking that indicates that the two PRCs are about 12 hours out of phase with each other. From Lewy et al. (56) with permission.

Therefore, the melatonin and light PRCs appear to be about 12 hours out of phase with each other (Figure 3) (56). This suggests that one function of endogenous melatonin may be to augment entrainment of the ECP by the light/dark cycle. When given to shift workers, the melatonin PRC has a similar shape, but the phase-shift magnitudes (as mentioned above) are ten times greater (53). Perhaps, melatonin and darkness potentiate each other's phase-shifting effects: a second function of melatonin may be to help the ECP discriminate between nighttime darkness and sporadic daytime naps, conferring more zeitgeber strength to the former (57). This may explain why melatonin is not produced when darkness occurs in the middle of the day.

MELATONIN TREATMENT OF WINTER DEPRESSION

Although morning light is a more effective antidepressant than evening light in winter depression (see above), this does not conclusively prove the PSH. It could be argued, for example, that people have a greater overall sensitivity to light in the morning. Therefore, it is necessary to demonstrate an association between symptomatic improvement and another agent that can cause phase shifts.

In the winter of 1996–1997, it was shown that melatonin (0.125 mg) given at CT 8 and CT 12 was more antidepressant than placebo, in a parallel study with five patients in each group (59). In the winter of 1997–1998, the dosing regimen was changed to 0.1 mg at CT 8, 10 and 12 compared to CT 0, 2 and 4. In addition, there was also a placebo group. This was also a parallel study and included 26 patients. As in the first study, most if not all of the responders were in the afternoon melatonin group. Morning melatonin caused a phase delay of the DLMO and afternoon melatonin caused a phase advance.

These patients seem to be exquisitely sensitive to the dose-dependent soporific side effect of melatonin. Consequently, the dose had to be lowered quite a bit in these patients. Since, according to the melatonin PRC, the earlier in the afternoon melatonin is given the greater is the resulting phase advance, a second or third dose was given to provide continuously elevated melatonin levels through the time of the endogenous onset. The thinking here was that this would create a single melatonin signal with an earlier onset and therefore maximize the phase-advance from a very small dose of exogenous melatonin.

CONCLUSIONS AND SUMMARY

With the advent of RIAs capable of measuring the melatonin onset (60), the DLMO has been increasingly utilized as a marker for the phase of the ECP. Measurement of melatonin not only helps assess circadian phase but also has been important in developing the use of bright light to assess light sensitivity, as well as to shift the ECP in the treatment of circadian phase disorders. Physiological doses of melatonin have been used to describe a PRC that is about 12 hours out of phase with the PRC to light. Melatonin treatment of winter depression may eventually provide conclusive proof of the PSH for this disorder. One function of endogenous melatonin may be to augment entrainment of the ECP by the light/dark cycle. Another function may be to serve as a nighttime darkness discriminator. The fact that melatonin is generally made only at night and can be suppressed by light have been the guiding principles behind these developments. The early studies were also made possible because of the highly accurate and sensitive GCMS assay for melatonin (61).

ACKNOWLEDGMENTS

We wish to thank the nursing staff of the OHSU GCRC and to acknowledge the assistance of Robert L. Sack, Neil L. Cutler, Vance K. Bauer, Saeed Ahmed, Katherine H. Thomas, Mary L. Blood, Jeanne M. Latham Jackson, Richard S. Boney, Neil R. Anderson and Aaron Clemons. Supported by Public Health Service research grants R01 MH55703, R01 AG15140, R01 MH56874, R01 MH40161, K02 MH00703, K02 MH01005 (RLS), PO1 AG10794 and M01 RR00334 (OHSU GCRC), and a grant from the National Alliance for Research on Schizophrenia and Depression.

REFERENCES

1. Lerner A.B., Case J.D., Takahashi Y., Lee T.H., and Mori N. Isolation of melatonin, a pineal factor that lightens melanocytes. J Am Chem Soc 80:2587, 1958.
2. Arendt J. Melatonin and the Mammalian Pineal Gland. London, Chapman & Hall, 1995.
3. Hoffman R.A. and Reiter R.J. Responses of some endocrine organs of female hamsters to pinealectomy and light. Life Sci 5:1147–1151, 1966.
4. Tamarkin L., Westrom W.K., Hamill A.I., and Goldman B.D. Effects of melatonin on the reproductive systems of male and female Syrian hamsters: a diurnal rhythm in sensitivity to melatonin. Endocrinology 99:1534–1541, 1976.

5. Goldman B.D. and Darrow J.M. The pineal gland and mammalian photoperiodism. Neuroendocrinology 37:386–396, 1983.
6. Gwinner E. and Benzinger I. Synchronization of a circadian rhythm in pinealectomized European starlings by injections of melatonin. J Comp Physiol 127:209–213, 1978.
7. Redman J., Armstrong S., and Ng K.T. Free-running activity rhythms in the rat: entrainment by melatonin. Science 219:1089–1091, 1983.
8. Underwood H. Circadian rhythms in lizards: phase response curve for melatonin. J Pineal Res 3:187–196, 1986.
9. Underwood H. and Harless M. Entrainment of the circadian activity rhythm of a lizard to melatonin injections. Physiol Behav 35:267–270, 1985.
10. Armstrong S.M. and Chesworth M.J. Melatonin phase shifts a mammalian circadian clock. In: Fundamentals and Clinics in Pineal Research. Edited by Trentini G.P., de Gaetani C., Pévet P. New York, Raven Press, pp. 195–198, 1987.
11. Lewy A.J., Wehr T.A., Goodwin F.K., Newsome D.A., and Markey S.P. Light suppresses melatonin secretion in humans. Science 210:1267–1269, 1980.
12. Lewy A.J., Wehr T.A., Goodwin F.K., Newsome D.A., and Rosenthal N.E. Manic-depressive patients may be supersensitive to light. Lancet i:383–384, 1981.
13. Lewy A.J., Nurnberger J.I., Wehr T.A., Pack D., Becker L.E., Powell R., and Newsome D.A. Supersensitivity to light: possible trait marker for manic-depressive illness. Am J Psychiatry 142:725–727, 1985a.
14. Czeisler C.A., Brown E.N., Ronda J.M., Kronauer R.E., Richardson G.S., and Freitag W.O. A clinical method to assess the endogenous circadian phase (ECP) of the deep circadian oscillator in man. Sleep Res 14:295, 1985.
15. Daan S. and Pittendrigh C.S. A functional analysis of circadian pacemakers in nocturnal rodents. II. The variability of phase response curves. J Comp Physiol 106:253–266, 1976.
16. Lewy A.J., Sack R.L., and Singer C.M. Assessment and treatment of chronobiologic disorders using plasma melatonin levels and bright light exposure: the clock-gate model and the phase response curve. Psychopharmacol Bull 20:561–565, 1984.
17. Lewy A.J., Sack R.L., and Singer C.M. Immediate and delayed effects of bright light on human melatonin production: shifting "dawn" and "dusk" shifts the dim light melatonin onset (DLMO). Ann NY Acad Sci 453:253–259, 1985b.
18. Daan S. and Lewy A.J. Scheduled exposure to daylight: a potential strategy to reduce "jet lag" following transmeridian flight. Psychopharmacol Bull 20:566–568, 1984.
19. Honma K. and Honma S. A human phase response curve for bright light pulses. Jap J Psychiatry 42:167–168, 1988.
20. Czeisler C.A., Kronauer R.E., Allan J.S., Duffy J.F., Jewett M.E., Brown E.N., and Ronda J.M. Bright light induction of strong (Type O) resetting of the human circadian pacemaker. Science 244:1328–1333, 1989.
21. Wever R.A. Light effects on human circadian rhythms. A review of recent Andechs experiments. J Biol Rhythms 4:161–186, 1989.
22. Minors D.S., Waterhouse J.M., and Wirz-Justice A. A human phase-response curve to light. Neurosci Lett 133:36–40, 1991.
23. Lewy A.J. Biochemistry and regulation of mammalian melatonin production. In: The Pineal Gland. Edited by Relkin R.M. New York, Elsevier, pp. 77–128, 1983.
24. Lewy A.J., Sack R.L., Fredrickson R.H., Reaves M., Denney D., and Zielske D.R. The use of bright light in the treatment of chronobiologic sleep and mood disorders: the phase-response curve. Psychopharmacol Bull 19:523–525, 1983.
25. Lewy A.J., Sack R.L., and Singer C.M. Melatonin, light and chronobiological disorders. In: Photoperiodism, Melatonin and the Pineal. Edited by Evered D., Clark S. London, Pitman, pp. 231–252, 1985c.
26. Mills J.N., Minors D.S., and Waterhouse J.M. Exogenous and endogenous influences on rhythms after sudden time shift. Ergonomics 21:755–761, 1978.
27. Czeisler C.A., Shanahan T.L., Klerman E.B., Martens H., Brotman D.J., Emans J.S., Klein T., and Rizzo J.F. Suppression of melatonin secretion in some blind patients by exposure to bright light. N Eng J Med 332:6–11, 1995.
28. Lewy A.J. and Sack R.L. Exogenous melatonin's phase shifting effects on the endogenous melatonin profile in sighted humans: a brief review and critique of the literature. J Biol Rhythms 12:595–603, 1997.

29. Lewy A.J. Human melatonin secretion (I): a marker for adrenergic function. In: Neurobiology of Mood Disorders. Edited by Post R.M., Ballenger J.C. Baltimore, Williams and Wilkins, pp. 207–213, 1984.
30. Illnerová H. and Vanecek J. Two-oscillator structure of the pacemaker controlling the circadian rhythm of *N*-acetyltransferase in the rat pineal gland. J Comp Physiol [A] 145:539–548, 1982.
31. Sack R.L., Hughes R.J., Parrott K., and Lewy A.J. Melatonin synthesis is sharply terminated at the beginning of the circadian day. Sleep 21 (suppl):216, 1998.
32. Shanahan T.L. and Czeisler C.A. Light exposure induces equivalent phase shifts of the endogenous circadian rhythms of circulating plasma melatonin and core body temperature in men. J Clin Endocr Metab 73:227–235, 1991.
33. Wehr T.A., Schwartz P.J., Turner E.H., Feldman-Naim S., Drake C.L., and Rosenthal N.E. Bimodal patterns of human melatonin secretion consistent with a two-oscillator model of regulation. Neurosci Lett 194:105–108, 1995.
34. Lewy A.J. and Sack R.L. The dim light melatonin onset (DLMO) as a marker for circadian phase position. Chronobiol Int 6:93–102, 1989.
35. Lewy A.J. Chronobiologic disorders, social cues and the light-dark cycle. Chronobiol Int 7:15–21, 1990.
36. Lewy A.J., Kern H.A., Rosenthal N.E., and Wehr T.A. Bright artificial light treatment of a manic-depressive patient with a seasonal mood cycle. Am J Psychiatry 139:1496–1498, 1982.
37. Rosenthal N.E., Sack D.A., Gillin J.C., Lewy A.J., Goodwin F.K., Davenport Y., Mueller P.S., Newsome D.A., and Wehr T.A. Seasonal affective disorder: a description of the syndrome and preliminary findings with light therapy. Arch Gen Psychiatry 41:72–80, 1984.
38. Kern H.E. and Lewy A.J. Corrections and additions to the history of light therapy and seasonal affective disorder [letter]. Arch Gen Psychiatry 47:90–91, 1990.
39. Wehr T.A. and Rosenthal N.E. Reply to: corrections and additions to the history of light therapy and seasonal affective disorder [letter]. Arch Gen Psychiatry 47:91, 1990.
40. Lewy A.J., Sack R.L., Miller S., and Hoban T.M. Antidepressant and circadian phase-shifting effects of light. Science 235:352–354, 1987a.
41. Sack R.L., Lewy A.J., White D.M., Singer C.M., Fireman M.J., and Vandiver R. Morning versus evening light treatment for winter depression; evidence that the therapeutic effects of light are mediated by circadian phase shifts. Arch Gen Psychiatry 47:343–351, 1990a.
42. Lewy A.J., Bauer V.K., Cutler N.L., Sack R.L., Ahmed S., Thomas K.H., Blood M.L., and Latham Jackson J.M. Morning versus evening light treatment of winter depressive patients. Arch Gen Psychiatry 55:890–896.
43. Avery D., Khan A., Dager S., Cohen S., Cox G., and Dunner D. Morning or evening bright light treatment of winter depression? The significance of hypersomnia. Biol Psychiatry 29:117–126, 1991.
44. Eastman C.I., Young M.A., Fogg L.F., Liu L., and Meaden P.M. Bright light treatment of winter depression: a placebo-controlled trial. Arch Gen Psychiatry 55:883–889, 1998.
45. Terman M., Terman J.S., and Ross D.C. A controlled trial of timed bright light and negative air ionization for treatment of winter depression. Arch Gen Psychiatry 55:875–882, 1998.
46. Lewy A.J., Sack R.L., Singer C.M., and White D.M. The phase shift hypothesis for bright light's therapeutic mechanism of action: theoretical considerations and experimental evidence. Psychopharmacol Bull 23:349–353, 1987b.
47. Lewy A.J., Sack R.L., Singer C.M., White D.A., and Hoban T.M. Winter depression: the phase angle between sleep and other circadian rhythms may be critical. In: Seasonal Affective Disorder. Edited by Thompson C., Silverstone T. London, CNS (Clinical Neuroscience), pp. 205–221, 1989.
48. Arendt J., Bojkowski C., Folkard S., Franey C., Marks V., Minors D., Waterhouse J., Wever R.A., Wildgruber C., and Wright J. Some effects of melatonin and the control of its secretion in humans, in Photoperiodism, Melatonin and the Pineal. Edited by Evered D., Clark S. London, Pitman, pp. 266–283, 1985.
49. Sack R.L., Lewy A.J., and Hoban T.M. Free-running melatonin rhythms in blind people: phase shifts with melatonin and triazolam administration. In: Temporal Disorder in Human Oscillatory Systems. Edited by Rensing L., an der Heiden U., Mackey M.C. Heidelberg, Springer-Verlag, pp. 219–224, 1987.
50. Sack R.L., Stevenson J., and Lewy A.J. Entrainment of a previously free-running blind human with melatonin administration. Sleep Res 19:404, 1990b.
51. Sack R.L., Lewy A.J., Blood M.L., Stevenson J., and Keith L.D. Melatonin administration to blind people: phase advances and entrainment. J Biol Rhythms 6:249–261, 1991.
52. Lewy A.J., Ahmed S., Jackson J.M.L., and Sack R.L. Melatonin shifts circadian rhythms according to a phase-response curve. Chronobiol Int 9:380–392, 1992.

53. Sack R.L. Blood M.L., and Lewy A.J. Melatonin administration promotes circadian adaptation to night-shift work. Sleep Res 23:509, 1994.
54. Zaidan R., Geoffriau M., Brun J., Taillard J., Bureau C., Chazot G., and Claustrat B. Melatonin is able to influence its secretion in humans: description of a phase-response curve. Neuroendocrinology 60:105–112, 1994.
55. Deacon S. and Arendt J. Melatonin-induced temperature suppression and its acute phase-shifting effects correlate in a dose-dependent manner in humans. Brain Res 688:77–85, 1995.
56. Lewy A.J., Bauer V.K., Ahmed S., Thomas K.H., Cutler N.L., Singer C.M., Moffit M.T., and Sack R.L. The human phase response curve (PRC) to melatonin is about 12 hours out of phase with the PRC to light. Chronobiol Int 15:71–83, 1998a.
57. Lewy A.J., Sack R.L., Blood M.L., Bauer V.K., Cutler N.L., and Thomas K.H. Melatonin marks circadian phase position and resets the endogenous circadian pacemaker in humans. In: Circadian Clocks and their Adjustment. Edited by Chadwick D.J., Ackrill K. New York, John Wiley, pp. 303–321, 1994.
58. Johnson C.H. An Atlas of Phase Response Curves for Circadian and Circatidal Rhythms. Nashville, TN: Department of Biology, Vanderbilt University, 1990.
59. Lewy A.J., Bauer V.K., Cutler N.L., and Sack R.L. Melatonin treatment of winter depression: a preliminary study. Psychiatry Res 77:57–61, 1998b.
60. Lewy A.J., Sack R.L., Boney R.S., Clemons A.A., Anderson N.R., Pen S.D., Bauer V.K., Cutler N.L., and Harker C.T. Assays for measuring the dim light melatonin onset (DLMO) in human plasma. Sleep Res 26:733, 1997.
61. Lewy A.J. and Markey S.P. Analysis of melatonin in human plasma by gas chromatography negative chemical ionization mass spectrometry. Science 201:741–743, 1978.

52

MELATONIN AND AGING

Fred W. Turek,[1] Phyllis Zee,[2] and Olivier Van Reeth[3]

[1] Department of Neurobiology and Physiology
Northwestern University
2153 North Campus Drive
Evanston, Illinois 60208
fturek @ nwu.edu
[2] Department of Neurology
Northwestern University Medical School
645 North Michigan Avenue—Suite #1058
Chicago, Illinois 60611
p-zee @ nwu.edu
[3] Center des Rhythmes Biologiques School of Medecine
Université Libre de Bruxelles
Hôpital Erasme
Route de Lennik 808
B-1070 Brussels
Belgium
ovanree @ resu1.ulb.ac.be

1. INTRODUCTION

Had there been a meeting on melatonin three decades after its discovery, and an assessment of its potential in 1988, perhaps many of the same topics would have been covered as are being addressed after four decades today. However, one feature about melatonin has clearly changed over the past 10 years. Whereas just a few years ago melatonin was only of interest (and even known by) a handful of scientific investigators intrigued by the control and functions of this "obscure molecule" produced by the pineal gland, today melatonin is not only widely known by most of the biomedical community, it has become, at least in the United States, almost a household name (31). Much of the popular attention surrounding melatonin derives from the claims/

Melatonin after Four Decades, edited by James Olcese.
Kluwer Academic / Plenum Publishers, New York, 2000.

possibilities that melatonin treatment may be an anti-aging substance; a fountain of youth for the baby-boomer generation. Whereas many of the claims for melatonin as an anti-aging substance have been greatly exaggerated (23,30,31), there are some indications that it may indeed be useful in the treatment of various human disorders, including age-related illnesses.

While some of the claims for melatonin as an anti-aging hormone are based on studies indicating it can have immunomodulatory and tumor suppressing effects, as well as it being a potent-free radical scavenger (1,2,8,18,20,22) much of the current scientific interest in melatonin and aging is based on many years of scientific investigation on the role of the pineal gland, and melatonin in its regulation of daily (circadian) rhythms, including the sleep-wake cycle (see other chapters in this volume). Thus, this brief review will focus on age-related changes in circadian rhythms and sleep and then discusses the possible role of melatonin as a cause or consequence of these changes.

2. AGE-RELATED CHANGES IN CIRCADIAN RHYTHMICITY

The significant reduction in amplitude and the increased daily and day-to-day variability are among the most characteristic age-related changes of circadian rhythmicity in rodents and humans alike (4,16,24,29,35,36,40). Decreased sensory input, deficits in the output pathways of the circadian system, or both, as well as changes in noncircadian homeostatic mechanisms during senescence, may represent some of the underlying mechanisms. However, the consistency of amplitude reduction among a number of humoral, physiological and behavioral variables may indicate changes in the circadian pacemaker system itself (24,36). It has been argued that amplitude is an important determinant of internal stability in the circadian system, and it may be a key variable during senescence (2). Recent studies have supported this possibility by showing a reduction in the amplitude of circadian rhythmicity at the level of the SCN in old rats (26).

Further evidence indicates that aging may interfere with the structural and functional integrity of the circadian pacemaker itself. General features of cell degeneration and selective loss of vasopressin and vasoactive intestinal polypeptide-immunoreactive neurons have been reported in the SCN region of old rodents (40,44). Moreover, aging is associated with altered circadian patterns of glucose utilization, receptor density, and neuronal firing in the rat SCN (26,41,42). Limited evidence indicates the presence of neuronal loss in the aging human SCN, particularly in patients with Alzheimer's disease, (16,40) although negative findings have been reported more recently by other investigators (33). Finally, age-related changes in the free-running period of circadian rhythmicity, believed to reflect an intrinsic property of the SCN, have been repeatedly reported in several rodent species and in a few human studies.

Several features characterize the interaction between the aging circadian system and its environment. Advanced phase angle of entrainment of circadian rhythms and altered rate of resynchronization or functional recovery after large shifts in the external light-dark cycle have been reported in rodent and human aging studies (5,17,24,35,45). In addition to being consistent with the age-related changes of the free-running period in some cases (45), these differences in entrainment and resynchronization may also suggest an altered ability of the aging circadian system to respond to major time-giving cues in the environment.

Indeed, recent studies in golden hamsters have demonstrated that the phase-shifting and entraining effects of several nonphotic and activity-inducing stimuli are significantly attenuated or absent during senescence (38,39). Furthermore, unusually large phase shifts of the circadian pacemaker were observed after old animals were exposed to light pulses during the mid-subjective night (25). In addition, aging can affect the photic induction of the immediate early gene (c-fos) product in the rodent SCN (28). The ability of increased light intensity in the external light-dark cycle to reverse some of the changes in the circadian sleep-wake-cycle of old rats (43) and the activity rhythm in old hamsters (14,43) further supports the existence of altered sensitivity to environmental timing signals during senescence. Moreover, the demonstration that external photic and nonphotic stimuli can rapidly reset the phase of the human circadian pacemaker (34,37) opens new possibilities for research on the potential changes of this response in the elderly.

3. MELATONIN, AGING, AND THE CIRCADIAN SYSTEM

Aging is associated with a number of changes in the morphology, physiology and biochemistry of the pineal gland resulting in a significant reduction of the nocturnal melatonin levels in rodents and humans alike (12,20,27). It has been hypothesized that the age-related disruption of this robust signal affects the integrity of circadian time structures and is a precursor of disease states (2,21). Recent studies indicate that decreased binding of melatonin to the SCN of old rats is correlated with disruption in overt circadian rhythmicity (43), whereas treatment with the melatonin agonist, S-20242, can partially reverse the age-related decrease in the amplitude of the circadian temperature rhythm (13). We have recently found that in some hamsters treatment with melatonin in the food increases the amplitude and cohesiveness of the rhythm of locomotor activity which often becomes fragmented and disorganized in old animals (Van Reeth, unpublished results). Further support for some linkage between melatonin and aging is the intriguing finding that increased longevity in rats after life-long food restriction is associated with improved pineal function and increased melatonin levels (20).

In addition to the possibility that a more robust melatonin signal somehow improves the integrity of the circadian clock system of old animals, the sleep-inducing effects of melatonin, while somewhat controversial (32) provide a second avenue for possible beneficial effects of melatonin in older humans. Disrupted sleep at an advanced age is a frequent complaint/problem in the elderly, which often takes the form of fragmentation of the periods of wake and sleep or a phase advance of the timing of sleep onset and offset (3). Thus any substance which enhances the overall temporal organization, causes phase shifts in circadian rhythmicity, and can act as a mild hypnotic would have the potential of enhancing the timing and duration of sleep. Since there is evidence that melatonin can have all of these effects, the possible anti-aging effects of melatonin could involve a number of different pathways.

The potential use of melatonin replacement for treating sleep-wake disorders in the elderly is a particularly attractive hypothesis since: 1) disturbed sleep becomes more prevalent with advanced age (3,6,19), 2) melatonin levels decline with age (35), and 3) melatonin levels are reported to be significantly lower in elderly insomniac patients than in age-matched controls (9). In a few studies using wrist activity as a marker of sustained sleep in elderly subjects complaining of insomnia, both low and high doses of melatonin were found to reduce motor activity at night (7,10,11,15). While such findings have raised the hope that melatonin treatment will be effective for improving sleep in the elderly, more rigorous studies involving polysomnographic sleep recordings are required before it can be concluded that melatonin is indeed an effective hypnotic for use in the elderly.

4. FUTURE PERSPECTIVES

Thus, although many of the exaggerated claims in the mass media and popular books for melatonin as a fountain-of-youth have little scientific support, there is a growing literature to indicate that there are some important links between the 24 hour rhythm of melatonin and age-related changes in physiology and behavior. Because melatonin can act as a signal to the brain and body providing information as to the time of day (and year), the potential effects of melatonin (or the lack of effect) on the aging of various physiological systems may be tied in with its properties of providing information to the organism about its overall temporal organization. Improving temporal organization in older humans is the goal of many pharmacological and non-pharmacological studies. Only time (i.e. more research), perhaps another decade, will tell whether or not melatonin or related molecules will be an important agent for attenuating age-related changes in sleep and circadian organization.

REFERENCES

1. Arendt, J. in Endocrinology (ed. DeGroot, L.J.) 432–444 Saunders, Philadelphia, 1994.
2. Armstrong, S.M. and Redman, J.R. Melatonin: A chronobiotic with anti-aging properties? Med Hypotheses 34:300–309, 1991.
3. Bliwise, D.L. in Principles and Practices of Sleep Medicine (eds. Kryger, M.H., Roth, T. and Dement, W.C.) 26–39 W.B. Saunders, Philadelphia, 1994.
4. Brock, M.A. Chronobiology and aging. J Am Geratrics Soc 39:74–91, 1991.
5. Buresova, M., Benesova, O., and Illnerova, H. Aging altersresynchronization of the circadian system in rats after a shift of the light-dark cycle. Experientia 46:75–77, 1990.
6. Dement, W.C., Miles, L., and Carskadon, M. "White Paper" on sleep and the elderly. J Am Geriatr Soc 30:25–50, 1982.
7. Garfinkel, D., Laudon, M., Nof, D., and Zisapel, N. Improvement of sleep quality in elderly people by controlled-release melatonin. Lancet 346:541–544, 1995.
8. Grad, B.R. and Rozencwaig, R. The role of melatonin and serotonin in aging: Update. Psychoneuroendocrinology 18:283–295, 1993.
9. Haimov, I., Laudon, M., Zisapel, N., Souroujon, M., Nof, D., Shlitner, A., Herer, P., Tzischinsky, P., and Lavie, P. Sleep disorders and melatonin rhythms in elderly people. British Med J 309:167, 1994.
10. Haimov, I. and Lavie P. Potential of melatonin replacement therapy in older patients with sleep disorders. Drugs Aging 7:75–78, 1995.
11. Haimov, I., Lavie, P., Laudon, M., Herer, P., and Zisapel, N. Melatonin replacement therapy of elderly insomniacs. Sleep 8:598–603, 1995.
12. Humbert, W. and Pevet, P. The decrease of pineal melatonin production with age. Ann NY Acad Sci 719:43–63, 1994.
13. Koster-Van Hoffen, G.C., Mirmiran, M., Bos, N.P., Witting, W., Delagrange, P., and Guardiola-Lemaitre, B. Effects of a novel melatonin analog on circadian rhythms of body temperature and activity in young, middle-aged, and old rats. Neurobiol Aging 14:565–569, 1993.
14. Labyak, S.E., Turek, F.W., Wallen, E.P., and Zee, P.C. The effects of bright light on age-related changes in the locomotor activity of Syrian hamsters. Am J Physiol 274:R830–R839, 1998.
15. Lockley, S., Tabandeh, H., Skene, D., Buttery, R., Bird, A., Defrance, R., and Arendt, J. Day-time naps and melatonin in blind people. Lancet 346:1491, 1995.

16. Mirmiran, M., Swaab, D.F., Kok, J.H., Hofman, M.A., Witting, W., and Van Gool, W.A. Circadian rhythms and the suprachiasmatic nucleus in perinatal development, aging and Alzheimer's disease. Progress in Brain Research 93:151–163, 1992.
17. Monk, T.H., Buysse, D.J., Reynolds, C.F., and Kupfer, D.J. Inducing jet lag in older people: adjusting to a 6-hour phase advance in routine. Exp Gerontol 28:119–133, 1993.
18. Penev, P.D. and Turek, F.W. The role of melatonin and serotonin in circadian rhythms and aging. Cur Opin Endocrinol Diab 2:169–176, 1995.
19. Prinz, P.N. Sleep and sleep disorders in older adults. J Clin Neurophysiol 12:139–146, 1995.
20. Reiter, R.J. The aging pineal gland and its physiological consequences. Bioessays 14:169–175, 1992.
21. Reiter, R.J. The pineal gland and melatonin in relation to aging: a summary of the theories and the data. Exp Gerontol 30:199–212, 1995.
22. Reiter, R.J., Poeggeler, B., Menendez-Pelaez, A., Chen, L.D., and Saarela, S. Melatonin as a free radical scavenger: Implications for aging and age-related diseases. Ann NY Acad Sci 719:1–12, 1994.
23. Reppert, S.M. and Weaver, D.R. Melatonin madness. Cell 83:1059–1062, 1995.
24. Richardson, G.S. in Handbook of the Biology of Aging (eds. Schneider, E.L. and Rowe, J.W.) 275–305 Academic Press, San Diego, 1990.
25. Rosenberg, R.S., Zee, P.C., and Turek, F.W. Phase response curves to light in young and old hamsters. Am J Physiol 261:R491–R495, 1991.
26. Satinoff, E., Li, H., Tcheng, T.K., Liu, C., McArthur, A.J., Medanic, M., and Gillette, M.U. Do the suprachiasmatic nuclei oscillate in old rats as they do in young ones? Am J Physiol 265:R1216–R1222, 1993.
27. Schmid, H.A. Decreased melatonin biosynthesis, calcium flux, pineal gland calcification and aging: a hypothetical framework. Gerontology 39:189–199, 1993.
28. Sutin, E.L., Dement, W.C., Heller, H.C., and Kilduff, T.S. Light-induced gene expression in the suprachiasmatic nucleus of young and aging rats. Neurobiol Aging 14:441–446, 1993.
29. Touitou, Y., Fevre, M., Bogdan, A., Reinberg, A., DePrins, j., Beck, H., and Touitou, C. Patterns of plasma melatonin with aging and mental condition: Stability of nyctohemeral rhythms and differences in seasonal variations. Acta Endocrinol 106:145–151, 1984.
30. Turek, F.W. Melatonin hype hard to swallow. Nature 379:295–296, 1996.
31. Turek, F.W. Melatonin: pathway from obscure molecule to international fame. Persp Biol Med 41:8–20, 1997.
32. Turek, F.W. and Czeisler, C.A. in Neurobiology of Sleep and Circadian Rhythms, Chapter 7 (eds. Turek, F.W. and Zee, P.C.) Marcel Dekker, Inc., New York, 1998.
33. Turek, F.W., Pénev, P., Zhang, Y., Van Reeth, O., Takahashi, J.S., and Zee, P.C. in Circadian Clocks and their Adjustment, Ciba Foundation Symposium 212–234 Pitman Press, London, 1994.
34. Van Cauter, E., Sturis, J., Byrne, M.M., Blackman, J.D., Leproult, R., Ofek, G., L'Hermite-Balériaux, M., Refetoff, S., Turek, F.W., and Van Reeth, O. Demonstration of rapid light-induced advances and delays of the human circadian clock using hormonal phase markers. Am J Physiol 266:E953–E963, 1994.
35. Van Coevorden, A., Mockel, J., Laurent, E., Kerkhofs, M., L'Hermite-Balériaux, M., Decoster, C., Nève, P., and Van Cauter, E. Neuroendocrine rhythms and sleep in aging. Am J Physiol. 260:E651–E661, 1991.
36. Van Gool, W.A. and Mirmiran, M. in Progress in Brain Research (ed. Swaab, D.F.) 255–277 Elsevier, New York, 1986.
37. Van Reeth, O., Sturis, J., Bryne, M.M., Blackman, J.D., L'Hermite-Balériaux, M., Leproult, R., Oliner, C., Refetoff, S., Turek, F.W., and Van Cauter, E. Nocturnal exercise phase-delays the circadian rhythms of melatonin and thyrotropin secretion in normal men. Am J Physiol 266:E964–E974, 1994.
38. Van Reeth, O., Zhang, Y., Reddy, A., Zee, P.C., and Turek, F.W. Aging alters the entraining effects of an activity-inducing stimulus on the circadian clock. Brain Res 607:286–292, 1993.
39. Van Reeth, O., Zhang, Y., Zee, P.C., and Turek, F.W. Aging alters feedback effects of the activity-rest cycle on the circadian clock. Am J Physiol 263:R981–R986, 1992.
40. Van Someren, E.J.W., Mirmiran, M., and Swaab, D.F. Nonpharmacological treatment of sleep and wake disturbances in aging and Alzheimer's disease: chronobiological perspectives. Behav Brain Res 57:235–253, 1993.
41. Weiland, N.G. and Wise, P.M. Aging progressively decreases the densities and alters the diurnal rhythms of alpha-1 adrenergic receptors in selected hypothalamic regions. Endocrinology 126:2392–2397, 1990.

42. Wise, P.M., Cohen, I.R., Weiland, N.G., and London, D.E. Aging alters the circadian rhythm of glucose utilization in the suprachiasmatic nucleus. Proc Natl Acad Sci USA 85:5305–5309, 1988.
43. Witting, W., Mirmiran, M., Bos, N.P.A., and Swaab, D.F. Effect of light intensity on diurnal sleep-wake distribution in young and old rats. Brain Res Bull 30:157–162, 1993.
44. Woods, W.H., Powell, E.W., Andrews, A., and Ford, C.W. Light and electron microscopic analysis of two divisions of the suprachiasmatic nucleus in the young and aged rat. Anat Rec 237:71–88, 1993.
45. Zee, P.C., Rosenberg, R.S., and Turek, F.W. Effects of aging on entrainment and rate of resynchronization of the circadian locomotor activity. Am J Physiol 263:1099–1103, 1992.

53

PHASE OF MELATONIN RHYTHM IN WINTER DEPRESSION

Arcady A. Putilov,* Galena S. Russkikh, and Konstantin V. Danilenko

Institute for Medical and Biological Cybernetics
Siberian Branch, RAMS
Novosibirsk, and International Scientific Center ARKTIKA
Far East Branch, RAS, Magadan, Russia

1. INTRODUCTION

Seasonal affective disorder (SAD), winter type, is a relatively recently described syndrome of recurrent episodes of fall-winter depression, fatigue, social withdrawal, oversleeping, overeating, carbohydrate craving and weight gain. Bright light treatment (LT) reverses these symptoms, else they disappear in spring (47).

Association of winter depression with reduced day length may suggest that photoperiodic time measurement plays a role in etiology of this disease. LT of the first patient with winter depression (25) was inspired by the facts that, in animals, the day length response is mediated by the circadian rhythm of melatonin (MLT) (14) and, in humans, the exposure to bright light is necessary to suppress MLT secretion (24). However, most investigations argue against the involvement of this hormone in SAD and LT. The reports of the efficacy of midday light (66) were considered to lend no support for the involvement of day length response in SAD pathogenesis and the mechanism by which LT works. MLT administered orally in the morning and early evening did not completely reverse the effects of LT, although it did reproduce the atypical depressive symptoms of SAD (social withdrawal, hyperphagia, appetite and weight increase, carbohydrate craving, hypersomnia, fatigability and reverse diurnal variations) (49). In another study (73), MLT given to SAD patients either in the morning or in the evening had no effect on depressive symptoms, while bright light did.

* e-mail: putilov@cyber.ma.nsc.ru.

Melatonin after Four Decades, edited by James Olcese.
Kluwer Academic / Plenum Publishers, New York, 2000.

The experiments with two beta-adrenergic blockers, which suppress MLT secretion, led to discrepant results. In an early study, atenolol did not reproduce the antidepressant effect of light in most SAD subjects treated in the afternoon. However, some subjects responded very well (50). In a more recently reported experiment (55), an antidepressant response was demonstrated in SAD patients treated in early morning with another beta-adrenergic blocker, propranolol. The subsequent study (34) provides additional support for the efficacy of timed beta blockade.

Moreover, some recent investigations of healthy subjects have drawn renewed interest to the MLT hypothesis. Wehr et al. (67,68) have shown that the duration of human MLT secretion responds to changes in photoperiod in ways that resemble the responses seen in animals. Spiegel et al. (58) have demonstrated that the duration of MLT secretion may be shortened or lengthened by sleep curtailment or extension, respectively. However, the seasonal changes in the circadian patterns of MLT and some other hormones heve not been found in winter SAD (3,69).

Since the extension of day length appears not to be critical for the antidepressant effect of bright light, another chronobiological response to light—correcting of abnormally phased circadian rhythms—was proposed as a possible mechanism of LT for SAD (27,28). According to the phase-shift hypothesis, endogenous circadian rhythms in most SAD patients are abnormally phase-delayed with respect to real time or sleep time, and LT in the morning can correct this phase disturbance by advance shift of circadian pacemaker.

Hypersomnia and morning fatigue have been proposed as markers of delayed circadian rhythms (26,62,74). In a number of investigations, the symptom of hypersomnia was found to be a predictor of clinical response to LT in SAD (2,21,22,35,37,59). Besides, the significant correlation between difficulty awaking and severity of depression was reported by Avery et al. (4). However, in some SAD populations the symptom of hypersomnia was not very common, and hypersomnic SAD subjects responded well to nighttime LT, although this treatment does not produce the advance of their circadian phase (74).

As has been noted by Avery et al. (5), a phase delay of circadian rhythms relative to sleep may explain why SAD subjects experience hypersomnia. When, under conditions of internal desynchronization, the temperature minimum is phase-delayed relative to sleep onset, the sleep duration is relatively long (9,77). However, the necessity of pretreatment phase delay and phase advance by morning light for clinical response might be questioned. The results of SAD studies designed to test the pretreatment circadian phase and its shift following LT are inconsistent. A phase delay of the MLT rhythm in depressed SAD patients was noted in a number of investigations (53,60,72,75). However, there were also reports of a normal phase (8,57,76). Similarly, a phase advance of the MLT rhythm following LT was reported by several (10,27,28,46,53,60), but not all groups (8,57,75,76).

In the studies of masked temperature rhythms in SAD subjects, neither baseline delays of phase, nor advances caused by LT were observed (13,23,41,51,57). However, in constant routine studies, both pretreatment phase delay and advance shift following morning LT were reported for main markers of circadian phase position: the rhythms of body temperature, cortisol and MLT (5,10). In another study using constant routine protocol (76), the temperature rhythm showed the tendency to delay in winter, and certain parameters of the rhythm were advanced by midday light.

The meta-analysis of data from different research centers shows that light is therapeutic at most times of day including hours when MLT is not secreted. However, the efficacy of early morning LT seems to be somewhat higher compared to the efficacy of evening LT, and the efficacy of these treatments seems to be higher than the efficacy of treatments scheduled in between (see review (74)). It was noted that evening LT does not potentiate the following response to morning LT, whereas morning light inhibits the following response to evening light (45,61). However, no differences in the antidepressant response were found in studies using a parallel design with random assignment to time of a.m. and p.m. treatment (11,32,40,75).

Since the phase-shifting effect of daytime bright light was reported in healthy subjects (18), it is possible that the human circadian pacemaker maintains the sensitivity to light throughout subjective day and, therefore, midday LT could advance the circadian phase. However, the findings of considerable improvement caused by day-by-day variable LT (1,32) may be considered as evidence against the phase-shift hypothesis. Besides, the causal link between the antidepressant response and phase shifts in winter depression was not supported satisfactorily (20,59,63). Even negative relations between advance shift of the temperature rhythm and clinical response were reported (5,13). Thus, the phase shifting nature of SAD and LT remains a controversial theory.

In this study we tested the hypothesis that abnormal phase of MLT rhythm underlies SAD and that in the majority of SAD patients LT works by advancing of MLT phase (27). We also summarize briefly the results of some other our studies to show that any simple pathophysiological model of SAD seems not to be adequate, and that several physiological mechanisms could be involved in regulating the mood of winter depressives.

2. METHODS

2.1. The Questionnaire Study

A total sample of 2183 adults who had lived in Turkmenia (38 deg North, n = 328), Novosibirsk (55 deg North, n = 761), Chukotka and Yakutia (61–66 deg North, n = 1094) was surveyed to examine the relations between self-assessed level of depression and types of seasonal and diurnal rhythms. In overall, these respondents were more likely to be female (54.4%) and older than 34 (54.3%). One of Novosibirsk subsamples (n = 150) was obtained in the process of recruiting female subjects with winter depression and nonseasonal female controls for our LT study in the hospital of the Siberian Branch of the Russian Academy of Medical Sciences near Novosibirsk.

2.1.1. Self-Assessment of Seasonality. The Seasonal Pattern Assessment Questionnaire (SPAQ) (48) was applied to 1) determine the patterns of seasonal variations in the symptoms of seasonal depression and 2) assign the respondents to the diagnostic groups: nonseasonals (N-SAD) and seasonals with subsyndromal SAD (S-SAD) or full-blown SAD (see criteria in (6,19)). The respondents were ask to circle all months during which they: feel the best and feel the worst, more energetic and active than usual and feel drowsy during the daytime, sleep less than usual and sleep more than usual, wake up too early and have difficulty falling asleep, lose the most weight and gain the most weight, eat less than usual and eat more than usual, socialize most and socialize least, etc. Besides, they were asked to assess the degree and severity of seasonal variations in sleep length, social activity, well-being, weight, appetite and energy level.

Figure 1 illustrates the relations between seasonal changes in scotoperiod (night-time plus twilight, hours) and seasonal variations in some neurovegetative depressive symptoms. The smoothed symptom curves represent the difference between percent of respondents who reported positive symptom and percent of those who reported negative symptom, i.e. early waking minus late waking (post/presleep insomnia), lose weight minus gain weight (under/overweightiness), sleep less minus sleep more (under/oversleeping), and feel more active minus feel more drowsy (hyper/hypoactivity). Moreover, the annual pattern of general well-being (feel best minus feel worst) is shown in Figure 1 to provide a comparison of this pattern with the patterns of neurovegetative symptoms.

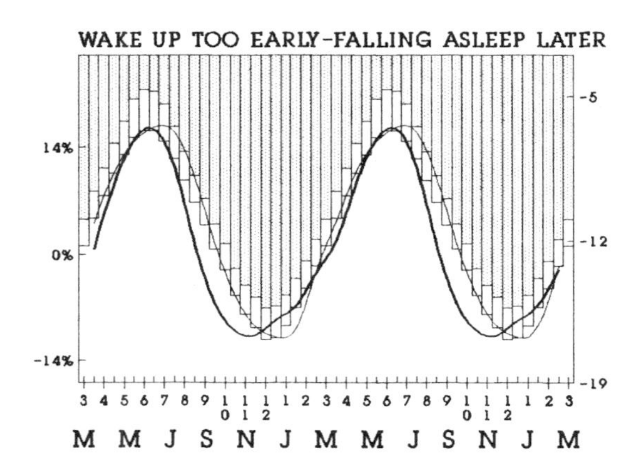

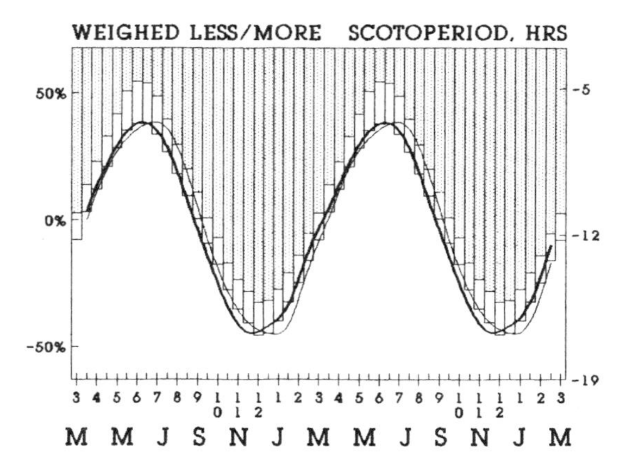

Figure 1. Annual rhythms of four typical neurovegetative depressive symptoms in winter seasonals (solid lines). All data are double-plotted. The scotoperiod data are shown with two-week interval. Right scale of Y-axis: duration of scotoperiod in hours. Filled columns show night length, open columns show duration of twilight. The symptom data (n = 324) were reported with one-month interval. Left scale of Y-axis: differences in percent of respondents reported a positive symptom (e.g. sleep less) and percent of respondents reported a negative symptom (e.g. sleep more) in a given month. The thin lines show seasonal variations in general well-being. For better fitting, the amplitude of these variations was adjusted to the amplitudes of seasonal variations in neurovegetative symptoms, and the scotoperiodic data were shifted on two weeks relative to the symptom data.

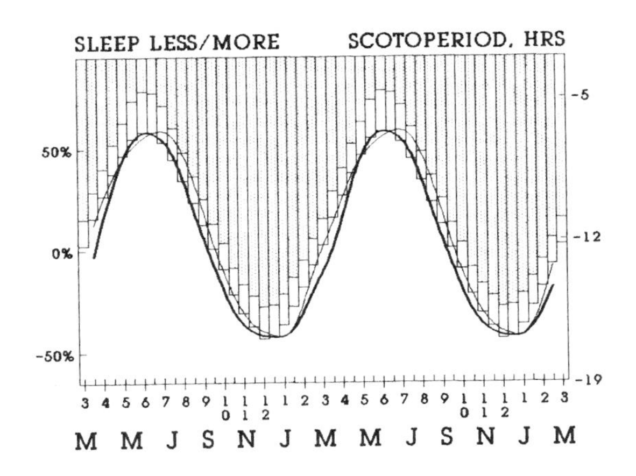

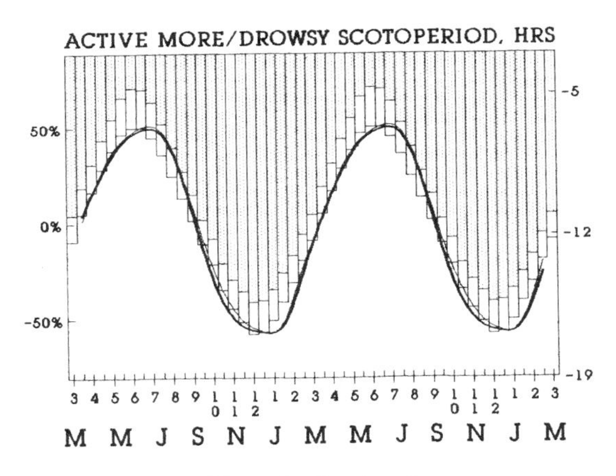

Figure 1. *Continued.*

The delays of seasonal variations in 6 depressive symptoms relative to the annual changes in photoperiod were calculated by two methods. First, by searching for maximal crosscorrelation between weekly data on photoperiod and row monthly data on symptom (P). Second, by averaging delays (p) for: 1) month of maximal rate of symptom relative to June (M), 2) month of minimal rate relative to December (m), 3) month of mean level upcrossing relative to March (C) and 4) month of mean level downcrossing relative to September (c). For the purpose of statistical comparison, the delays for every symptom were obtained in 24 subsamples or less (the delays were not calculated when maximal coefficient of crosscorrelation was lower than +0.82, $p > 0.001$ for $n = 12$ months). The mean delays shown Figure 2 were calculated by averaging the delays either over all subsamples (All) or within every of 5 groups of subsamples: #1–6 subsamples of residents of Turkmenia including summer seasonals ($n = 51$), winter seasonals ($n = 20$), and 4 subsamples of nonseasonals: males younger 35 or older ($n = 44$ and 47, respectively) and females younger 35 or older ($n = 77$ and 67, respectively); #2—nonseasonals from Novosibirsk divided on 4 subgroups according to their age and gender ($n = 118$, 145, 148 and 87, respectively); #3–6 subsamples of winter seasonals from Novosibirsk: females screened for LT study (44 were younger than 35 and 28 were 35 or older), and other winter seasonals divided on 4 groups according to their age and

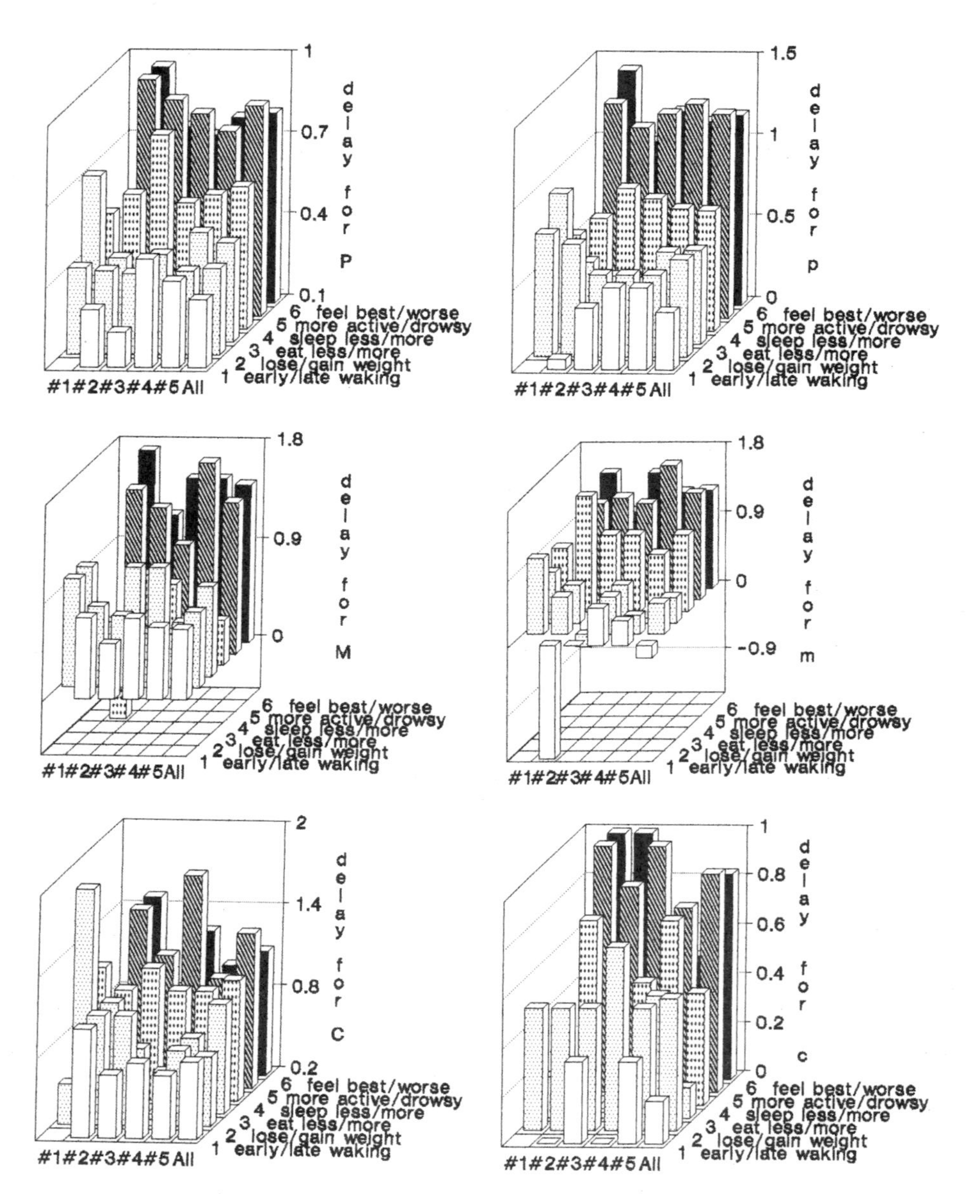

Figure 2. Delays of annual rhythms of six depressive symptoms relative to the annual light cycle. P—delays for maximal crosscorrelation between weekly data on day length and monthly data on depressive symptoms. p—delays averaged over the following four phases of annual rhythm: M—delays for month of maximal rate of symptom relative to June, m—delays for month of minimal rate relative to December, C—delays for month of mean level upcrossing relative to March, c—delays for month of mean level downcrossing relative to September. ##1–5—delays for separate groups of subsamples, All—delays for all subsamples.

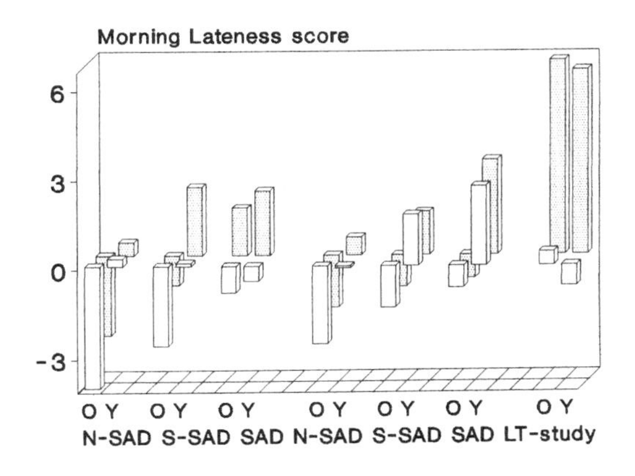

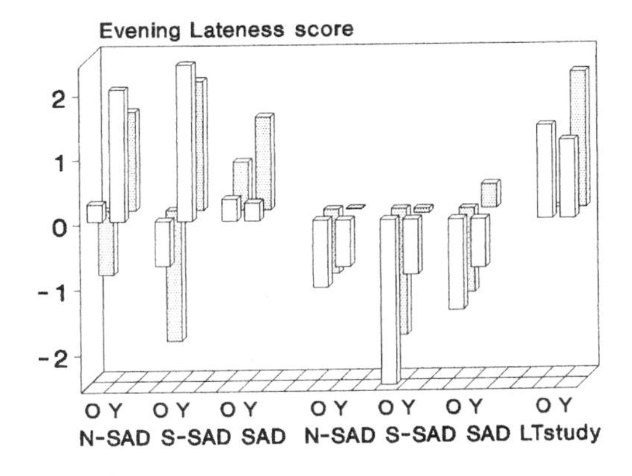

Figure 3. Scores on M- and E-scales (Morning- and Evening Lateness). Positive scores suggest a delay of the sleep-wake cycle. Filled columns represent depressed respondents from the questionnaire study or patients with winter depression from the LT-study, open columns represent non-depressed respondents or healthy controls, respectively. O—35 year of age or older, Y—younger than 35. N-SAD, S-SAD and SAD subgroups on the left represent male respondents of the questionnaire study with diagnoses No-SAD, Subsyndromal SAD and SAD; on the right—female respondents of N-SAD, S-SAD and SAD subgroups.

gender (n = 17, 13, 35 and 18, respectively); #4—nonseasonals from Yakutia and Chukotka divided on 4 age-gender groups (n = 148, 279, 149 and 258, respectively); and #5—winter seasonals from Yakutia and Chukotka divided on 4 age-gender groups (n = 18, 29, 43 and 58, respectively).

2.1.2. Self-Assessment of the Sleep-Wake Pattern and Level of Depression. Individual traits of the sleep-wake cycle were self-assessed with the 40-item Sleep-Wake Pattern Assessment Questionnaire (SWPAQ) (38). This instrument includes two morningness-eveningness scales which can be defined briefly as levels of morning and evening wakefulness (12-item M-scale and 8-item E-scale). The positive scores indicate the tendencies toward late awakening on M-scale (Morning Lateness) and late bedtimes on E-scales (Evening Lateness). The 20-item Center for Epidemiologic Studies—Depression Scale (CES-D) (44) was used to group respondents into those with lower and higher current level of depressive symptoms (the scores of 16 or more may indicate clinical depression (7,44). Figure 3 gives M- and E-scores in 28 subgroups of respondents: the rightist 4 columns represent the subjects screened for LT-study (younger and older female nonseasonals and younger and older female winter seasonals), the other 24 columns represent the rest of the questionnaire sample: more and less depressed male and female respondents of younger and older age grouped in N-SAD, S-SAD and SAD diagnostic categories. To clarify the effects of seasonality and

Table 1. Results of 4-Way ANOVAs for morning and evening lateness

	Morning lateness score				Evening lateness score		
Factor	F	p	Df	Factor	F	p	Df
1 Gender	2.12	0.141	1/2026	1 Gender	25.30	0.000	1/2026
2 Age	61.80	0.000	1/2026	2 Age	40.50	0.000	1/2026
3 Depression	6.62	0.009	1/2026	3 Depression	0.25	0.621	1/2026
4 Diagnosis	13.15	0.000	2/2026	4 Diagnosis	1.17	0.311	2/2026

Notes. Level of significance (p) for all interactions was higher than 0.05. Gender: Male/Female; Age: younger than 35/older; Depression: CES-D score <16/>15; Diagnosis: N-SAD/S-SAD/SAD.

mood state on diurnal type, the M- and E-scores in the latter 24 subsamples were compared using 4-way ANOVA with 4 grouping factors (level of depression, diagnostic category, age and gender; Table 1).

2.2. The LT-Study

This set of investigations was designed to determine whether bright light improves both psychic and physiological functions in SAD. In total, 61 female patients with mean age 33.6 and 36 female controls with mean age 36.1 participated in 1-week trials of LT (2,500 lux of white light for 2 hours daily). The subjects were divided on a.m. and p.m. subgroups treated in the morning (from 8:00) or in the second part of the day (from 16:00 or later).

The therapeutic effects of LT were evaluated by KVD with the 21-item Hamilton Rating Scale for Depression (HRSD) (17) and self-rated with the 29-item Structured Interview Guide for the Hamilton Depression Rating Scale: Seasonal Affective Disorder (SIGH-SAD). The SIGH-SAD includes the HRSD and 8-item Addendum concerning atypical depressive symptoms (71). To define the recovery on HRSD, we used the outcome criteria of Terman et al. (61): at least 50% reduction of the depression score till a final level under 8. The analogous criteria were chosen to determine the recovery on SIGH-SAD: at least 50% drop of score till a final level under 16.

2.2.1. Measurement of Physiological Responses to Light. In every subject we examined the signs of such physiological effects of light as advance of circadian phase, increase in energy expenditure and activation of sympatho-adrenal system. Only few biological variables (i.e. body weight, and daily variations of axillary temperature and heart rate) were measured in all subjects, while in more detail circadian rhythms, metabolism and sympathetic reactivity were investigated in the subsamples of at least 20 patients with SAD and 20 age-matched controls (12,39,41,43,54). In particular, the levels of serum MLT, cortisol and some other hormones were measured in a subsample of 23 patients and 20 controls (Figures 4 and 5). Moreover, the indicators of the fourth physiological effect, intensification of non-rapid eye movement sleep (non-REMS), were studied in a subsample of 21 patients and 10 controls (36,42). Patients were grouped according to whether or not they had an expected physiological change. A subject was included in a group with expected change if more than a half of available physiological indexes showed the change in a direction of non-REMS intensification, circadian phase advance, increase in energy expenditure or activation of sympathoadrenal system.

Table 2. Results of 5-Way and 4-Way ANOVAs for serum melatonin

	All subjects				Patients		
Factor	F	p	Df	Factor	F	p	Df
1 Diagnosis	4.33	0.042	1/35	1 Time of LT	4.31	0.049	1/19
2 M-type				2 Response			
3 E-type	4.51	0.038	1/35	3 Condition			
4 Condition				4 Time of day	8.92	0.000	4/76
5 Time of day	13.73	0.000	4/140	1 × 2	3.31	0.082	1/19
2 × 5	2.63	0.036	4/140	3 × 4	3.45	0.011	4/76
3 × 5	2.60	0.037	4/140	1 × 2 × 4	3.77	0.007	4/76
4 × 5	3.28	0.013	4/140	1 × 2 × 3 × 4	3.56	0.010	4/76

Notes. Level of significance (p) for other interactions was higher than 0.1. Diagnosis: Controls/Patients; M-type or E-type: Positive/Negative score; Condition: Before/After LT; Time of day: Morning/Midday/Afternoon/ Evening/Midnight. Time of LT: a.m./p.m.; Response: Nonresponders/Responders (The patients of responder group showed the reduction of both HRSD and SIGH-SAD scores by at least 50% to levels of less than 8 on HRSD and less than 16 on SIGH-SAD).

2.2.2. Measurement of MLT and Cortisol. At a day before and a day after LT an indwelling venous catheter was inserted into a forearm vein at 8 a.m. and kept patent with heparinized normal saline solution till midnight or 4 a.m. the following day. Blood samples were drawn every 4 hr from 8 a.m. either until midnight (the majority of subjects) or until 4 a.m. (7 patients of the morning subgroup). Fourteen patients and 14 controls were restudied in summer. MLT was determined by a direct radioimmunoassay (15). The antibodies from Stockgrand Ltd, G/S/704–6483 (Guilford, England), were used at an initial dilution of 1:3000. Cortisol was measured according to a concurrent immunometric method based on enhanced luminescence (70). The assay was performed using commercially available kits (LAN.0801) from Amerlite Diagnostic Ltd (Amersham, England).

The effects of LT on hormonal levels in patients and controls was evaluated using a five-way ANOVA, with three independent between-subject factors (diagnosis, M- and E-types) and two within-subject factors (condition and time of day). A four-way ANOVA was employed to patients' data to evaluate the interaction between the effects of time of LT, clinical response, condition and time of day (Table 2).

For analysis of phases of MLT and cortisol rhythms, the individual values were expressed as percentage of deviation from the individual daily mean (the mean was calculated by averaging over 5 time points). Diurnal rhythms of MLT in subjects of four M-&E-types are illustrated in Figure 4.

Circadian phase position was measured indirectly as an evening-morning difference: a difference between relative variable values obtained before and after sleep (at 24:00 and 8:00, respectively). We suggested that lower values correspond to later phases of MLT and cortisol rhythms (Figure 5; Table 3).

3. RESULTS

3.1. Results of the Questionnaire Study

Figure 1 demonstrates that most of winter seasonals reported annual variations in such bipolar neurovegetative symptoms as under/overweightiness, under/oversleeping and hyper/hypoactivity. Only one symptom, post/presleep insomnia that could be

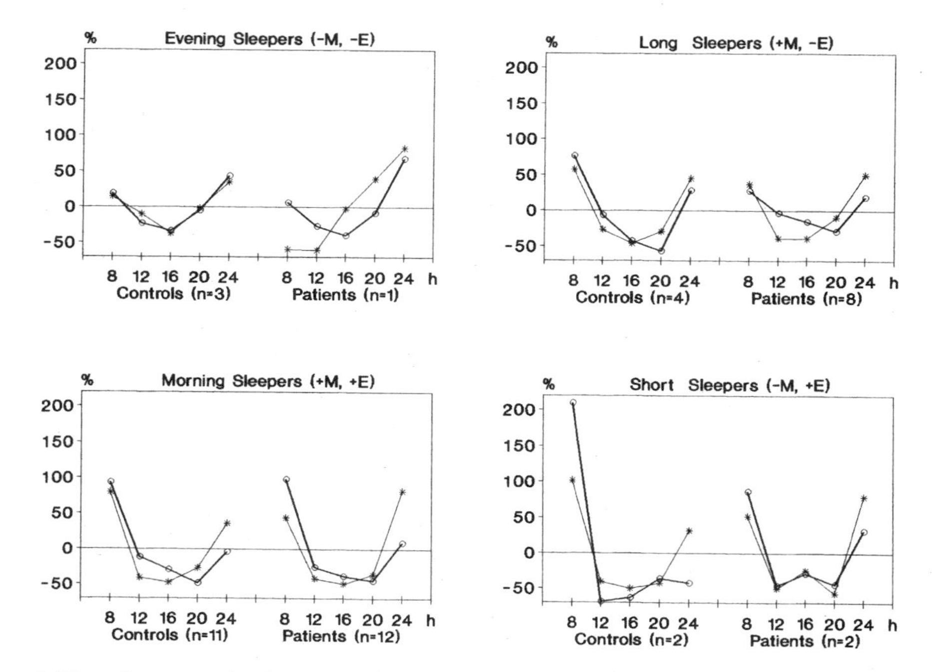

Figure 4. Diurnal patterns of melatonin in subjects of different diurnal types. Solid lines show patterns before light treatment, thin lines show patterns after light treatment.

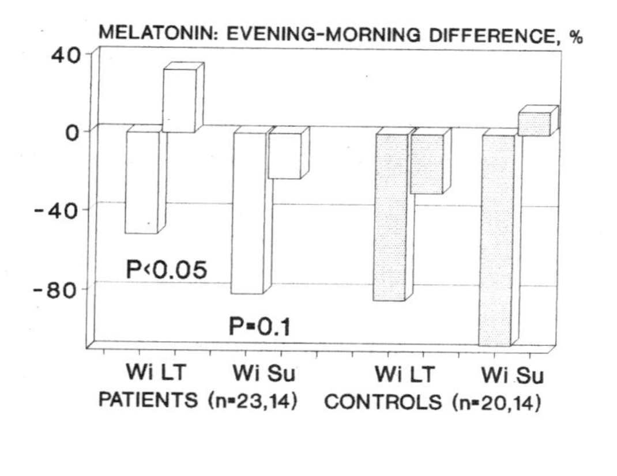

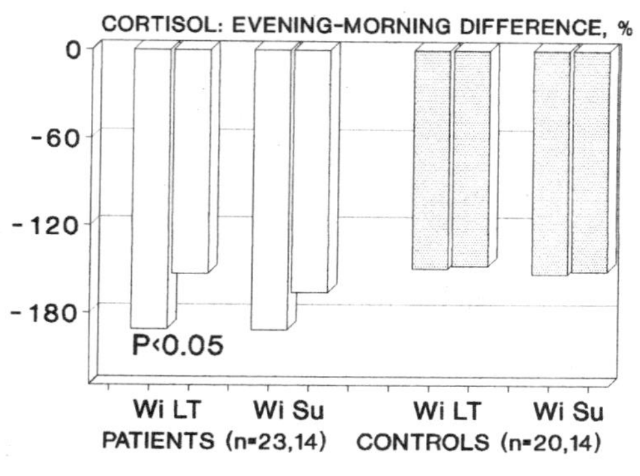

Figure 5. Evening-morning differences for melatonin and cortisol. The significant differences between pre- and posttreatment values are indicated by p-values (Student's paired t-test).

Table 3. Evening-morning difference for serum melatonin before/after a.m. or p.m. light treatment

Groups	All treatments	A.M. treatment	P.M. treatment
Controls, n = Mean (SD)	20 –84.4 / –30.2 (177.3) / (174.8)	10 –109.6 / –20.5 (132.4) / (123.9)	10 –59.3 / –39.9 (217.8) / (221.3)
Patients, n = Mean (SD) Paired t-test	23 –51.7 / –32.5 (128.2) / (132.3) t = 2.67, p < 0.05	13 –49.1 / 63.5 (135.4) / (156.6) t = 2.44, p < 0.05	10 –55.0 / –7.7 (125.3) / (83.3)
Nonresponders, n = Mean (SD) Paired t-test	8 –47.3 / 64.2 (123.1) / (193.5)	3 80.9 / 205.9 (80.3) / (257.8)	5 –124.3 / –20.7 (59.7) / (90.8) t = 3.60, p < 0.05
Responders, n = Mean (SD) Paired t-test	15 –54.0 / 15.6 (134.9) / (88.9) t = 2.01, p < 0.1	10 –88.1 / 20.7 (125.2) / (95.5) t = 3.17, p < 0.05	5 14.1 / 5.2 (140.5) / (83.4)

Note. t-values (Student paired t-test) are not shown if the level of significance for a difference between pre- and post-treatment values was found to be 0.1 or higher.

related to delaying/advancing of sleep timing, was much less common in comparison with other neurovegetative symptoms. The annual patterns were very similar in N-SAD, S-SAD and SAD. Only the amplitude parameter was considerably different in these diagnostic groups.

3.1.1. Delays of the Annual Rhythms of Depressive Symptoms. Although the seasonal variations in depressive symptoms closely follow the variations in day length, most of the symptoms lag behind the seasonal changes of daylight and darkness. Figure 2 shows that only two phases of one of the neurovegetative symptoms, namely the phase of fall downcrossing of the mean level (c) and the phase of winter minimum (m) for the symptom of post/presleep insomnia (=early/late waking), correspond directly with the seasonal decrease of daylight. All other phases appear to be delayed one to six weeks relative to the changes in day length. Besides, the differences between similar phases of different symptoms as well as the differences between different phases of the same symptom often reach a statistically significant level. These results suggest that, although the photoperiod appears to be the main seasonal timer for depressive symptoms, the modulating influence of some other environmental factors may account for the observed specific of the annual patterns of these symptoms. Winter cold may produce the increase in sleep duration and, due to that, the winter minimum of the corresponding symptom may be shifted towards January/February (the months of the lowest temperatures). The specific effects of summer heat could include the decrease of appetite leading to the delay of summer minimum of the eating symptoms on July/August (the months of highest temperatures). Besides, especially in the subsamples from the southernmost location (Turkmenia, #1), both summer heat and winter cold + winter photoperiod could affect several symptoms in a similar way, i.e. they both may cause such problems as late waking, daytime drowsiness and bad feeling. When the impacts of summer and winter seasons were similar, the annual patterns of the symptoms of post/presleep insomnia, hyper/hypoactivity and good/bad mood were bimodal and, as a result, the maximal crosscorrelations between the annual curves of symptom and scotoperiod were lower than +0.82 (due to that the data on these 3 symptoms in Turkmenian subsamples are not shown in Figure 2). In contrast, the effects of winter and summer environment on the symptoms related to sleep length and metabolism were always opposite and, therefore, all maximal crosscorrelations between the annual rhythms of these symptoms and seasonal changes in day length were highly significant (+0.99 in the whole sample).

In general, the findings on seasonality of neurovegetative depressive symptoms raise a question about multi-component nature of physiological dysfunction in SAD. We suggest that, depending on the particular underlying dysregulation, the MLT rhythm may be of more or less importance for control of seasonal variations in a certain symptom. Such chronobiological mechanisms as phase resetting and daytime measurement could be primarily responsible for the symptoms closely related to circadian and metabolic dysfunctions in winter depressives, while the disturbances in sleep homeostasis and arousal may be associated with both chronobiological and non-chronobiological mechanisms including thermoregulation.

3.1.2. Delay of the Sleep-Wake Cycle. Significant differences on scores of evening lateness were found neither between nonseasonals and seasonals in the questionnaire sample, nor between healthy controls and patients with winter depression in the LT-study. By contrast, the scores of morning lateness were significantly higher in seasonals compared to nonseasonals (Student's unpaired t-test, $p < 0.0001$ for both studies). Table 1 and Figure 3 indicate that high current level of depression and, especially, high degree of seasonality could account for the elevation of M-scores in SAD. The symptom of late awakening appears to be caused by both an increase in sleep need and a delay of circadian phase. The lack of significant association between E-score and seasonal depression may be explained by contradictions between the actions of two physiological factors on sleep onset: the increased sleep need produces an advance of bed time, while the delay of circadian phase leads to a delay of bed time.

3.2. Results of LT-Study

LT caused clinically and statistically significant reduction of depressive scores. Time of LT was not important for antidepressant response.

3.2.1. Physiological Effects of LT. The reduction of HRSD and SIGH-SAD scores was lower than 60% in all 4 groups without expected physiological changes in sleep homeostasis, metabolism, circadian or sympatho-adrenal systems. In contrast, the groups with expected changes had mean score reductions higher than 70%. The associations between different physiological responses to LT were not strong. Only the direction of change in non-REMS homeostasis was found to be associated significantly with the direction of change in metabolic rate (Chi-square = 4.89, $p < 0.05$, $n = 21$). Additionally, an insignificant association between metabolic activation and phase advance was noted (Chi-square = 3.32, $p = 0.1$, $n = 61$).

Thus, despite rather weak positive interrelations between studied physiological effects of light, each of these effects contributed to reduction of depressive symptoms. Since any of biological responses and their sum was related positively to clinical efficacy of LT, the observed physiological effects appear to be additive.

3.2.2. Effects of LT on MLT and Cortisol. Although the diurnal means of MLT and cortisol did not change significantly throughout the study, the daytime MLT levels (at 12:00 and 16:00) were significantly higher in patients compared to controls in wintertime. This difference disappeared after LT and in summer (See Figures 1–4 in (12)). Moreover, a five-way ANOVA indicates that the mean MLT levels were higher in patients compared to controls (Table 2).

Both hormonal variables underwent significant diurnal variations with a pattern similar in patients and controls (see Figure 1 in (11), and Figures 1–4 in (12)). Figure 4 and Table 2 show that the circadian pattern of serum MLT reflected the sleep-wake pattern. Even among SAD patients, we found the morning type subjects (evening sleepers) with phase-advanced MLT rhythm and the short sleepers (habitually larks in the morning and owls in the evening) with low daytime MLT levels.

Evening-morning differences for MLT in patients were similar to that in controls throughout the study (Figure 5). The Evening-morning difference for cortisol was lower in patients compared to controls before ($t = -2.765$, $p < 0.01$), but not after LT ($t = -0.310$, $p > 0.1$) or in summer ($t = -0.708$, $p > 0.1$). The phases of both rhythms did not change significantly in controls, but the MLT phase in controls showed the same tendency as did the patients' phase (Figures 4 and 5; Table 3). An interaction between time of day and condition was significant for MLT both in the whole group of subjects and in the patients' group alone ((Table 2). This result indicates an advance shift of the circadian phase following LT (Figures 4 and 5). Moreover, the significant interaction between time of LT and condition was revealed by 4-way ANOVA of cortisol in patients ($F = 5.72$, $Df = 1, 19$, $p < 0.05$). This interaction reflects the decrease of morning levels of cortisol after morning, but not after afternoon LT (assumably, due to an advance shift of the cortisol rhythm). As can be seen in Figure 5, LT and change of season affected the evening-morning differences in a similar way.

It may be generalized that no difference between patients and controls on the timing of MLT secretion was observed in our sample, but patients more likely responded to LT by advancing circadian phase.

For the patients group, the significant interaction between diurnal profile of MLT, time of LT and response was revealed by the 4-way ANOVA. Moreover, a triple interaction (response × time of LT × time of day) as well as the interaction of all four factors (condition × response × time of LT × time of day) reached statistically significant levels (Table 2).

The relative values of MLT and cortisol at 08:00 and 24:00 were similar in control and patient groups. However, compared to the whole control group, responders to morning light had significantly higher cortisol levels at 08:00. Besides, the significant positive correlation was found between HRSD and relative cortisol levels at 24:00 in patients treated in the afternoon, and, in patients treated in the morning, higher baseline HRSD and SIGH-SAD determined increase of relative value of cortisol at 08:00. In contrast to nonresponders to morning light, nonresponders to afternoon light showed significantly lower MLT levels at 24:00 and significantly higher MLT levels at 08:00. Light seems to work worse when patients with advanced MLT phase are treated in the morning, and when patients with delayed phase are treated in the afternoon (see also Figure 2 in (52)).

In general, only morning LT, and then only in patients, was able to produce a significant advance of circadian phase. Phase shifts were significant in responders and the same tendency was noted in nonresponders. Neither in patients, nor in controls were the phase changes following afternoon LT an significant. However, the significant decrease of MLT levels at 08:00, indicating advance shift of the rhythm, was observed when the nonresponder subgroup was analyzed separately. The results show that, in the majority of patients treated in the afternoon, the antidepressant response occurred despite the lack of phase shift of circadian rhythms, whereas the advance shifts were observed in both responders and nonresponders treated in the morning.

4. DISCUSSION

The results of our questionnaire study may be considered as supporting the idea of close link between seasonal changes in photoperiod and seasonal variations in the symptoms of winter SAD. However, we also found considerable differences between annual phases of different depressive symptoms. The annual photoperiodic cycle seems not to be the only important determinant for seasonal variations in neurovegetative and psychic symptoms. Compared to the symptoms of mood and activity, the seasonal variations in the symptoms related to sleep phase and rate of metabolism corresponded more directly to the seasonal cycles of daylight and darkness. It is likely that the studied depressive symptoms seem to reflect the disturbances in several systems of physiological regulation.

Moreover, the findings of the LT-study suggest the multi-component nature of the physiological response to light. We suggest that bright light treats the depressive symptoms via normalization in distinct systems of physiological regulation and that the MLT rhythm may be more or less involved in the mechanisms underlying different therapeutic effects of LT. Our findings seem to provide an explanation for the contradicting results of investigations on SAD physiology. Most experiments are designed to test only one physiological abnormality, while the therapeutic action of bright light could be mediated by at least four physiological effects (shifts of circadian phase, increase of non-REMS pressure, rate of metabolism and sympatho-adrenal activity). Each of these effects need not be necessarily observed in the vast majority of patients (40).

At least two physiological tendencies, the increase in sleep need and circadian phase delay, could account for such features of the sleep-wake cycle of winter depressives as hypersomnia and late awakening. The winter seasonals experience serious problems with morning sleep termination, but, presumably, due to the opposite effects of sleep need and phase delay on sleep onset, they do not show differences from nonseasonals on bed times. Since the recent observations (33,56,65) showed the close link between time of MLT secretion and sleep onset, it could be suggested that use of the onset of MLT secretion as a marker of circadian phase position (26,27,28) may lead to underestimation of circadian phase differences between winter depressives and healthy subjects.

Our LT-study does not provide solid evidence that winter depression is triggered by a seasonal phase delay of the MLT rhythm. However, we observed the tendency for seasonal changes of MLT phase position in SAD subjects and healthy controls. Besides, such a parameter of the circadian phase of cortisol rhythms as evening-morning difference was found to indicate winter phase delay in SAD, and the circadian phase for cortisol was related to depth of depression. Finally, a pretreatment MLT phase was found to predict a preferential improvement to morning or afternoon light. The early morning decay and evening rise of MLT predicted worse response to morning LT, whereas the late morning decay and evening rise predicted worse response to afternoon LT.

Our results also suggest that the advance of the pacemaker phase could be necessary for the excellent antidepressant response to light. We found that the response of patients with SAD to the phase shifting action of morning light is enhanced compared to healthy subjects. In this respect, we replicated the result of Lewy et al. (28,30) who reported that patients with SAD did not differ much from controls on baseline MLT phase position (which was nonsignificantly delayed), but more pronounced differences from controls were found for the extent of advance shift of circadian phase of MLT rhythm following morning LT. Since several reports (16,31,64,68) suggest that the MLT secretion in patients with SAD is more sensitive to light than in normal subjects, the increased sensitivity to light rather than the initial phase delay may explain the phase shifts in SAD patients.

In conclusion, a question about role of MLT and circadian phase in the mechanisms of LT for SAD is still open. Our findings demonstrate that, despite lack of solid evidence for involvement of circadian pacemaker in pathogenesis of SAD, both an individual circadian phase and a direction of phase shift seem to be associated with the antidepressant action of light, and that the circadian phase-advance is one of at least four physiological effects of light that could participate in improvement of mood in winter depression.

ACKNOWLEDGMENTS

The authors appreciate very much the assistance of Vladislav Palchikov, Dmytry Zolotarev, Tatjana Neschumova, Oleg Vasiljev, Andrey Abelev, Valery Kozaruk and Andrey Samsonov in collecting data, and the assistance of Dmitriy Putilov in data analysis. We are grateful to Alexndra Schurgaja, Stanislav Schergin and Boris Churin for the opportunity to use the hospital services for this study. We also thank Prof. Sven Ebbesson, Prof. Lawrence Duffy, Dr. Brian Barnes and Ms Andree Porchet for help in MLT assay (supported by the Alaska-Siberia Medical Exchange Program).

REFERENCES

1. Anderson, J.L., Vasile, R.G., Bloomingdale, K.L., and Schildkraut, J.J. SAD: Varied schedule phototherapy and catecholamines. Amer. Psychiatr. Assoc., 141st Annual Meet., Program and Abs., #166, 1988.
2. Avery, D.H., Khan, A., Dager, S.R., Cohen, S., Cox, G.B., and Dunner, D.L. Morning or evening bright light treatment of winter depression? The significance of hypersomnia. Biol. Psychiatry 29:117–126, 1991.
3. Avery, D.H., Dahl, K., Eder, D.N., Larsen, L.H., Vitiello, M.V., Prinz, P.N., Brengelmann, G.L., Savage, M.V., and Kenny, M.A. Is PSI increased M-E in winter depression? Soc. Light Treatment Biol. Rhythms, 5th Meet. Abs., p. 5, 1993.
4. Avery, D.H., Bolte, M.A., and Eder, D. Difficulty awakening as asymptom of winter depression. Soc. Light Treatment Biol. Rhythms, 6th Meet. Abs., p. 21, 1994.
5. Avery, D.H., Dahl, K., Savage, M.V., Brenglemann, G.L., Larson, L.H., Kenny, M.A., Eder, D.N., Vitiello, M.V., and Prinz, P.N. Circadian temperature and cortisol rhythms during a constant routine are phase-delayed in hypersomnic winter depression. Biol. Psychiatry 41:1109–1123, 1997.
6. Booker, J.M. and Hellekson, C.J. Prevalence of seasonal affective disorder in Alaska. Amer. J. Psychiat. 149:1176–1182, 1992.
7. Boyd, J., Weissman, M.M., Thompson, W.D., and Myers, J.K. Screening for depression in a community sample: Understanding the discrepancies between depression symptom and diagnostic scales. Arch. Gen. Psychiatry 39:1195–1200, 1982.
8. Checkley, S.A., Murphy, D.G.M., Abbas, M., Marks, M., Winton, F., Palazidou, E., Murphy, D.M., Franey, C., and Arendt, J. Melatonin rhythms in seasonal affective disorder. Brit. J. Psychiatr. 163:332–337, 1993.
9. Czeisler, C.A., Weitzman, E.D., Moore-Ede, M.C., Zimmerman, J.C., and Knauer, R.S. Human sleep: Its duration and organization depend on its circadian phase. Science 210:1264–1267, 1980.

10. Dahl, K., Avery, D.H., Lewy, A.J., Savage, M.V., Brengelmann, G.L., Larsen, L.H., Vitiello, M.V., and Prinz, P.N. Dim light melatonin onset and circadian temperature during a constant routine in hypersomnic winter depression. Acta Psychiatr. Scand. 88:60–66, 1993.
11. Danilenko, K.V. and Putilov, A.A. Diurnal and seasonal variations in cortisol, prolactin, TSH and thyroid hormones in women with and without seasonal affective disorder. J. Interdisc. Cycle Res. 24:185–196, 1993.
12. Danilenko, K.V., Putilov, A.A., Russkikh, G.S., Duffy, L.K., and Ebbesson, S.O.E. Diurnal and seasonal variations in melatonin and serotonin in women with seasonal affective disorder. Arc. Med. Res. 53:137–145, 1994.
13. Eastman, C.I., Gallo, L.C., Lahmeyer, H.W., and Fogg, L.F. The circadian rhythm of temperature during light treatment for winter depression. Biol. Psychiatry 34:210–220, 1993.
14. Elliott, J.A. and Goldman, B.D. Seasonal reproduction: Photoperiodism and biological clocks. In: Adler N.T. (Ed.). Neuroendocrinology of Reproduction. Plenum Press, New York, pp. 377–423, 1981.
15. Fraser, S., Cowen, P., Franklin, M., Franey, C., and Arendt, J. Direct radioimmunoassay for melatonin in plasma. Clin. Chem. 29:396–397, 1983.
16. Gaddy, J.R., Stewart, K.T., Byrne, B., Doghramji, K., Rollag, M.D., and Brainard, G.C. Light-induced plasma melatonin suppression in seasonal affective disorder. Prog. Neuropsychopharmacol. Biol. Psychiatry 14:563–568, 1990.
17. Hamilton, M. Development of a rating scale for primary depressive illness. Br. J. Soc. Clin. Psychol. 6:278–296, 1967.
18. Jewett, M.E., Rimmer, D.W., Duffy, J.F., Klerman, E.B., Kronauer, R.E., and Czeisler, C.A. Human circadian pacemaker is sensitive to light throughout subjective day without evidence of transients. Am. J. Physiol. 273(5 Pt 2):R1800-R1809, 1997.
19. Kasper, S., Wehr, T.A., Bartko, J.J., Gaist, P.A., and Rosenthal, N.E. Epidemiological findings of seasonal changes in mood and behavior: A telephone survey of Montgomery County, Maryland. Arch. Gen. Psychiat. 46:823–833, 1989.
20. Kurtz, J., Bauer, M.S., and Poland, R. A test of the phase-shift hypothesis of SAD. Amer. Psychiatr. Assoc., 150th Annual Meet., San Diego, Syllabus & Proc., #453, 1997.
21. Lam, R.W., Buchanan, A., Mador, J.A., and Corral, M.R. Hypersomnia and morning light therapy for winter depression. Biol. Psychiatry 31:1062–1064, 1992.
22. Lam, R.W. Morning light therapy for winter depression: Predictors of response. Acta Psychiatr. Scand. 89:97–101, 1994.
23. Levendosky, A.A., Josef-Vanderpool, J.R., Hardin, T., Sorek, E., and Rosenthal, N.E. Core body temperature in patients with seasonal affective disorder and normal controls in summer and winter. Biol. Psychiatry 29:524–534, 1991.
24. Lewy, A.J., Wehr, T.A., Goodwin, F.K., Newsome, D.A., and Markey, S.P. Light suppresses melatonin secretion in humans. Science 210:1267–1269, 1980.
25. Lewy, A.J., Kern, H.A., Rosenthal, N.E., and Wehr, T.A. Bright artificial light treatment of a manic-depressive patient with a seasonal mood cycle. Am. J. Psychiat. 139:1496–1498, 1982.
26. Lewy, A.J. and Sack, R.L. Minireview: Light therapy and psychiatry. Proc. Soc. Exp. Biol. Med. 183:11–18, 1986.
27. Lewy, A.J., Sack, R.L., Miller, L.S., and Hoban, T.M. Antidepressant and circadian phase-shifting effect of light. Science 235:352–354, 1987.
28. Lewy, A.J., Sack, R.L., Singer, C.M., White, D.M., and Hoban, T.M. Winter depression and the phase-shift hypothesis for bright light's therapeutic effects: History, theory, and experimental evidence. J. Biol. Rhythms 3:121–134, 1988.
29. Lewy, A.J. and Sack, R.L. The dim light melatonin onset as a marker for circadian phase position. Chronobiol. Int. 6:93–102, 1989.
30. Lewy, A.J., Sack, R.L., Bauer, V.K., and Cutler, N.L. Melatonin as treatment for winter depression. Amer. Psychiatr. Assoc., 150th Annual Meet., San Diego, Syllabus & Proc., #74B, 1997.
31. McIntryre, I.M., Norman, T.R., Burrows, G.D., and Armstrong, S.M. Melatonin supersensivity to dim light in seasonal affective disorder. Lancet 335:488, 1990.
32. Meesters, Y., Jansen, J.H.C., Beersma, D.G.M., Bouhuys, A.L., and van den Hoofdakker, R.H. Light therapy for seasonal affective disorder: The effects of timing. Br. J. Psychiatr. 166:607–612, 1995.
33. Nakagawa, H., Sack, R.L., and Lewy, A.J. Sleep propensity free-runs with the temperature, melatonin and cortisol rhythms in a totally blind person. Sleep 15:330–336, 1992.
34. Norman, C.C. and Schlager, D.S. Early morning, short-acting beta-blockers as treatment for winter depression. Amer. Psychiatr. Assoc., 150th Annual Meet., San Diego, Syllabus & Proc., #210, 1997.

35. Oren, D.A., Jacobsen, F.M., Wehr, T.A., Cameron, D.L., and Rosenthal, N.E. Predictors of response to phototherapy in seasonal affective disorder. Compr. Psychiatry 33:111–114, 1992.
36. Palchikov, V.E., Zolotarev, D.Y., Danilenko, K.V., and Putilov, A.A. Effects of season and of bright light administered at different times of day on sleep EEG and mood in patients with seasonal affective disorder. Biol. Rhythm Res. 28:166–184, 1997.
37. Partonen, T. Effects of morning light treatment on subjective sleepiness and mood in winter depression. J. Affect. Dis. 30:47–56, 1994.
38. Putilov, A.A. A questionnaire for self-assessment of individual traits of sleep-wake cycle. (in Russian) Bull. Siberian Branch, USSR Acad. Med. Sci. 1:22–25, 1990.
39. Putilov, A.A. "Owls", "Larks" and Others: About Clocks Inside Us and Their Influence on Our Health and Character. Novosibirsk: Novosibirsk University Press; Moscow: Publishing House "Soverschenstvo", 1997.
40. Putilov, A.A. Multi-component physiological response mediates therapeutic benefits of bright light in winter seasonal affective disorder. Biol. Rhythm Res. 29:367–386, 1998.
41. Putilov, A.A., Danilenko, K.V., Volf, N.V., Cherepanova, V.A., Palchikov, V.E., Neschumova, T.V., Zolotarev, D.Y., and Senkova, N.I. Chronophysiological aspects of light treatment for seasonal affective disorder: Siberian studies. Light Treatment Biol. Rhythms 3:43–45, 1991.
42. Putilov, A.A., Palchikov, V.E., Zolotarev, D.Y., and Danilenko, K.V. Sleep architecture in seasonal affective disorder throughout the year. Sleep Res. 22:159, 1993.
43. Putilov, A.A., Danilenko, K.V., Russkikh, G.S., and Duffy, L.K. Phase typing of patients with seasonal affective disorder: A test for the phase shift hypothesis. Biol. Rhythm Res. 27:431–451, 1996.
44. Radloff, L.S. The CES-D scale: A self-report depression scale for research in the general population. Appl. Psychol. Measurement 1:385–401, 1977.
45. Rafferty, B., Terman, M., Terman, J.S., and Reme, C.E. Does morning light prevent evening light effect? A statistical model for morning/evening crossover studies. Soc. Light Treatment Biol. Rhythms, 2nd Meet. Abs., p. 18, 1990.
46. Rice, J., Mayor, J., Tucker, H.A., and Bielski, R.J. Effect of light therapy on salivary melatonin in seasonal affective disorder. Psychiat. Res. 56:221–228, 1995.
47. Rosenthal, N.E., Sack, D.A., Gillin, J.C., Lewy, A.J., Goodwin, R.K., Davenport, Y., Mueller, P.S., Newsome, D.A., and Wehr, T.A. Seasonal affective disorder: A description of the syndrome and preliminary findings with light therapy. Arch. Gen. Psychiat. 41:72–80, 1984.
48. Rosenthal, N.E., Bradt, G.H., and Wehr, T.A. Seasonal Pattern Assessment Questionnaire, National Institute of Mental Health, Bethesda, MD, 1984.
49. Rosenthal, N.E., Sack, D.A., Jacobsen, F.M., Parry, B.L., Arendt, J., Tamarkin, L., and Wehr, T.A. Melatonin in seasonal affective disorder and phototherapy. J. Neural. Trans. (Suppl.) 21:257–267, 1986.
50. Rosenthal, N.E., Jacobsen, F.M., Sack, D.A., Arendt, J., James, S.P., Parry, B.L., and Wehr, T.A. Atenolol in seasonal affective disorder: A test of melatonin hypothesis. Am. J. Psychiatry 145:52–56, 1988.
51. Rosenthal, N.E., Levendosky, A.A., Skwerer, R.G., Josef-Vanderpool, J.R., Kelly, K.A., Hardin, T., Kasper, S., Della Bella, P., and Wehr, T.A. Effects of light treatment on core body temperature in seasonal affective disorder. Biol. Psychiatry 27:39–50, 1990.
52. Russkikh, G.S., Danilenko, K.V., Putilov, A.A., and Duffy, L.K. Diurnal pattern of melatonin secretion and preferential response to morning or afternoon light treatment. In: Biol. Effects of Light 1995. Walter de Gruyter & Co., Berlin—New York, pp. 418–420, 1996.
53. Sack, R.L., Lewy, A.J., White, D.M., Singer, C.M., Fireman, M.J., and Vandiver, R. Morning vs evening light treatment for winter depression: Evidence that the therapeutic effects of light are mediated by circadian phase position. Arch. Gen. Psychiatry 47:343–351, 1990.
54. Schergin, S.M., Danilenko, K.V., and Putilov, A.A. Biological and psychic effects of bright light in seasonal affective disorder. Biol. Effects of Light 1995. Walter de Gruyter & Co., Berlin—New York, pp. 409–411, 1996.
55. Schlager, D. Early-morning administration of short-acting beta blockers for treatment of winter depression. Am. J. Psychiat. 151:1383–1385, 1994.
56. Shochat, T., Luboshitzky, R., and Lavie, P. Nocturnal melatonin onset is phase locked to the primary sleep gate. Am. J. Physiol. 273(1 Pt 2):R364–370, 1997.
57. Skwerer, R.G., Jacobson, F.M., Dunkan, C.C., Kelly, K.A., Sack, D.A., Tamarkin, L., Gaist, P.A., Kasper, S., and Rosenthal, N.E. Neurobiology of seasonal affective disorder and phototherapy. J. Biol. Rhythms 3:135–154, 1988.

58. Spiegel, K., Leproult, R., l'Hermite-Baleriaux, M., and Van Cauter, E. Effect of sleep restriction and sleep extension on the 24-h profiles of melatonin and body temperature. 6th Meet. Soc. Res. Biol. Rhythms, Amelia, Island FL, p. 36, 1998.
59. Terman, M. Problems and prospects for use of bright light as a therapeutic intervention. In: L. Wetterberg (Ed.). Light and Biol. Rhythms in Man, Pergamon Press, Oxford. pp. 421–436, 1993.
60. Terman, M., Terman, J.S., Quitkin, F.M., Cooper, T.B., Lo, E.S., Gorman, J.M., Stewart, J.W., and McGrath, P.J. Response of the melatonin cycle to phototherapy for seasonal affective disorder. J. Neural. Transm. 72:145–165, 1988.
61. Terman, M., Terman, J.S., Quitkin, F.M., McGrath, P.J., Stewart, J.W., and Rafferty, B. Light therapy for seasonal affective disorder: A review of efficacy. Neuropsychopharmacol. 2:1–22, 1989.
62. Terman, M., Terman, J.S., Lewy, A.J., and Dijk, D.-J. Light treatment for sleep phase and duration disturbances. Light Treatment Biol. Rhythms 7:7–14, 1994.
63. Terman, M. and Terman, J.S. Phase shifts in melatonin and sleep under light therapy for winter depression. Soc. Light Treatment Biol. Rhythms, 7th Meet. Abs. p. 15, 1995.
64. Thompson, C., Stinson, D., and Smith, A. Seasonal affective disorder and season-dependent abnormalities of melatonin suppression by light. Lancet 336:703–706, 1990.
65. Tzischinsky, O., Shlitner, A., and Lavie, P. The association between the nocturnal sleep gate and the nocturnal onset of urinary 6-sulphatoxymelatonin. J. Biol. Rhythms 8:199–209, 1993.
66. Wehr, T.A., Sack, D.A., Jacobsen, F., Tamarkin, L., Arendt, J., and Rosenthal, N.E. Phototherapy of seasonal affective disorder: Time of day and suppression of melatonin are not critical for antidepressant effects. Arch. Gen. Psychiatry 145:52–56, 1986.
67. Wehr, T.A., Moul, D.E., Barbato, G., Giesen, H., Seidel, J.A., Barker, C., and Bender, C. Conservation of photoperiod-responsive mechanisms in humans. Am. J. Physiol. 265:R846-R857, 1993a.
68. Wehr, T.A., Schwartz, P.J., Turner, E.H., Feldman-Naim, S., Drake, C.L., and Rosenthal, N.E. Bimodal patterns of human melatonin secretion consistent with a two-oscillator model of regulation. Neuroscience Letters 194:105–108, 1995.
69. Wehr, T.A., Schwartz, P.J., Turner, E.H., Drake, C.L., and Rosenthal, N.E. Summer-winter differences in duration of nocturnal melatonin secretion in SAD patients and healthy controls. Soc. Light Treatment Biol. Rhythms, 7th Meet. Abs., p. 9, 1995.
70. Whitehead, T.P., Thorpe, G.H.G., Carter, T.J.N., Groucutt, C., and Kricka, L.J. Enhanced luminescence procedure for sensitive determination of peroxidase-labelled conjugates in immunoassay. Nature 305:158–159, 1983.
71. Williams, J.B.W., Link, M.J., Rosenthal, N.E., and Terman, M. Structured Interview Guide for the Hamilton Depression Rating Scale: Seasonal Affective Disorder Version (SIGH-SAD). New York State Psychiatric Institute, New York, 1988.
72. Winton, F., Corn, T., Huson, L., Franey, C., Arendt, J., and Checkley, S. Effects of light treatment upon mood and melatonin in patients with seasonal affective disorder. Psychol. Med. 19:585–590, 1989.
73. Wirz-Justice, A., Graw, P., Krauchi, K., Gisin, B., Arendt, J., Aldhous, M., and Poldinger, W. Morning or night-time melatonin is ineffective in seasonal affective disorder. J. Psychiat. Res. 24:129–137, 1990.
74. Wirz-Justice, A. and Anderson, J.L. Morning light exposure for the treatment of winter depression: The one true light therapy? Psychopharmacol. Bull. 26:511–520, 1990.
75. Wirz-Justice, A., Graw, P., Krauchi, K., Gisin, B., Jochum, A., Arendt, J., Fish, H.-U., Buddeberg, C., and Poldinger, W. Light therapy in seasonal affective disorder is independent of time of day or circadian phase. Arch. Gen. Psychiatry 50:929–937, 1993.
76. Wirz-Justice, A., Krauchi, K., Graw, P., Arendt, J., English, J., Hetsch, C., Haug, H.-J., Leonhardt, G., and Brunner, D.P. Circadian rhythms of core body temperature and salivary melatonin in winter SAD before and after midday light. Soc. Light Treatment Biol. Rhythms, 6th Meet. Abs., p. 12, 1994.
77. Zulley, J., Wever, R., and Aschoff, J. The dependence of onset and duration of sleep on the circadian rhythm of rectal temperature. Pflugers Arch. 391:314–318, 1981.

54

INFLUENCE OF LOW-FREQUENCY MAGNETIC FIELD OF DIFFERENT CHARACTERISTICS ON SERUM MELATONIN CONCENTRATIONS IN HUMANS

Michal Karasek,[1] Marta Woldanska-Okonska,[2] Jan Czernicki,[3] Krystyna Zylinska,[4] and Jacek Swietoslawski[1]

[1] Laboratory of Electron Microscopy, Chair of Pathomorphology, Medical University of Lodz
[2] Division of Rehabilitation Medicine, Regional Hospital in Sieradz
[3] Department of Rehabilitation Medicine, Military Medical Academy, Lodz Poland
[4] Department of Experimental Endocrinology and Hormone Diagnostics, Institute of Endocrinology, Medical University of Lodz, Poland

1. INTRODUCTION

There is substantial evidence that magnetic field (MF) influences melatonin secretion in animals (see 5). However, data on its influence on human melatonin levels are scarce, and contradictory (3,4,6–8). On the other hand, because of its many beneficial effects, such as anti-inflammatory and analgesic action, improvement of soft tissue regeneration processes, vasodilatory action or oxygen utilization and tissue respiration, very low-frequency MF is used in physiotherapy of some neurological diseases and overloading syndromes of locomotor system (1,2).

Since the data on the influence on melatonin levels, both in humans and animals are inconsistent, and in our previous studies (6) we suggested that differences among various studies may depend on different characteristics of applied MF, we decided to compare the effects of MF of different parameters in subjects with low back pain syndrome before and after chronic exposure to low-frequency MF generated by two different apparatuses used in standard physiotherapy.

Melatonin after Four Decades, edited by James Olcese.
Kluwer Academic / Plenum Publishers, New York, 2000.

2. MATERIAL AND METHODS

The study was performed in 11 men with low back pain syndrome, who were admitted to the Division of Medical Rehabilitation of the Regional Hospital in Sieradz. The patients were divided into two groups. Group 1 consisted of 6 men (mean age 42.9 years; range 32–55) exposed to a pulsating MF (2.9 mT, 40 Hz, square impulse shape, bipolar) generated by Magnetronic MF 10 apparatus for three weeks (5 days per week, at 10:00 h, 20 min per day) applied as a coil in lower back region. Group 2 consisted of 5 men (mean age 44.8 years; range 41–47) exposed to a pulsating MF (0.025–0.08 mT, 200 Hz, complex saw-like impulse shape, bipolar) generated by Quatronic MRS 2000 apparatus ("magnetic bed") for three weeks (5 days per week, twice a day at 08:00 h and 13:00 h for 8 min each) applied for the whole body in patients laying in horizontal position. The study was performed in spring.

Diurnal serum melatonin profiles were estimated a day before exposure to MF (baseline), and the day after the last exposure. Each subject served as his own control. On the day before and during blood sampling, the period of darkness in the patients' room lasted from 22:00 h to 06:00 h. Blood samples were collected at 08:00, 12:00, 16:00, 20:00, 24:00, 02:00, 04:00 and 08:00 h; the nighttime samples were taken under dim red light. All blood samples were allowed to clot for 45 min; serum was removed by centrifugation and stored at –20°C until assayed. Melatonin concentration was measured using RIA kit (DGR Instr. GmbH, Marburg, Cat. No. IH RH 29301): intra assay CV 8%, inter assay CV 14.8%. The data were statistically analyzed using paired Student's t test.

The study was approved by the Regional Committee for Studies with Human subjects. The experimental protocol was explained to each patient and informed consent was obtained.

3. RESULTS

Chronic exposure to 2.9 mT, 40 Hz MF caused significant depression in nocturnal melatonin rise in patients of group 1 (Figure 1A), whereas 0.08 mT, 200 Hz MF did not influence melatonin levels in patients of group 2 (Figure 1B).

4. DISCUSSION

It has been suggested that considerable differences among various studies on the influence of MF on melatonin secretion may depend on different experimental paradigms, including certain characteristics of applied MF (e.g., field strength, frequency, duration, applied vector, etc.), acute or chronic exposure, differences in exposure time and duration, and factors that could interfere with the results of MF studies (e.g., vibration, noise, synchronization or desynchronization with geomagnetic field) (5,6).

The results of the present study seem to support the hypothesis that pineal response to MF may depend on field strength, frequency, impulse shape, and application system, because in the experimental conditions used in this study there were differences only in the above mentioned parameters. Remaining parameters, such as time of the year, housing, photoperiod, and duration of the exposure were the same in both groups of patients of similar age.

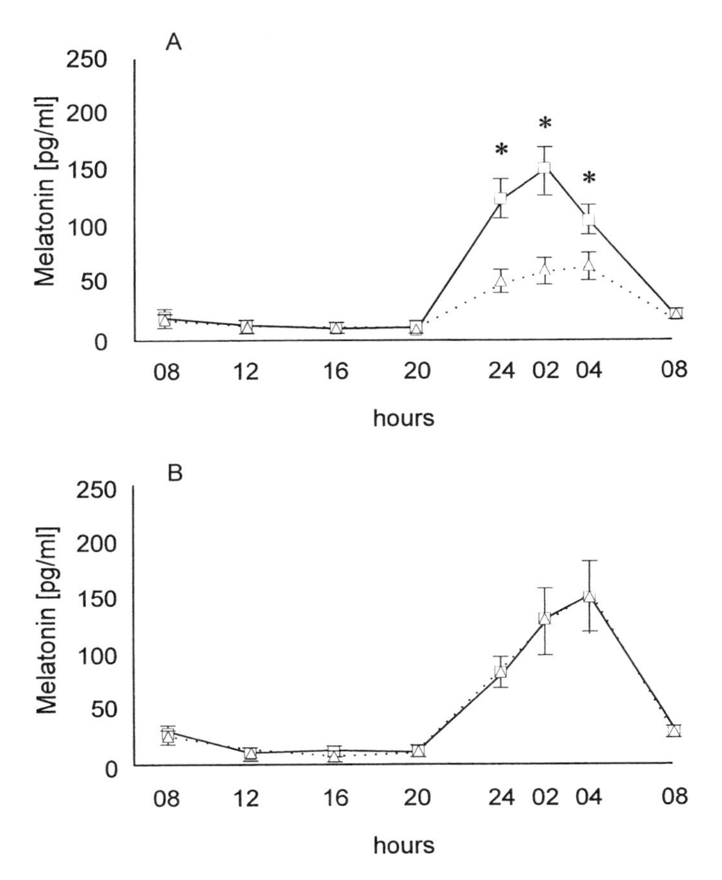

Figure 1. Diurnal serum melatonin concentrations in patients exposed to 2.9 mT, 40 Hz magnetic field (A) or 0.08 mT, 200 Hz magnetic field (B) before (solid line) and after (dotted line) exposure. Data are expressed as means ± SEM; * $-p < 0.05$.

ACKNOWLEDGMENTS

The study was supported by grants from the Medical University of Lodz (No. 503-129-0), and from Military Medical Academy of Lodz.

REFERENCES

1. Basset C.A. Beneficial effect of electromagnetic fields. J Cell Biochem 4:387–393, 1993.
2. Fisher G. Grundlagen der Quanten-Therapie. Hecataeus Verlagsanstalt, Tiesenberg, 1996.
3. Graham C., Cook M.R., Riffle D.W., Gerkovich M.M., and Cohen H.D. Nocturnal melatonin levels in human volunteers exposed to intermittent 60 Hz magnetic fields. Bioelectromagnetics 17:263–273, 1996.
4. Graham C., Cook M.R., and Riffle D.W. Human melatonin during continuous magnetic field exposure. Bioelectromagnetics 18:166–171, 1997.
5. John T.M., Liu G.Y., and Brown G.M. Electromagnetic field exposure and indoleamine metabolism: an overview. Front Horm Res 21:42–50, 1996.

6. Karasek M., Woldanska-Okonska M., Czernicki J., Zylinska K., and Swietoslawski J. Chronic exposure to 2.9 mT, 40 Hz magnetic field reduces melatonin concentrations in humans. J Pineal Res 1998, in press.
7. Selmaoui B., Lambrozo J., and Touitou Y. Magnetic fields and pineal function in humans: evaluation of nocturnal acute exposure to extremely low frequency magnetic fields on serum melatonin and urinary 6-hydroxymelatonin circadian rhythms. Life Sci 58:1539–1549, 1996.
8. Wilson B.W., Wright C.W., Morris J.E., Buschbom R.L., Brown D.P., Miller D.L., Sommers-Flannigan R., and Anderson L.E. Evidence for an effect of ELF electromagnetic fields on human melatonin function. J Pineal Res 9:259–269, 1990.

CIRCADIAN SERUM MELATONIN PROFILES IN PATIENTS WITH VERY LARGE GOITRE BEFORE AND AFTER SURGERY

Preliminary Report

Aleksander Stankiewicz,[1] Krzysztof Kuzdak,[2] Krystyna Zylinska,[3] Elżbieta Bandurska-Stankiewicz,[1] Jacek Swietoslawski,[4] and Michal Karasek[4]

[1] Regional Hospital in Olsztyn
[2] Department of Surgical Endocrinology
[3] Department of Experimental Endocrinology and Hormone Diagnostics
Institute of Endocrinology
[4] Laboratory of Electron Microscopy
Chair of Pathomorphology
Medical University of Lodz
Lodz, Poland

1. INTRODUCTION

Surgical removal of very large goitre may traumatize adjacent anatomical structures, such as superior cervical ganglia that play a very important role in the synthesis of melatonin, and in consequence, may alter melatonin secretion. To test this hypothesis we decided to study diurnal serum melatonin profiles in patients with very large goitre before and after the surgery.

2. MATERIAL AND METHODS

The study was performed in 8 women (mean age—46.3 ± 1.8 years; range 40–56 years) with very large non-toxic nodular goitre (mean thyroid volume—135.4 ± 30.9 cm^3; range 82.6–326.7 cm^3) admitted to Regional Hospital in Olsztyn to undergo strumectomy.

Melatonin after Four Decades, edited by James Olcese.
Kluwer Academic / Plenum Publishers, New York, 2000.

Diurnal serum melatonin profiles were estimated 2 days before the operation and 10 days after the surgery. Each subject served as her own control. One day before and during blood sampling the period of darkness in patients' room lasted from 21:00 to 07:00 h. Blood samples were collected at 08:00, 12:00, 16:00, 20:00, 22:00, 24:00, 02:00, 04:00, 06:00 and 08:00 h; the nighttime samples were taken under dim red light. All blood samples were allowed to clot for 45 min, serum was removed after centrifugation, and stored at –20°C until assayed. Melatonin concentration was measured using RIA kit (DRG Inst. GmbH, Marburg; cat. No. IH RE 29301, sensitivity 3.5 pg/ml, intra assay CV—8%, inter assay CV—14.8%). The samples from all subjects (pre- and post-assay) run together. The data were statistically analyzed using paired Student's t test.

The study was approved by the Regional Committee for Studies with Human Subjects. The experimental protocol was explained to each patient, and informed consent was obtained.

3. RESULTS

Nocturnal serum melatonin concentrations (at 24, 02, and 04 hours) were significantly higher after the surgery than before operation (Figure 1).

4. DISCUSSION

A relationship between the pineal gland and the thyroid has been suggested in many reports (3,5). Especially, the involvement of melatonin in the control of proliferation of thyroid follicular cells has been shown in experimental animal models (2,3). The data on pineal—thyroid relationship in humans are, however, scarce. No changes were seen in melatonin levels in both hypothyroidism and hyperthyroidism (4). A decrease in nocturnal melatonin concentrations was observed in the patients with recurrent non-toxic nodular goitre in comparison with the control group of healthy women (1).

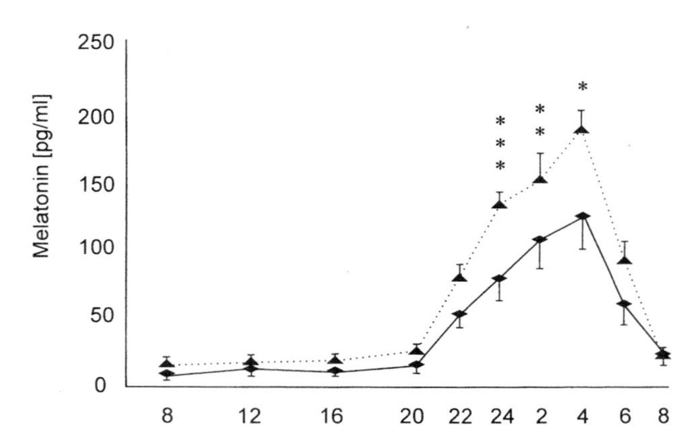

Figure 1. Diurnal serum melatonin profiles in patients with very large goitre before (solid line) and after (dotted line) surgery. Data are expressed as mean ± SEM; * $-p < 0.05$, ** $-p < 0.01$, *** $-p < 0.001$.

We expected that manipulations during surgical removal of very large goitre might alter melatonin secretion via possible lesions of superior cervical ganglia during the surgery. Surprisingly, in all patients nocturnal melatonin concentrations were significantly higher after the operation than before the surgery. One of the possible explanations of this finding is that very large goitre may compress the superior cervical ganglia altering their function. It is also possible that the presence of goitre in some way affects melatonin secretion. In both our present studies and Kuzdak (1) decreased nocturnal melatonin concentrations were observed.

ACKNOWLEDGMENTS

The study was supported by a grant from Medical University of Lodz (No. 503-129-0).

REFERENCES

1. Kuzdak K. The role of endocrine and autocrine growth factors after subtotal restrumectomy in non-toxic nodular goitre—special reference to risk of goitre relapse. Endokrynol Pol—Pol J Endocrinol 46, suppl 2/2:59–68, 1995.
2. Lewinski A. Evidence for pineal gland inhibition of thyroid growth: contribution to the hypothesis of a negative feedback between the thyroid and the pineal gland. Adv Pineal Res 1:167–176, 1986.
3. Lewinski A. Some aspects of the pineal—thyroid interrelationship and their possible involvement in the regulation function and growth of these two glands. Adv Pineal Res 4:175–188, 1990.
4. Soszynski P., Zgliczynski S., and Pucilowska J. The circadian rhythm of melatonin in hypothyroidism and hyperthyriodism. Acta Endocrinol (Copenh.) 119:240–244, 1988.
5. Vriend J. Pineal—thyroid interactions. Pineal Res Rev 1:183–206, 1983.

56

ADVANCED IMMUNOASSAYS FOR THE DIRECT DETERMINATION OF MELATONIN IN HUMAN SERUM AND CULTURE MEDIA

Matthias Schumacher, Anita Nanninga, Richard Werner, and James Olcese

Institute for Hormone and Fertility Research
University of Hamburg

INTRODUCTION

The study of the role of melatonin (MEL) in the human and in various experimental cell systems requires sensitive and accurate assays for the measurement of MEL. In recent times various immunoassays were described in the literature and several commercial immunoassay kits are on the market. Although the assays have improved with time, the measurement in human plasma/serum is still associated with complications, since day-time levels are rather low (<10 pg/ml), and there are numerous other factors in plasma or serum beside the MEL, which interfere in the assay, and which show a high individual variation. Therefore accurate measurements of MEL in plasma/serum require an extraction of MEL prior to assay, which reduces non-specific matrix effects, which, however, is cost- and time consuming.

In the present study a very sensitive and specific radioimmunoassay (RIA) for the direct measurement of MEL in human plasma is described. The variable interference of plasma factors were kept low by combining a highly specific antiserum with a new, polar ^{125}I-tracer, in which the MEL is bridged to ^{125}I-glycylhistidin (GH) over a short spacer at the indole nitrogen (N_i). The presence of PEG and plasma in the incubation mixture reduced the individual matrix effects to a minimum. In addition, analogous to the ^{125}I-RIA-tracer, we have prepared a biotinylated MEL derivative, which allows ultrasensitive measurements in a competitive ELISA system. This assay works well for cell culture media and C18-extracted samples.

Melatonin after Four Decades, edited by James Olcese.
Kluwer Academic / Plenum Publishers, New York, 2000.

METHODS

The MEL-BSA-conjugate for immunization was prepared according to Grota et al. (1). Antibodies were developed in rabbits by injecting 0.5 mg conjugate in Freund's Adjuvant at 6–8 weeks intervals. N_i-Carboxypropyl-MEL (CP-MEL) was prepared by alkylating MEL at N_i with 3-bromopropionic acid. For the preparation of the ^{125}I-MEL-tracer CP-MEL was coupled to glycylhistidine (GH) via the mixed anhydride procedure. The GH-CP-MEL was ^{125}I-labeled by the chloroamine-T procedure. The ^{125}I-GH-CP-MEL was purified by TLC on Empore-silica (Analytichem) using ethylacetate:methanol:acetic acid:H_2O (50:50:2:2) as the mobile phase. MEL-free plasma was prepared by stripping pooled plasma with octadecyl silica (C18 columns). The 400 µl-incubation mixture for the RIA contained 200 µl plasma sample or 200 µl C18-plasma as standard matrix. Standards (0.5–100 pg), tracer (20,000–30,000 cpm) and antiserum (anti-MEL-3; pre-precipitated; final dilution 1:240,000) were dissolved in RIA-buffer (PBS) supplemented with 6% polyethylene glycol (PEG) 6000 and 65% C18 plasma. Incubations were run overnight at 4°C. The biotinylated tracer for the ELISA was prepared by coupling CP-MEL to N-biotinoyl-1,8-diamino-3,6-dioxaoctane (Boehringer, Mannheim) over the mixed anhydride. The incubation mixture for the ELISA consisted of 50 µl sample (C18-extract or cell culture medium) or 50 µl MEL standard (0.1–20 pg in sample matrix), 50 µl biotinylated MEL (0.7 pg) and 100 µl antiserum (1:100,000) in PBS + 0.2% BSA, 5 mM EDTA 0.05% Tween 20.

RESULTS

RIA

Human plasma contains factors beside MEL, which interfere in the direct RIA and which show a high individual variation. As illustrated in Figure 1, the assay of 12 individual plasma samples (freed from MEL by C18-extraction) showed a variation in tracer binding of 37.3% associated with a general low binding. The presence of 3% PEG in the incubation mixture decreased the individual variation to only 6.9% and increased the tracer binding about two-fold. The smallest detectable MEL concentration in the RIA (difference from zero-standard by two standard deviations) was 0.3 pg/tube corresponding to 1.5 pg/ml plasma. A typical standard curve is shown in Figure 2. Intra-assay variations for 15, 55 and 150 pg/ml were 10.2, 7.1 and 9.4%, respectively. Inter-assay variations for these concentrations were 15.2, 8.8 and 11.8%, respectively. The antiserum shows a very high specificity for MEL as indicated in Table 1. In the RIA, MEL added to individual plasma samples in different concentrations could be quantitatively recovered. Day time levels (late afternoon) were in general below 5 pg/ml. The measurement of the circadian patterns of plasma MEL concentrations by the direct RIA and the RIA with prior C18-extraction of plasma samples showed very similar results for both assay variants (correlation: $r = 1.0$). The validation of the direct RIA by gas chromatographic-mass spectrometry (GCMS) showed a nice correlation of $r = 0.95$ (Figure 3). The use of the less polar ^{125}I-2-iodo-MEL (2) as tracer in the direct RIA often resulted in an overestimation of individual plasma MEL concentrations particularly of the low day time levels (data not shown).

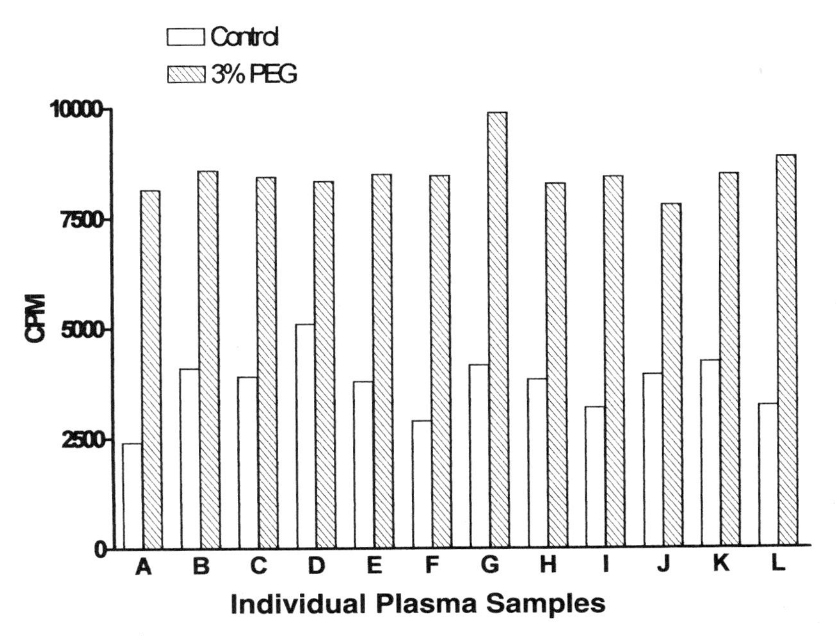

Figure 1. Effect of polyethylene glycol (PEG) on the measurement of melatonin-free individual plasma samples. Individual samples were freed of melatonin by stripping with octadecylsilica (C18) and assayed in the direct RIA with and without 3% PEG 6000. Mean of all samples (each in duplicate): 4350 ± 1620 (SD) for control; 8932 ± 618 (SD) for PEG.

ELISA

The smallest detectable dose in the ELISA is below 0.1 pg/well. A typical standard curve is shown in Figure 2. The intra-assay variation for the assay in cell culture medium (without extraction) is between 4 and 9%. In cell culture media and C18-extracted samples the recovery is usually 100%, provided that standards and samples contain a similar matrix and received the same treatment (data not shown).

Table 1. Cross-reactions of various indole derivatives measured in the RIA system

Indole derivative	Cross-reaction (%)
Melatonin	100.0
6-Hydroxy-melatonin	0.800
5-Methoxy-tryptamine	0.030
N-Acetyl-5-hydroxy-tryptamine	0.010
6-Sulfatoxymelatonin	0.020
5-Methoxy-tryptophol	<0.008
5-Methoxy-DL-tryptophan	<0.005
Tryptamine	<0.005
L-Tryptophan	<0.005
N-Acetyl-L-tryptophan	<0.005
5-Hydroxy-tryptamine	<0.005

Pooled human Plasma was spiked with different concentrations of indole derivatives. Cross-reactions were calculated from molar concentrations, which produced a 50% -displacement of the ^{125}I-tracer.

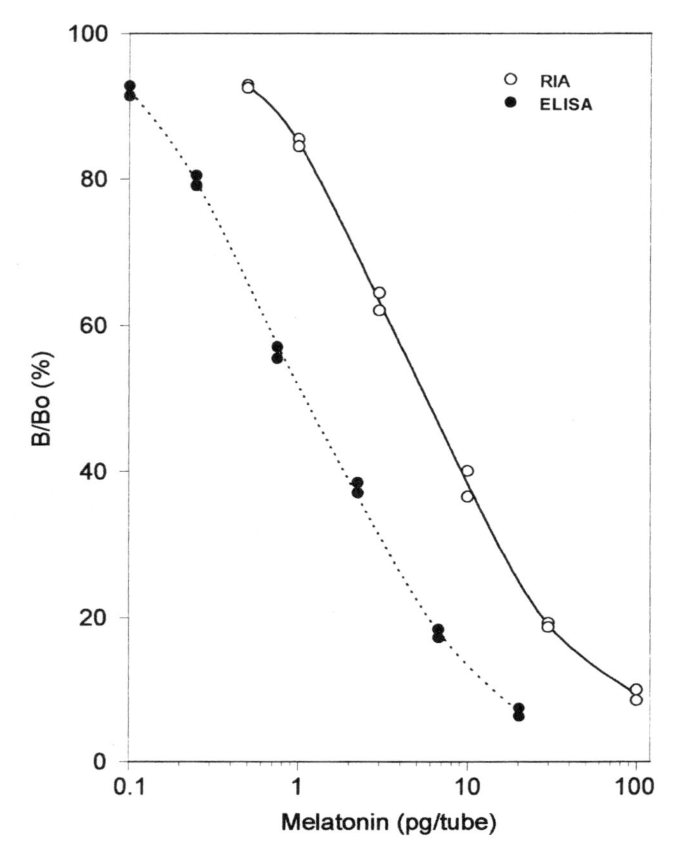

Figure 2. Typical melatonin standard curves obtained by RIA and ELISA.

In the ELISA, plasma or serum directly added to the incubation mixture caused a drastic decrease in the optical density measured (data not shown).

CONCLUSION

An excellent antiserum (as the one used here) is an absolute prerequisite for the specific and sensitive measurement of physiological concentrations of plasma MEL. However, the accuracy of direct measurements is strongly influenced by individual plasma factors, which act in a way distinct from common cross-reactions. In the present work the observed individual variations in interference by these plasma factors could be kept very low by the use of a polar ^{125}I-tracer and the supplementation of the incubation mixture with PEG and additional plasma. The accuracy of this direct RIA is documented by the high correlation of assay results with those obtained by the C18-extraction RIA and by GCMS. Later studies have shown that under the RIA conditions described here human serum and porcine serum behaved very similarly with respect to non-specific matrix effects. Therefore, a more recent version of the direct

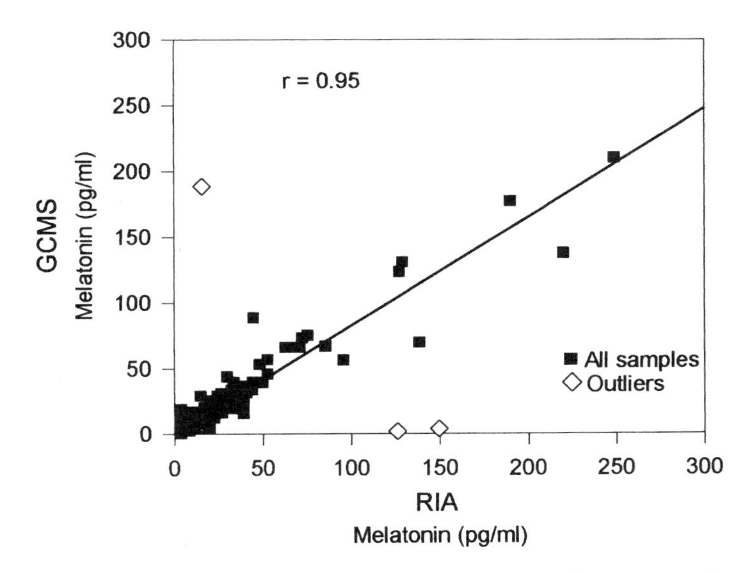

Figure 3. Comparison of gas chromatographic-mass spectrometry (GCMS) and RIA. Melatonin concentrations of 93 plasma samples were measured. The correlation coefficient is based on 90 samples, i.e. excludes the 3 outliers. Data kindly provided by Dr. R. Sack, Oregon Health Science University, Portland, Oregon.

RIA is an assay for MEL in human serum, which uses C18-porcine serum as standard matrix and buffer supplement. Efforts are ongoing to make the ELISA suitable for direct measurement of MEL in serum, too.

For the direct measurement of melatonin in human serum we have developed highly specific antibodies in rabbits against a melatonin-BSA-conjugate, in which the melatonin was bridged to the protein at the C2-position via the Mannich reaction (1). This coupling strategy is advantageous over others, since it results in the production of antibodies which do not accept any changes in the sidechains of melatonin, but tolerate modifications at or near C2. Initially we used 2-Iodo-melatonin (2) as tracer, which was bound well by the antiserum and which produced sensitive standard curves. However, its use in the direct assay of serum samples was usually associated with complications, presumably due to its highly hydrophobic nature. Therefore we have constructed a polar tracer molecule, in which melatonin at the N_i-position was coupled to Gly-His over a short spacer, followed by iodination of the latter via the chloroamine T-procedure. The combination of this tracer with the above antiserum allowed the direct measurement of the very low day time serum levels (detection limit 1.5 pg/ml). Variations in individual serum matrix effects could be abolished by the addition of C18-stripped porcine serum and polyethylene glycol to the incubation mixture. Measurements by RIA and GCMS showed a good correlation ($r = 0.95$).

By analogy to the RIA we have developed a competitive ELISA using the same antiserum and a biotinylated tracer molecule, in which the melatonin was linked to a biotin derivative at the N_i-position. Microtiter plates were coated with a second antibody (goat anti rabbit) and streptavidin-HRP and the chromogen TMB were used in the detection system. This assay is routinely used for the measurement of melatonin in culture media and extracted samples of other sources in a concentration range of 0.1 to 20 pg/well.

REFERENCES

Grota L.J. and Brown G.M. 1974 Antibodies to indolealkylamines: serotonin and melatonin. Can. J. Biochem. 52:196–202.

Vakkuri O., Lamsa E., Rahkamaa E., Ruotsalainen H., and Leppaluoto J. 1984 Iodinated melatonin: preparation and characterization of the molecular structure by mass and 1H NMR spectroscopy. Anal. Biochem. 142:284–289.

INDEX